# Methods in Molecular Biology



For further volumes:
http://www.springer.com/series/7651

For over 35 years, biological scientists have come to rely on the research protocols and methodologies in the critically acclaimed *Methods in Molecular Biology* series. The series was the first to introduce the step-by-step protocols approach that has become the standard in all biomedical protocol publishing. Each protocol is provided in readily-reproducible step-by-step fashion, opening with an introductory overview, a list of the materials and reagents needed to complete the experiment, and followed by a detailed procedure that is supported with a helpful notes section offering tips and tricks of the trade as well as troubleshooting advice. These hallmark features were introduced by series editor Dr. John Walker and constitute the key ingredient in each and every volume of the *Methods in Molecular Biology* series. Tested and trusted, comprehensive and reliable, all protocols from the series are indexed in PubMed.

# The Plant Endoplasmic Reticulum

## Methods and Protocols

### Second Edition

Edited by

**Verena Kriechbaumer**

*Biological and Medical Sciences, Oxford Brookes University, Oxford, UK*

Humana Press

*Editor*
Verena Kriechbaumer
Biological and Medical Sciences
Oxford Brookes University
Oxford, UK

ISSN 1064-3745 ISSN 1940-6029 (electronic)
Methods in Molecular Biology
ISBN 978-1-0716-3709-8 ISBN 978-1-0716-3710-4 (eBook)
https://doi.org/10.1007/978-1-0716-3710-4



This Humana imprint is published by the registered company Springer Science+Business Media, LLC, part of Springer Nature.
The registered company address is: 1 New York Plaza, New York, NY 10004, U.S.A.

Paper in this product is recyclable.

## Dedication

*In memoriam Prof Chris Hawes: "Don't ask why—ask how!"*

This book is dedicated to Prof Chris Hawes who inspired so many of us as a wonderful colleague and friend, amazing mentor, and brilliant and passionate scientist. Nobody could phrase this better than Chris himself: "I have just loved plants all my life."

Like so many things, editing this book was not the same without you.

# Preface

The endoplasmic reticulum (ER) is a dynamic and multifaceted organelle, carrying out a multitude of structural, biosynthetic, and metabolic functions. Spanning the whole of the cell cortex, the ER network consist of tubules and cisternae. The structure of the ER has long been claimed as important for ER functionality and productivity but a convincing link between structure and function of the ER has not been established [1].

The ER forms the first compartment in the secretory pathway and is a major factory for protein and lipid synthesis, assembly, quality control, and export, and it is also involved in many other functions including co- and post-translational glycosylation [2, 3], growth hormone regulation [4], calcium homeostasis [5], immunity [6], biogenesis of peroxisomes [7, 8], and specialist functions such as oil [9, 10] and protein body formation [10]. More recently, the ER has been linked to autophagy and the generation of the phagophore membrane [11, 12]. The ER can also be used to produce molecules of importance to industry such as antibodies and valuable chemicals and supports a range of enzymatic activities including hosting enzymes involved in the biosynthesis of hormones such as auxin [1, 13].

ER-Golgi interactions have been extensively studied, but interactions of the ER with other organelles such as the nucleus [14], actin [15–18], chloroplasts [2, 19, 20], or plasma membrane [16, 21] are moving more and more in the spotlight [22].

Here, we present a range of different techniques and state-of-the-art approaches to study the structure and function of the plant ER. These comprise modern microscopy techniques by fluorescence and electron microscopy, software protocols for analyzing ER network structure, and methods to purify and analyze ER membrane structure and to study protein glycosylation.

*Oxford, UK* *Verena Kriechbaumer*

### References


1. Kriechbaumer V, Brandizzi F (2020) The plant endoplasmic reticulum: an organized chaos of tubules and sheets with multiple functions. J Microsc 280
2. Stefano G, Hawes C, Brandizzi F (2014) ER – the key to the highway. Curr Opin Plant Biol 22:30–38
3. Hawes C, Kiviniemi P, Kriechbaumer V (2015) The endoplasmic reticulum: a dynamic and well-connected organelle. J Integr Plant Biol 57(1):50–62
4. Friml J, Jones AR (2010) Endoplasmic reticulum: the rising compartment in auxin biology. Plant Physiol 154(2):458–462
5. Hong B et al (1999) Identification of a calmodulin-regulated Ca2+-ATPase in the endoplasmic reticulum. Plant Physiol 119(4):1165–1176
6. Nakano RT et al (2014) ER bodies in plants of the Brassicales order: biogenesis and association with innate immunity. Front Plant Sci 5:73
7. Sparkes IA et al (2009) The plant endoplasmic reticulum: a cell-wide web. Biochem J 423(2):145–155
8. Barton K, Mathur N, Mathur J (2013) Simultaneous live-imaging of peroxisomes and the ER in plant cells suggests contiguity but no luminal continuity between the two organelles. Front Physiol 4:196

9. Herman EM (2008) Endoplasmic reticulum bodies: solving the insoluble. Curr Opin Plant Biol 11(6): 672–679
10. Huang AH (1996) Oleosins and oil bodies in seeds and other organs. Plant Physiol 110(4):1055–1061
11. Wang P et al (2016) Arabidopsis NAP1 regulates the formation of autophagosomes. Curr Biol 26(15): 2060–2069
12. Le Bars R et al (2014) ATG5 defines a phagophore domain connected to the endoplasmic reticulum during autophagosome formation in plants. Nat Commun 5:4121
13. Kriechbaumer V, Botchway SW, Hawes C (2016) Localization and interactions between Arabidopsis auxin biosynthetic enzymes in the TAA/YUC-dependent pathway. J Exp Bot 67(14):4195–4207
14. Brandizzi F et al (2003) ER quality control can lead to retrograde transport from the ER lumen to the cytosol and the nucleoplasm in plants. Plant J 34(3):269–281
15. Pain C et al (2022) intER-ACTINg: the structure and dynamics of ER and actin are interlinked. J Microsc 291:105
16. Wang P et al (2014) The plant cytoskeleton, NET3C, and VAP27 mediate the link between the plasma membrane and endoplasmic reticulum. Curr Biol 24(12):1397–1405
17. Wang P, Hussey PJ (2017) NETWORKED 3B: a novel protein in the actin cytoskeleton-endoplasmic reticulum interaction. J Exp Bot 68(7):1441–1450
18. Zang J et al (2021) A novel plant actin-microtubule bridging complex regulates cytoskeletal and ER structure at ER-PM contact sites. Curr Biol 31(6):1251–1260.e4
19. Mathur J (2021) Organelle extensions in plant cells. Plant Physiol 185(3):593–607
20. Mathur J et al (2022) The ER Is a common mediator for the behavior and interactions of other organelles. Front Plant Sci 13:846970
21. Wang P et al (2018) Characterization of proteins localized to plant ER-PM contact sites. Methods Mol Biol 1691:23–31
22. Wang P et al (2023) Keep in contact: multiple roles of endoplasmic reticulum-membrane contact sites and the organelle interaction network in plants. New Phytol 238(2):482–499

# Contents

# Contributors

CARLOS AGUILAR • *Department of Biological and Environmental Science, University of Jyväskylä, Jyväskylä, Finland*
JING AN • *Centre for Plant Sciences, School of Biology, Faculty of Biological Sciences, University of Leeds, Leeds, UK*
ELSA ARCALÍS • *Department of Applied Genetics and Cell Biology, University of Natural Resources and Life Sciences, Vienna, Austria*
LISA YASMIN ASSECK • *Centre for Plant Molecular Biology, Developmental Genetics, University of Tübingen, Tübingen, Germany*
BENJI BATEMAN • *Central Laser Facility, Science and Technology Facilities Council, Oxon, UK*
KAREN BELL • *Institute of Molecular Plant Sciences, School of Biological Sciences, University of Edinburgh, Edinburgh, UK*
KENNETH W. BERENDZEN • *Centre for Plant Molecular Biology, University of Tübingen, Tübingen, Germany*
VICTOR M. BOLANOS-GARCIA • *Department of Biological and Medical Sciences, Faculty of Health and Life Sciences, Oxford Brookes University, Gipsy Lane, Headington, Oxford, UK*
STANLEY W. BOTCHWAY • *Central Laser Facility, Science and Technology Facilities Council (STFC) Rutherford Appleton Laboratory, Research Complex at Harwell, Didcot, UK*
FEDERICA BRANDIZZI • *MSU-DOE Plant Research Lab and Plant Biology Department Michigan State University, East Lansing, MI, USA; Department of Plant Biology, Michigan State University, East Lansing, MI, USA; Great Lakes Bioenergy Research Center, Michigan State University, East Lansing, MI, USA*
EMILY BREEZE • *School of Life Sciences, University of Warwick, Coventry, UK*
ALESSIA CANDEO • *Dipartimento di Fisica, Politecnico di Milano, Milan, Italy; Central Laser Facility, Research Complex at Harwell, Science and Technology Facilities Council, Rutherford Appleton Laboratory, Harwell, Didcot, Oxford, UK*
JURGEN DENECKE • *Centre for Plant Sciences, School of Biology, Faculty of Biological Sciences, University of Leeds, Leeds, UK*
MATTEO DORA • *Applied Mathematics and Computational Biology, Ecole Normale Supé rieure, Paris, France*
NATASHA DZIMITROWICZ • *School of Life Sciences, University of Warwick, Coventry, UK*
NADINE FIELD • *Department of Biological and Medical Sciences, Oxford Brookes University, Oxford, UK*
LAETITIA FOUILLEN • *CNRS-University of Bordeaux, UMR 5200 Membrane Biogenesis Laboratory, INRAe Bordeaux Aquitaine, Villenave d'Ornon, France*
MARK FRICKER • *Department of Biology, University of Oxford, Oxford, UK*
LORENZO FRIGERIO • *School of Life Sciences, University of Warwick, Coventry, UK*
PUJA PUSPA GHOSH • *Laboratory of Plant Development & Interactions, Department of Molecular & Cellular Biology, University of Guelph, Guelph, ON, Canada*
KATJA GRAUMANN • *Department of Biological and Medical Sciences, Oxford Brookes University, Oxford, UK*

CHRISTOPHER GREFEN • *Molecular & Cellular Botany, Ruhr-University Bochum, Bochum, Germany*
LAWRENCE R. GRIFFING • *3258 TAMU, College Station, TX, USA*
CHRIS HAWES • *Endomembrane Structure and Function Research Group, Department of Biological and Medical Sciences, Oxford Brookes University, Oxford, UK*
DAVID HOLCMAN • *Applied Mathematics and Computational Biology, Ecole Normale Supé rieure, Paris, France; Churchill College, Cambridge University, Cambridge, UK*
PATRICK J. HUSSEY • *Department of Biosciences, Durham University, Durham, UK*
VERENA IBL • *Department of Applied Genetics and Cell Biology, University of Natural Resources and Life Sciences, Vienna, Austria; Molecular Systems Biology, Faculty of Life Sciences, University of Vienna, Vienna, Austria*
MAIKE KITTELMANN • *Cell and Developmental Biology, Biological and Medical Sciences, Oxford Brookes University, Oxford, UK*
JÜRGEN KLEINE-VEHN • *Institute of Biology II, Chair group of Molecular Plant Physiology (MoPP), University of Freiburg, Freiburg, Germany; Center for Integrative Biological Signalling Studies (CIBSS), University of Freiburg, Freiburg, Germany*
KIRSTEN KNOX • *School of Molecular Biosciences, College of Medical, Veterinary and Life Sciences, University of Glasgow, Glasgow, UK*
DAE KWAN KO • *MSU-DOE Plant Research Lab, Michigan State University, East Lansing, MI, USA; Department of Plant Biology, Michigan State University, East Lansing, MI, USA; Great Lakes Bioenergy Research Center, Michigan State University, East Lansing, MI, USA*
VERENA KRIECHBAUMER • *Department of Biological and Medical Sciences, Oxford Brookes University, Oxford, UK; Endomembrane Structure and Function Research Group, Department of Biological and Medical Sciences, Oxford Brookes University, Oxford, UK*
INGEBORG LANG • *Department of Functional and Evolutionary Ecology, Faculty of Life Sciences, University of Vienna, Vienna, Austria*
LIFAN LI • *National Key Laboratory for Germplasm Innovation & Utilization of Horticultural Crops, College of Horticulture and Forestry Sciences, Huazhong Agricultural University, Wuhan, China; Hubei Hongshan Laboratory, Wuhan, China*
LILLY MANETA-PEYRET • *CNRS-University of Bordeaux, UMR 5200 Membrane Biogenesis Laboratory, INRAe Bordeaux Aquitaine, Villenave d'Ornon, France*
JAIDEEP MATHUR • *Laboratory of Plant Development & Interactions, Department of Molecular & Cellular Biology, University of Guelph, Guelph, ON, Canada*
DIETMAR MEHLHORN • *Molecular & Cellular Botany, Ruhr-University Bochum, Bochum, Germany*
ANDREAS J. MEYER • *INRES-Chemical Signalling, University of Bonn, Bonn, Germany*
PATRICK MOREAU • *CNRS-University of Bordeaux, UMR 5200 Membrane Biogenesis Laboratory, INRAe Bordeaux Aquitaine, Villenave d'Ornon, France; Bordeaux Imaging Center, UMS 3420 CNRS, US004 INSERM, University of Bordeaux, Bordeaux, France*
SEINAB NOURA • *Institute of Biology II, Chair group of Molecular Plant Physiology (MoPP), University of Freiburg, Freiburg, Germany; Center for Integrative Biological Signalling Studies (CIBSS), University of Freiburg, Freiburg, Germany*
KARL OPARKA • *Institute of Molecular Plant Sciences, School of Biological Sciences, University of Edinburgh, Edinburgh, UK*

CHARLOTTE PAIN • *Endomembrane Structure and Function Research Group, Biological and Medical Sciences, Oxford Brookes University, Oxford, UK; Department of Biological and Medical Sciences, Oxford Brookes University, Oxford, UK; Endomembrane Structure and Function Research Group, Department of Biological and Medical Sciences, Oxford Brookes University, Oxford, UK*

FRÉDÉRIC PAQUIN-LEFEBVRE • *Applied Mathematics and Computational Biology, Ecole Normale Supérieure, Paris, France*

HARRIET T. PARSONS • *Department of Biochemistry, Cambridge University, Cambridge, UK*

EMANUELA PEDRAZZINI • *Istituto di Biologia e Biotecnologia Agraria, Consiglio Nazionale delle Ricerche, Milan, Italy*

JENNY PETERS • *Department of Applied Genetics and Cell Biology, University of Natural Resources and Life Sciences, Vienna, Austria*

CRISTINA RUBERTI • *MSU-DOE Plant Research Lab and Plant Biology Department Michigan State University, East Lansing, MI, USA*

MICHELLE SCHLÖßER • *INRES-Chemical Signalling, University of Bonn, Bonn, Germany*

JENNIFER SCHOBERER • *Department of Applied Genetics and Cell Biology, University of Natural Resources and Life Sciences, Vienna, Austria*

YUN-JI SHIN • *Department of Applied Genetics and Cell Biology, University of Natural Resources and Life Sciences, Vienna, Austria*

IMOGEN SPARKES • *School of Biological Sciences, University of Bristol, Bristol, UK*

TATIANA SPATOLA ROSSI • *Endomembrane Structure and Function Research Group, Department of Biological and Medical Sciences, Oxford Brookes University, Oxford, UK*

EVA STOGER • *Department of Applied Genetics and Cell Biology, University of Natural Resources and Life Sciences, Vienna, Austria*

RICHARD STRASSER • *Department of Applied Genetics and Cell Biology, University of Natural Resources and Life Sciences, Vienna, Austria*

JIAQI SUN • *The Key Laboratory of Plant Development and Environmental Adaptation Biology, Ministry of Education, School of Life Sciences, Shandong University, Qingdao, China*

CHRISTOPHER TYNAN • *Central Laser Facility, Science and Technology Facilities Council (STFC) Rutherford Appleton Laboratory, Research Complex at Harwell, Didcot, UK*

JOSÉ M. UGALDE • *INRES-Chemical Signalling, University of Bonn, Bonn, Germany*

ULRIKE VAVRA • *Department of Applied Genetics and Cell Biology, University of Natural Resources and Life Sciences, Vienna, Austria*

CHRISTIANE VEIT • *Department of Applied Genetics and Cell Biology, University of Natural Resources and Life Sciences, Vienna, Austria*

ALESSANDRO VITALE • *Istituto di Biologia e Biotecnologia Agraria, Consiglio Nazionale delle Ricerche, Milan, Italy*

SASCHA WAIDMANN • *Institute of Biology II, Chair group of Molecular Plant Physiology (MoPP), University of Freiburg, Freiburg, Germany*

NIKLAS WALLMEROTH • *Centre for Plant Molecular Biology, University of Tübingen, Tübingen, Germany; Centre for Plant Molecular Biology, Developmental Genetics, University of Tübingen, Tübingen, Germany*

PENGWEI WANG • *College of Horticulture and Forestry Sciences, Huazhong Agricultural University, Wuhan, Hubei Province, China; National Key Laboratory for Germplasm Innovation & Utilization of Horticultural Crops, College of Horticulture and Forestry Sciences, Huazhong Agricultural University, Wuhan, China; Hubei Hongshan Laboratory, Wuhan, China*

ANDY WARD • *Central Laser Facility, Science and Technology Facilities Council, Oxon, UK*

RHIANNON R. WHITE • *Biosciences, University of Exeter, Exeter, UK*

MARYAM ALSADAT ZEKRI • *Department of Functional and Evolutionary Ecology, Faculty of Life Sciences, University of Vienna, Vienna, Austria*

TONG ZHANG • *National Key Laboratory for Germplasm Innovation & Utilization of Horticultural Crops, College of Horticulture and Forestry Sciences, Huazhong Agricultural University, Wuhan, China; Hubei Hongshan Laboratory, Wuhan, China*

HUANQUAN ZHENG • *Department of Biology, McGill University, Montreal, Quebec, Canada*

# Chapter 1

# Make It Shine: Labelling the ER for Light and Fluorescence Microscopy


## Chris Hawes, Pengwei Wang, and Verena Kriechbaumer



## Abstract

The ER is a highly dynamic network of tubules and membrane cisternae. Hence, imaging this organelle in its native and mobile state is of great importance. Here we describe methods of labelling the native plant ER using fluorescent proteins and lipid dyes as well as methods for immunolabelling on plant tissue.



**Key words** Endoplasmic reticulum, Labelling, Fluorescent protein, Stable, Transient expression, Immunofluorescence, Fluorescent dyes


## 1 Introduction

The endoplasmic reticulum (ER) forms a dynamic, continually changing network of tubules and membrane cisternae that ramify throughout the cytoplasm of cells. As such, in order to appreciate its true nature, it is often necessary to image the organelle in its native state in living cells. For convenience, it is often best to think of the ER as two interconnected populations of tubules and cisternae. Firstly, there is the geometrical cortical network that overlies the cortical cytoskeleton and is connected to the plasma membrane at specific contact points [1]. Secondly, cytoplasmic ER often rapidly streams and transverses the vacuolar lumen via transvacuolar strands.

Two strategies are available for imaging the ER in its native form. A number of different probes have been used to directly label the ER in living cells. Most of them are not 100% specific for the ER membrane and may label other organelles at varying concentrations and incubation times. Two probes with different emission wavelengths, which are relatively easy to use on a range of plant tissues, are the rhodamine B hexyl ester [2] and 3,3′-dihexylocarbocyanine iodide (DiOC6, green emission) [3, 4]. Expression of fluorescent protein constructs that are targeted to the ER is another efficient


Verena Kriechbaumer (ed.), *The Plant Endoplasmic Reticulum: Methods and Protocols*, Methods in Molecular Biology, vol. 2772, https://doi.org/10.1007/978-1-0716-3710-4_1,

strategy for in vivo labelling. It is possible to use labelled proteins or peptide sequences that are targeted to the ER membrane [5, 6]; although on occasions, these can perturb the membrane and alter the structure of the ER network [5, 7]. Alternatively, a simple construct comprising a signal sequence and fluorescent protein with a KDEL or HDEL retrieval motif spliced to the C terminus is sufficient to label the lumen of the ER and is often the construct of choice [8, 9].

Immunocytochemistry is an important technique for locating native proteins and confirming the validity of the location of fluorescent protein constructs. Many different preparative techniques for the immunofluorescence labelling of plant proteins have been described over the years, including production of single cells by enzyme digestion (the root squash technique), cryo-sectioning, wax or resin embedding, and sectioning [10]. Here we describe a modification of the root squash technique to permeabilize root tip cells and the freeze shatter technique developed by Wasteneys et al. [11], which permits the physical rupture of *Arabidopsis* tissues and cells, permitting antibody penetration, and is particularly suitable for *Arabidopsis* seedling tissues.

In this chapter, we describe the following methods:

(a) *Agrobacterium*-mediated transient protein expression in tobacco epidermal leaf cells.

(b) Stable protein expression in tobacco.

(c) Stable protein expression in *Arabidopsis.*

(d) Immunofluorescence in *Arabidopsis* roots.

(e) Freeze shattering and immunofluorescence.

(f) The use of lipid dyes Rhodamine B hexyl ester or $DiOC_6$.

## 2 Materials

### 2.1 Solutions and Equipment

#### 2.1.1 For Agrobacterium-Mediated Transient Protein Expression in Tobacco Epidermal Leaf Cells

1. YEB medium: 5 g/L of beef extract, 1 g/L of yeast extract, 5 g/L of sucrose, 0.5 g/L of $MgSO_4{\cdot}7H_2O$.
2. Infiltration buffer: 50 mM MES, 2 mM $Na_3PO_4{\cdot}12H_2O$, 0.1 mM acetosyringone and 5 mg/mL glucose.
3. Water bath.
4. Nanodrop spectrophotometer (or equivalent) to determine optical density of bacterial culture.

#### 2.1.2 For Stable Protein Expression in Tobacco

1. Sterilization solution: 1:1 hypochlorite solution: $dH_2O$, 0.01% (v/v) Tween 20.
2. Plates with shooting medium: 2.15 g/L Murashige and Skoog salts, pH 5.2 (without IAA, kinetin or sucrose; MP Biomedicals Inc.), 0.8% (w/v) agar, 3.0% (w/v) sucrose, 0.1 mg/L indole

butyric acid (1.0 mg/mL stock), 0.8 mg/L 6-benzylaminopurine (1.0 mg/mL stock), 0.1 mg/L carbenicillin, 0.2 mg/L ticarcillin/clavulanic acid (Ducheva) and suitable selection for the binary vector carrying your FP fusion construct. Dissolve Murashige and Skoog salts and sucrose in ultrapure deionized water. Add stock solutions of plant growth regulators. Adjust the pH to 5.2 and autoclave. When the medium has cooled to 50 °C, add filter-sterilized antibiotics and pour into Petri dishes. Plates are kept at 4 °C in the dark.

3. Plates with rooting medium: 2.15 g/L Murashige and Skoog salts, 0.8% (w/v) agar, 3.0% (w/v) sucrose, 0.5 mg/L indole butyric acid (1 mg/mL stock), 0.1 mg/L carbenicillin, 0.2 mg/L ticarcillin/clavulanic acid (Ducheva).

#### 2.1.3 For Stable Protein Expression in Arabidopsis

1. LB broth: Tryptone 10 g/L, NaCl 10 g/L yeast extract 5 g/L.
2. Dipping buffer: 5% sucrose, 50 μl/L Silwet L-77 in $dH_2O$.

#### 2.1.4 For Immunofluorescence in Arabidopsis Root Cells

1. MBS ester (m-maleimidobenzoyl-N-hydroxysuccinimide).
2. Fixative: paraformaldehyde (PFA) 4%, glutaraldehyde 0.5%, EGTA 5 mM (pH 8.0), PIPES 50 mM (pH 7.0), $MgSO_4$ 2 mM, 0.01% Triton X-100 0.01% (*see* **Note 1**).
3. PBS buffer (pH 7.4).
4. Blocking buffer: PBS supplemented with 2% BSA.
5. Permeable buffer: PBS supplemented with 0.1% Triton X-100.
6. Dricelase solution: Dricelase (2%) in PBS.
7. Proteinase inhibitors: PMSF, leupeptine and pepstatine A.
8. Liquid nitrogen.
9. Metal block and a hammer.
10. Glass Microscope slide.
11. Plastic sieves.
12. Vectashield.

#### 2.1.5 For Freeze Shattering and Immunofluorescence

1. MBS ester (m-maleimidobenzoyl-N-hydroxysuccinimide).
2. PBS buffer.
3. Blocking buffer: PBS supplemented with 2% BSA.
4. Permeable buffer: PBS supplemented with 0.1% Triton X-100.
5. Driselase solution: Driselase (2%) in PBS.
6. Proteinase inhibitors: PMSF, leupeptine and pepstatine A.
7. Liquid nitrogen.
8. Metal block and a hammer.
9. Glass microscope slide.

10. Plastic sieves.
11. Vectashield or equivalent anti-fade mountant.

*2.1.6 For Lipid Dyes Rhodamine B Hexyl Ester or DiOC$_6$*

1. Rhodamine B hexyl ester solution: stock solution1 mM in DMSO, working solution 1 μM in water.
2. DiOC$_6$ (molecular probes): working solution: 1.8 mM in water.
3. 2 mL Eppendorf tubes.

### 2.2 Antibodies for Immunofluorescence

Various suppliers are possible for both primary and secondary antibodies.

1. Polyclonal mouse antibody against VAP27-1 [12].
2. Polyclonal mouse antibody against NET3C [1].
3. Polyclonal Rabbit antibody against BIP2 (Agriserum).
4. Polyclonal Rabbit antibody against HDEL (Agriserum).
5. Goat-anti-rabbit FITC conjugated antibody (Jackson Immuno or other supplier).
6. Goat-anti-mouse TRITC conjugated antibody (Jackson Immuno or other supplier).

### 2.3 Microscopy

1. Upright or inverted laser scanning or spinning disc confocal microscope, TIRF microscope, super resolution fluorescence microscope.

## 3 Methods

1. Agrobacterium-mediated transient expression in tobacco epidermal leaf cells.
2. Stable protein expression in tobacco,
3. Stable protein expression in *Arabidopsis*,
4. Immunofluorescence in *Arabidopsis* root.
5. Freeze shattering and immunofluorescence.
6. Labelling of ER with lipid dyes (Table 1).

### 3.1 Agrobacterium-Mediated Transient Expression in Tobacco Epidermal Leaf Cells

1. For agrobacterium-mediated transient expression, 5-week-old tobacco (*Nicotiana tabacum* SR1 cv Petit Havana) plants grown in the greenhouse are used. Transient expression is carried out according to [13]. Alternatively, *N. benthamiana* plants are also suitable for transient expression experiments.
2. Construct a suitable expression vector using standard molecular biology techniques such as conventional Cut & Paste cloning [14] or Gateway cloning [15, 16].

**Table 1**
**Some suggested constructs suitable for imaging the ER in epidermal cells**

| Construct | Comments | References |
|---|---|---|
| SS-FP-HDEL | Labels ER lumen | [8] |
| Reticulon-FP | Labels membrane of ER tubules and cisternal rims. Over-expression can constrict tubules | [5] |
| Calnexin-TMD-FP | Membrane marker can induce cisternae on high expression | [6] |
| Derlin-FP | Labels membrane of whole ER network | [7] |

3. Introduce the expression vector into agrobacterium strain GV3101 by heat shock.
4. Inoculate transformants into 5 mL of YEB medium with the appropriate antibiotic for the bacterial vector used as well as 25 mg/L rifampicin.
5. After overnight shaking at 28 °C, pellet 1 mL of the bacterial culture by centrifugation at 2000 *g* for 5 min at room temperature.
6. Wash the pellet twice with 1 mL of infiltration buffer, and then resuspended in 1 mL of infiltration buffer.
7. Initially dilute the bacterial suspension to a final $OD_{600}$ of 0.1. If expression is successful, then different $OD_{600}$ values can be tested to improve the expression results or to reduce expression levels, such as $OD_{600}$ of 0.05 or 0.01.
8. Carefully press the suspension through the stomata on the lower epidermal surface using a 1 mL syringe (*see* **Note 2**). The surface infiltrated appears darker at this point and should be outlined with a thin black marker pen to aid removal of correct segments of leaf for microscopy (for details, *see* Fig. 1).
9. Inoculated plants are then incubated under normal growth conditions for 48–72 h depending on the proteins expressed. This will have to be determined experimentally by checking the expression after, e.g., 2, 3, and 4 days after infiltration.
10. Excised a 0.5×0.5 cm piece of the infiltrated leaf, and mount in a drop of water, and observed with a microscope (*see* **Note 3** and Subheading 3.7).
11. Images can be recorded, e.g., using a laser scanning confocal microscope with ×63 or ×100 high numerical aperture oil or water immersion objective lenses. For imaging of combinations of the green fluorescent (GFP) and red fluorescent protein (RFP), samples should be excited using 488 and 543 nM laser lines in multitrack mode preferably using line switching.

**Fig. 1** Infiltration of tobacco epidermal leaf cells with *Agrobacterium tumefaciens*. (**a**) A hole is punched into a leaf section using a 100 μL-pipette tip. (**b**) The bacterial suspension culture is carefully pressed into the leaf by covering the hole with a 1 mL syringe on the lower epidermal site and a finger on the upper side. (**c**) As much as possible of that leaf section is filled with the culture. (**d**) The infiltrated parts will show up darker now. (**e**) Mark the infiltrated area with a pen. (**f**) Infiltrate as many leaf sections as required for the constructs tested

Images can be analyzed with proprietary software from the confocal manufacturer, with commercial image analysis software or freeware such as Fiji ImageJ (Fig. 2).

### *3.2 Stable Transformation of Tobacco from Transiently Expressing Tobacco Cells [13]*

1. Infiltrate at least two leaves on two different tobacco plants according to Subheading 3.1, and check the expression of the fusion protein by microscopy. If expression is low, then repeat infiltration until transformation levels are satisfactory and at least 70% of the cells in the field of view on the microscope are expressing the fusion protein.
2. Remove infiltrated leaves from the plant using scissors, and immerse them completely into sterile 500 mL beakers containing sterilization solution to remove trace amounts of *Agrobacterium* and other microbes from the leaf surface.

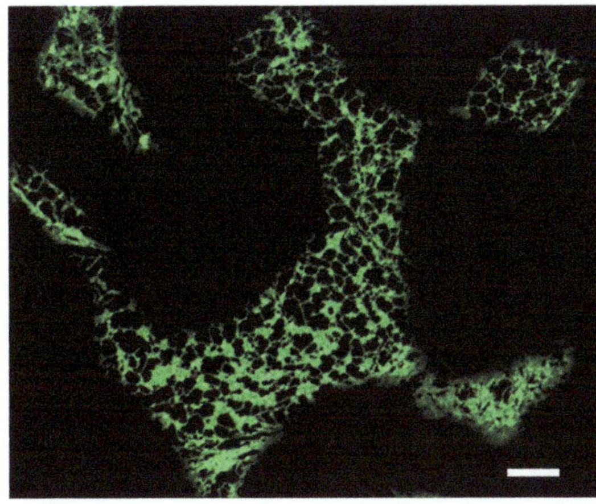

**Fig. 2** GFP-labelled ER in tobacco leaf epidermal cells. The ER network is labelled with the ER membrane localized protein TAR2 [20] fused to GFP as a fluorophore and visualized using confocal microscopy. Scale bar = 10 μm

3. Carefully agitate the leaves for 8 min, and then rinse them in a series of three sterile beakers containing sterile water (*see* **Note 4**).
4. Using scissors, cut the leaves into pieces approximately 2 × 2 cm avoiding the midrib (*see* **Note 5**).
5. Place the leaf squares onto agar plates of shooting medium plus the corresponding antibiotics for the transformation vector used, and incubate at 25 °C, 16 h light, 8 h dark, until shoots appear. This will take between 3 and 4 weeks.
6. Check regularly for contamination, and if apparent, transfer uncontaminated pieces to plates of fresh shooting medium.
7. When shoots are present, transfer them from the tissue with sterile forceps to plates with rooting medium. Roots will appear within 7–10 days.
8. Check again with the microscope for expression as there may be many shoots to deal with.
9. If the plants are expressing well, transfer them to Phytatrays or other suitable growth containers containing 0.5% Murashige and Skoog agar to develop until they are large enough to be transferred to soil.

### 3.3 Transformation of Arabidopsis Using the Floral Dip Method

1. A 5 mL liquid preculture, inoculated from single colony of agrobacterium, carrying the appropriate binary vector, is grown with vigorous agitation (180 rpm) over night at 28 °C with the appropriate antibiotics.
2. 100 mL of LB medium is inoculated with 1 mL of the preculture and shaken over night at 28 °C and 180 rpm.

3. Decant culture into two 50 mL Falcon tubes.
4. Pellet the cells by centrifugation at 4 °C with 4000 rpm for 10 min. Discard the supernatant, and wash the resulting pellets by resuspension in 10 mL infiltration medium.
5. The pelleting step is repeated, and the two pellets are combined into one Falcon tube in a total of 50 mL of dipping buffer.
6. Remove siliques from the inflorescence shoots from 2–4 plants of *Arabidopsis thaliana* (wild type Col-0) so only unfertilized flowers remain.
7. Dip the shoots upside down into the Falcon tubes containing the *Agrobacterium* suspension and soaked for 30 s.
8. After dipping, seal the plants with transparent plastic bags or cling film for the next 24 h, and then remove the cover and return plants to their normal growing conditions (*see* **Note 6**).
9. After 2–3 weeks, the plants should not be watered anymore to allow the seeds to ripen. Plants with seeing heads should be bagged to stop loss of material or siliques grown through Aracon containers (Arasystem) to collect any loose seed.
10. Collect the seeds that are brown and dry (approximately after about 6 weeks).
11. Selection of transformed *Arabidopsis* plants is carried out by germinating seeds collected from transformed plants on MS plates containing 1% sucrose and the appropriate antibiotics or herbicide depending on the vector used.

### 3.4 Immunofluorescence in Arabidopsis Roots

This protocol is designed for immunolabelling of root tips, a similar method, and also can be used for fixing cells from suspension culture (e.g., BY2 cells).

1. *Arabidopsis* seedlings are grown vertically in Petri dishes for 5–7 days before fixation. The tips region (about 1 cm) are collected and transferred into a plastic sieve (Fig. 3).
2. The fixative is prepared fresh each time. To make 10 mL, 0.4 g of PFA is added to 5 mL of dd$H_2O$ in a falcon tube. The mixture is warmed to 65 °C in a water bath, and drops (about 20 μL per 10 mL) of 0.1 M NaOH solution are added. The tube is agitated periodically until the PFA is dissolved. Cool the solution to room temperature, and add remaining buffer components. Bring the final volume to 10 mL with dd$H_2O$. Finally, add Triton X-100 to 0.01% v/v (*see* **Note 7**).
3. Fix roots for 90 min, and then wash in PBS buffer 3 times for 5 min each.
4. Incubate roots in a Driselase solution (2% w/v) for 7 min at room temperature. Driselase solution is stored at −20 °C. For each 1 mL solution, add 10 μL of PMSF (100 mM), 1 μL of

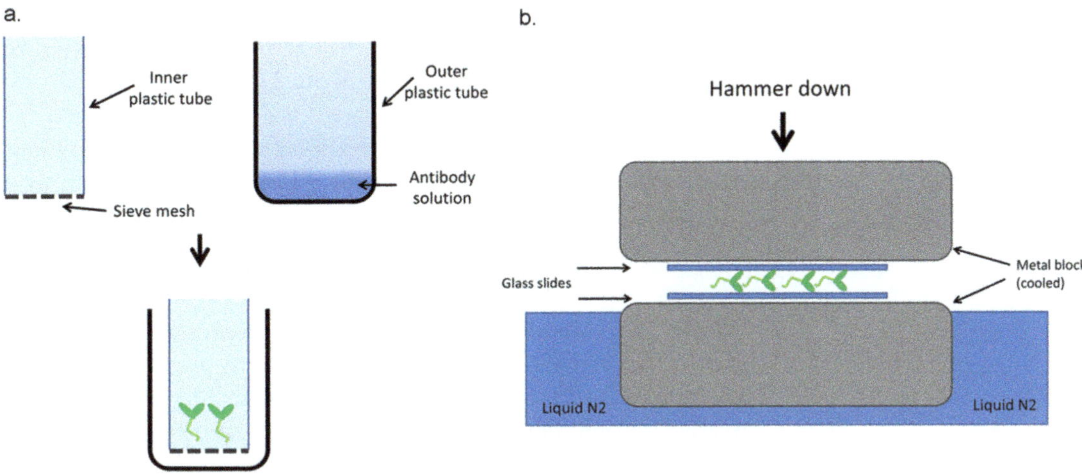


**Fig. 3** Experiment setup for immune-labelling (**a**) and freeze shattering (**b**)

leupeptine (10 mg/mL), and 1 μL of pepstatin A (1 mg/mL) immediately before use (*see* **Note 8**). The protease inhibitors are there to block proteinases in the Driselase solution.

5. Quickly wash roots in PBS 3 times, and incubate them for 15 min in PBS containing 0.1% Triton X-100 to make cells permeable.
6. Quickly wash roots in PBS 3 times, and leave them in blocking buffer for 1 h.
7. Incubate roots in primary antibody (diluted in blocking buffer; *see* **Note 9**) for 3–6 h at room temperature or overnight at 4 °C.
8. Wash roots in PBS buffer 3 times for 30 min each, and incubate them in secondary antibody solution for 1–3 h. Antibodies (e.g., FITC or TRITC conjugated) with minimal cross activity should be used for dual labelling; this is especially important for using primary antibodies raised in two close related species (e.g., rat and mouse; *see* **Note 10**).
9. Wash roots in PBS buffer 3 times for 30 min each (*see* **Note 11**), and mount the slide using Vectashield. Samples are now ready for microscopy.
10. In most cases, ER bodies, a spindle-like ER-derived structure, can be seen in wild-type *Arabidopsis* root cells. This can affect the visibility of the ER network (Fig. 3). To overcome this problem, an *Arabidopsis* mutant without ER bodies (*nai1*, [17]) can be used (Fig. 3).
11. If plants expressing GFP-HDEL are used for immunocytochemistry, the GFP signal can withstand fixation; therefore, it can be used as a marker to test the co-localization of any protein of interest. As an example here, we use *Arabidopsis* (*nai1*) roots expressing GFP-HDEL, immuno-stained with

anti-NET3C (an ER-PM contact site protein). The result indicates the endogenous NET3C localized to punctate structure that co-aligned with the ER (*see* **Note 12**).

### 3.5 Freeze Shattering and Immunofluorescence

This protocol is designed for immunolabelling of whole plant tissue, ideally for leaf epidermal cells [11]. It is based around using physical pressure on frozen material, after fixation, to fracture the cuticles and cell walls, thus enabling the easy penetration of antibodies.

1. Whole *Arabidopsis* plantlets or excised leaves (10 days old) grown in Petri dishes can be used; they are pre-incubated for 15 min in MBS ester (100 μM) solution. Transfer plants to fixative (supplemented with 100 μM MBS ester) as described in Subheading 3.4 for 1 h.
2. After incubation, wash samples twice in PBS to remove all fixative. Transfer seedlings onto a glass slide, and dry them using tissue paper. Put another glass slide on top to make a "sandwich" (Fig. 3).
3. Rapidly freeze the "sandwich" by immersion in liquid nitrogen, and place it in between two metal blocks that are also cooled in liquid nitrogen (Fig. 3). Knock the top metal block gently with a hammer; this will shatter the *Arabidopsis* seedlings into small pieces of 1 mM × 1 mM (*see* **Note 13**).
4. Use cold tweezers to transfer plant pieces to a plastic sieve (Fig. 3). Then follow the protocol in Subheading 3.4 (**Step 5–9**) for antibody incubation.
5. Mount the slide using Vectashield or Citifluor antifadent mountants. Samples are now ready for microscopy (Fig. 4).

### 3.6 Lipophilic Dyes Rhodamine B Hexyl Ester or DiOC$_6$ for Live Imaging of the ER

1. Whole *Arabidopsis* seedlings 7–10 days after germination are transferred to an Eppendorf tube containing the staining solution (Rhodamine B hexyl ester, 1 μM; or DiOC$_6$, 1.8 mM) and incubated for 15 min for Rhodamine B or 10–30 min for DiOC$_6$ (*see* **Note 14**).
2. After incubation in the dye, transfer the seedlings to a fresh Eppendorf tube containing water to wash off excess staining solution.
3. Samples stained with Rhodamine B hexyl ester can be imaged with a 514 nM line of an argon ion laser using a 458/514 dichroic mirror, and the subsequent emission can be detected using 470–500 nM and 560–615 nM band-pass filters (Fig. 5).
4. Samples stained with DiOC$_6$ should be imaged with a 488 nM line of an argon ion laser, and emission can be detected at 492–629 nM.

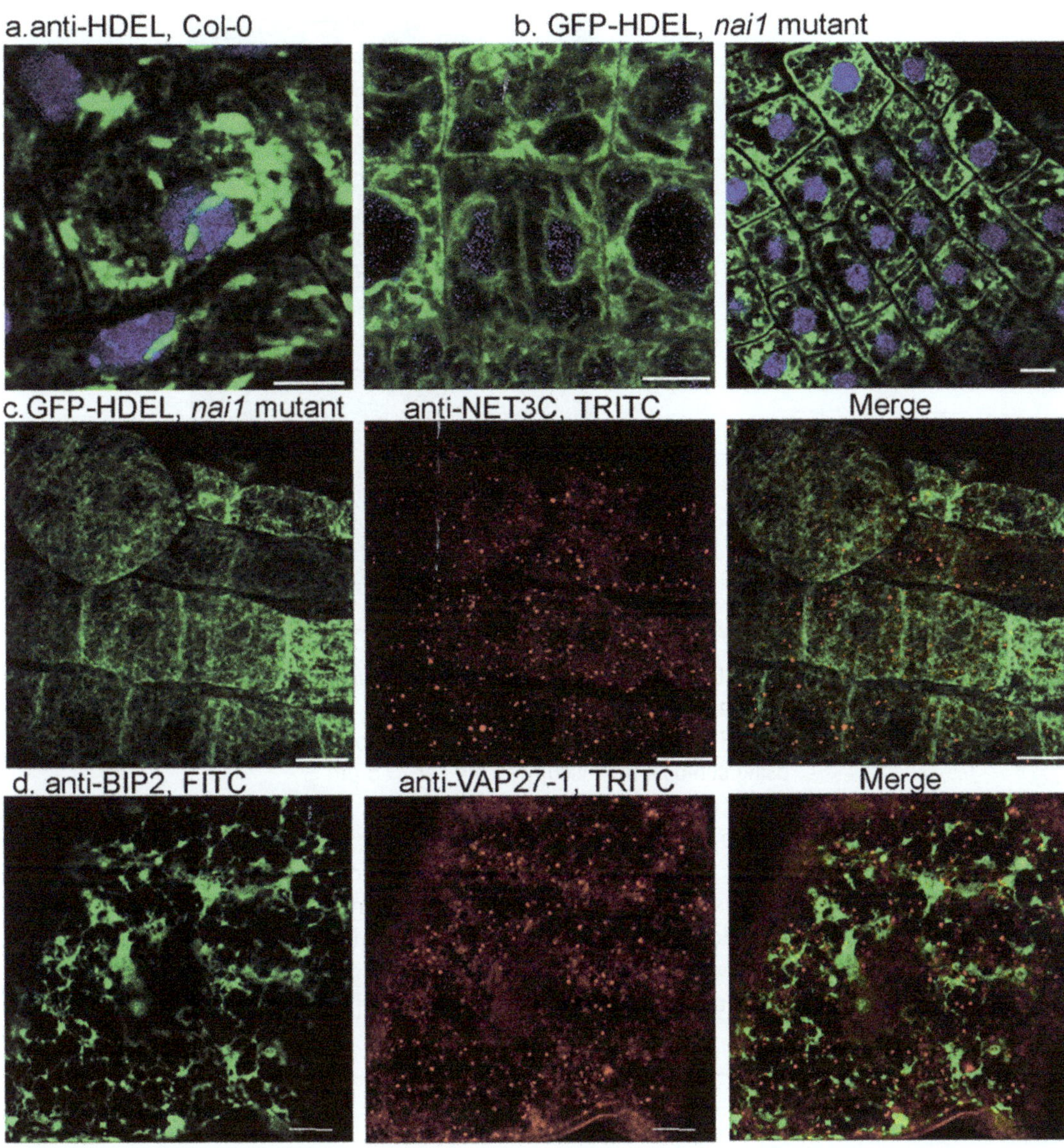


**Fig. 4** Immunolabelling of the ER network from different cell types. (**a**) The ER network is labelled with anti-HDEL in combination with a FITC-conjugated secondary antibody; ER bodies can be identified throughout the cell. (**b**) *Arabidopsis* mutant (*nai 1*) transformed with GFP-HDEL do not produce ER bodies. The GFP-HDEL labelled ER network is still strong after fixation. The ER network from root meristem cells and a dividing cell at the anaphase is shown. For both (**a**) and (**b**), nucleus are stained with DAPI. **(c)** GFP-HDEL expressing root cells (*nai 1*) are fixed and stained with anti-NET3C in combination with a TRITC-conjugated secondary antibody. As demonstrated, NET3C labelled punctate structures are localized to the ER network. (**d**) After freeze shattering, *Arabidopsis* leaf epidermal cells are stained with anti-BIP2 (ER) and anti-VAP27–1 (ER and ER-PM contact sites) antibodies. The ER structure is much clear and defined in leaf cells than in root tips

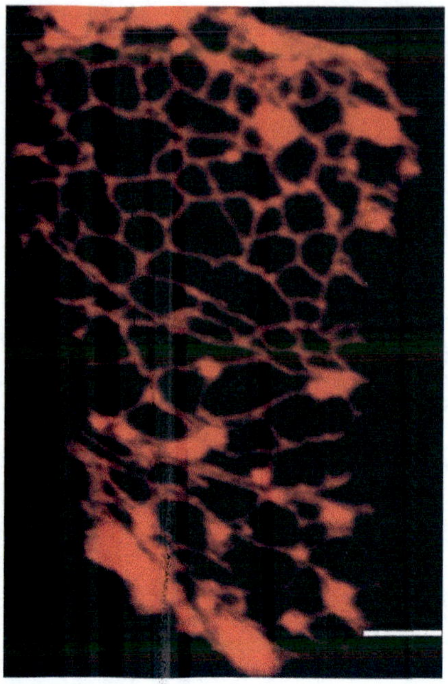

**Fig. 5** Confocal image of *Arabidopsis reticulon6* mutant dyed with Rhodamine B hexyl ester
The ER network is stained with the dye Rhodamine B hexyl ester and visualized using confocal microscopy. Scale bar = 5 μm

### 3.7 Microscopy

While it is perfectly feasible to image fluorescent endoplasmic reticulum with a conventional wide-field epifluorescence microscope, the use of a point scanning or spinning disc confocal is recommended [18]. For cortical ER in epidermal cells of living tissue, it is also possible to obtain excellent images with a TIRF (total internal reflection fluorescence microscopy) [19].

## 4 Notes

1. Concentrated stock solution of PIPES, EGTA, and $MgSO_4$ can be made, aliquoted, and stored at −20 °C.
2. Try to avoid large veins. A small hole may be punched into the lower epidermis to aid infiltration.
3. For use with an inverted microscope, it can be useful to stick coverslips to the slide with a tape such as electrical tape, in order to prevent movement of the coverslip and specimen during observation of the specimen.
4. Make sure the leaves are not damaged by hypochlorite, and shorten the exposure time to a minimum of 5 min if necessary. Incubations longer than 8 min will be detrimental to the tissue, resulting in cell death.

5. From this step on throughout the following tissue culture steps, proper sterile techniques need to be applied.
6. The same plants can be dipped again about 5–7 days after the first dipping, which can dramatically increase the transformation efficiency. Note: do not cut any siliques for the second dipping as they might already have been successfully transformed in the first round.
7. The concentration of Triton X-100 is critical here, especially for fixing fragile ER network. High levels of detergent can destroy the membrane structure and produce fragmented ER.
8. After enzyme treatment, handle the roots gently to avoid loss of tips.
9. Dilution factors vary for different antibodies; 1:100–200 is recommended to start with. For the experiments described here, about 200–300 μL of antibody solution was sufficient to cover root in the plastic sieve.
10. For negative controls, omit the primary antibody, and the secondary antibody only sample should not produce any signal. Likewise with antipeptide antibodies, the antibody in the serum can be titrated out through addition of the peptide.
11. For nuclear staining, PBS supplemented with 10 ng/mL DAPI or Hoechst can be used after the second wash.
12. The ER network is cisternalized in cells from root tips, and these cells are small. Therefore, it can often be difficult to resolve the ER membrane from the cytoplasmic signals.
13. Only a gently tap with the hammer is required. Wear eye protection when handling liquid nitrogen.
14. Low concentrations of $DiOC_6$ will label mitochondria.

## References


1. Wang P, Hawkins TJ, Richardson C, Cummins I, Deeks MJ, Sparkes I, Hawes C, Hussey PJ (2014) The plant cytoskeleton, NET3C, and VAP27 mediate the link between the plasma membrane and endoplasmic reticulum. Curr Biol 24:1397–1405
2. Boevink P, Santa Cruz S, Hawes C, Harris N, Oparka KJ (1996) Virus mediated delivery of the green fluorescent protein to the endoplasmic reticulum of plant cell. Plant J 10:35–941
3. Terasaki M, Song J, Wong JR, Weiss MJ, Chen LB (1984) Localization of endoplasmic reticulum in living and glutaraldehyde-fixed cells with fluorescent dyes. Cell 38:101–108
4. Quader H, Schnepf E (1986) Endoplasmic reticulum and cytoplasmic streaming: fluorescence microscopical observations in adaxial epidermis cells of onion bulb scales. Protoplasma 131:50–252
5. Sparkes I, Tolley N, Aller I, Svozil J, Osterrieder A, Botchway S, Mueller C, Frigerio L, Hawes C (2010) Five Arabidopsis reticulon isoforms share endoplasmic reticulum location, topology, and membrane-shaping properties. Plant Cell 22:1333–1343
6. Huang LQ, Franklin AE, Hoffman NE (1993) Primary structure and characterization of an *Arabidopsis thaliana* calnexin-like protein. J Biol Chem 268:6560–6566
7. Ivanov S, Harrison MJ (2014) A set of fluorescent protein-based markers expressed from constitutive and arbuscular mycorrhiza-inducible promoters to label organelles,

membranes and cytoskeletal elements in *Medicago truncatula*. Plant J 80:1151–1163
8. Lee H, Sparkes I, Gattolin S, Dzimitrowicz N, Roberts LM, Hawes C, Frigerio L (2013) An Arabidopsis reticulon and the atlastin homologue RHD3-like2 act together in shaping the tubular endoplasmic reticulum. New Phytol 197:481–419
9. Denecke J, De Rycke R, Botterman J (1992) Plant and mammalian sorting signals for protein retention in the endoplasmic reticulum contain a conserved epitope. EMBO J 11: 2345–2355
10. Gomord V, Denmat LA, Fitchette-Lainé AC, Satiat-Jeunemaitre B, Hawes C, Faye L (1997) The C-terminal HDEL sequence is sufficient for retention of secretory proteins in the endoplasmic reticulum (ER) but promotes vacuolar targeting of proteins that escape the ER. Plant J 11:313–325
11. Satiat-Jeunemaitre B, Hawes C (2001) Chapter 10: immunocytochemistry for light microscopy. In: Hawes C, Satiat-Jeunemaitre B (eds) Plant cell biology: practical approach. Oxford University Press, New York, pp 207–233
12. Wasteneys G (1997) Freeze shattering: a simple and effective method for permeabilizing higher plant cell walls. J Microsc 188:51–61
13. Wang P, Richardson C, Hawkins TJ, Sparkes I, Hawes C, Hussey PJ (2016) Plant VAP27 proteins: domain characterization, intracellular localization and role in plant development. New Phytol 210:1311–1326
14. Sparkes IA, Runions J, Kearns A, Hawes C (2006) Rapid, transient expression of fluorescent fusion proteins in tobacco plants and generation of stably transformed plants. Nat Protoc 1:2019–2025
15. Kriechbaumer V, Shaw R, Mukherjee J, Bowsher CG, Harrison AM, Abell BM (2009) Subcellular distribution of tail-anchored proteins in Arabidopsis. Traffic 10:1753–1764
16. Karimi M, De Meyer B, Hilson P (2005) Molecular cloning in plant cells. Trends Plant Sci 10:103–105
17. Kriechbaumer V, Botchway SW, Slade SE, Knox K, Frigerio L, Oparka K, Hawes C (2015) Reticulomics: protein-protein interaction studies with two plasmodesmata-localized reticulon family proteins identify binding partners enriched at plasmodesmata, endoplasmic reticulum, and the plasma membrane. Plant Physiol 169:1933–1945
18. Matsushima R, Fukao Y, Nishimura M, Hara-Nishimura I (2004) NAI1 gene encodes a basic-helix-loop-helix-type putative transcription factor that regulates the formation of an endoplasmic reticulum-derived structure, the ER body. Plant Cell 16:1536–1549
19. Pawley JB (2006) Handbook of biological confocal microscopy. Springer, New York
20. Kriechbaumer V, Botchway SW, Hawes C (2016) Localization and interactions between Arabidopsis auxin biosynthetic enzymes in the TAA/YUC-dependent pathway. J Exp Bot 67: 4195–4207

# Chapter 2

# Electron Microscopy Techniques for 3D Plant ER Imaging


## Charlotte Pain and Maike Kittelmann



## Abstract

The endoplasmic reticulum (ER) forms an extensive network in plant cells. In leaf cells and vacuolated root cells it is mainly restricted to the cortex, whereas in the root meristem the cortical and cytoplasmic ER takes up a large volume throughout the entire cell. Only 3D electron microscopy provides sufficient resolution to understand the spatial organization of the ER in the root. Here we present two protocols for 3D EM imaging of the ER across a range of scales. For large-scale ER structure analysis, we describe selective ER staining with ZIO that allows for automated or semi-automated ER segmentation. For smaller regions of ER, we describe high-pressure freezing, which enables almost instantaneous fixation of plant tissues but without organelle specific staining. These fixation and staining techniques are suitable for a range of imaging modalities, including serial sections, array tomography, serial block face-scanning electron microscopy (SBF-SEM), or focused ion beam (FIB) SEM.



**Key words** 3D EM, Serial block face SEM, Endoplasmic reticulum, ZIO, High-pressure freezing, Automated segmentation


## 1 Introduction

The plant endoplasmic reticulum has a very distinct morphology of tubular and cisternal domains [1]. However, how they are generated and maintained and how alterations in ER morphology alter ER function remain to be fully understood in all eukaryotic cells. In plants, the physical nature of ER exit sites and their connection to the Golgi apparatus [2], ER function in cell plate formation during cell division [3], and the characterization of anchoring points of the ER with the plasma membrane [4] are still under investigation and require high-resolution imaging in 3D. Light microscope resolution is usually not sufficient to analyze these ultrastructural details, especially not in small *Arabidopsis* root cells. Electron microscopy has therefore been the imaging technique of choice.

Numerous approaches can be applied to interrogate 3D ER architecture across a range of scales. Historically, subregions of the ER could only be studied by serial sectioning, tomography, or


Verena Kriechbaumer (ed.), *The Plant Endoplasmic Reticulum: Methods and Protocols*, Methods in Molecular Biology, vol. 2772, https://doi.org/10.1007/978-1-0716-3710-4_2,

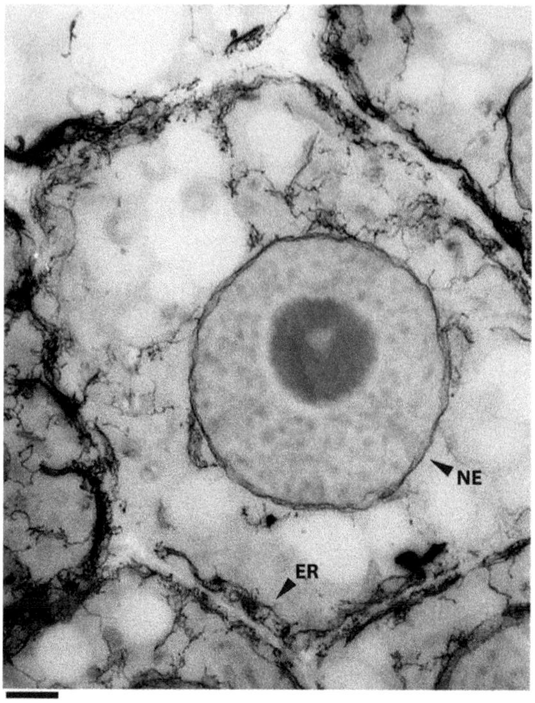


**Fig. 1** High-voltage electron micrograph of a 1 μm thick ZIO impregnated *Zea* root tip meristem cell. The cortical ER and the nuclear envelop (NE) are heavily stained. Scale bar = 2 μm

high-voltage electron microscopy of thick sections (Fig. 1). While these techniques allow for more flexibility, including post staining and the ability to re-use samples, these approaches are either extremely laborious or would only depict a very small area of the ER. These methods are not readily applied to study whole-cell ER architecture. Instead, techniques such as SBF-SEM, FIB-SEM, and Array tomography are preferred as they allow the acquisition of hundreds of images in the Z axis, of a fairly large area in an automated or semi-automated fashion. Especially SBF-SEM has recently become more widely used for imaging of plant tissue [5].

FIB-SEM and SBF-SEM are both destructive techniques where the surface of the resin-embedded sample is scanned, and then, a layer is removed by either ion beam milling (removing a layer of few nanometers) or diamond knife sectioning (sections of at least 20 nm). Both techniques require strong pre-embedding staining to allow sufficient signal detection without the electron beam penetrating too far into the sample.

Array tomography involves the collection of serial sections onto slides or automated with the ATUMtome (RMC) [6] onto tape that is mounted onto wafers. Sections can then be post-stained for additional contrast, immunolabelled, or otherwise post-processed and imaged repeatedly with desired settings and region of interest (ROI) in a high-resolution field emission SEM.

Depending on which technique is most appropriate for a given application—ER subregion analysis or whole-cell ER architecture analysis—different fixation methods and stains are recommended. To allow automated or semi-automated segmentation of the complex ER network in a 3D image stack, tissue- or organelle-specific staining is essential. Automated algorithms including thresholding, region growth, or watershed then allow unbiased 3D modelling [5].

There are two main ways to achieve selective staining of the ER:

1. Enzyme cytochemistry, based on heavy metals as osmium or lead salts reacting locally with the substrates of the targeted enzyme. For example, the inosine 5′-diphosphatase (IDPase) shows specific dictyosome staining [7] in the green alga, *Gloeomonas kupfferi*. In animal cells, a horseradish peroxidase can be targeted to the ER by fusion with the KDEL retention signal [8]. A variety of other enzymes for plant enzyme cytochemistry were reviewed by Sexton and Hall already in 1991 [9]. More recently, miniSOG [10] and APEX [11] have been developed as genetically encoded tags allowing also correlative microscopy. However, both techniques are technically challenging and risk sample damage due to the unique procedures. APEX staining requires an extended period of exposure to $H_2O_2$. Higher concentrations are sometimes required for high contrast staining and can have unwanted effects on cellular structure [12]. miniSOG on the other hand requires sample exposure to intense blue light and a stream of pure oxygen to be blown over the sample for complete staining, a process which is technically challenging and requires specialist equipment [10].
2. Endomembrane-specific staining techniques where zinc-iodine or potassium ferryicyanide guides osmium reduction to the membrane and cisternae of the ER [13, 14] or a combination of uranyl acetate, copper, and lead citrate that leads to selective membrane and polysaccharide contrast [15]. So far, the zinc-iodide-osmium technique (ZIO) is the most reliable and selective method for the ER. After prefixation in aldehydes, the tissue is incubated in a mixture of zinc-iodide and osmium and then embedded in resin. The ER membrane as well as ER lumen, Golgi bodies, and the nuclear envelop is heavily stained, whereas other organelles show no or only very little staining (Fig. 2).

ER network characteristics can then be quantified not only in 2D but also in 3D. Measurements of the surface area and volume are standard options in 3D rendering programs. Algorithms that are able to distinguish between sheets and tubules and allow the quantification of tubule lengths and sheet size have yet to be developed but will in the future allow the analysis of ER network composition and changes in mutant backgrounds.

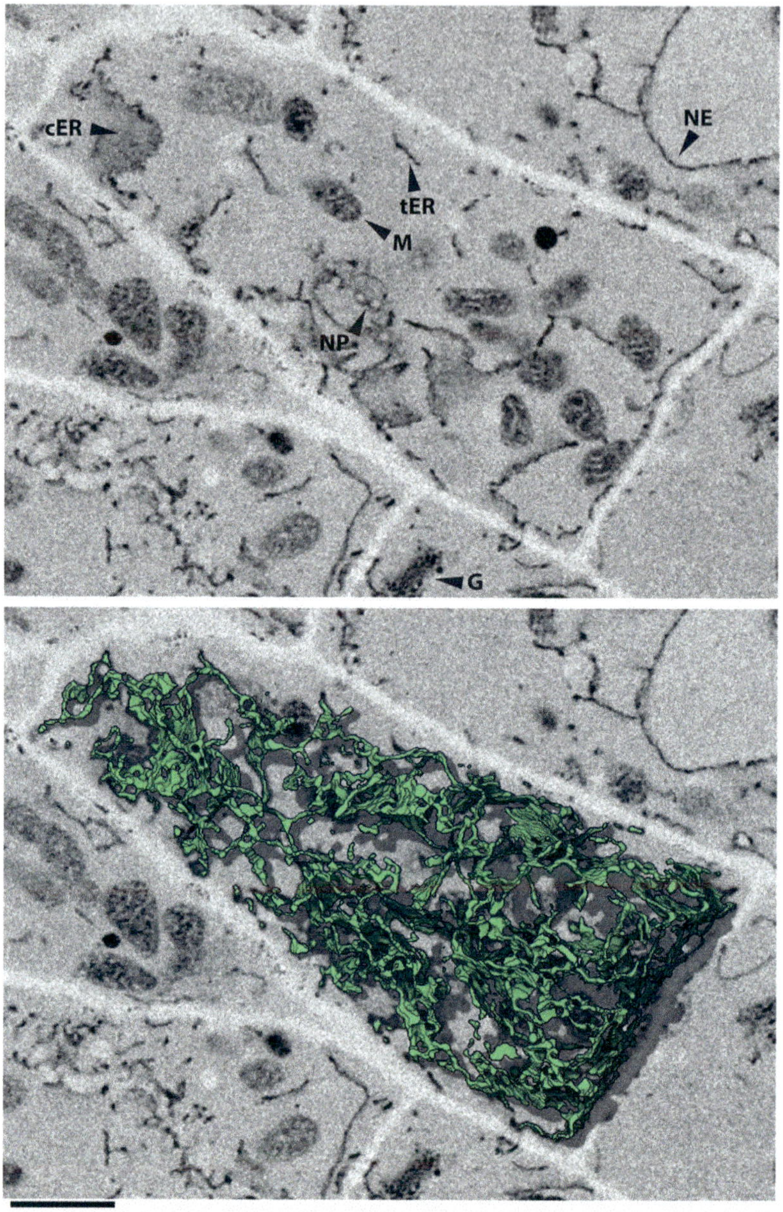


**Fig. 2** *Arabidopsis* root meristem cell stained with the ZIO technique
The endomembrane system including the cisternal ER (cER), tubular ER (tER), nuclear envelop (NE) with clearly visible nuclear pores (NP), and Golgi bodies (G) is heavily stained (top). Mitochondria (M) show variable staining as well. This staining technique allows a semi-automated rendering of the ER network (green, bottom). Scale bar = 1 μm

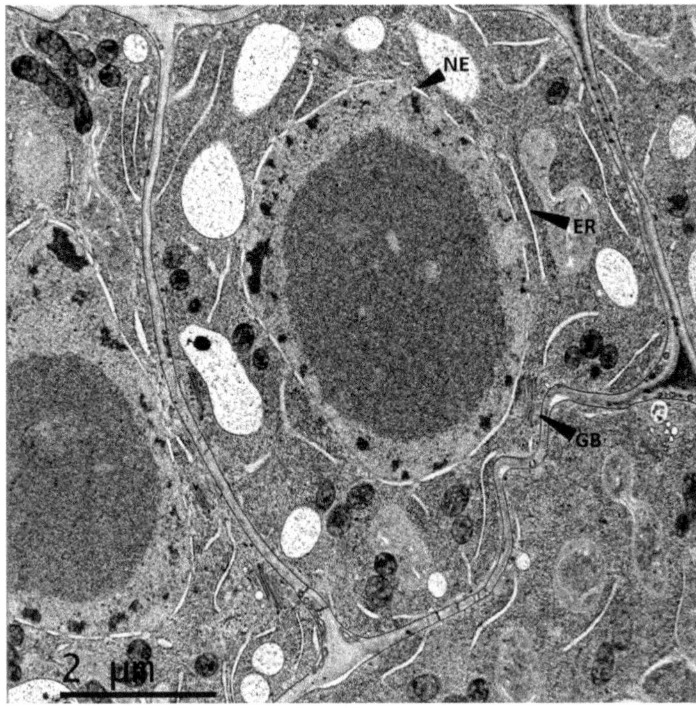


**Fig. 3** Example of an *Arabidopsis* root cell fixed using the HPF technique and stained with osmium tetroxide and uranyl acetate. The endomembrane system, including the nuclear envelope (NE), endoplasmic reticulum (ER), and Golgi body (GB), is easily observable. The lumen of these membranes is not stained, unlike cells stained with ZIO

Should automated segmentation not be a requirement or smaller, specific areas of the ER, such as ER-organelle contact sites, be the goal of study, an alternative fixation method may be more suitable. High-pressure freezing (HPF) allows for the rapid freezing of relatively thick sections of samples (up to 700 μm) [16] with limited ice crystal damage. HPF has clear advantages regarding ultrastructural preservation over room temperature chemical fixation. Samples are initially "cryofixed" to stabilize cellular components in their native state. Subsequently, chemical infiltration at low temperatures (freeze substitution) allows fixation and crosslinking with minimal shrinkage and loss of soluble components [17, 18] (Fig. 3). However, it requires the use of large volumes of liquid nitrogen and specialized equipment, including a high-pressure freezer and freeze substitution machine. The low temperatures of the freeze substitution process (−90 °C) require the use of solvents like acetone instead of aqueous solutions. Therefore, staining techniques like ZIO are difficult and often do not lead to a similar result as classic room temperature fixations and lack specificity to the ER and Golgi bodies. However, freeze substitution with osmium tetroxide and uranyl acetate results in excellent staining of the lipid bilayer of organelle membranes such as the ER and Golgi body.

Automated segmentation of high contrast samples is well established; however, recent developments in artificial intelligence may soon provide alternative means for EM image segmentation. Deep learning approaches using neural networks have been applied to segment either specific organelles [19] or all organelles present in a cell [20, 21]. In addition, large EM datasets are now available as training datasets for AI, e.g., CEM500K [22].

Here we present two techniques for sample preparation of plant tissues for 3D microscopy techniques:

1. Chemical fixation followed by ZIO staining.
2. High-pressure freezing followed by freeze substitution with osmium and uranyl acetate.

## 2 Materials

### 2.1 Materials for Chemical Fixation and ZIO Staining

1. Fixative: 1% paraformaldehyde (*see* **Note 1**), 1% glutaraldehyde, 2% sucrose, 50 mg calcium chloride, drop of Brij.35 surfactant (*see* **Note 2**) in 0.1 M sodium cacodylate buffer (NaCac, pH 6.9) (*see* **Note 3**).
2. Zinc-Iodide: Add 3 g zinc to 20 mL of deionized water, add 0.5 g of resublimed iodine (*see* **Note 4**), stir for 5 min, and then filter through No. 1 filter paper.
3. Osmium: prepare 2% osmium tetroxide in deionized water (*see* **Note 5**). Mix 2% osmium and filtered zinc-iodide 1:1 just before adding to the sample.
4. EtOH dilution series and absolute EtOH dried over a molecular sieve.
5. Spurr resin, hard (*see* **Note 6**).

#### 2.1.1 Materials for High-Pressure Freezing and Freeze Substitution

1. 7–10-day-old *Arabidopsis* seedlings grown on ½ MS.
2. High-pressure freezing and freeze substitution machines, in this case the Baltec HMP 010 high-pressure freezer and Leica EM AFS2.
3. For high-pressure freezing: planchettes (0.2 mM and 0.3 mM), 0.1 M sucrose, 100 mg/mL lecithin in chloroform, microdissecting scissors, tweezers, hexadecane, dissecting microscope.
4. Freeze substitution cocktail: 2% osmium prepared in 100% acetone from crystals and 0.1% uranyl acetate (prepared initially as a 5% solution in methanol).
5. Liquid nitrogen (*see* **Note 7**).
6. Containers for liquid nitrogen (*see* **Note 8**).
7. Cryovials.

8. En bloc staining: 2% osmium prepared in 100% acetone from crystals, toothpicks.
9. For washes: 100% acetone.
10. Taab 812 hard resin (*see* **Note 9**).

## 3 Methods

### *3.1 Chemical Fixation and ZIO Staining of* Arabidopsis *Seedling for 3D EM*

1. Prepare fixative and zinc iodide.
2. Place 1-week-old seedling or cut off root and leaves (*see* **Note 10**) into small glass vial with fixative, and apply mild vacuum for 1 min by placing open vial into vacuum pump (*see* **Note 11**).
3. Close glass vials, and leave on rocker for 1 h at room temperature to fix.
4. Wash seedlings 2× with 0.1 M NaCac buffer for 10 min and then 1× with deionized water for 10 min.
5. Mix zinc iodide and 2% osmium tetroxide 1:1 to make an appropriate volume of ZIO depending on the number of samples and size of fixation vials. Remove water from samples and add ZIO mix.
6. Leave closed vials for incubation for 4 h for roots or at least 12 h for leaf tissue to obtain sufficient staining (*see* **Note 12**).
7. Wash quickly 3× with deionized water to remove precipitates and ZIO solution.
8. Wash 2× for 10 min with deionized water.
9. If seedlings are still intact, cut off roots to allow better dehydration and resin infiltration.
10. Dehydrate tissue in ethanol series for at least 15 min per dilution in 10%, 30%, 50%, 70%, 90%, and 100% EtOH. Dehydrate 2× 30 min in absolute EtOH.
11. Infiltrate with Spurr resin (*see* **Note 13**) for at least 1 h per dilution in 10%, 30%, and at least 3 h in 50%, 70%, and 90% and 3× in 100% Spurr (*see* **Note 14**). Mix with fresh 100% Spurr resin, and pour into flat embedding dishes, or place samples into embedding molds, and polymerize for 12 h at 70 °C (*see* **Note 15**).

### *3.2 High-Pressure Freezing of* Arabidopsis *Seedlings for 3D EM*

1. Prepare planchettes with lecithin: On the day of the experiment, prepare a mixture of 100 mg/mL lecithin in chloroform in the fume hood. Pipette the lecithin solution onto the center of the planchette and the part of the lid that will be in contact with the sample. Allow the solution to evaporate, leaving a thin layer on the planchette.

2. Perform test shots: ensure the high-pressure freezing equipment is working correctly by performing three test shots on test samples, which can be two 0.3 mM planchettes placed back to back.
3. Load the planchettes: pipette cryoprotectant (0.1 M sucrose) into the 0.2 mM planchette. Using the microdissecting scissors and tweezers, under a dissecting microscope, cut a small piece of root tip that will fit full inside the planchette, and transfer to the planchette using the tweezers. Each planchette can hold 2–4 *Arabidopsis* roots. Once the roots have been transferred, ensure there are no bubbles or air spaces in the planchette, which could cause cavitation and damage the sample. Place the 0.3 planchette flat side down as lid on the sample planchette using tweezers, and transfer the planchette sandwich into the sample holder.
4. Perform the shot: insert the sample holder into the machine, and perform the high-pressure freezing shot. Immediately transfer the sample to liquid nitrogen to prevent defrosting of the sample. If possible, separate the planchettes, and transfer the planchette containing the sample to a labelled cryovial submerged in liquid nitrogen (*see* **Note 16**).
5. Prepare freeze substitution cocktail: in a fume hood, prepare the freeze substitution cocktail of 2% osmium and 0.1% uranyl acetate in 100% acetone by dissolving osmium tetroxide crystals in acetone and then adding the appropriate volume of uranyl acetate in methanol. Separate this cocktail into enough cryotubes to hold all the samples created, and then freeze in liquid nitrogen. Transfer planchettes to these frozen vials under liquid nitrogen. Place the precooled lid on, but do not seal tightly to prevent pressure accumulation from nitrogen gas.
6. Perform freeze substitution: transfer samples to the freeze substitution machine. Set up the freeze substitution program as follows: soak at −90 °C for 12 h, ramp −90 °C to −85 °C for 6 h, ramp −85 °C to −20 °C for 48 h, soak at −20 °C for 12 h, and ramp −20 °C to 10 °C for 12 h (*see* **Note 17**).
7. En bloc staining: once the sample is at room temperature, remove from the freeze substitution machine (*see* **Note 18**). In a fume hood, wash the sample in 100% acetone 3 times, and try to remove the sample from the planchette using the disposable toothpicks. The planchette can then be discarded. Incubate the samples in 2% osmium in acetone for 1 h. Wash the samples in 100% acetone 3 times.
8. Infiltrate with resin: mix samples with 30% resin to acetone (TAAB 812) and rock at room temperature for 1 h. Repeat overnight with 50% resin. The next day, replace with 70% resin and rock at room temperature for 24 h. Transfer samples to

100% resin and keep rocking at room temperature during the day. At the end of the day transfer to fresh resin and leave in the fridge overnight. Repeat this process the next day.

9. Embed in fresh resin: On the final day, embed the samples in fresh resin in appropriate embedding dishes, and then polymerize at 70 °C for at least 12 h.

## 4 Notes

1. 10% stock paraformaldehyde: mix 2 g of PFA in 20 mL deionized water, and heat up to 70 °C while stirring. When reaching target temperature, slowly add drops of 1 M NaOH until PFA is completely dissolved. Store in fridge.
2. Brij.35 surfactant is used to enhance access of the fixative to the hydrophobic leaf surface. Other or no surfactant may be used and may result in similarly good fixation results.
3. 0.2 M stock sodium cacodylate buffer: mix 400 mL of deionized water with 21.4 g sodium cacodylate, adjust pH with HCl, and add deionized water to final volume of 500 mL. Dilute 1:1 with deionized water for 0.1 M buffer.
4. Resublimed iodine currently available from most companies seems to cause precipitates when used in the previously published ratio with zinc (15). We reduced the iodine to 0.5 g per 3 g zinc and found that the precipitates disappeared almost completely, and the ZIO solution after incubation showed significantly less precipitates.
5. If preparing osmium tetroxide solution from crystalline osmium tetroxide, leave overnight for the crystals to dissolve completely. Sonication speeds up the process if needed.
6. Depending on the country, Spurr resin is now available again with ERL 4221D instead of ERL 4206, or a similarly low-viscosity resin is offered as a substitute.
7. Liquid nitrogen is required for the freeze substitution and high-pressure freezing machines and to keep samples and tools used to manipulate samples cool.
8. Any containers that can be used to temporarily hold liquid nitrogen are suitable, for example, polystyrene ice boxes.
9. TAAB 812 resin is recommended for the breadth of possible applications, including thin-section TEM and SBF-SEM.
10. For *Arabidopsis* seedling roots, one can leave the seedlings intact for at least the initial fixation step to allow easier handling. For larger tissues and plants, cut off tissue of interest to allow quicker diffusion of fixative into the tissue.

11. Air is extruded from the intercellular space, thus allowing the fixative to enter the leaves more easily for quicker fixation. If only the roots are of interest, this procedure is not necessary.
12. Depending on the tissue, those incubation times vary. It is a good idea to run an initial experiment with three different time points to determine the ideal incubation time to obtain enough staining but prevent cytoplasmic precipitates.
13. We prefer Spurr resin due to its low viscosity and thus quicker and better infiltration properties through the cell walls. Other resins might be similarly suitable but potentially require longer infiltration times.
14. Resin infiltration times can be shortened by mild centrifugation to about half an hour each step; however, the root tips are very delicate at this stage, and damage may occur when trying to detach the roots from the tube wall.
15. To label the sample, we print fixation date, strain, and other important information in Arial font size 3, and place the paper into the mold or embedding dish with the sample. Test if the ink is stable in Spurr resin before using the label. Color printed labels usually do not work; therefore, use the black and white print option.
16. It is recommended that cryovials are perforated 3–4 times in the upper third of the vial and once in the lid using a hot needle. This allows for liquid nitrogen to flow over the sample while the tube is closed without allowing samples to escape into the liquid nitrogen. These vials can be reused multiple times. Multiple samples can be stored in such a way in liquid nitrogen until the freeze substitution cocktail has been prepared.
17. Ensure that the lids on the sample vials are tightened before the second temperature ramp once all the liquid nitrogen has evaporated, otherwise the freeze substitution cocktail might evaporate at the end of the programme.
18. From this point in the protocol onward, it is essential that samples never fall fully dry as this can cause damage to samples.

## References


1. Friedman JR, Voeltz GK (2011) The ER in 3D: a multifunctional dynamic membrane network. Trends Cell Biol 12:709–717
2. Hawes C et al (2008) The plant ER-Golgi interface. Traffic 10:1571–1580
3. Seguí-Simarro JM et al (2004) Electron tomographic analysis of somatic cell plate formation in meristematic cells of Arabidopsis preserved by high-pressure freezing. Plant Cell 16:836–856
4. Hamada T et al (2014) Microtubules contribute to tubule elongation and anchoring of endoplasmic reticulum, resulting in high network complexity in Arabidopsis. Plant Physiol 166:1869–1876
5. Kittelmann M, Hawes C, Hughes L (2016) Serial block face scanning electron microscopy

and the reconstruction of plant cell membrane systems. J Microsc 263:200–211

6. Hayworth KJ, Morgan JL, Schalek R, Berger DR, Hildebrand DG, Lichtman JW (2014) Imaging ATUM ultrathin section libraries with WaferMapper: a multi-scale approach to EM reconstruction of neural circuits. Fronti Neural Circuits 8(68):1–18
7. Domozych DS (1989) The endomembrane system and mechanism of membrane flow in the green alga, Gloeomonas kupfferi (Volvocales, Chlorophyta) II. A cytochemical analysis. Protoplasma 149:108–119
8. Puhka M et al (2007) Endoplasmic reticulum remains continuous and undergoes sheet-to-tubule transformation during cell division in mammalian cells. J Cell Biol 179:895–909
9. Sexton R, Hall JL (1991) Enzyme cytochemistry. In: Hawes C, Hall JL (eds) Electron mocroscopy of plant cells. Academic Press, London, pp 67–84
10. Shu X et al (2011) A genetically encoded tag for correlated light and electron microscopy of intact cells, tissues, and organisms. PLoS Biol 4:e1001041
11. Martell JD et al (2012) Engineered ascorbate peroxidase as a genetically encoded reporter for electron microscopy. Nat Biotechnol 30(11): 1143–1148
12. Li H et al (2022) A validated set of ascorbate peroxidase-based organelle markers for electron microscopy of saccharomyces cerevisiae. mSphere 7(4):1–12
13. Barlow PW, Hawes C, Horne JC (1984) Structure of amyloplasts and endoplasmic reticulum in the root caps of Lepidium sativum and Zea mays observed after selective membrane staining and by high-voltage electron microscopy. Planta 160:363–371
14. Helper PK (1981) The structure of the endoplasmic reticulum revealed by osmium tetroxide-potassium ferricyanide staining. Eur J Cell Biol 26:102–111
15. Hawes CR, Horne LC (1983) Staining plant cells for thick sectioning: uranyl acetate, copper and lead citrate impregnation. Biol Cell 48: 207–210
16. Sartori N, Richter K, Dubochet J (1993) Vitrification depth can be increased more than 10-fold by high-pressure freezing. J Microsc 172(1):55–61
17. Hawes P et al (2007) Rapid freeze-substitution preserves membranes in high-pressure frozen tissue culture cells. J Microsc 226(2):182–189
18. Gilkey JC, Staehelin LA (1986) Advances in ultrarapid freezing for the preservation of cellular ultrastructure. J Electron Microsc Tech 3(2):177–210
19. Spiers H et al (2021) Deep learning for automatic segmentation of the nuclear envelope in electron microscopy data, trained with volunteer segmentations. Traffic 22(7):240–253
20. Khadangi A, Boudier T, Rajagopal V (2020) EM-net: deep learning for electron microscopy image segmentation. In: 25th International Conference on Pattern Recognition (ICPR). IEEE, Milan, pp 31–38
21. Gallusser B et al (2023) Deep neural network automated segmentation of cellular structures in volume electron microscopy. J Cell Biol 222(2):e202208005
22. Conrad R, Narayan K (2021) CEM500K, a large-scale heterogeneous unlabeled cellular electron microscopy image dataset for deep learning. elife 10:e65894

# Chapter 3

# Studying Plant ER-PM Contact Site Localized Proteins Using Microscopy

**Lifan Li, Tong Zhang, Patrick J. Hussey, and Pengwei Wang**

## Abstract

As in most eukaryotic cells, the plant endoplasmic reticulum (ER) network is physically linked to the plasma membrane (PM), forming ER-PM contact sites (EPCS). The protein complex required for maintaining the EPCS is composed of ER integral membrane proteins (e.g., VAP27, synaptotagmins), PM-associated proteins (e.g., NET3C), and the cytoskeleton. Here, we describe methods for studying EPCS structures and identifying possible EPCS-associated proteins. These include using artificially constructed reporters, GFP tagged protein expression followed by image analysis, and immunogold labelling at the ultrastructural level. In combination, these methods can be used to identify the location of putative EPCS proteins, which can aid in predicting their potential subcellular function.

**Key words** Endoplasmic reticulum, ER-PM contact sites, *N. benthamiana*, Chi-square analysis, Immunogold TEM, VAP27 family

## 1 Introduction

The cortical endoplasmic reticulum (ER) in plants forms a dynamic geometrical network. These dynamics are controlled, in part, by the cortical actin network [1] and several members of the myosin XII family [2, 3]. Another interesting feature of the ER network is that it forms physical interactions with various membrane compartments and organelles [4, 5], and the best characterized to date is the ER-PM contact/anchor site (EPCS) [6–12]. Contact sites can be identified using both light and electron microscopy, but their function in plants is not fully understood, possibly due to the lack of molecular markers and the difficulty in resolving their structures using conventional microscopy.

ER-PM contact sites can be distinguished at the ultrastructural level. Morphologically, EPCS can be defined as a region where the two membranes are less than 10–30 nM apart and where ribosomes are excluded (as suggested in our study [6] and studies conducted

Verena Kriechbaumer (ed.), *The Plant Endoplasmic Reticulum: Methods and Protocols*, Methods in Molecular Biology, vol. 2772, https://doi.org/10.1007/978-1-0716-3710-4_3,

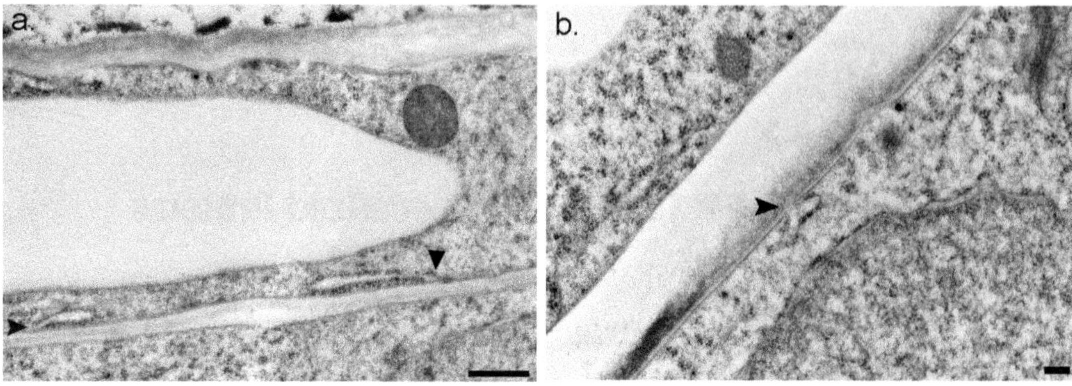


**Fig. 1** Transmission electron microscopy image of a cell from Arabidopsis hypocotyl (**a**) and *N. benthamiana* leaf epidermal cells (**b**). The ER-PM contact site (arrowhead) can be observed. ER membrane and plasma membrane are connected but not fused to each other; the average space between the membranes is less than 10 nM (scale bar = 100 nM)

in yeast). At the EPCS, PM and ER membranes contact each other but are not fused (Fig. 1). Our understanding of plant EPCS at the molecular level has increased dramatically over the last few years. Various proteins have been reported to be associated with the EPCS [6–12], and this has allowed studies to be carried out on the function of plant EPCS (e.g., cargo transport, signal communication, and cytoskeletal reorganization).

Fluorescent protein fusions coupled with their transient protein expression in leaf epidermal cells (*N. benthamiana* or *N. tabacum*) is a widely accepted method for studying protein localization [13], and this is the most straightforward approach for studying putative EPCS localized proteins in plants. However, due to the nature of light microcopy, EPCS structures can be difficult to resolve, and protein overexpression, in some cases, can cause problems in ascertaining the endogenous localization. Therefore, additional confirmation is required, such as freeze shattering followed by immunofluorescence, which is important in determining the localization of the native protein. Furthermore, immunogold labelling can also be used to characterize the sites, and we routinely use Chi-square analysis to ensure that gold particles found at the EPCS are specifically localized and enriched.

Moreover, some artificially designed reporters are also widely used to study proteins localized to EPCS, such as MAPPER [11, 12, 14–17] and LiMETER [18]. The MAPPER construct (membrane-attached peripheral ER) consists of an ER transmembrane domain, a cytosolic linker, and a PM-targeting motif (Fig. 2), which non-selectively labels EPCS under low-expression conditions [14]. LiMETER (light-inducible membrane-tethered peripheral ER) contains a flexible linker and a cytoplasmic LOV2-Jα domain derived from *Avena sativa* phototropin 1 [19, 20]. In the resting

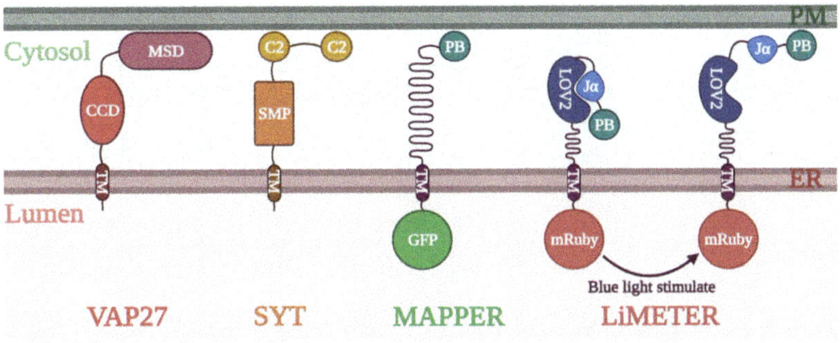


**Fig. 2** Diagramatic illustrations of different EPCS localized proteins

state, the Jα helix docks to the LOV2 domain while caging the PB domain to prevent its targeting to the PM. After blue light activation, the PB domain is exposed to the cytosol and associated with the PM. Both reporters were originally developed for mammalian systems but can also be used in plants.

In summary, we present several methods that can be used to confirm the EPCS localization of candidate proteins. These include conventional light and electron microscopy approaches, immunocytochemical studies, and the use of artificially designed reporters.

## 2 Materials

### 2.1 Plant and Microscopy Material

1. *N. benthamiana* or *N. tabacum* grown in plant growth rooms with a 16-h day length (25 °C) and 8-h dark (18 °C) regime. Plants around 5 weeks old are ideal for transient expression studies.
2. Wild-type Arabidopsis (Col-0) or genetically transformed lines can be grown either on compost or 1/2MS medium (1% sucrose, 0.7% plant agar) in a growth chamber with a 16-h light (22 °C) and 8-h dark (18 °C) regime.
3. Low melting point agarose.
4. FM4-64 membrane dye.
5. Glass slides and cover slips.
6. Razor blade.

### 2.2 Plasmids and Agrobacterium-Mediated Transformations

1. Endoplasmic reticulum markers: GFP/RFP-HDEL [6].
2. ER-PM contact site markers: VAP27–1-YFP [6, 8], GFP-MAPPER [14], and YFP/mRuby -LiMETER (unpublished).
3. Suppressor of post-transcriptional gene silencing (PTGS): pCB301-p19 [22].
4. Agrobacterium strain: GV3101::pMP90.

**2.3 Immunogold Labeling**

1. MonoStep Lowicryl HM20 (Agar Scientific).
2. Mouse polyclonal antibody raised against NET3C, NAP1 [6, 23] or protein of interest.
3. 5 or 10 nM gold-conjugated goat anti-mouse antibody (British Biocell International, Cardiff, UK).
4. Blocking buffer: PBS supplemented with BAS, 1% w/v.
5. BSAc buffer: PBS buffer supplemented with acetylated BSA, 0.1% w/v.
6. Phosphate buffered saline (PBS).

**2.4 Computer Programs Required**

1. Image J.
2. Adobe Photoshop.

## 3 Methods

**3.1 Live Cell Imaging and Quick Identification of Immobile ER-PM Contact Sites**

The method described here can be used for a brief determination of putative EPCS localized proteins using fluorescent protein fusions and bacteria infiltration-mediated transient expression.

1. Follow the leaf infiltration protocol (*see* Chap. 1), and infiltrate with *Agrobacterium* (GV3101) transformed with the relevant construct, e.g., VAP27-1-YFP (*see* **Notes 2** and **3**).
2. Normally 48–60 h after infiltration, dissect a leaf segment (25–50 $mM^2$), and mount on a slide with a cover slip. To avoid sample drifting during live imaging, low melting agarose can be used here instead of water for mounting (*see* **Note 3**), or tape can be used to stick coverslips to the slide.
3. Using a laser scanning confocal microscope, capture a time series over 3–5 min (frame size 512 × 512; 2 times line average; no time interval between every image).
4. Make sure samples do not drift during the image capture series, and choose three frames at different time points, and assign them a different color. For example, use red for the image at 0 s, green for 15 s, and cyan for 30 s (Fig. 3).
5. With Adobe Photoshop, overlay three images on top of each other, and merge them into one picture. Any persistent sites over the time period should be colored magenta, and the color for each point should still be distinguished for mobile structures.
6. As shown in Fig. 3a, a time series of VAP27-YFP over 30 s, each image represents a time point and is pseudo-colored to represent the different time frames. A merged picture indicates that the ER-PM contact sites are persistent (magenta), while the

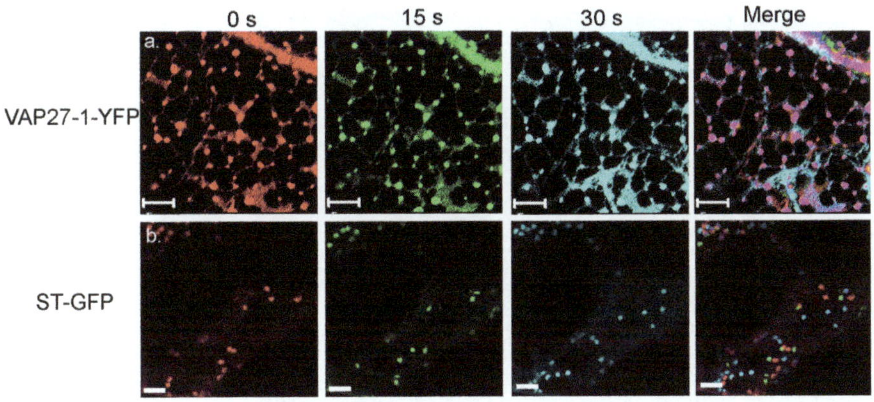


**Fig. 3** Persistency analysis of VAP27–1-YFP labelled EPCS (**a**) and ST-GFP labelled Golgi bodies (**b**). Images representing time points (0 s, 15 s, and 30 s) are pseudo-colored, and structures colored magenta indicate they are persistent during the time frame, such as VAP27-1 labelled EPCS (**a**). In contrast, when the same analysis is applied to moving Golgi bodies (**b**), no persistent structures can be identified (scale bar = 5 μm)

rest of the ER network remodels. In contrast, when the same analysis is applied to moving Golgi bodies, no persistent structures can be identified (Fig. 3b).

7. Persistent structures that appear to be localized to stable ER nodes are likely to be ER-PM contact site associated and can be analyzed using the procedures detailed in Subheadings 3.2, 3.3 and 3.4.
8. In addition, protein co-expression studies with EPCS markers (e.g., VAP27, NET3C) or artificially designed reporters (Subheading 3.3) can be used to determine EPCS localization. But care should be taken, as our study suggested the EPCS in plants are likely to be represented by different protein complexes (e.g., SYT1), which do not necessarily overlap [24].

### *3.2 ImageJ Analysis for Signal Distribution*

By definition, protein complexes of EPCS must localize in between the ER and PM. Using fluorescent protein markers, we should be able to distinguish the distribution of the signals from the ER, PM, and EPCS localized proteins.

1. *Agrobacterium*-mediated transformation and sample preparation are the same as in Subheading 3.1.
2. Dissect a leaf segment ($5 \times 5$ mM$^2$), and incubate it in water solution containing FM4-64 (8 μM) for 10 min. This will stain the plasma membrane without the labelling of endocytic vesicles.
3. Set the focal plane in the middle of the cell, and take images of the cell periphery, where cytoplasm, plasma membrane, and cell wall can be distinguished (*see* **Note 4**).

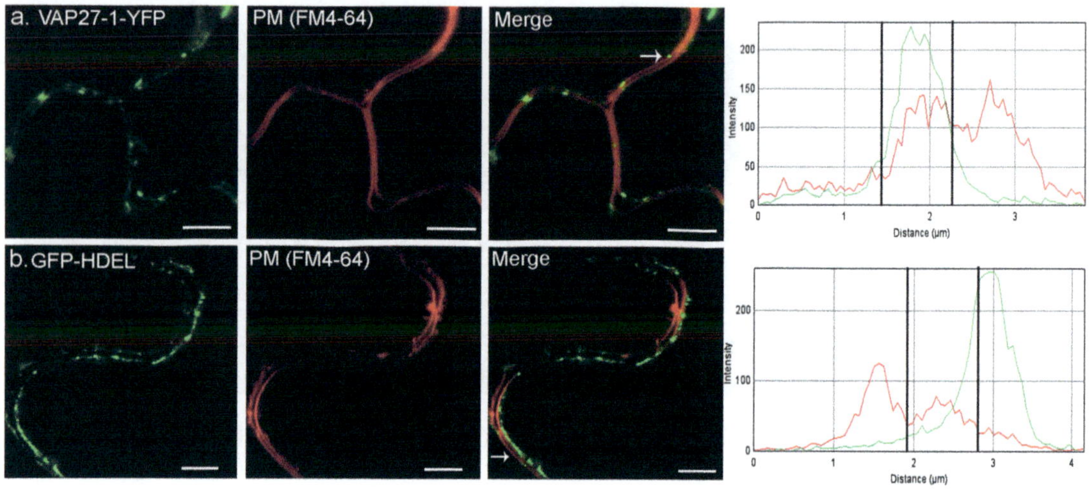


**Fig. 4** Signal distribution of EPCS (**a**) and the ER network (**b**) in relation to the plasma membrane. A selected area (arrow) was analyzed and the signal from VAP27-1 labeled EPCS (green) overlaps with PM (red), whereas the signal from ER network (GFP-HDEL) is distinct from the PM (scale bar = 5 μm)

4. Open the merged picture using ImageJ, and draw a line where ER-PM contact sites can be identified (Fig. 4a, arrow).
5. In the Plugins option, go to Graphics, and then choose RGB Profile Plot. An image that indicates signal distribution should appear. The signal from ER network, ER-PM contact sites, and plasma membrane should be identifiable. For example, the signal from EPCS structures and PM appears to be overlapped, while ER network/cytoplasmic labelling is separated from PM (Fig. 4b, arrow).
6. This only gives a rough idea of the signal distribution, and it largely depends on the quality of the image. Further analysis is required (*see* Subheading 3.4).

### *3.3 Labelling EPCS Using Artificially Designed Reporters*

1. LiMETER and MAPPER can be used to label EPCS through their transient expression in tobacco leaf epidermal cells. Since the overexpression of some ER integral membrane proteins may change the morphology of ER, it is necessary to determine the optimal expression conditions from the start (*see* **Notes 1** and **2**).
2. Both 448 nM and 514 nM lasers can excite the LiMETER, while the 514 nM laser is less effective (Fig. 5a). In the resting state, LiMETER localizes to the entire ER network evenly. In the excited state, LiMETER is recruited to immobile punctate structures that resemble EPCS.
3. The labelling of EPCS by LiMETER is rapid and reversible. First, select a region of interest, and turn on the 488 nM laser using the time-lapse imaging mode, and EPCSs become visible

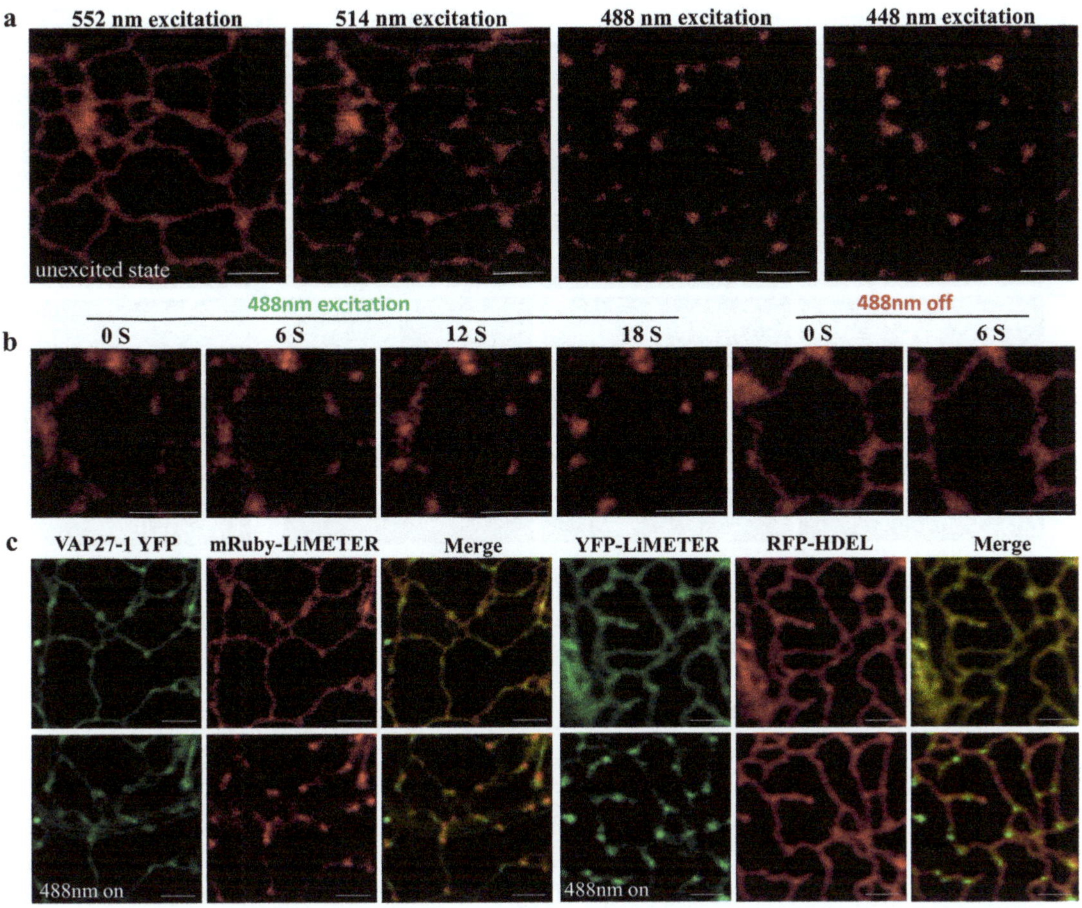


**Fig. 5** Labelling EPCS using artificially designed reporter. (**a**) mRuby-LiMETER was activated by the 458 nM and 488 nM lasers more effectively, while no activation was found with a 552 nM laser. (**b**) Time-lapse imaging showed that LiMETER could rapidly be activated and recruited to EPCS when 488 nM was turned on, while it redistributed to the ER network once the 488 nM laser was turned off. (**c**) VAP27-1 YFP and mRuby-LiMETER co-localized and at EPCS after excitation, and the LiMETER-labelled EPCSs are structurally distinct from the ER network that was labelled by RFP-HDEL (scale bar = 2 μm)

immediately (Fig. 5b). After 30 s of excitation, approximately 95% of EPCS reach a stable state that do not disappear within 15 s (*see* **Note 5**). Once the 488 nM laser is turned off, the signal of LiMETER labelled EPCS quickly disappears. This process can be repeated multiple times as long as the fluorescent proteins tagged to LiMETER do not get photobleached.

4. Therefore, these reporters are going to be useful in determining the EPCS localization of a candidate protein by using transient co-expression. For example, VAP271-YFP and photoactivated LiMETER were found to co-localize at EPCS, while little EPCS colocalization was found between LiMETER and RFP-HDEL (Fig. 5c).

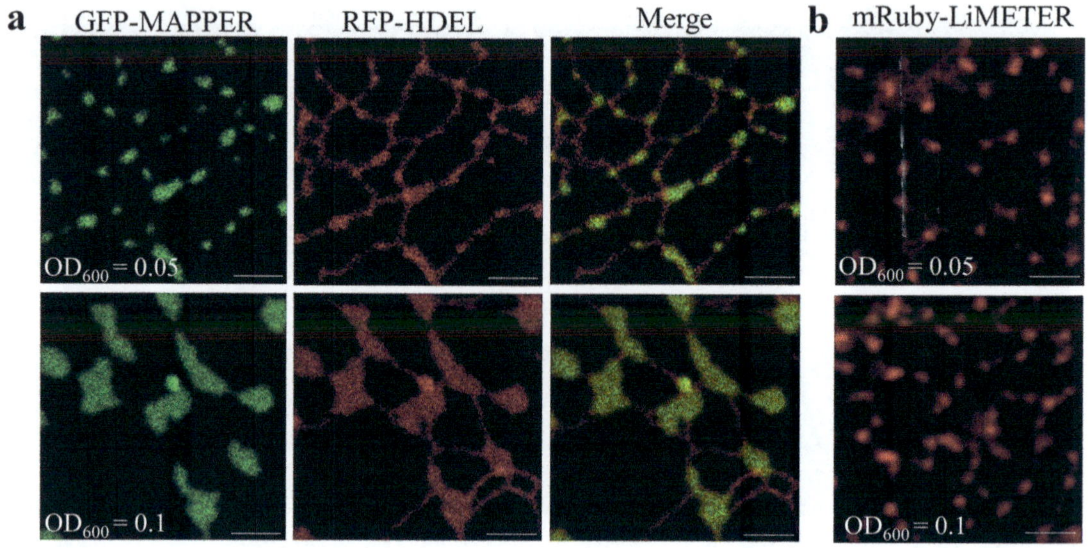


**Fig. 6** Transient over-expression of EPCS reporters affects ER and EPCS morphology. (**a**) The overexpression of MAPPAER increases the area of EPCS and affects the overall morphology of the ER. (**b**) The overexpression of LiMETER increases the number of EPCS (scale bar = 2 μm)

5. As for the MAPPER reporter, this especially labels EPCS when its expression level is low. However, high expression of MAPPER does cause EPCS expansion and does alter the morphology of the ER network (Fig. 6a).
6. In contrast, the overexpression of LiMETER has little influence on the morphology of EPCS and ER in general but may affect the density of EPCS (Fig. 6b).
7. Therefore, we do not recommend that these reporters are used to compare the number and the structure of EPCS in a transient expression system; they are only useful to determine co-localization at EPCS with candidate proteins (*see* **Note 6**).

### 3.4 Immunogold TEM and Chi-Square Analysis of Gold Particle Enrichment

Protein localization at the ultrastructural level can be achieved by immunogold labelling. This approach, in combination with Methods, Subheadings 3.1, 3.2 and 3.3, can provide convincing results for determining the location of proteins and proteins of the ER-PM contact site.

1. Ideally, high-pressure freezing with freeze substitution should be used for this purpose. It faithfully preserves cell structure especially the plasma membrane, which is important for the correct location of EPCS.
2. The protocol is adapted from [25] with some modifications. Before high-pressure freezing, the distal 1–2 mM tips of the Arabidopsis roots (7 days after germination) were excised with a razor blade in 20% BSA.

3. The samples were freeze-substituted in anhydrous acetone containing 0.25% (v/v) glutaraldehyde and 0.1% (w/v) uranyl acetate for 48 h at −80 °C; then, the temperature was slowly raised to −50 °C over a 30-h period.
4. After several rinses in anhydrous acetone at −50 °C, infiltration continued at −50 °C into MonoStep Lowicryl HM20 (Agar Scientific) by increasing the concentration of resin to acetone stepwise (12 h in 22, 33, and 66% and then three times in 100% 96 h). Final embedding and UV polymerization were carried out at −50 °C for 48 h followed by a slow warming to 20 °C; polymerization then continued for a further 24 h.
5. Ultrathin sections (50–70 nM) are prepared according to [21] and transferred onto Formvar-coated nickel grids.
6. Incubate sections for 20 min at room temperature in a PBS buffer supplemented with glycine (0.1% w/v), and then incubate for a further 30 mins in a blocking buffer containing 1% BSA.
7. Wash the grids in BSAc buffer 3 times for 5 min.
8. In a humid chamber, incubate the sections for a further 30 min with a primary antibody, which is diluted in blocking buffer (dilution from 1:50 to 1:200 depends on the antibody; *see* **Note 8**).
9. After washing in BSAc buffer (3 × 5 min), incubate the sections in 5–10 nM gold labelled goat anti-mouse antibody diluted in blocking buffer (1:20) for 30 min (*see* **Note 7**).
10. With the TEM, the structure of ER-PM contact sites can be identified (Fig. 5). The closely associated gold particle indicates that the protein that the primary antibody recognizes is localized to these sites (*see* **Note 8**).
11. Next, a Chi-square test can be used to test if the gold particle labelling enriched at the selected region is genuine or random.
12. For ER-PM contact sites, count the number of gold particles in an area 50 nM on either side of the plasma membrane. Similarly, count the number of gold particles at the EPCS (Fig. 7).
13. From these measurements, the area of EPCS in proportion to the PM, the density of gold particles in EPCS, and the expected number for a random distribution (total gold number × the proportion of EPCS area) can be calculated (Fig. 7).
14. Finally, the p-value can be calculated based on the real distribution of gold particles compared to the expected number from a random distribution. This can be achieved using the online program in http://graphpad.com/quickcalcs/chisquared1.cfm.

a. b.

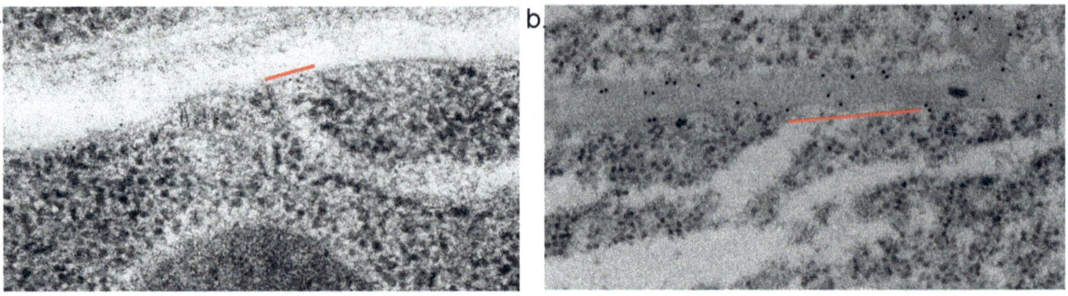

a.

| | Total Numbers | Surface area (μm2) | Proportion to total area | Density (per μm2) | Expected for random distribution | P-Value |
|---|---|---|---|---|---|---|
| PM associated | 4 | 0.076 | 9.2% | 92 | 6.36 | P = 0.002* |
| ER-PM contact sites associated | 3 | 0.007 | 90.8% | 428.6 | 0.36 | |

b.

| | Total Numbers | Surface area (μm2) | Proportion to total area | Density (per μm2) | Expected for random distribution | P-Value |
|---|---|---|---|---|---|---|
| PM associated | 12 | 0.064 | 26.6% | 265.6 | 12.5 | P = 0.783* |
| ER-PM contact sites associated | 5 | 0.017 | 73.4% | 294.1 | 4.5 | |

**Fig. 7** Immunogold labelling of Arabidopsis root tips. (**a**) With NET3C antibody gold particles are located at the ER-PM contact sites and the PM. The region of ER-PM contact sites is highlighted in red. The p-value of 0.002 from Chi-square analysis indicates that the labelling of NET3C positive gold is enriched at the EPCS compared to the PM. (**b**) When a NAP1 (actin binding protein) antibody is used as a control, it is evenly distributed on the PM. The calculated p-value is much larger than 0.05, indicating that any association with EPCS is a random event. The Chi-square calculation for PM and EPCS is shown in a table (scale bar = 100 nM)

15. As shown in Fig. 7, the real association (Fig. 7a) and random association (Fig. 7b) can be distinguished based on their p values, and a number below 0.05 indicates the enrichment of gold particles in the selected area.

## 4 Notes

1. When infiltrating the leaf epidermis of *N. benthamiana,* the viralp19 suppressor [22] can be used to increase protein expression and signal intensity. However, it cannot be applied to *N. tabacum* as it causes lethality.
2. The ideal optical density of GV3101 ($OD_{600}$) used in infiltration depends on the construct that will be transformed. Normally, an $OD_{600}$ of around 0.05–0.2 is used, the principle being that the lowest OD is used, which can result in a useable level of protein expression.
3. Prolonged imaging on the same leaf segment should be avoided; the mechanical pressure from cover slips can induce cellular stress. Normally, samples should be changed at least every 20 min.
4. Such procedures can also be carried out on most proprietary confocal and image analysis programs.

5. The time required for EPCS to stabilize after excitation depends on the wavelength and intensity of the laser. The higher the energy given, the shorter the time it takes to reach stability.
6. If the aim of a study is to compare the abundance or dynamics of EPCS at different conditions, using a stable transgenic line is preferred.
7. For each grid, approximately 10uL of antibody solution can be applied.
8. Negative controls are required. To do this, the secondary antibody can be applied directly without primary antibody incubation.

## Acknowledgments

The project was supported by NSFC grants (no. 32261160371, 92254307,91854102) and the Foundation of Hubei Hongshan Laboratory (2021hszd016) to P.W. We thank Prof. Chris Hawes (Oxford Brookes University) and Dr. Christine Richardson (Durham University) for their contributions to the previous edition of this chapter.

### References

1. Boevink P, Oparka K, Santa Cruz S et al (1998) Stacks on tracks: the plant Golgi apparatus traffics on an actin/ER network. Plant J 15: 441–447
2. Sparkes I, Runions J, Hawes C et al (2009) Movement and remodeling of the endoplasmic reticulum in nondividing cells of tobacco leaves. Plant Cell 21:3937–3949
3. Avisar D, Abu-Abied M, Belausov E et al (2009) A comparative study of the involvement of 17 Arabidopsis myosin family members on the motility of Golgi and other organelles. Plant Physiol 150:700–709
4. Gao H, Metz J, Teanby NA et al (2016) In vivo quantification of peroxisome tethering to chloroplasts in tobacco epidermal cells using optical tweezers. Plant Physiol 170:263–272
5. Stefano G, Hawes C, Brandizzi F (2014) ER – the key to the highway. Curr Opin Plant Biol 22C:30–38
6. Wang P, Hawkins TJ, Richardson C et al (2014) The plant cytoskeleton, NET3C, and VAP27 mediate the link between the plasma membrane and endoplasmic reticulum. Curr Biol 24:1397–1405
7. Wang P, Richardson C, Hawkins TJ et al (2016) Plant VAP27 proteins: domain characterization, intracellular localization and role in plant development. New Phytol 210:1311–1326
8. Levy A, Zheng JY, Lazarowitz SG (2015) Synaptotagmin SYTA forms ER-plasma membrane junctions that are recruited to plasmodesmata for plant virus movement. Curr Biol 25:2018–2025
9. Brault ML, Petit JD, Immel F et al (2019) Multiple C2 domains and transmembrane region proteins (MCTPs) tether membranes at plasmodesmata. EMBO Rep 20:e47182
10. Zang J, Klemm S, Pain C et al (2021) A novel plant actin-microtubule bridging complex regulates cytoskeletal and ER structure at ER-PM contact sites. Curr Biol 31:1251–1260
11. Ruiz-Lopez N, Pérez-Sancho J, Del Valle AE et al (2021) Synaptotagmins at the endoplasmic reticulum-plasma membrane contact sites maintain diacylglycerol homeostasis during abiotic stress. Plant Cell 33:2431–2453
12. Codjoe JM, Richardson RA, Elizabeth SH et al (2022) Unbiased proteomic and forward genetic screens reveal that mechanosensitive

ion 1 channel MSL10 functions at ER-plasma membrane contact sites in Arabidopsis thaliana. elife 11:e80501

13. Sparkes I, Runions J, Kearns A et al (2006) Rapid, transient expression of fluorescent fusion proteins in tobacco plants and generation of stably transformed plants. Nat Protoc 1: 2019–2025
14. Chang C, Hsieh T, Yang T et al (2013) Feedback regulation of receptor-induced $Ca^{2+}$ signaling mediated by E-Syt1 and Nir2 at endoplasmic reticulum-plasma membrane junctions. Cell Rep 5:813–825
15. Poteser M, Leitinger G, Pritz E et al (2016) Live-cell imaging of ER-PM contact architecture by a novel TIRFM approach reveals extension of junctions in response to store-operated $Ca^{2+}$-entry. Sci Rep 6:35656
16. Henry C, Carreras-Sureda A, Demaurex N (2022) Enforced tethering elongates the cortical endoplasmic reticulum and limits store-operated $Ca^{2+}$ entry. J Cell Sci 135:jcs259313
17. Lee E, Vanneste S, Pérez-Sancho J et al (2019) Ionic stress enhances ER-PM connectivity via phosphoinositide-associated SYT1 contact site expansion in Arabidopsis. PNAS 116:1420–1429
18. Jing J, He L, Sun A et al (2015) Proteomic mapping of ER–PM junctions identifies STIMATE as a regulator of $Ca^{2+}$ influx. Nat Cell Biol 17:1339–1347
19. Harper SM, Neil LC, Gardner KH (2003) Structural basis of a phototropin light switch. Science 301:1541–1544
20. Wu YI, Frey D, Lungu OI et al (2009) A genetically encoded photoactivatable Rac controls the motility of living cells. Nature 461: 104–108
21. Deeks MJ, Calcutt JR, Ingle EK et al (2012) A superfamily of actin-binding proteins at the actin-membrane nexus of higher plants. Curr Biol 22:1595–1600
22. Voinnet O, Rivas S, Mestre P et al (2003) An enhanced transient expression system in plants based on suppression of gene silencing by the p19 protein of tomato bushy stunt virus. Plant J 33:949–956
23. Wang P, Richardson C, Hawes C et al (2016) Arabidopsis NAP1 regulates the formation of autophagosomes. Curr Biol 26(15): 2060–2069
24. Siao W, Wang P, Voigt B et al (2016) Arabidopsis SYT1 maintains stability of cortical ER networks and VAP27-1 enriched ER-PM contact sites. J Exp Bot 67:6161–6171
25. Fiserova J, Goldberg MW (2011) Immunoelectron microscopy of cryofixed freeze-substituted Saccharomyces cerevisiae. Methods Mol Biol 657:191–204

# Chapter 4

# Preparation and Imaging of Specialized ER Using Super-Resolution and TEM Techniques

Karen Bell, Karl Oparka, and Kirsten Knox

## Abstract

The plant endoplasmic reticulum (ER) forms several specialized structures. These include the sieve element reticulum (SER) and the desmotubule formed as the ER passes through plasmodesmata. Imaging both of these structures has been inhibited by the resolution limits of light microscopy and their relatively inaccessible locations, combined with the fragile nature of the ER. Here we describe methods to view desmotubules in live cells under 3D-structured illumination microscopy (3D-SIM) and methods to fix and prepare phloem tissue for both 3D-SIM and transmission electron microscopy (TEM), which preserve the fragile structure and allow the detailed imaging of the SER.

**Key words** 3D-SIM, BY2, Endoplasmic reticulum, Desmotubule, Sieve-element reticulum, Imaging, OMX, ZIO

## 1 Introduction

The limit of lateral resolution in light microscopy was determined more than 100 years ago [1]. The diffraction limit of light means that objects closer together than 200 nM cannot be fully resolved, instead appearing blurred. This limit remained largely unchallenged, despite the development of confocal laser scanning microscopy (CLSM). Theoretically, CLSM can produce images below the diffraction limit, but this is not generally seen with biological samples, principally as the pinhole needs to be smaller than the Airy pattern [2]. This results in most of the superfluous out-of-focus light being discarded but also has the unwanted consequence of losing a significant portion of the in-focus emission. With biological specimens, the fluorescence is often too weak, or labile, to sustain a detectable signal when emission light is discarded. Therefore, in practice, the pinhole is used at an aperture larger than the Airy pattern, and as such any chance of significantly increased lateral resolution is lost.

Verena Kriechbaumer (ed.), *The Plant Endoplasmic Reticulum: Methods and Protocols*, Methods in Molecular Biology, vol. 2772, https://doi.org/10.1007/978-1-0716-3710-4_4,

Structured illumination microscopy (SIM) offers an alternative method to gain greater resolution while avoiding discarding the desired emitted light [3]. Optimally conducted, SIM can increase the lateral resolution to 100 nM and the axial resolution to 200 nM [4]. The technique relies on using spatially patterned excitation light. By introducing structure to the excitation light, and subtracting that known value from the emitted light pattern, it is possible to gain new information about the unknown sample. Exciting a fluorescently labelled sample with such structured light results in a Moiré interference fringe, created by the two different fine patterns being superimposed. As the illumination pattern is pre-determined, the Moiré Fringe will describe the unknown structure of the sample, thus accessing super-resolution data otherwise unobtainable [3].

Due to the directionality of the light, and to gain a complete image, it is necessary to shift the phase of the light through five patterns and then the orientation angle, which is usually shifted three times by at least 60° [5]. Post-imaging processing then produces the final image from 15 raw images from each z-plane.

The 3D-SIM imaging system was developed commercially by Applied Precision Inc. (Washington, USA), and the platform was named Optical Microscope eXperimental (OMX). The OMX is designed to maximize physical stability and photon budget with the aim of producing a platform suited to both rapid live-cell imaging and super-resolution of fixed material [5–7]. The basic components of the OMX system are described in [5], and the current Deltavision OMX Blaze model specification is available on the GE Healthcare website (http://www.gelifesciences.com).

In this chapter, we present protocols that describe methods of preparing fixed and live plant cells in order to visualize specialized ER components. Each protocol presents its own challenges concerning image optimization. Elements of these protocols have been described elsewhere [8–10], but here, we describe methods optimized for the imaging of specialized ER structures.

Acquiring high-resolution 3D-SIM images from fixed tissue requires that the samples are labelled with a probe that is specific and highly photostable and with a high quantum yield. The tissue must be carefully fixed in order to preserve as much of the fine morphological structure as possible.

Imaging live cells requires a highly stable fluorophore for labelling structures that will not move during the timescale required to capture an image. During live-cell imaging, speed is of the essence in order to avoid any blur caused by components streaming within the cells. In this regard, the OMX system offers several advantages: simultaneous recording of up to four separate channels or, in cases where there may be some emission overlap between fluorophores, sequential imaging combined with rapid shuttering and fast, stable focusing, resulting in delays of less than 1–2 ms. This means that together with its fully integrated electronic control, fast and precise 3D high-resolution image capture is possible [5, 6]. Despite this, many live cells will suffer bleaching during the image capture, so it is important to optimize the setup to minimize bleaching, although this may come at the cost of losing some of the resolution.

The sieve element reticulum (SER) is a specialized ER system in the phloem of higher plants [11] and has long posed difficulties when imaging, as the sieve elements are located deep within tissues, beyond the working distance of most objectives. Fixation and sectioning are therefore a good way to visualize this structure (Fig. 1e, f). However, extreme care must be taken that the fixation method chosen does not disrupt the delicate structure.

In addition, tobacco-derived BY2 cell lines make good model systems as they are readily transformable, both transiently and stably with fluorescent reporters, and often divide in long chains

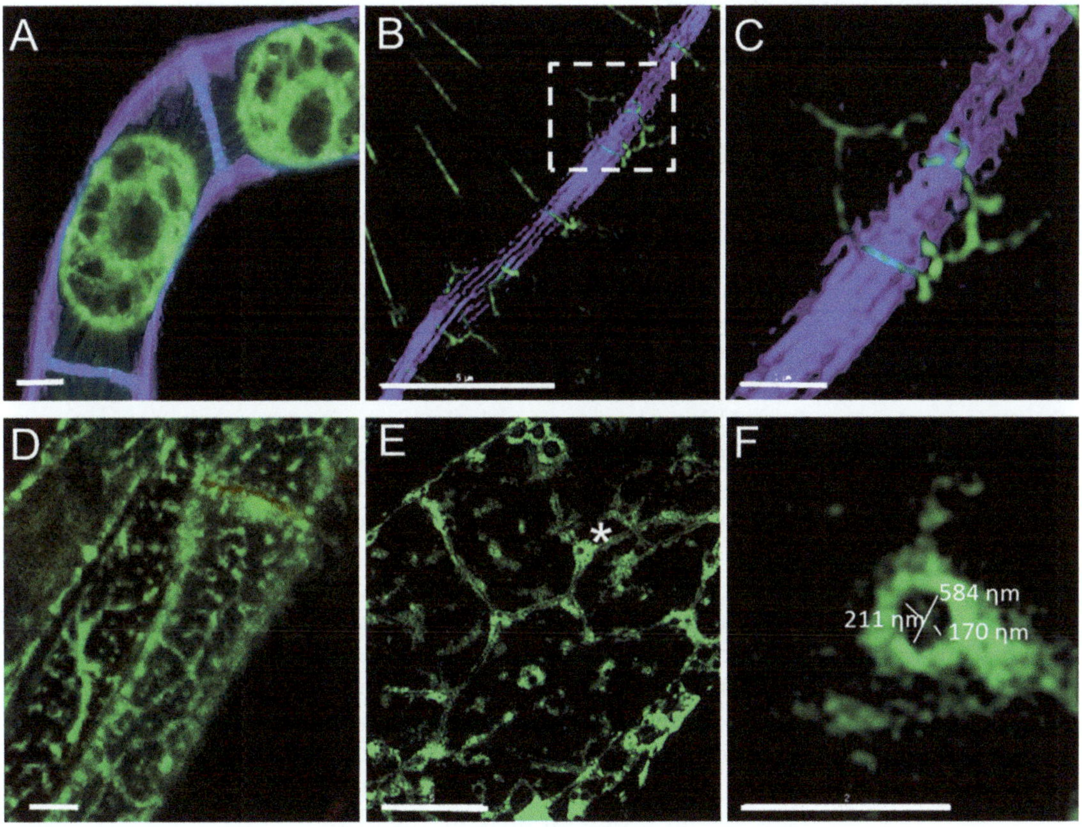


**Fig. 1** (**a**) Confocal image of plasmolyzed BY2 cells, showing ER labelled by RTN6-GFP (green) and the cell wall labelled with calcofluor white (blue); scale 10 μm. (**b**) 3D-SIM image of the cross wall between two BY2 cells; scale 5 μm. (**c**) Detailed view of the boxed area in B, showing the narrowing of the ER as it forms the desmotubule (arrow, scale 1 μm). (**d**) Confocal image of sieve elements showing the SER labelled by pSEO2: GFP-HDEL; scale 5 μm. (**e**) 3D-SIM image showing structural details of the reticulum not resolved by the confocal scale 5 μm. (**f**) detailed view of a region of SER similar to the region marked (*) in (**e**); scale 2 μm

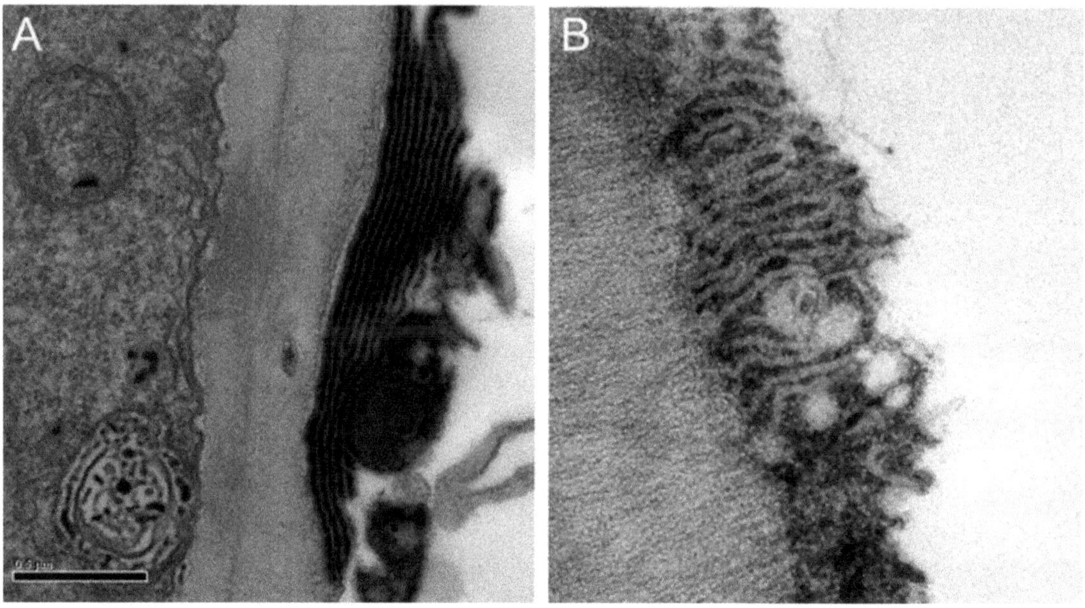


**Fig. 2** TEM images taken at the sieve element and companion cell walls of tobacco petioles, stained with ZIO. (**a**) The SER adjacent to the cell wall was strongly labelled, showing detailed structure of the SER cisternae. (**b**) An alternate angle section of the SER, with ZIO clearly enhancing the contrast between the lumen and the membranes. Scale 0.5 μm

along one plane, allowing single cell layers to be imaged [12]. Close to the cell wall, the ER is tightly appressed as it passes through the plasmodesmata, forming a desmotubule [13]. At just 15 nM wide, this structure is well beyond the resolution limit of a standard confocal, but 3D-SIM allows greater resolution [14] (Fig. 1b, c).

Although super-resolution imaging can resolve structures beyond the diffraction limit, it cannot compete with transmission electron microscopy for the magnification of fine structures. To observe such fine structures requires contrast enhancement, and this is usually achieved by using specific electron dense heavy-metal based stains. In this chapter, we describe a protocol for zinc-iodide osmium tetroxide (ZIO) staining of the SER (see Fig. 2 for representative images). The impregnation of tissue with ZIO was initially used to stain autonomic nerve fibers for visualization using light microscopy and subsequently TEM [15]. Notable for forming electron dense accumulations between double-membraned structures, it was soon adopted for the study of the ER in general [16] and by our group for imaging the SER.

## 2 Materials

### 2.1 Preparation of Phloem Tissue for 3D-SIM

1. Fixative: 50 mM 1,4-piperazinediethanesulfonic acid (PIPES) pH 6.9, 2 mM EGTA, 1% (w/v) bovine serum albumin (BSA) in $dH_2O$. Add 4% (v/v) formaldehyde and 0.25% glutaraldehyde (*see* **Note 1**).
2. Stabilizing solution: 5% (w/v) Phytoagar in $ddH_2O$ (*see* **Note 2**).
3. Wash Buffer: 2 mM EGTA, 1% BSA, 50 mM PIPES, pH 6.9 in $ddH_2O$.
4. PBS: Phosphate buffered saline, 1 mM $KH_2PO_4$, 155 mM NaCl, and 10 mM $Na_2HPO_4$ in $ddH_2O$, pH 7.4.
5. Calcofluor White solution: 10 μg/mL in $dH_2O$.

### 2.2 Growth and Preparation of BY2 Cells for Live Imaging

1. BY2 growth media: 0.43% MS basal salts media, 3% sucrose, 2 μg/mL 2,4-Dichlorophenoxyacetic acid in $ddH_2O$. Sterilize by autoclaving.
2. 1 M Mannitol: dissolved in $dH_2O$.
3. 170 μg/mL Calcofluor White Stock Solution: dissolved in ethanol. Stored in the dark at −20 °C.
4. 100 mM $DiOC_6$ (3,3′-dihexyloxacarbocyanine iodide) stock solution: dissolved in DMSO. Stored in the dark at −20 °C.

### 2.3 ZIO Staining

1. Fixative: 3% (v/v) glutaraldehyde in 0.1 M sodium cacodylate (*see* **Note 3**).
2. 0.2 M sodium cacodylate: In a fume cupboard, add 21.4 g sodium cacodylate to 400 mL $dH_2O$. Mix well, and add $dH_2O$ to a final volume of 500 mL. Add 0.1 M HCl dropwise until the solution reaches pH 7.3. For wash buffer, dilute 1:1 with $dH_2O$.
3. 1% Osmium tetroxide in 0.1 M sodium cacodylate: Add a 0.25 g glass vial of osmium tetroxide to a glass DURAN bottle and screw lid on. Shake to break vial, and then add 12.5 mL $dH_2O$; mix gently on a rocking platform. Mix equal volumes of osmium tetroxide and sodium cacodylate buffer.
4. Zinc iodide: 3 g zinc powder and 1 g resublimed iodine to 20 mL $dH_2O$; stir for 5 min, and then filter. Leave to mature in flow hood for 4 h (*see* **Note 4**).
5. ZIO: Combine equal volumes of zinc iodide with 2% osmium tetroxide in a small aliquot, mix, and use immediately.
6. Epon812.
7. 1:1 ratio of Epon812:propylene oxide.
8. 2:1 ratio of Epon812 and propylene oxide.

## 3 Methods

### 3.1 Preparation of Phloem Tissue

1. Cut the stem or petiole of a 35–55-day-old tobacco plant expressing the desired fluorescent reporter (in this case, pSEO2.GFP-HDEL [17]), and transfer immediately to a glass beaker containing the fixative solution, and submerge the cut end. Trim approximately 5 mM from the end of the stem to avoid air blocks. Allow to transpire in an illuminated fume hood for 1 h (*see* **Note 5**).
2. Trim the tissue, and add to molten stabilizer cooled to 40 °C. Submerge sample fully by gently pressing with a blunt instrument. Remove any air bubbles, and carefully orient the tissue to allow for the desired angle of sectioning, e.g., positioned at 90° to the surface for transverse sections or at 180° for longitudinal sections.
3. When set, trim around the sample to form a block, and then section using a vibrating microtome on a medium-fast setting to obtain sections of 100 μm.
4. Place sections into wash buffer contained in a 5 mL petri dish. Wash by replacing the buffer 3 times, 10 min for each wash.
5. Rinse with PBS and stain with calcofluor white (at 10 μg/mL) for 1 min at room temperature. Rinse sections well with $dH_2O$.
6. Mount sections directly on a number 1.5 coverslip with a drop of Citifluor AF1 antifade medium. Apply gentle pressure to spread the mountant, and remove air bubbles before sealing with nail varnish.

### 3.2 Preparing BY2 Cells for Live 3D-SIM Imaging

1. In advance, culture BY2 cell lines in 50 mL Erlenmeyer flasks with sterile Murashige and Skoog basal salts media (*see* **Note 6**).
2. Aliquot 1 mL of cell suspension to an Eppendorf tube, and stain cell walls with calcofluor white at a final concentration of 3.5 μg/mL. Stain for 5 min at room temperature before rinsing cells twice with fresh media. Optional: if imaging wild-type cells, use $DiOC_6$ to stain the ER at a final concentration of 50 μM (*see* **Note 7**). Incubate for 10 min, and then rinse twice with fresh media.
3. Induce plasmolysis by pipetting off 450 μL of media and adding 450 μL 1 M mannitol. Incubate for 10 min at room temperature before gently inverting the tube to resuspend the cells (*see* **Note 8**).
4. Pipette 40 μL of cells onto a very clean microscope slide before carefully placing the coverslip. Using a folded paper towel, apply gentle, even pressure to remove excess media from the slide (*see* **Note 9**). Carefully wipe all edges, making sure the coverslip does not become smudged. Seal with nail varnish.

### 3.3 3D-SIM Imaging with an OMX DeltaVision Blaze

1. Locate ideal cells on the auxiliary microscope (PersonalDV Deltavision), which has stage coordinates synchronized with the OMX. Use the lowest-level brightfield possible to view candidate cells, and mark their position on the slide using the point visiting tool. Expose all candidate cells briefly to check that both the calcofluor white stain and the GFP reporter (or $DIOC_6$) are sufficiently bright (*see* **Note 10**).
2. Transfer the slide to the OMX, and apply immersion oil (*see* **Note 11**). Using the Point List, find the marked cells, and then center the cell in the image using the spiral mosaic function.
3. Empirically determine the lowest laser power required to achieve optimal intensities of between 1000 and 3000 counts in a raw image acquired by a 15-bit dynamic range Edge sCMOS camera. Also, optimize the shortest exposure times for each channel, typically between 100 and 200 ms (*see* **Note 12**).
4. Following optimization, acquire image stacks of appropriate cells. Set the start and end positions for a minimal z-stack quickly to avoid excess light exposure.
5. Using the SoftWorx 6.0 alignment tool, adjust the images from the separate channels, and then reconstruct 3D super-resolution image stacks using softWorx 6.0 with channel specific OTFs and Wiener filter settings of 0.002 (*see* **Note 13**).

### 3.4 ZIO Staining of the SER

1. Cut the stem or petiole of a 35–55-day-old tobacco plant expressing the desired fluorescent marker (in this case, pSEO2.GFP-HDEL [17]), and transfer immediately to the fixative solution, and submerge the cut end. Trim approximately 5 mM from the end of the stem to avoid air blocks. Allow to transpire in an illuminated fume hood for 1 h (*see* **Note 5**).
2. Remove the leaf, and chop the petiole into 5 mM by 5 mM sections.
3. Wash twice in 0.1 M sodium cacodylate for 10 min each wash.
4. Wash twice in $dH_2O$ for 10 min each wash.
5. Incubate in ZIO solution at room temperature for 4 h in a sealed vial with gentle agitation on a rotating wheel.
6. Wash twice in $dH_2O$ for 10 min each wash.
7. Dehydrate tissue in an ethanol series: submerge for 15 min per solution in 50%, 70%, and 95% ethanol before two 15 min incubations in 100% ethanol.
8. Infiltrate dehydrated tissue with a 1:1 ratio of Epon812 and propylene oxide at room temperature for 2 h.

9. Continue infiltration overnight in a 2:1 ratio of Epon812: propylene oxide at 58 °C.
10. Incubate in 100% Epon812 at room temperature for 1 h. Replace the Epon812 and incubate for one further hour.
11. Embed in flat bed molds, for 48 h at 58 °C.
12. Check orientation and quality of the tissue by cutting semi-thin sections (0.5–1 μm) using a glass knife. Stain sections briefly with toluidine blue for accurate visualization on a bright-field microscope.
13. Following identification of an appropriate area of a block, cut ultrathin sections (60 nM) with a diamond knife, and float onto $dH_2O$. Mount sections on an EM grid by touching the dull face of the grid to the middle of a section without breaking the surface tension of the water. The sections will readily adhere to a clean grid.
14. Stain sections with uranyl acetate and lead citrate by first wetting the grids in $dH_2O$ (*see* **Note 14**), and then float section side down for 45 min on a drop of uranyl acetate (*see* **Note 15**). Wash by dipping in $dH_20$, manipulating the grid by its edge using fine forceps for 1 min, and repeat the wash in fresh $dH_2O$. Invert the grid onto a drop of lead citrate for counterstaining, for 10 min. Wash again in $dH_2O$ for 1 min, and then repeat for a further 1 min in fresh $dH_2O$.
15. Air-dry before imaging with the TEM.

## 4 Notes

1. The fixative solution should be prepared freshly on the day of use. However, the PIPES, EGTA, and BSA stock solutions may be prepared in advance and stored on the bench following autoclave sterilization. Glutaraldehyde should be EM-grade and formaldehyde from a methanol-free solution. Fixatives provided in single-use glass ampules provide the best source, and any unused can be decanted into a glass bottle and sorted for up to 1 month, glutaraldehyde at 4 °C and formaldehyde at room temperature.
2. Phytoagar is a specialist agar used for plant tissue culture. With a high gel strength at a 5% solution, this matches the mechanical properties of petiole tissue and thus supports the tissue well during sectioning. The concentration may require optimization when used as a stabilizer for other tissues types to avoid the agar pulling away from the tissue or disintegrating.

3. To preserve the fine ultrastructure of the cell in ultrathin TEM sections, it is necessary to increase the glutaraldehyde content of the fixative to 3%.
4. The solution should turn a light-straw color during the maturation process.
5. Due to transpiration, the fixative is drawn up into the xylem, from where it moves laterally to the phloem and other tissues, providing a gentle delivery system, which avoids mechanical damage of the phloem cells.
6. Incubate liquid BY2 cell cultures at 28 °C, in the dark, shaking at 140 rpm. Sub-culture cells weekly using a 1:40 dilution. If the cells are stably transformed with a fluorescent marker for the ER such as RTN6-GFP, ensure that the signal is strong and relatively homogeneous in the line of choice. Best results are usually obtained with cells, which were subbed 3–4 days prior to imaging.
7. $DiOC_6$ will stain the ER, mitochondria, and vesicle membranes. It is however, extremely phototoxic so the cells must be kept in the dark once stained and illumination minimized to reduce ER damage.
8. Plasmolysis allows clearer imaging of the desmotubules as the membrane is retracted away from the cell wall, and the Hechtian strands can be clearly labelled either by RTN6-GFP or $DiOC_6$.
9. To achieve optimal imaging, the cells need to be as close to the coverslip as possible.
10. It is vital to keep all light exposure as brief as possible prior to the SIM imaging so as to minimize any pre-imaging photobleaching or phototoxicity from $DiOC_6$.
11. The type of immersion oil should be matched as closely as possible to the refractive Index of the sample to minimize spherical aberrations. To achieve the best combination, in theory, the cells should be mounted in a high concentration of glycerol/antifade mountant, but this was found to affect the ER structure, so cells were mounted in media.
12. It is advisable to carry out the optimization on a sacrificial cell—by the time the process is complete, the GFP is likely to have bleached sufficiently to prohibit good image capture during the 3D-SIM process.
13. The microscope must be routinely calibrated by measuring channel-specific optical transfer functions (OTFs) to optimize both lateral and axial image resolutions. Alignment tools should also be regularly calibrated based on alignment parameters obtained from calibration measurements with 100 nM diameter TetraSpeck beads.

14. Pre-wetting the EM grids minimizes potential artifacts by reducing air-stain contact [18].
15. Spot stain on to a plastic Petri dish and keep covered to reduce contamination or artifacts from dust, evaporation, and $CO_2$ (particularly important for lead citrate).

## References

1. Abbe E (1873) Beiträge zur Theorie des Mikroskops und der mikroskopischen Wahrnehmung. Arkiv Mikroskop Anat 9:413–468
2. Wilson T (1995) The role of the pinhole in confocal imaging system. In: Pawley JB (ed) And book of biological confocal microscopy. Plenum Press, New York, pp 167–182
3. Gustafson MGL (2000) Surpassing the lateral resolution limit by a factor of two using structured illumination microscopy. J Microsc 198:82–87
4. Schermelleh L, Carlton PM, Haase S et al (2008) Subdiffraction multicolour imaging of the nuclear peripheray with 3D structured illumination microscopy. Science 320:1332–1336
5. Dobbie IM, King E, Partion RM et al (2011) OMX: a new platform for multimodal, multi-channel wide-field imaging. Cold Spring Harb Protoc 8:899–909
6. Kner P, Chhun BB, Griffis ER et al (2009) Super-resolution video microscopy of live cells by structured illumination. Nat Methods 6: 339–342
7. Fitzgibbon J, Bell K, King E, Oparka K (2010) Super-resolution imaging of plasmodesmata using 3-dimensional structured illumination microscopy. Plant Physiol 153:1453–1463
8. Bell K, Mitchell S, Paultre D, Posch M, Oparka K (2013) Correlative imaging of fluorescent proteins in resin-embedded plant material. Plant Physiol 161:1595–1603
9. Bell K, Oparka K (2015) Preparative methods for imaging plasmodesmata at super-resolution. Methods Mol Biol 1217:67–79
10. Bell K, Oparka K, Knox K (2016) Super-resolution imaging of live BY2 cells using 3D-structured illumination. Microscopy Bio-protocol 6(1):e1697
11. Evert RF (1990) Dicotyledons. In: Behnke H-D, Sjolund RD (eds) Sieve elements: comparative structure, induction and development. Springer, Berlin, pp 103–137
12. Nagata T, Nemoto Y, Hasezawa S (1992) Tobacco BY-2 cell line as the "HeLa" cell in the cell biology of higher plants. Int Rev Cytol 132:1–30
13. Overall RL, Blackman LM (1996) A model of the macro-molecular structure of plasmodesmata. Trends Plant Sci 1:307–331
14. Knox K, Wang P, Kriechbaumer V, Tilsner J, Frigerio L, Sparkes I, Hawes C, Oparka K (2015) Putting the squeeze on plasmodesmata: a role for reticulons in primary plasmodesmata formation. Plant Physiol 168:1563–1572
15. Maillet M (1968) Etude critique des fixations au tétraoxyde d'osmium-iodure. Bull Assoc Anat 79:233–394
16. Barlow PW, Hawes C, Horne JC (1984) Structure of amyloplasts and endoplasmic reticulum in the root caps of Lepidium sativum and Zea mays observed after selective membrane staining and by high-voltage electron microscopy. Planta 160:363–371
17. Knoblauch M, Peters WS (2010) Münch, morphology, microfluidics: our structural problem with the phloem. Plant Cell Environ 33:1439–1452
18. Hayat MA (1968) The principles and techniques of electron microscopy, vol 1. Van Nostt and Reinold, New York

# Chapter 5

# Quantitation of ER Morphology and Dynamics

**Mark Fricker, Emily Breeze, Charlotte Pain, Verena Kriechbaumer, Carlos Aguilar, José M. Ugalde, and Andreas J. Meyer**

## Abstract

The plant endoplasmic reticulum forms a network of tubules connected by three-way junctions or sheet-like cisternae. Although the network is three-dimensional, in many plant cells, it is constrained to thin volume sandwiched between the vacuole and plasma membrane, effectively restricting it to a 2-D planar network. The structure of the network, and the morphology of the tubules and cisternae can be automatically extracted following intensity-independent edge-enhancement and various segmentation techniques to give an initial pixel-based skeleton, which is then converted to a graph representation. ER dynamics can be determined using optical flow techniques from computer vision or persistency analysis. Collectively, this approach yields a wealth of quantitative metrics for ER structure and can be used to describe the effects of pharmacological treatments or genetic manipulation. The software is publicly available.

**Key words** Quantitative confocal imaging, Endoplasmic reticulum morphology, Endoplasmic reticulum dynamics, Persistency analysis, Network analysis, Phase congruency, Reticulon, Lunapark, ER cisternae

## 1 Introduction

The endoplasmic reticulum (ER) forms a complex and dynamic network of tubules and sheet-like cisternae that ramify throughout the cytoplasm [1]. In this chapter, we describe AnalyzER v2 ER network analysis program that is designed to quantify:

(i) The length, width, morphology, and protein distribution along ER tubules following extraction of a single-pixel wide skeleton (Fig. 1a, b).

(ii) The degree and branch angles at three-way junctions (nodes) in the tubular network (Fig. 1a, b).

(iii) The size, shape, and protein distribution in cisternae, with the latter estimated using gray-level co-occurrence (GLCM) texture metrics (Fig. 1c).

Verena Kriechbaumer (ed.), *The Plant Endoplasmic Reticulum: Methods and Protocols*, Methods in Molecular Biology, vol. 2772, https://doi.org/10.1007/978-1-0716-3710-4_5,

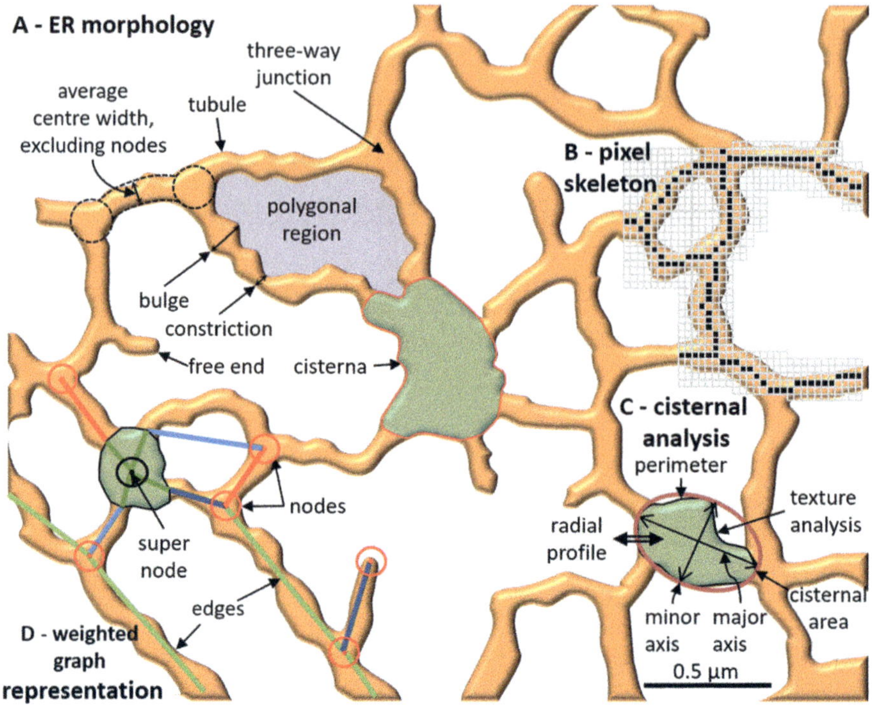


**Fig. 1** Schematic outline of the terminology used to describe the ER morphology. (**a**) Representation of the main features of the ER including tubules, cisternae, enclosed polygonal regions, three-way junctions, and free ends. Measurements of the tubule width may average over the "center" region, excluding the junctions, or involve more detailed analysis of local bulges and constrictions. (**b**) The image of the network is converted to a single-pixel wide skeleton following automated ridge enhancement and segmentation. (**c**) The cisternae (and polygonal regions) are analyzed separately to determine their area and various shape metrics such as the major and minor axis of a bounding ellipse. (**d**) Weighted graph representation: Elements within the pixel skeleton are classified as nodes (junctions and free ends) connected by edges that inherit the length and width of the underlying tubules. Cisternae are represented as a super-node that is connected to each tubule incident on the boundary

(iv) The topological organization of the tubular and cisternal network determined using graph-theoretic metrics (Fig. 1d).

(v) The distribution of immobile nodes, tubules, and cisternae using persistency mapping.

(vi) The local speed, direction, coherence, divergence, and curl of movement of tubules and cisternae using optical flow.

(vii) The size and shape of the polygonal regions enclosed by the network.

(viii) Physiological responses measured using ratiometric reporters.

The AnalyzER package is implemented in MATLAB® (www.Mathworks.com) and available as a MATLAB® app or as a stand-alone package. The program was originally designed to quantify ER organization in plant epidermal cells where the ER is confined to a very thin layer of cytoplasm appressed to the periclinal cell wall as a planar, 2-D network. It is a development of a tubule morphology analysis program [2], which was used to quantify constrictions in ER tubular network caused by reticulon overexpression in single-channel (*x*,y) images [3]. AnalyzER extends the analysis to a complete quantitation of all aspects of ER structure and dynamics for multi-channel 4D time series [4]. This includes multiple improvements and expansion to the processing, segmentation, and graph extraction algorithms, comparison with ground-truth data to validate the automated network structure, introduction of time domain metrics for persistency and optical flow, and introduction of cisternal substructure analysis, including texture features, perimeter and radial profiles, and extension of the graph-theoretic metrics. All aspects of the analyses are handled through a single graphical user interface (GUI) to provide an integrated platform. A complete manual for the program is available at https://doi.org/10.5281/zenodo.6386982.

The simplest method to identify the ER automatically is an intensity-based segmentation of the fluorescence image to give a binary image, with ones representing the ER structure and zeros for the background. However, the resultant binary image is critically dependent on the value for the threshold used, and it is rare that a single threshold provides adequate segmentation without either losing dimmer structures if it is set too high or artificially expanding and fusing adjacent regions if it is set too low (see Bouchekhima et al. [5]). Thus, the approach adopted here exploits additional intensity-independent information over a range of scales and orientations to enhance the network structure, prior to segmentation as a single-pixel wide skeleton (Fig. 1b). The skeleton is then used as a template to interrogate the image locally to provide an estimate of the relative amount of fluorescent probe present.

The labelling intensity also provides an indication of the tubule width, with a number of assumptions. The expected width of the ER tubule is 50–70 nM [1] or less, which is below the resolution of the confocal microscope, but can just be resolved with super-resolution techniques, such as stimulated emission depletion microscopy (STED, [6]), although this is not routine for dynamic imaging. For most laboratories, access to super-resolution techniques may be limited, necessitating the development of approaches that can be used on a routine basis with existing systems. Nevertheless, with additional assumptions about the distribution of the fluorescent lumenal marker and the point spread function (psf) of the microscope, the approximate width of the ER can be estimated, even if this is below the resolution limit of the microscope system [7]. In this approach, the skeleton is used as a template to interrogate the image locally to provide an estimate of the relative amount of fluorescent probe present, and the width is inferred from the integrated intensity signal from estimates of the profile of intensity with distance normal to the skeleton. We describe some basic routines to estimate the relative tubule width (effectively convolved with the psf) and also introduce measures using the intensity values to infer the approximate sub-resolution tubule width, along with calibration against ER tubules and cisternae measured from serial block-face scanning electron microscopy (SBF-SEM) [4].

Topological measures of the ER network structure can also be extracted following conversion of the pixel skeleton to a weighted, undirected graph, where nodes represent junction points and edges represent the tubules that connect them [5, 8, 9] (see Fig. 1d). A wide range of metrics can be calculated using such a graph representation, such as node degree or betweenness centrality of nodes or edges, although it is not yet clear how these may relate to underlying mechanism or function. Nevertheless, unlike morphological measurements, the topology of the network is less sensitive to the resolution of the imaging system as it reflects the connectivity of the ER rather than the physical size of the components [5].

### 1.1 Outline of the Main Steps in the ER Analysis

The ER Network Analysis progresses through a number of parallel threads that are designed to extract different information from the underlying image to characterize the tubular network, cisternae, and enclosed polygonal regions of cytoplasm. A flow diagram of the overall sequence for morphological measurements is shown in Figs. 2 and dynamics in Fig. 3. The principles and processing details are discussed in more depth in Subheading 3.1.

## 2 Materials

The AnalyzER program was written in MATLAB (The Mathworks, Nantick, MA) and is packaged in a single compiled executable file for distribution as a stand-alone package or as MATLAB app that runs within the MATLAB 2021a environment or later. In the case of the app, the AnalyzER package requires the MATLAB image processing toolbox, computer vision toolbox and statistics, and machine learning toolbox. The software and future updates, manual, and tutorial are available from Zenodo:

doi: 10.5281/zenodo.6386982,

web link: https://zenodo.org/record/6386983).

The software has been tested on Windows 10 and 11 and requires a minimum screen resolution of 1600 × 900. In addition, for the stand-alone version, an appropriate version of the MATLAB Compiler Runtime (MCR) is required to install the set of shared libraries that enables execution of the compiled MATLAB application. The MCR should automatically download from MathWorks

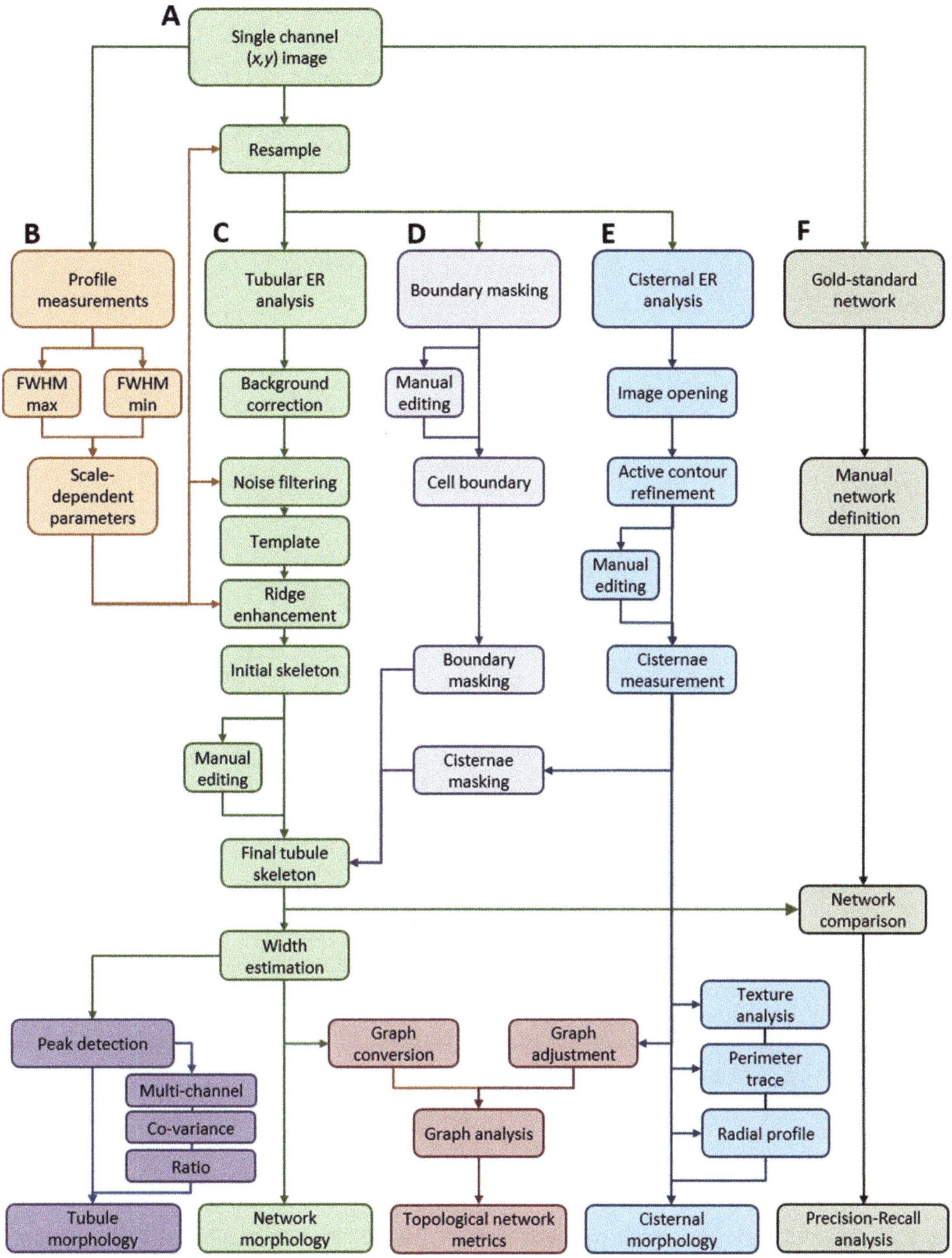


**Fig. 2** Flow diagram showing the main steps in the morphological ER analysis. (**a**) The starting point is typically a single channel confocal fluorescence of the ER labelled with a fluorescent protein. (**b**) The minimum and maximum tubule diameters are estimated manually from line transects to resample the image and standardize all the subsequent processing parameters; (**c**) ER tubules are segmented following filtering and enhancement

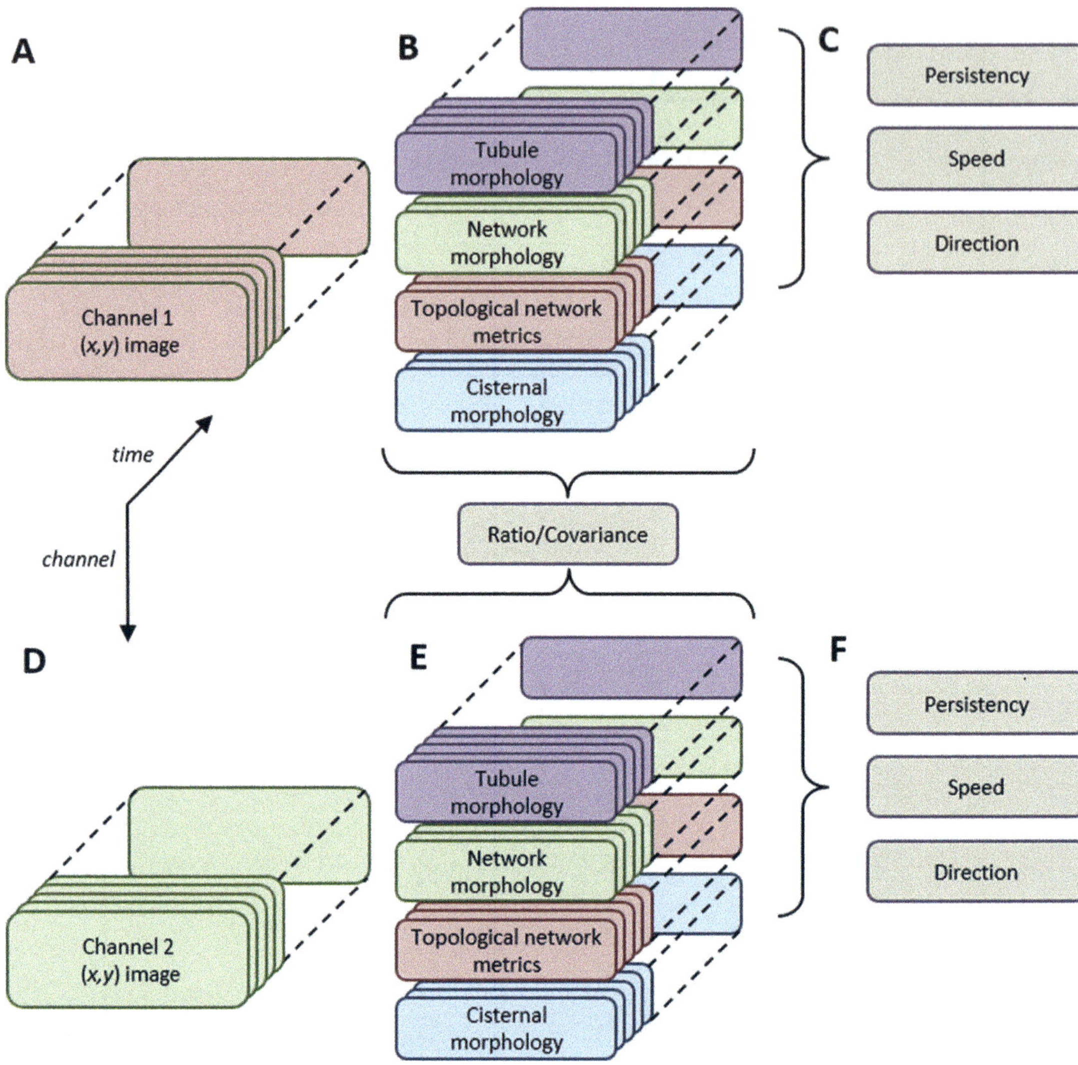


**Fig. 3** Flow diagram showing the main steps in multi-channel time series analysis
Time series images for one or more channels (**a**, **d**) are processed to extract morphological metrics (**b**, **e**) that can be analyzed using ratiometric or covariance techniques. In addition, ER dynamics can be assessed from optical flow techniques to measure speed and direction or persistency analysis to identify fixed points in the network that might correspond to ER-PM contact sites, for example

**Fig. 2** (continued) steps to give a single-pixel wide skeleton. This provides basic morphological information on the length and width of the tubules and can be interrogated further to examine individual tubule morphology. (**d**) The analysis can be constrained to a particular cell or sub-cellular region by masking the image; (**e**) ER cisternae are detected independently using image opening followed by active contour refinement; (**f**) the performance of the automated segmentation approaches can be compared to a manually defined "gold-standard" pixel skeleton

when the program is installed for the first time. Alternatively, the MCR can be downloaded separately from the MathWorks Website:

http://www.mathworks.com/products/compiler/mcr

A number of additional files needed to run the full suite of programs may also be installed at the same time as the main program. For example, the Bio-Formats package [10] has been designed to read in images from different microscope manufacturers and store them in a standardized format. Full details are available on the open microscopy website:

https://bio-formats.readthedocs.io/en/v7.0.0/

The bioformats_packages.jar program needs to be available on the search path or installation directory of the MATLAB programs. The bioformats_package.jar is available from:

https://www.openmicroscopy.org/bio-formats/downloads/

The latest version of Java needs to be installed, and is available from:

http://www.java.com/en/

Output of images at full resolution uses *export_fig.m* originally written by Oliver Woodford (2008–2014) and now maintained by Yair Altman (2015–). When exporting to vector format (PDF or EPS), this function requires Ghostscript, which can be downloaded from:

http://www.ghostscript.com

When exporting images to eps, *export_fig* additionally requires pdftops, from the XpdfReader suite of functions. This can be downloaded from:

https://www.xpdfreader.com/pdftops-man.html

## 3 Methods

### 3.1 Extraction of ER Network Structure

#### 3.1.1 Loading Images

Single-channel or multi-channel confocal images or time series are imported into the software from a wide variety of microscope manufacturers using the Bio-Formats program [10] running in MATLAB and called from the AnalyzER GUI. The channel order can be assigned and the image rotated or cropped as required.

#### 3.1.2 Setting the Scale Parameters

Images of the ER may have been collected at different pixel resolutions, depending on the microscope settings, and may span different width scales depending on the experimental treatment and genetic background. If direct physical estimates of the tubule width are required, the pixel size needs to be reduced below the optimal Nyquist sampling value (to ~20 nM) using a super-resolution technique and to reduce downstream discretization errors due to pixelation. However, for intensity-based prediction of the width, confocal or Airyscan imaging is sufficient using a 1.2 NA water immersion lens or a 1.4NA oil lens and a zoom of around 3–4, to give a pixel sampling in the 40–80 nM range. To standardize all the subsequent processing steps (*see* **Note 1**), it is useful to define the expected minimum and maximum width of the tubular components manually using a transect drawn on the image.

1. Draw a two-point transect on the image across a tubule that, by eye, appears to be close to the smallest tubule diameter. A graph of the transect profile is automatically displayed and the full width at half-maximum peak height (FWHM) calculated to give $FWHM_{min}$. This value would normally be rounded down to the nearest odd integer.
2. Repeat the transect profile across the largest diameter tubule present to give the maximum tubule diameter ($FWHM_{max}$). This number should be rounded up to the nearest odd integer.
3. Values for FWHM displayed in the interface are given in pixels, while the peak intensities are given in normalized units ranging from 0 to 1.

#### 3.1.3 Background Measurement and Correction

Accurate measurements of the fluorescence signal from the ER typically require correction for the instrument dark current, amplifier offset, and background signal for each channel and image in the time series. A number of options to automatically or manually correct the background contribution are provided, including:

1. *Subtraction of a single measured value:* A constant value is measured from a ROI defined on the image and is subtracted for each channel and frame. This is appropriate for removing instrument black-level offsets or very diffuse fluorescence.
2. *Image opening or top-hat*: The image is processed with a gray-scale opening function using a disk-shaped kernel with the radius set by $FWHM_{max}$ (i.e., twice the size of the largest tubular feature expected). This removes any features smaller than 2 × $FWHM_{max}$ and provides an estimate of the local background around each pixel. The opened image is subtracted from the original to correct for the local background. This method may be useful if there is some out-of-focus blur in the image or an amount of signal from another compartment, such as the cytoplasm. However, it is less useful if there are sheet-like cisternal regions larger than 2 × $FWHM_{max}$, as these remain after the opening operation and are then subtracted from the image, distorting the pixel intensities in the neighboring regions.

3. *Subtraction of a surface fit to the local background:* The local minima across the image are detected and a smooth surface fit to points in the 10–90% interval of the local minimum distribution. The surface is converted to an image and subtracted from the original.
4. *Subtraction of a low-pass filtered image:* The image is filtered using a Gaussian kernel with a large radius sufficient to remove all the high-frequency information in the image. The standard deviation for the Gaussian kernel is calculated as $FWHM_{max}$ or approximately twice the size of the largest tubule diameter. The low-pass image is subtracted from the original and the image re-normalized. This approach also suffers if there are cisternal regions present.

#### 3.1.4 Filtering the Image to Improve Signal-to-Noise Ratio

Most simple noise reduction algorithms use isotropic kernels and smooth the noise in the image equally in all directions. This is not desirable when analyzing the ER, as the tubule boundaries will become blurred. Instead, a number of adaptive anisotropic filters are provided that smooth within the tubular network structures, but do not spread across boundaries. In addition, the image intensities can subsequently be rescaled using local contrast-limited adaptive histogram equalization (CLAHE, [11]) (*see* **Notes 2** and **3**).

1. *Coherence filtering*: This applies an anisotropic diffusion filter written by Dirk-Jan Kroon [12] that smooths regions of low variance, but avoids blurring of object boundaries.
2. *Guided filtering:* This uses an edge-preserving smoothing filter that is "guided" by the structure of the underlying image [13].

#### 3.1.5 Setting a Template Image

While each channel in the time series can be analyzed separately, it is appropriate to define a single image or image combination for each time point to identify the ER structure. Typically, this template can combine multiple channels, for example, labelling the lumen and a tagged reticulon, as an average or maximum projection across the channels to ensure full segmentation of the ER structure.

#### 3.1.6 Defining a Mask and Cell Boundary

In some instances, it may be appropriate to define a mask to exclude regions that should not be analyzed or to restrict the analysis to a specific region of the cell or tissue. Several approaches are available to segment the main fluorescent structures automatically or manually:

1. Automatic segmentation using a threshold determined by Otsu's method [14] that minimizes the intraclass variance of the foreground and background distributions. Multiple partitions are possible, and the segmentation can select different partition levels.

2. Adaptive thresholding using local first-order image statistics around each pixel using Bradley's method [15].
3. Manual segmentation using a threshold determined from a region of interest (ROI) placed on the background. The threshold is calculated as mean + 2 × SD (*see* **Note 4**).
4. Manual delineation by drawing a boundary on the image. If this option is selected, a separate window opens to give access to a number of drawing tools. The edited binary image is automatically saved alongside the other parameters used for processing the image and can be re-applied or edited further every time the image is processed.

Once the relevant region(s) of ER is segmented, the boundary is defined using a "shrunken" convex hull. With a shrink factor set to zero, the boundary will map exactly onto the convex hull. As the shrink factor is increased, the boundary snaps back to follow concave sections. The area enclosed is important as it is used to define the analysis area for density measurements. Typical values for shrinkage are 0.3–0.5. An additional option is also given to erode the boundary by some number of pixels to avoid dim regions at the edge of the sampled region where the cell moves out of focus.

### 3.1.7 Enhancing the Tubular Elements

A number of options can be used to improve the relative contrast of tubular ER elements prior to segmentation, by using kernels designed to pick out "ridge" like features that are detected over a range of scales and angles (*see* **Note 5**). These include:

1. *Phase congruency*: This uses the phase-congruency approach developed by Peter Kovesi [16, 17], which provides contrast-invariant ridge detection over a range of scales and angles. The MATLAB implementation [17] provides a number of outputs, including the level of phase congruency as a measure of the edge strength, and also the "feature type," calculated as the weighted mean phase angle at every point in the image. A value for the feature type of pi/2 corresponds to a bright line, 0 corresponds to a step, and -pi/2 is a dark line. The feature type has proved to be one of the most robust and reliable outputs for subsequent segmentation, as all ridges, irrespective of their original intensity are identified with almost equal strength in the feature type image. The "Feature Type" measure is re-normalized to the range [0 1].
2. *Frangi vesselness filter:* This calls the MATLAB implementation of the classic Frangi "vesselness" filter [18] written by Marc Schrijver and Dirk-Jan Kroon and available from the MathWorks website (http://uk.mathworks.com/matlabcentral/fileexchange/24409-hessian-based-frangi-vesselness-filter). This gives a strong response for bright features where the second-order derivative of the intensity image (Hessian)

shows a strong anisotropy. It can perform well for tubular elements across a range of scales but leaves holes at the junctions, and only partially compensates for differences in intensity in the original image, which can lead to breaks in the skeleton during the subsequent segmentation step.

3. *Ridge enhancement.* Applies a version of the second-order anisotropic Gaussian kernel originally proposed by Meijering [19] as part of their "Neuriteness" detector. This uses a slightly flattened second-order Gaussian kernel at a range of scales and angles to give better discrimination of ridge-like structures, but does not include any additional information related to the anisotropy of the ridge.
4. *Anisotropic ridge enhancement.* This applies the multi-scale ridge detector developed by Lopez-Molina [20], which uses anisotropic second-order Gaussian kernels. These can be configured flexibly in terms of size, orientation, and anisotropy. This gives good ridge enhancement but still retains the variation in local intensity along the tubules that can make subsequent segmentation more difficult.

The result of the enhancement steps can also be smoothed using coherence or guided filtering.

### 3.1.8 Skeletonization

The aim of the skeletonization step is to convert the enhanced image to a single-pixel wide skeleton along the centerline of the tubule ridges. Nevertheless, this is only an approximation to the true centerline due to pixel discretization errors. There are two main approaches that can be used, namely, *hysteresis thresholding*, which uses the intensity information and some degree of local pixel connectivity to provide an initial binary image that is then thinned to give a single pixel skeleton, and *watershed thresholding*, which follows connected ridges, irrespective of their absolute intensity, and automatically generates a single-pixel skeleton.

1. *Hysteresis thresholding*: This starts with seed pixels above a user-defined upper threshold and then propagates the initial segmentation as long as pixels remain above a user-defined lower threshold. The resulting binary image is then thinned to give a single-pixel wide skeleton. The value of the lower threshold is critical – too high and the network becomes disconnected; too low and large blocks of the image are included in the resultant binary image that may fuse separate tubules into a single object. Consequently, when this block is thinned, the skeleton does not map onto the ridge centerlines. In addition, as the thinning process is not guided by the intensities in the enhanced image, the skeleton does not necessarily converge on the expected pattern at junction points. Using an h-minimum transform suppresses all minima in the image below a given depth.

Regional minima are connected components of pixels with a constant intensity value within a set of boundary pixels that are greater than the specified depth. This smooths out regions with low fluctuations in intensity but then sets these regions as local minima to ensure that the surrounding ridges will be segmented individually and do not spread into the basin, even if the absolute values are above the lower hysteresis threshold.

2. *Watershed thresholding*: this applies a watershed segmentation and then extracts the watershed lines as the pixel skeleton. The watershed method is good at segmenting the centerline of the ridges and can handle variations in intensity well. However, it does not include any tubules that have a free end and has a tendency to over-segment regions with noise. Over-segmentation can be avoided by including additional steps that suppress regions with small-intensity fluctuations using the h-minimum transform. In addition, setting the local minima to zero ensures that adjacent structures are separated. Free-ending veins can be included using an additional hysteresis segmentation with non-maximal suppression.

If the automated methods to delineate the pixel skeleton are still incorrect, the skeleton can be edited manually in an additional window equipped with basic drawing tools.

### 3.2 Validation of Parameter Settings Against a Ground Truth

The sequence of steps leading to the single-pixel skeleton (Fig. 4a, b) can be validated against a manually defined ground-truth image (Fig. 4c). The manual skeleton is thinned and masked in parallel to the automated skeleton and the cisternae excluded (see below). A number of true positives ($T_P$), true negatives ($T_N$), false positives ($F_P$), and false negatives ($F_N$) are calculated, allowing a tolerance of half $\text{FWHM}_{\text{min}}$ [21]. Performance is assessed using precision recall (P-R) analysis in preference to receiver operating characteristic (ROC) plots, as the former is better suited to imbalanced data sets, where the number of $T_N$ from the background is expected to be much greater than $T_P$ from the skeleton [22]. Precision (*P) is* calculated as $= \frac{T_P}{T_P + F_P}$, recall ($R$) as $R = \frac{T_P}{+F_N}$, and overall performance assessed with the Dice similarity coefficient or $F_1$ score as the harmonic mean of precision and recall, $F_1 = 2 \times \frac{P \times R}{P + R}$.

The validation can be visually assessed by mapping the results back on to the pixel skeleton with green representing $T_P$, red $F_N$, and blue $F_P$ (Fig. 4d).

### 3.3 Estimation of ER Tubule Width

Once the pixel skeleton has been segmented satisfactorily, the first step in the analysis is to estimate the tubule width. Unfortunately, the average ER tubule diameter is below the theoretical resolution of most confocal microscopes and at the resolution of current live-cell super-resolution techniques such as STED [6], so direct physical estimates of the width are challenging. The methods described

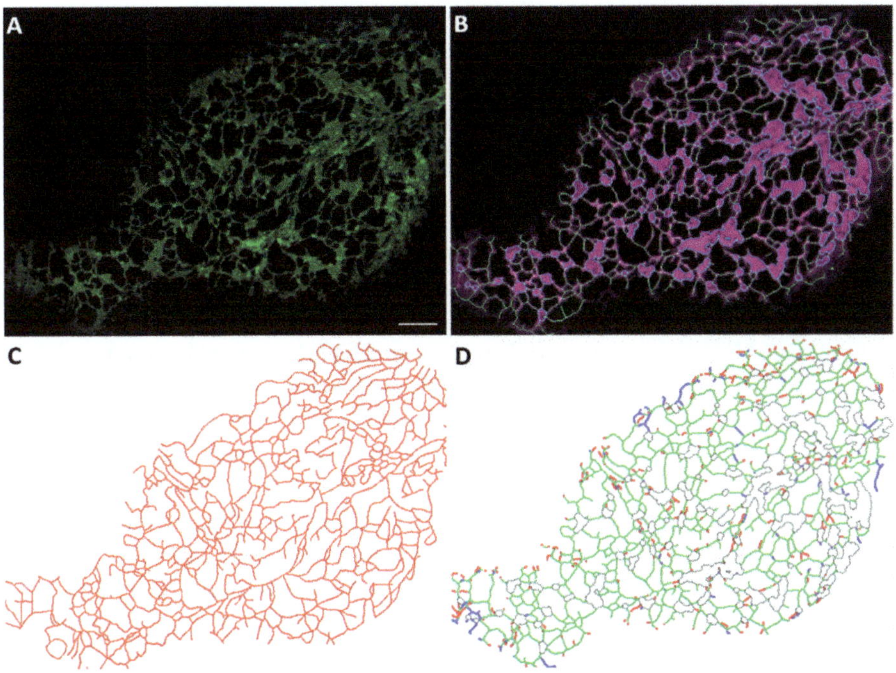


**Fig. 4** Comparison between the automated skeleton extraction and a manual gold standard network
A single image of the ER network (**a**) is processed to extract the pixel skeleton (green) overlaid on the original image (magenta), with the excluded cisternae in blue (**b**). Separately, a gold standard ground-truth network is manually defined (**c**) and compared with the automatic network, with green representing true positives, red false negatives, and blue false positives. Scale bar = 5 μm

here typically return a value of the true width convolved with the psf of the microscope. In addition, the typical pixel spacing leads to significant digitization errors, even with oversampling. A number of different approaches are available that all provide some information on the tubule diameter including:

1. *50% threshold distance*: This estimates the local FWHM from the original tubule image for each pixel in the skeleton. The peak height is estimated from the original intensity, while the distance is estimated from the distance transform of the pixel skeleton. The 50% threshold is estimated from where the pixel intensity falls below half the peak intensity, assuming a local background of zero.
2. *Maximum-gradient granulometry*: The intensity image is subject to a series of image openings (erosion followed by dilation) that successively remove structures as the size of the opening kernel ($s$) exceeds the underlying object. This results in an intermediate ($x$,$y$,$s$) image, where $s$ increases with the size of the disk-shaped kernel to a maximum of $FWHM_{max}$. The intensity of each pixel initially decreases slowly with $s$ as the kernel samples more of the object but then reduces

dramatically once the boundary of the object is reached, and the kernel only samples the background. The transition point for any pixel is determined from the maximum (negative) gradient of the granulometry curve. This approach constrains the width to integer pixel values and also suffers from the digital approximation of small kernels to a true disk-shaped kernel (*see* also **Note 6**).

3. *Integrated intensity granulometry*: This approach follows the same methodology as the maximum-gradient granulometry method, but rather than extract a specific size threshold, the integrated intensity under the granulometry curve is calculated. An additional option is available to use the $FWHM_{max} + 1$ kernel as a local background correction. The integrated intensity provides a more nuanced interrogation of the local ER structure, but cannot be directly related to the physical tubule width without additional assumptions about the relationship between fluorescence intensity and sampled volume (or full deconvolution with an experimentally measured psf). Nevertheless, this approach does help with estimation of relative tubule widths, even if they are sub-resolution objects, provided it is assumed that the fluorescent probe is evenly distributed throughout the ER, and the ER is within the sampling volume of the confocal defined by the psf. A conversion factor can be estimated to convert relative fluorescence to width if it is possible to correlate the physical width, measured from transect profiles, with the intensity.

In the case of integrated intensity granulometry, the simplest way to estimate the empirical intensity calibration factor is to take the average intensity for a cisternal sheet ($I_s$), measured using the profile tool. Typically, this is about 30–50% of the maximum intensity and represents the signal expected for a sheet-like structure completely filling the ($x$,$y$) plane of the psf with the same thickness as the tubules even though it is sub-resolution (Fig. 5). We have to assume that the fluorescent probe is evenly distributed throughout the ER, there is a linear relationship between signal and the amount of probe, and the ER is fully within the sampling volume of the confocal defined by the psf [23]. The estimated sheet thickness ($T_s$) of 40 nM is derived from serial block-face SEM (SBF-SEM), and the size of the lateral psf ($psf_{xy}$) of 140 nM is used to relate the measured intensity to the volume of ER that a fully filled psf would yield when sampling sheets of defined thickness ($V_s = \pi psf_{xy}^2 T_s$) (Fig. 5). The radius of the tubules ($r_t$) is estimated by assuming the volume of the tubule sampled by the same stylized psf ($V_t = \pi r_t^2 psf_{xy}$) compared to the sheet, scaled as the ratio of the intensities from the tubule and sheet ($I_t/I_s$). Thus, the radius of the tubule ($r_t$) was estimated as $r_t = \sqrt{(I_t/I_s)\left(psf_{xy} T_s/4\right)}$ (Fig. 5).

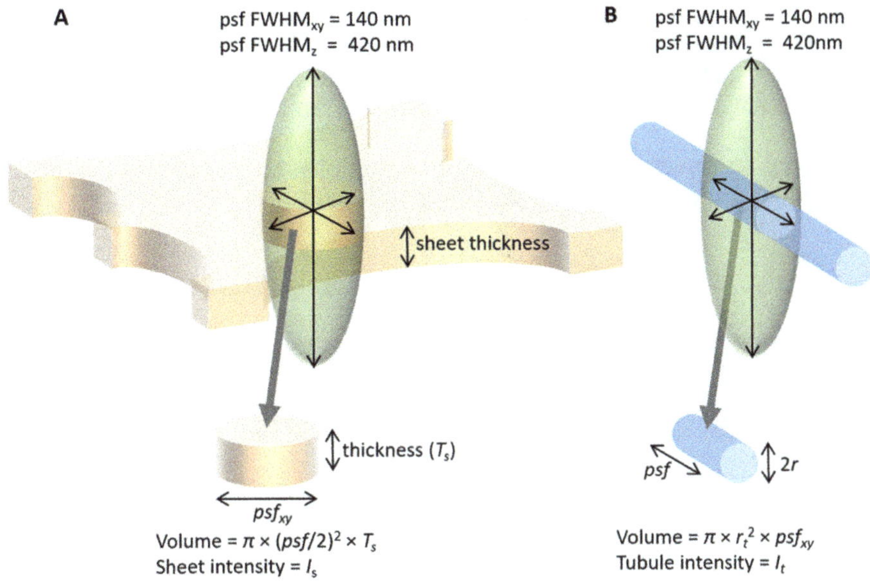


$$r_t = \sqrt{\frac{I_t \times psf_{xy} \times T_s}{I_s \times 4}}$$

**Fig. 5** Estimation of sub-resolution tubule widths from intensity measurements
(**a**) Representation of an idealized point spread function sampling a homogeneous cisternal sheet with an assumed thickness of 40 nM (based on serial-block face SEM measurements). The intensity of the signal can be attributed to a defined sampling volume and used to calculate the relative volume of an ER tubule extending through the same psf sampling volume (**b**). Assuming cylindrical geometry, the radius of the tubule can be calculated

The principle behind the integrated intensity measurements and calibration approach can be illustrated empirically using a simulated model of ER tubules with different widths (Fig. 6a, b) that are convolved with an excitation and emission psf, modelled as anisotropic 3-D Gaussian [23] blurring functions (Fig. 6c, d). The performance of the three different width estimators can be assessed by analysis of a single confocal plane to represent a single (*x,y*) image (Fig. 7e) and shows that all measures perform well above about twice the theoretical lateral FWHM of the psf (140 nM for the Zeiss Airyscan), but the 50% distance measure and the maximum-gradient granulometry measure both converge to a fixed value close to the psf for sub-resolution objects as expected (Fig. 6g). By contrast, the integrated intensity measure provides near linear performance even with sub-resolution objects (Fig. 6g).

A number of metrics are associated with each edge including the Euclidean length of the underlying pixel skeleton, the average width for the entire tubule, and the average width of the tubule excluding the region around the nodes (termed the "center width" – see Fig. 1), which provides a more accurate measure of the tubule diameter itself.

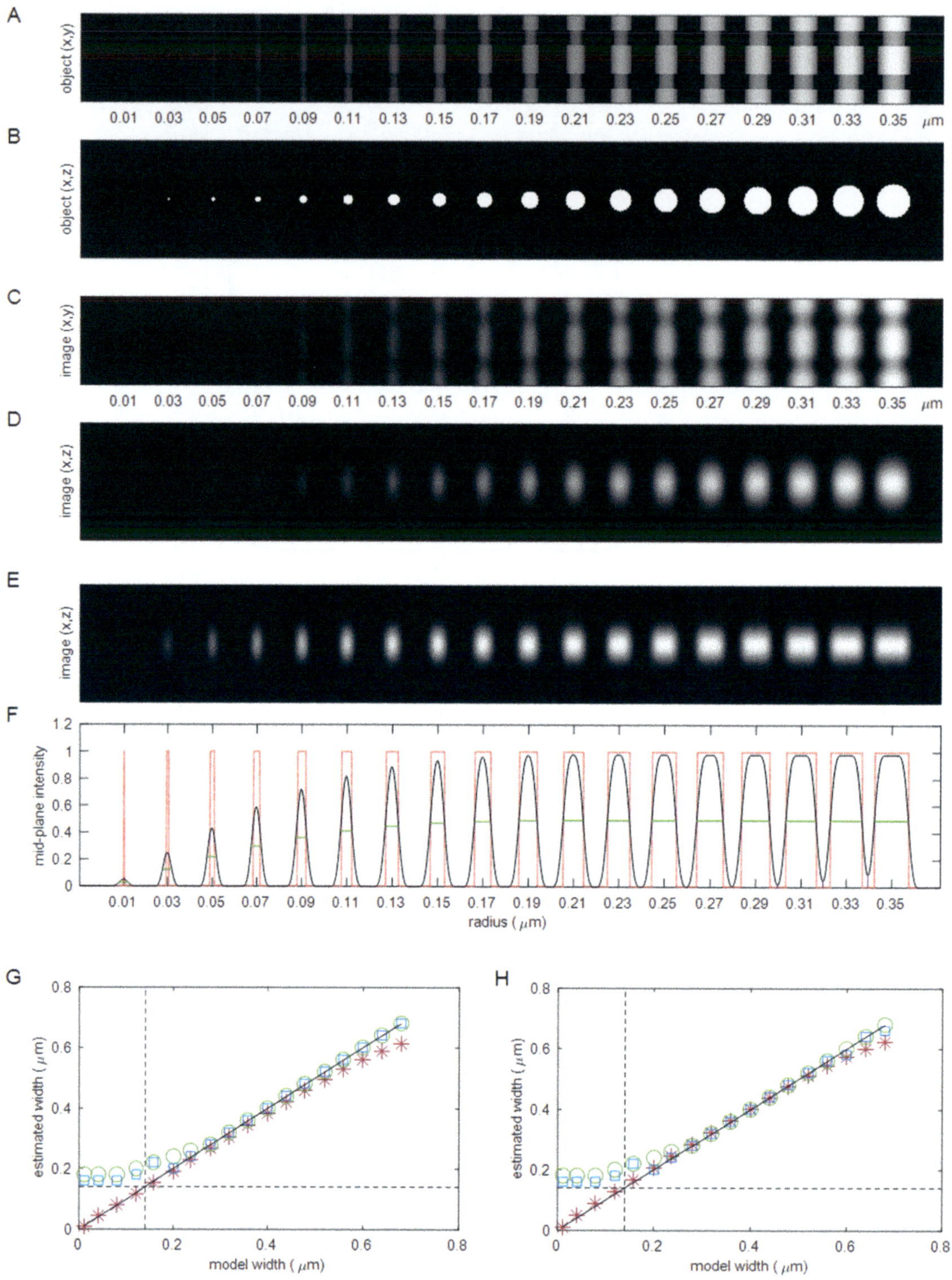


**Fig. 6** Validation of approaches on model tubules

A set of ER tubules with increasing width were simulated as cylinders and viewed as an average (*x,y*) projection (**a**, **c**) or a single (*x,z*) section (**b**, **d**) before (**a**, **b**) and after (**c**, **d**) blurring with simulated anisotropic 3-D Gaussian excitation and emission point spread functions (FWHM 0.14 μm × 0.14 μm × 0.42 μm in *x,y* and *z*, respectively). The simulated results for scanning a single confocal section are illustrated in (**e**). The intensity

Results for a typical ER network are presented in Fig. 7, which shows the original network (Fig. 7a); the "Feature-Type" mean phase angle from the phase congruency analysis that enhances the tubules (Fig. 7b); the segmented cisternae following a morphological top-hat filter to remove the tubules, refined by active contours (Fig. 7c); the intervening polygons color-coded by the $\log_{10}$ area (Fig. 7d); the color-coded width of the pixel-skeleton with the cell boundary marked in yellow and the cisternal boundary overlaid in red (Fig. 7e); the graph representation, including the "super-nodes" in the cisternae that ensure connectivity across the whole network (Fig. 7f); the local speed, color-coded on a rainbow scale between zero in blue and 2.5 μm $s^{-1}$ in red (Fig. 7g); and the persistency of the tubules(green) and cisternae (magenta) (Fig. 7h). Such images provide a useful visual check that the segmentation is working as expected. For example, the estimated width (using the integrated intensity) should match the subjective variation in intensity along different tubules, while the delineation of the sample (cell) boundary and cisternae should again match a subjective view of the structures. In this time series, there are also three cytoplasmic streams, recognized by notional cisternae with a maximum speed above a threshold of 1.3 μm $s^{-1}$ (black arrows in Fig. 7g).

### 3.4 Measurement of Tubule Morphology

Under some conditions or in certain genetic backgrounds, such as overexpression of reticulons [3], the ER tubules show fluctuations in diameter along their length (Fig. 8a). The number, size, and distribution of these bulges and constrictions can be analyzed using the graph representation of the network to extract the intensity profile, integrated normal to the tube axis, from each individual tubule (Fig. 8b). The method uses a peak-finding algorithm, initially to find the bulges (peaks) on the original intensity profile, and then the constrictions (troughs) on an inverted intensity profile (Fig. 8b). The process operates as follows:

**Fig. 6** (continued) profile across the mid-section of the original tubules and after blurring (black) is shown in (**f**), with the horizontal green lines representing the FWHM and the red box profile the true cross-section. The reduction in intensity for sub-resolution tubules is apparent. (**g**) shows the performance of the three different estimates of tubule width. The 50% distance measure (○, green) and the maximum gradient granulometry (□, blue) perform well above about twice the psf (indicated by dashed lines) but return a value close to the expected FWHM of the psf for sub-resolution objects. The integrated intensity measure (*, red) gives a near linear response throughout the range of tubule widths examined unless the tubes overlap at large diameters. Similar results are obtained using a faster granulometry method that includes decomposition of the circular structuring element into linear approximations (*see* **Note 6**), along with an empirical correction of −1 pixel to compensate for oversampling by rectangular kernels (**h**)

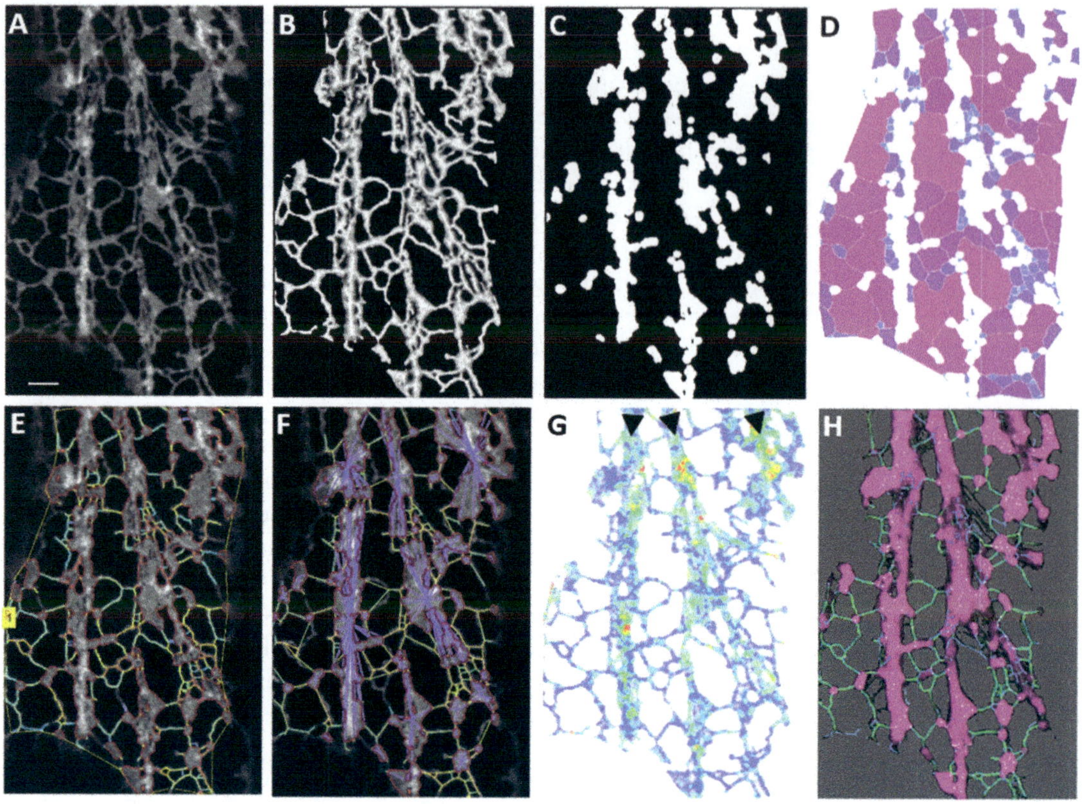


**Fig. 7** Stages in the analysis of the ER morphology
(**a**) Original single (*x,y*) image of GFP-HDEL targeted to the ER and imaged with a Zeiss Airyscan confocal. Scale bar = 5 μm. (**b**) Enhanced image of the tubules using the normalized feature-type image from phase-congruency analysis. (**c**) Automated segmentation of ER cisternae using image opening and active contours. (**d**) Enclosed polygon regions color-coded by $\log_{10}$ area. (**e**) Pseudo-color-coded representation of the tubule width ranging from blue ($FWHM_{min}$) to red ($FWHM_{max}$), superimposed on the filtered image with the sample boundary and cisternal outlines overlaid in yellow and red, respectively, to visually inspect the performance of the overall segmentation and width analysis method. (**f**) Conversion of the weighted pixel skeleton to a weighted graph. Junctions and free ends are represented as nodes linked by edges that have a vector of properties associated with them, including tubule width, length, and average intensity. Cisternae are represented as a "super-node" connected to all the incident tubules on the boundary. (**g**) Color-coded map of speed estimated from optical flow, with red/yellow representing speeds of around 1–2 μm $s^{-1}$. (**h**) Color-coded map of persistency for tubules (green) and cisternae (magenta) where more saturated colors represent regions that are relatively immobile

1. Scan the pixel intensities along the length of each tubule starting at the nodes (which are set to 1). The position and intensity of each peak that is greater than a user-defined minimum peak height and minimum peak prominence are recorded. An additional minimum separation condition is imposed to prevent noisy adjacent pixels from being considered as separate peaks. This separation defaults to $FWHM_{min}$.

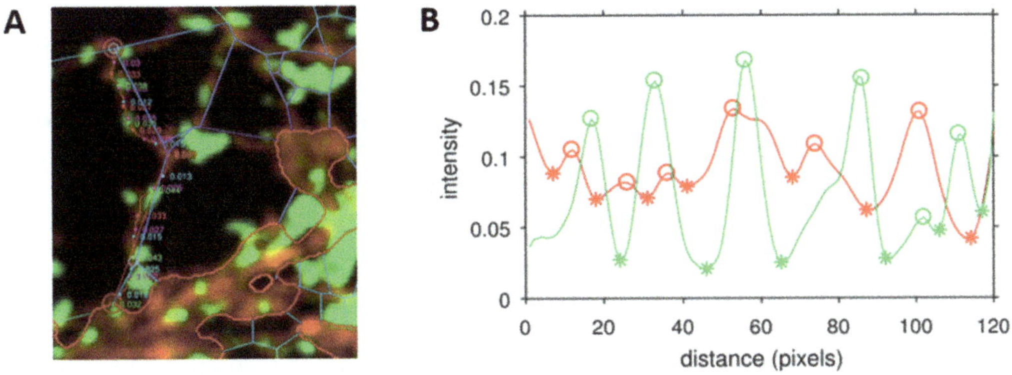


**Fig. 8** Tubule morphology
(**a**) Overexpression of certain proteins, such as reticulons (red), can lead to constrictions in the tubule width, visible as a decrease in intensity of a luminal marker (green), interspersed with tubule bulges that lack reticulons. (**b**) Peaks (open circles) and troughs (asterisks) along each tubule are automatically identified along the profile for each channel and also highlighted on the image in red and magenta, respectively, for the red reticulon marker and green and cyan, respectively, for the luminal HDEL marker

2. A similar scan is used on the inverted tubule profile to detect the "constrictions."
3. Detected peaks and troughs for each channel are superimposed on the original image to visually inspect performance of the algorithm and adjust the parameters if necessary (Fig. 8a).
4. Once the location of the bulges and constrictions has been determined, the width parameters at each point are extracted along with the separation between the peaks.
5. The summary statistics for the bulges and constrictions for all tubules are output to Excel.

### 3.5 Segmentation and Measurement of the Cisternal Regions

In addition to the tubular-reticulum, the ER may also include sheet-like regions called cisternae or regions of closely appressed tubules that are difficult to separate. The methods used to segment the tubular ER do not work with the cisternae, so a different set of methods are available to segment these structures independently. These include:

1. *Automatic segmentation* uses an intensity-based threshold to pick out the brightest structures. An automatic multi-threshold is used to partition pixel intensities into three bins, with optional selection of the threshold level that gives the best segmentation.
2. *Image opening* calculates a modified image after grayscale opening to remove all objects smaller than $FWHM_{max}$ and then converts the resultant image to a binary mask using an automatic threshold. This is more effective at detecting larger sheet-

like regions of the ER, but the resultant binary image tends to give blob-like features that extend beyond the original cisternal region.

3. *Active contour* performs the same image opening step as 2 above but then "shrinks" the features detected down to match the underlying intensity profile using an active contour algorithm.

If no satisfactory segmentation of the features can be achieved with the automatic settings, the cisternae can be defined manually. If this option is selected, the image editing window opens again to give access to the drawing tools. As with the boundary mask, the edited binary image is automatically saved and can be re-applied or edited further every time the image is processed.

A number of morphology measurements of the cisternae are provided including the area, area density, major and minor axis lengths, perimeter, roughness, and solidity (*see* **Notes 7** and **8**).

The cisternal regions are punched out of the pixel skeleton so that they do not contribute to analysis of the tubular elements. During the subsequent analysis steps for the tubular network, each cisternal region is represented as a set of linear connections radiating out from the centroid of the cisterna to connect with the tubules incident on the boundary. This ensures the overall connectivity of the network is retained.

The results for the cisternal segmentation using image opening and active contour refinement are shown in Fig. 8c. The modification to the color-coded pixel skeleton is shown in Fig. 8e and the graph representation in Fig. 8f where the cisternae are represented as a super-node connected to each of the tubules incident on the cisternal boundary.

### 3.6 Cisternal Texture Metrics

The variation in intensities within the cisternal region are characterized using texture analysis from the normalized gray-level co-occurrence matrix (GLCM) [24–26].

1. For each pixel in the segmented cisternal region, the GLCM examines pixels at a distance set by an integer multiple of the target FWHM to maximize the likelihood of revealing texture from appressed tubule regions (*see* **Note 9**). Results are aggregated in four cardinal directions (N, NE, E, SE) with symmetry to complete the angular sampling and ensure results are independent of orientation.

2. Results are added to an accumulator array (Fig. 9a) where the row ($i$) and column ($j$) indices correspond to the intensity value of the target pixel and the neighbor, grouped into a number of equal intensity bins to capture key differences in intensity without rendering the GLCM too sparse. The resultant GLCM is normalized to give a probability $p(i,j)$ of co-occurrence for each intensity pair.

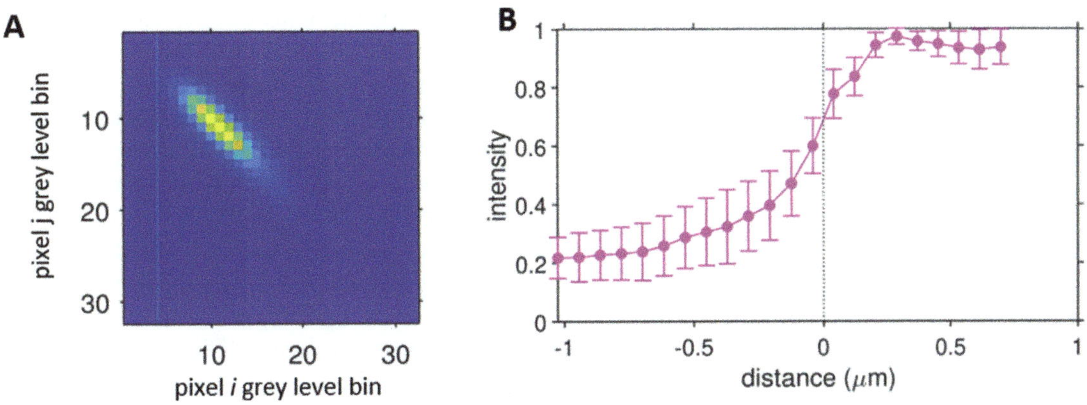


**Fig. 9** Cisternal texture metrics and protein profiles
(**a**) The variation in cisternal sub-structure can be probed using a set of texture metrics derived from the gray-level co-occurrence matrix (GLCM). For each pixel in the image, the intensity is compared to its neighbors at a fixed distance away at 45° intervals and the results accumulated into a 2D GLCM as a count for each pairwise intensity combination. Four orthogonal metrics (contrast, correlation, homogeneity, and energy) are calculated from the dispersion of the points in the GLCM. (**b**) A second measure of substructure in the cisternae is achieved from a radial profile normal to the cisternal boundary. Here the pixel intensities are summed at increasing distances outside and inside from the boundary to build up an average profile. If, for example, the target protein is specifically localized to the cisternal edge, it would be apparent as a peak in the profile on the boundary

3. Choice of the number of intensity bins and spacing is known to affect texture features [26]; thus, to standardize results [26], images also have to be collected under identical conditions and GLCM constructed using a fixed number of intensity bins over the full-intensity range (default 32).
4. Four orthogonal measures are calculated from the GLCM, either individually for each cisterna or as an a single accumulated GLCM for all cisternae (which effectively weights the contribution of each cisterna to the overall statistic by the number of pixels):
   (a) Contrast, $\sum_{i,j}|i-j|^2 p(i,j)$, measures the local variation in the GLCM.
   (b) Correlation, $\sum_{i,j}\frac{(i-\mu i)(j-\mu j)p(i,j)}{\sigma_i\sigma_j}$, where $\mu i$ and $\mu j$ were the weighted mean intensities and $\sigma i$ and $\sigma j$ their standard deviation. This provides a measure of the joint-probability distribution.
   (c) Energy, $\sum_{i,j}p(i,j)^2$, measures the sum of the squares.
   (d) Homogeneity, $\sum_{i,j}\frac{p(i,j)}{1+|i-j|}$, measures how close elements in the GLCM are to the diagonal.

Correlation is scaled from [−1 1], while energy and homogeneity are scaled between [0 1]. An idealized cisternal sheet would have a value of 1 for correlation and homogeneity and zero for energy. The contrast scales from 0 to $(\text{nbins} - 1)^2$ but is normalized here by $(\text{nbins} - 1)^2$ as all images are processed with the same number of bins, to constrain the range to [0 1], where an idealized sheet would have a value of zero (*see* **Note 10**).

The distribution of fluorescence intensities across the cisternal boundary was determined by averaging intensities in incrementing (external) and decrementing (internal) integer radial bins, determined from the Euclidean distance transform (EDT), normal to the edge of each cisternal region (Fig. 9b).

### 3.7 Measurement of Enclosed Polygonal Regions

The polygons enclosed by the ER network are automatically extracted as the inverse of the pixel-skeleton and cisternal areas and various morphological measures calculated, including the area, major and minor axis, circularity, roughness, and solidity in a similar manner to the cisternal measurements. All polygons should be included in area density measurements, but truncated polygons on the boundary of the sampled region should be excluded from shape metrics.

### 3.8 Analysis of ER Organization Using Graph-Theoretic Metrics

Topological measures of the ER structure are calculated by converting the pixel skeleton to a weighted graph representation (*see* **Note 11**). For the tubular regions of the skeleton, the nodes are defined at the junctions between the ER tubules, or the end-points of free tubules, which are connected by straight edges that preserve the topology of the network. If any ER cisternae have been identified, they are represented as a "super-node" positioned at the intensity-weighted centroid of the cisternae, which is connected to all the tubules that are incident on the feature boundary by edges that are given an arbitrary value equal to the average or maximum width value.

### 3.9 Measurement of Movement Using Optical Flow

The ER in plants is highly dynamic, with particularly rapid movement along actin bundles in cytoplasmic streams in the sub-cortical cytoplasm and trans-vacuolar strands [27]. Here we use the Farnebäck algorithm [28] based on local quadratic polynomial expansions to approximate image intensities over a given neighborhood, with weightings based on the strength of the signal and the distance from the central pixel. The algorithm also includes a coarse-to-fine pyramidal scale-space iteration to improve the a priori estimate of the initial displacement field, which allows the method to handle larger displacements. The average velocity for tubules and cisternae is calculated as the scalar mean to give an estimate of the total amount of movement irrespective of direction, while the net directed movement is estimated from the vector mean. The maximum speed is also recorded but potentially quite sensitive to noise.

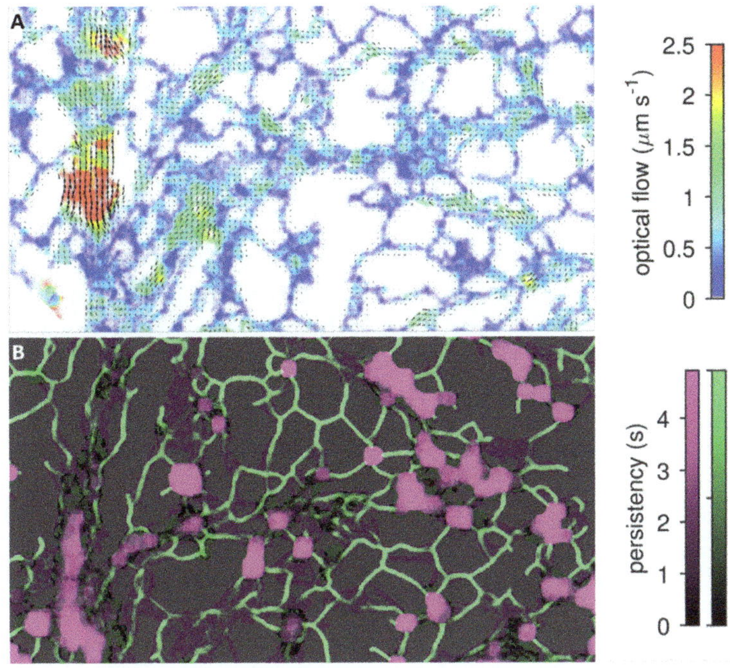


**Fig. 10** Endoplasmic reticulum dynamics
(**a**) The local speed and direction of macroscopic movement can be probed using optical flow techniques and presented as a color-coded map of the magnitude with the associated quiver plot showing direction. Second moments of the flow, such as curl and divergence, can also be calculated. (**b**) Persistency mapping provides a complementary method to highlight elements in the network that remain static and may be associated with other structural features, such as ER-plasma membrane contact sites

Flow coherence is calculated as the vector mean divided by scalar mean and gives a ratio of 1 for fully directed flow and 0 for random movement. The local optical flow is color-coded and mapped back onto the image for inspection, with the option to toggle on or off the quiver plots (arrows) showing the estimated direction of movement (Fig. 10a). Divergence and curl are also calculated from the velocity field and can be mapped back onto the image if required. Summary metrics for optical flow direction are calculated using circular statistics [29].

### 3.10 Persistency Mapping of Static Elements

In the original approach to persistency mapping [9], tubule and cisternal persistency were estimated from the difference between two images with a lag of 5–8 frames (8–14 s), followed by a Boolean AND operation between time points to only include features present in both. Results were normalized to the total segmented area present in all frames (Boolean OR operation on the image stack). Here we implement two approaches:

1. The first method follows the approach developed by I. Sparkes [9] using the difference in the intensity image over a user-defined time period, masked by a threshold binary image from both time points.
2. The second uses the segmented skeleton and cisternal regions directly. Binary images of the skeleton and cisterna are dilated by $FWHM_{min}/2$ and analyzed separately or in combination, by either summing over a defined time window, giving a graded estimate of persistency, or by differencing between the start and end of the time window, giving a binary estimate of persistency.
3. In either case, results are displayed as a color-coded persistency map, summed over the entire time course, and are also aggregated for each edge and cisterna and included in their vector of properties (Fig. 10b).
4. Nodes that were persistent over longer periods that might transiently be associated with tubules or cisternae moving across the anchor points [8, 30] were identified following a 1-D median filter in time across the entire time course followed by Gaussian smoothing in *xy* with $\sigma = FWHM_{min}/2$, normalization to [0 1], and extraction of the position of local maxima above an intensity threshold of 0.3–0.5.

## 4 Notes

1. The value of $FWHM_{min}$ is used to calculate a resampling factor needed to ensure that the minimum apparent tubule width is at least 5 pixels wide to reduce pixelation errors later on. This is based on a pixel size of 40–80 nM and a typical Airyscan or confocal lateral FWHM of 140–250 nM. Likewise, $FWHM_{max}$ is used to determine the number of scales to use in the subsequent steps to ensure both that the largest tubules are correctly segmented and also that structures above this limit are identified as cisternae.
2. CLAHE visually helps the user see detail across the whole image but is only needed if the subsequent segmentation step is dependent on thresholding based on image intensities rather than phase congruency approaches, which are intensity independent.
3. The filtering and contrast adjustment are applied to aid segmentation of the pixel skeleton. Quantitative measurements always refer back to the original image intensities.
4. Care is needed with this step as the automatic thresholding operation may introduce breaks in some dim tubules, rendering those tubules unavailable for consideration in subsequent processing and analysis steps. Trying different parameter settings may resolve this issue; alternatively, there is the option to manually edit the mask.

5. In the case of multi-scale filters, the number of scales is initially set by the ratio of $FWHM_{max}/FWHM_{min}$ with a minimum bound of 3. However, this can be overridden by the user. This may be necessary to ensure that larger features are also processed prior to segmentation. The number of orientations is fixed at 6. In the case of the second-order anisotropic Lopez-Molina filters, the anisotropy is set at 1.3.
6. An additional option is available to increase the speed of the granulometry by using decomposition of the disk-shaped structuring element into a set of linear elements. This requires an empirically determined correction of −1 pixel to the estimated width to compensate for the oversampling with essentially square kernel approximations (Fig. 6h).
7. Rapidly moving cytoplasmic streams are rarely resolved into their constituent tubules and cisternae but tend to get lumped together into one larger cisterna during the segmentation. This can lead to errors in the cisternal metrics unless they are removed from the analysis using a filtering step. This typically involves setting a maximum speed threshold to separate out streams from ordinary cisternae. We use a default of 1.3 $\mu m\ s^{-1}$ as a starting point.
8. Cisternae that are touching the boundary of the sampled region are likely to be truncated. Thus, they should be excluded from any cisternal shape metrics in post-capture data analysis. Conversely, they can be included in area density measurements and cisternal texture analysis.
9. Sampling at one FWHM radius may be useful to detect the minimum between two objects at the Nyquist sampling for the psf and tubule diameter. However, more texture, such as fenestrations, is likely to be detectable at a greater spacing. We currently use a default of 4.
10. A number of metrics, including the cisternal texture metrics, are bounded ranges and unlikely to conform to a normal distribution favored by most subsequent statistical approaches. Thus, data may require transformation, typically using arcsine, log or logit transforms, or analysis using non-parametric statistics.
11. At present, it is not clear which graph theoretic measurements are most useful, in part because the network is essentially planar, so metrics such as node degree, clustering, and assortativity are highly constrained, and also because entire network is sub-sampled to a varying degree by the geometry of the intersection between the cell and the scan plane. Nevertheless, such metrics may provide a fingerprint able to detect changes in ER organization [31].

## Acknowledgments

Funding is gratefully acknowledged from the DFG in the framework of the priority program SPP1710 (AJM), the Human Frontier Science Program (RGP0053/2012, MDF), the Leverhulme Trust (RPG-2015-437, MDF), and BBSRC (BB/W007126/1, MDF, EB).

### References

1. Westrate LM, Lee JE, Prinz WA, Voeltz GK (2015) Form follows function: the importance of endoplasmic reticulum shape. Annu Rev Biochem 84:791–811
2. Fricker M, Heaton L, Jones N, Obara B, Müller SJ, Meyer AJ (2018) Quantitation of ER structure and function. In: Hawes C, Kriechbaumer V (eds) The plant endoplasmic reticulum: methods and protocols. Springer, New York
3. Breeze E, Dzimitrowicz N, Kriechbaumer V, Brooks R, Botchway SW, Brady JP, Hawes C, Dixon AM, Schnell JR, Fricker MD, Frigerio L (2016) A C-terminal amphipathic helix is necessary for the in vivo tubule-shaping function of a plant reticulon. Proc Natl Acad Sci U S A 113:10902–10907
4. Pain C, Kriechbaumer V, Kittelmann M, Hawes C, Fricker M (2019) Quantitative analysis of plant ER architecture and dynamics. Nat Commun 10:984
5. Bouchekhima AN, Frigerio L, Kirkilionis M (2009) Geometric quantification of the plant endoplasmic reticulum. J Microsc 234:158–172
6. Hein B, Willig KI, Hell SW (2008) Stimulated emission depletion (STED) nanoscopy of a fluorescent protein-labeled organelle inside a living cell. Proc Natl Acad Sci 105:14271–14276
7. Streekstra GJ, van Pelt J (2002) Analysis of tubular structures in three-dimensional confocal images. Netw Comput Neural Syst 13:381–395
8. Lin CP, Zhang YW, Sparkes I, Ashwin P (2014) Structure and dynamics of ER: minimal networks and biophysical constraints. Biophys J 107:763–772
9. Sparkes I, Runions J, Hawes C, Griffing L (2009) Movement and remodeling of the endoplasmic reticulum in nondividing cells of tobacco leaves. Plant Cell 21:3937–3949
10. Linkert M, Rueden CT, Allan C, Burel JM, Moore W, Patterson A, Loranger B, Moore J, Neves C, Macdonald D, Tarkowska A, Sticco C, Hill E, Rossner M, Eliceiri KW, Swedlow JR (2010) Metadata matters: access to image data in the real world. J Cell Biol 189: 777–782
11. Pizer SM, Amburn EP, Austin JD, Cromartie R, Geselowitz A, Greer T, ter Haar Romeny B, Zimmerman JB, Zuiderveld K (1987) Adaptive histogram equalization and its variations. Comput Vis Graphics Image Process 39:355–368
12. Kroon DJ, Slump CH, Maal TJ (2010) Optimized anisotropic rotational invariant diffusion scheme on cone-beam CT. Med Image Comput Comput Assist Interv 13:221–228
13. He KM, Sun J, Tang XO (2013) Guided image filtering. IEEE Trans Pattern Anal Mach Intell 35:1397–1409
14. Otsu N (1979) Threshold selection method from gray-level histograms. IEEE Trans Syst Man Cybern 9:62–66
15. Bradley D, Roth G (2007) Adaptive thresholding using the integral image. J Graph Tools 12: 13–21
16. Kovesi PD (1999) Image features from phase congruency. Videre 1:1–26
17. Kovesi PD (2000) MATLAB and Octave functions for computer vision and image processing. http://www.csse.uwa.edu.au/~pk/research/matlabfns/
18. Frangi A, Niessen W, Vincken K, Viergever M (1998) Multiscale vessel enhancement filtering. In: Wells W, Colchester A, Delp S (eds) Medical image computing and computer-assisted interventation. Springer, Berlin/Heidelberg
19. Meijering E, Jacob M, Sarria JCF, Steiner P, Hirling H, Unser M (2004) Design and validation of a tool for neurite tracing and analysis in fluorescence microscopy images. Cytometry 58A:167–176
20. Lopez-Molina C, Vidal-Diez de Ulzurrun G, Baetens JM, Van den Bulcke J, De Baets B (2015) Unsupervised ridge detection using second order anisotropic Gaussian kernels. Signal Process 116:55–67

21. Lopez-Molina C, De Baets B, Bustince H (2013) Quantitative error measures for edge detection. Pattern Recogn 46:1125–1139
22. Saito T, Rehmsmeier M (2015) The precision-recall plot is more informative than the ROC plot when evaluating binary classifiers on imbalanced datasets. PLoS One 10:e0118432
23. Zhang B, Zerubia J, Olivo-Marin J-C (2007) Gaussian approximations of fluorescence microscope point-spread function models. Appl Opt 46:1819–1829
24. Haralick RM (1979) Statistical and structural approaches to texture. Proc IEEE 67:786–804
25. Haralick RM, Shanmugam K, Dinstein IH (1973) Textural features for image classification. IEEE Trans Syst Man Cybern 3
26. Brynolfsson P, Nilsson D, Torheim T, Asklund T, Karlsson CT, Trygg J, Nyholm T, Garpebring A (2017) Haralick texture features from apparent diffusion coefficient (ADC) MRI images depend on imaging and pre-processing parameters. Sci Rep 7:4041
27. Pain C, Kriechbaumer V (2020) Defining the dance: quantification and classification of endoplasmic reticulum dynamics. J Exp Bot 71:1757–1762
28. Farnebäck G (2003) Two-frame motion estimation based on polynomial expansion. In: Bigun J, Gustavsson T (eds) Image analysis: 13th Scandinavian conference, SCIA 2003 Halmstad, Sweden, June 29–July 2, 2003 proceedings. Springer, Berlin/Heidelberg
29. Berens P (2009) CircStat: a MATLAB toolbox for circular statistics. J Stat Softw 31:1–21
30. Lin C, White RR, Sparkes I, Ashwin P (2017) Modeling endoplasmic reticulum network maintenance in a plant cell. Biophys J 113: 214–222
31. Lu M, Christensen CN, Weber JM, Konno T, Läubli NF, Scherer KM, Avezov E, Lio P, Lapkin AA, Kaminski Schierle GS, Kaminski CF (2023) ERnet: a tool for the semantic segmentation and quantitative analysis of endoplasmic reticulum topology. Nat Methods 20:569–579

# Chapter 6

# Long-Term Imaging of Endoplasmic Reticulum Morphology in Embryos During Seed Germination


## Natasha Dzimitrowicz, Emily Breeze, and Lorenzo Frigerio



## Abstract

Imaging plant embryos at the cellular level over time is technically challenging, since the embryo, once its protective seed coat is removed, must be kept viable and unstressed on a microscope slide for the duration of the experiment. Here we describe a procedure and suitable apparatus for the visualization, over several days, of changes in endoplasmic reticulum (ER) morphology associated with the process of germination in *Arabidopsis thaliana* seeds. Moreover, we also present a user-friendly image analysis tool, which enables subtle perturbations in the ER network to be measured.



**Key words** Confocal microscopy, Imaging chamber, Embryo, Germination


## 1 Introduction

Seed germination can be defined as the series of events occurring between the uptake of water by the dry seed (imbibition) and the elongation of the embryonic axis which, in angiosperms, culminates in the emergence of the radicle through the seed coat [1, 2]. Upon imbibition, the seed rapidly commences metabolic activities with a high concomitant demand for de novo protein and lipid biosynthesis, thus placing a considerable burden on the endoplasmic reticulum (ER), which continues throughout germination [1, 3]. Since the time taken for germination and post-germination growth can vary from hours to weeks, depending on the plant species and/or environmental conditions, imaging changes in cellular morphology associated with this process is challenging. Moreover, it is not possible to obtain confocal microscopy images through the seed coat, which must, therefore, be removed prior to imaging. Here we describe a time series of ER morphology changes associated with the germination of *Arabidopsis thaliana* embryos over a period of 8 days, from seed imbibition through to fully emerged seedling. By first removing the seed coat and then


Verena Kriechbaumer (ed.), *The Plant Endoplasmic Reticulum: Methods and Protocols*, Methods in Molecular Biology, vol. 2772,
https://doi.org/10.1007/978-1-0716-3710-4_6,

immobilizing the embryos in sterile Murashige and Skoog (MS) media, containing appropriate antibiotics, in an imaging chamber, we were able to obtain live confocal images of cellular events occurring during germination. It is possible to embed several Arabidopsis embryos in a single imaging chamber, enabling multiple replicates to be imaged under identical conditions, and also to apply the described methodology to other plant species, depending on the size of the embryos. We were further able to analyze the images post-acquisition in order to assess the subtle temporal differences in the ER network.

## 2 Materials

Prepare all solutions using sterile ultrapure water.

1. Murashige and Skoog (MS) medium: Weigh 0.44 g Murashige and Skoog basal salt mixture and 1 g bacteriological agar into a suitable, autoclavable, container. Add water to a volume of 90 mL and mix. Adjust pH to 5.7 ± 0.1 with KOH or HCl. Add additional water to bring the total volume to 100 mL. Sterilize the medium by autoclaving. Use immediately, or store at 65 °C. Appropriate antibiotics can be added immediately prior to use.
2. Arabidopsis seeds transformed with the desired fluorescent markers or fusion proteins. Here we used stable lines carrying the ER luminal marker GFP-HDEL [4].
3. CoverWell™ Imaging chamber (1 × 20 mM) (supplier Grace Biolabs). A reusable press-to-seal silicone chamber that forms a removable enclosure designed to stabilize and support free-flowing samples (Fig. 1).

## 3 Methods

All procedures should be carried out at room temperature unless otherwise stated.

### *3.1 Preparation of Embryos for Imaging*

1. Add 50–100 dry seeds to a 2 mL round-bottomed Eppendorf tube, and carefully add 1 mL sterile water. Incubate tube at 4 °C for 1 h in the dark.
2. Using a wide-bore pipette tip (or a cut yellow tip), transfer some of the seeds to a clean petri dish (*see* **Note 1**), and view under a stereomicroscope with appropriate lighting and magnification.
3. Hold an individual seed against its longitudinal axis with fine forceps. Carefully cut the seed coat (including testa and

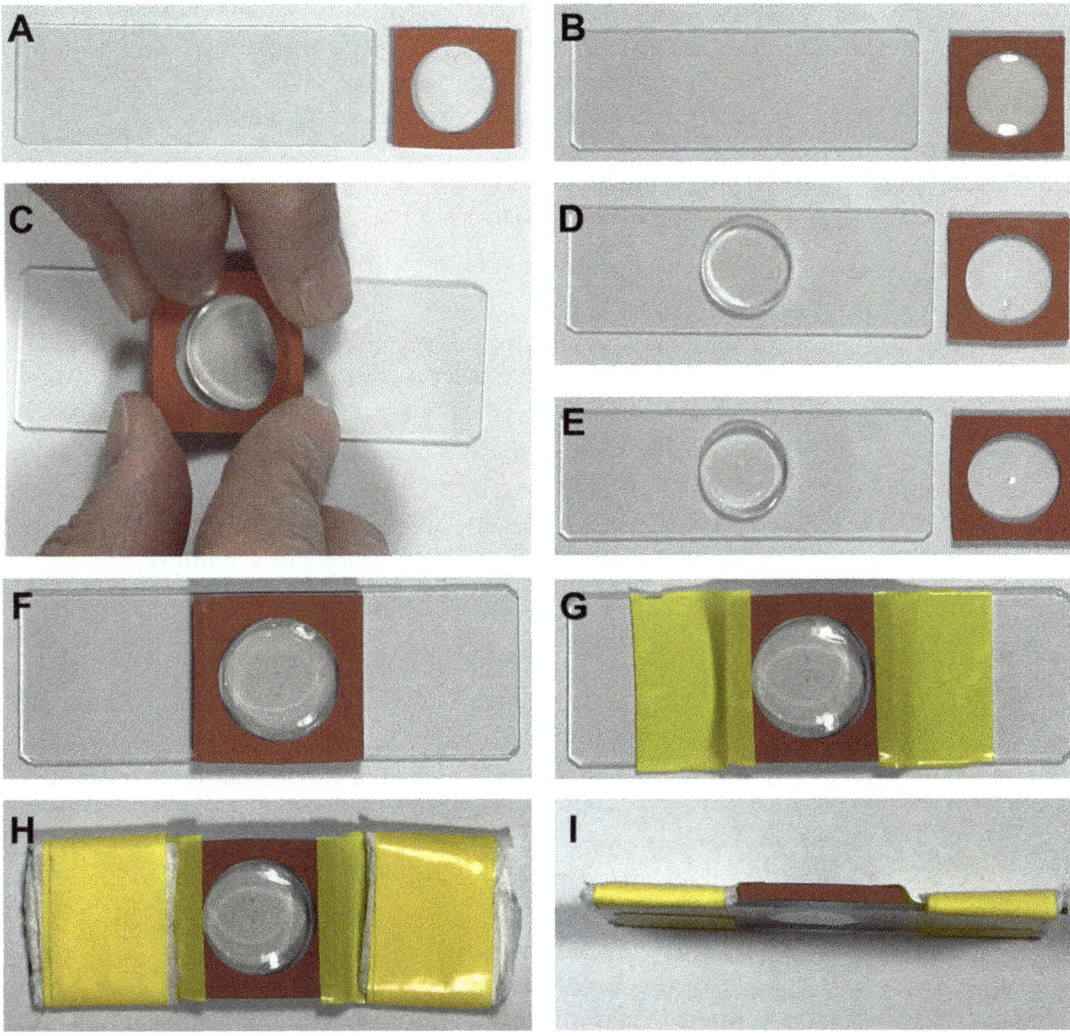


**Fig. 1** Assembly of the imaging chamber
The clean imaging chamber (**a**) is filled with MS agar media containing appropriate antibiotics (**b**) and allowed to set before carefully being turned out onto a sterile microscope slide (**c**, **d**). Uncoated embryos, prepared previously and stored in water, are gently pipetted into the chamber (**e**) and covered with the agar disk (**f**) and the chamber securely taped to the slide (**g**). The height of the assembled apparatus is adjusted, as required with tissue and further tape (**h**, **i**). All steps should be performed under sterile conditions. Arrow indicates position of uncoated embryo

endosperm) along its length using a syringe needle (e.g., U100 insulin needle), starting at the end of the seed closest to where the cotyledons join the radicle.

4. Using forceps, gently and smoothly squeeze the seed at the end furthest from the cut (i.e., closest to the tip of the cotyledons and radicles) to release the intact embryo out of the seed coat through the created opening (*see* **Note 2**).

5. Using a syringe needle, carefully transfer the uncoated embryo to a 2 mL round-bottomed Eppendorf tube, and store in a minimal amount of sterile water while further embryos are collected, for up to a maximum of 60 min.

### 3.2 Preparation of Imaging Chamber

The following steps should be carried out inside a laminar flow hood cabinet. For all stages see Fig. 1.

1. Prior to use, thoroughly clean the imaging chamber with 70% ethanol (Fig. 1a).
2. With a sterile pipette tip, add 900 μL MS agar at ~60 °C to a CoverWell™ imaging chamber (or equivalent), and allow to set (Fig. 1b).
3. Carefully transfer the disc of agar onto a clean, sterile, microscope slide (Fig. 1c, d) (*see* **Note 3**).
4. Using a sterile wide-bore pipette, transfer 3–4 uncoated embryos into the imaging chamber in a minimal amount of water (Fig. 1e) (*see* **Note 4**).
5. Carefully invert the imaging chamber, and place it onto the agar disc on top of the embryos in the chamber (Fig. 1f) (*see* **Note 5**). Remove excess water with a tissue. Once mounted, the embryos should not move freely under the agar.
6. Fix the prepared imaging chamber to a clean microscope slide, and secure with electrical insulation tape or equivalent, ensuring that the embryos are clearly visible (Fig. 1g). Fold two suitably sized pieces of tissue paper, and affix to the outer edges of the slide with additional tape, to equalize the thickness of the whole apparatus (Fig.1g, h).
7. Place the assembled chamber into a growth cabinet under the desired environmental conditions. Remove the apparatus at the desired time points and image under a confocal microscope with the appropriate settings for the desired fluorophores (Fig. 2). After imaging, remove any residual immersion oil from the slide with ethanol. Between imaging sessions, return the chamber to the growth cabinet.

### 3.3 Image Analysis

The following macro is designed to estimate the proportion of ER sheets to tubules from a confocal image (Fig. 3). The macro runs in the freely available ImageJ-based open source processing package Fiji [5] using the Bio-Formats plugin [6] and so requires the latest version of this software to be installed prior to use. It also requires a representative image of the ER containing both sheets and tubules in order to train the segmentation classifier.

1. Open the relevant confocal image file in Fiji (*see* **Note 6**), and set the background of the binary image to black (Edit> Options> Colours…).

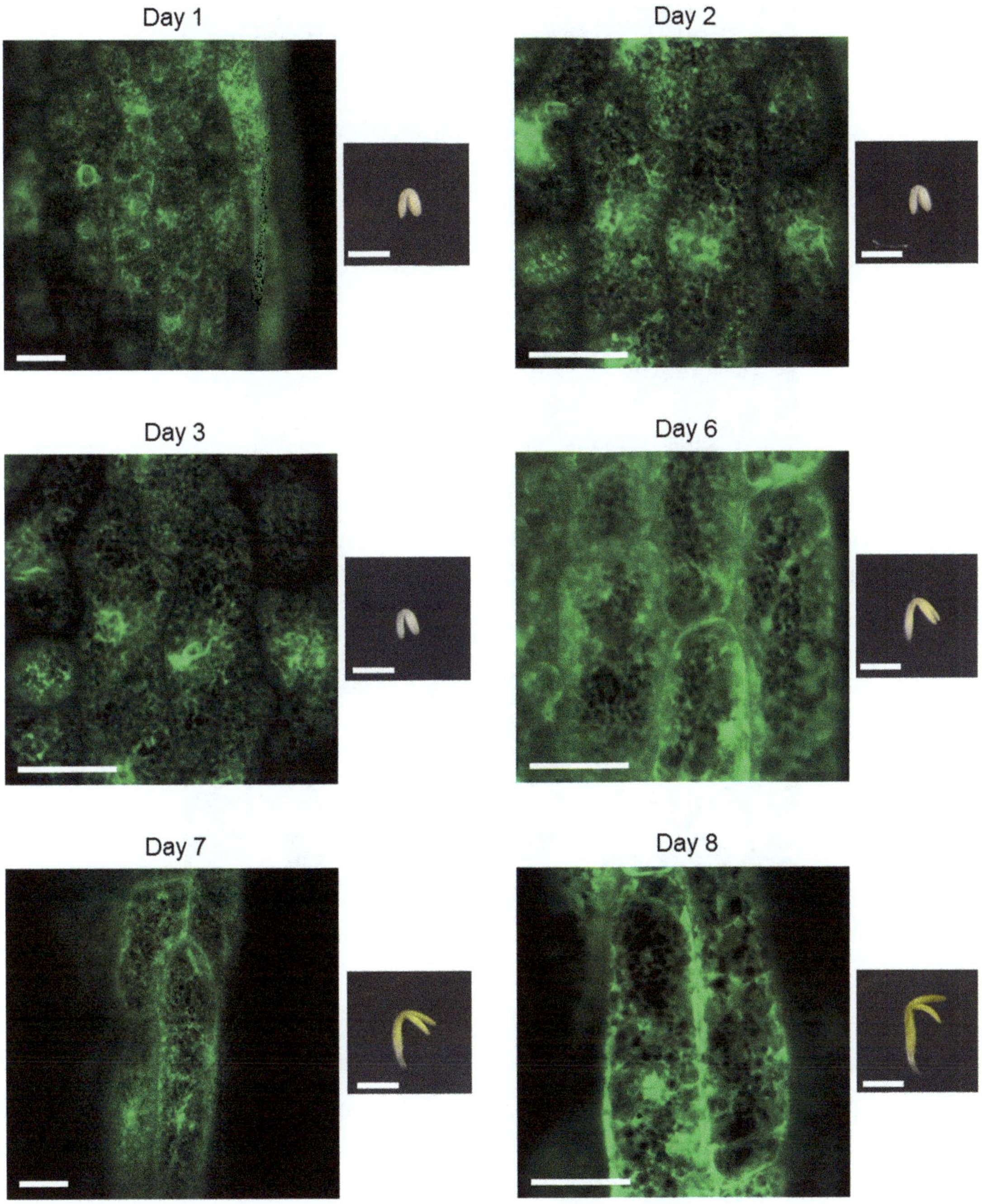


**Fig. 2** Changes in ER morphology during germination
Confocal images of the ER, as labelled with the luminal marker GFP-HDEL, obtained from a single germinating embryo in an imaging chamber with images obtained daily for a total of 8 days, together with the corresponding images of the whole embryo. Confocal image scale bars = 10 μm. Whole embryo image scale bars = 0.5 mM

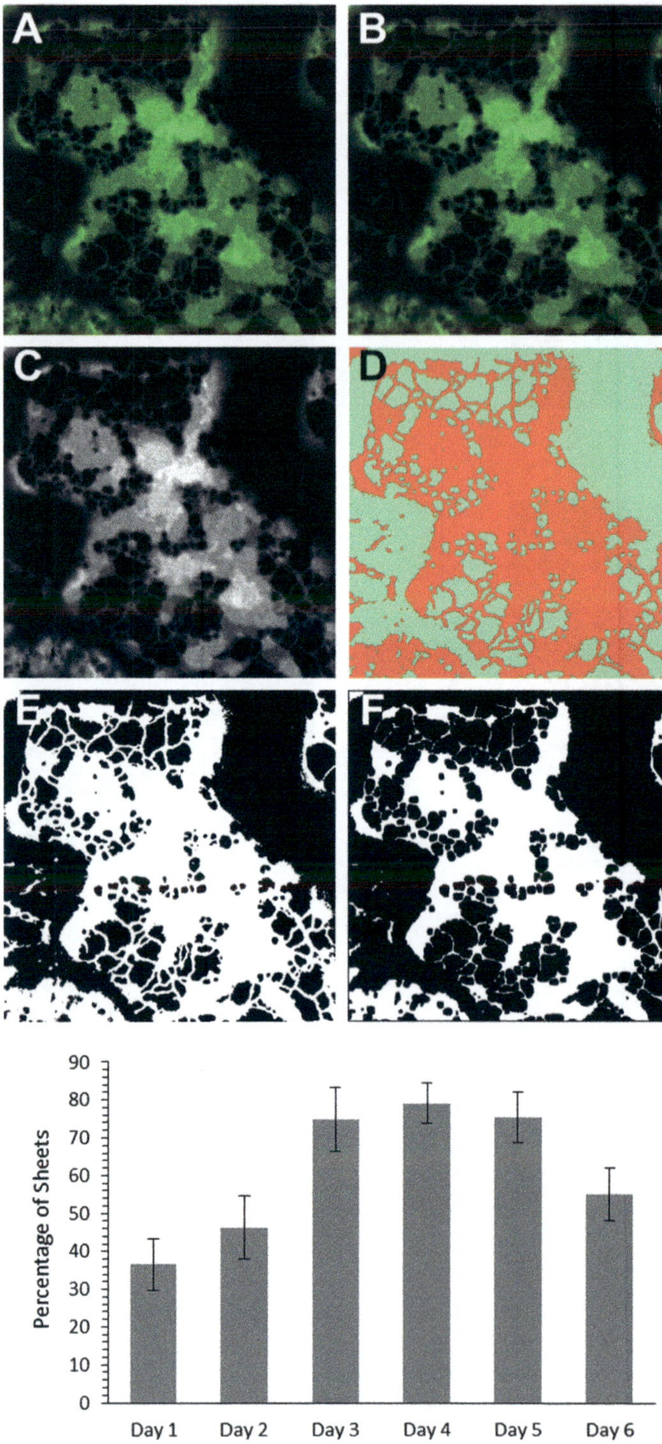


**Fig. 3** A Fiji-based macro for the estimation of ER sheet abundance post-image acquisition
A confocal image of the ER network in a germinating embryo was analyzed by a macro written for the open-source platform Fiji [5]. (**a**) Original (unmanipulated) confocal image. (**b**) Median-filtered image. (**c**) Contrast-enhanced image. (**d**) Trainable Weka Classified image. (**e**) Binary image (total). (**f**) Eroded binary image. (**g**) Proportion of the total ER network identified as sheets by our image analysis macro from Days 1 to 6 of embryo germination

2. Apply a median filter (radius 2 pixels) to reduce the salt and pepper noise (Process> Filters> Median).
3. Normalize the image by setting the saturation levels to 0.5% (Process> Enhance> Contrast...).
4. Segment the image to create a binary image of the total ER network by applying the Trainable Weka Classifier (v3.2.5) function (Plugins> Segmentation> Trainable Weka Segmentation) [7, 8]. Load a representative image of the ER containing both tubules and sheets. Use the "Freehand line" selection tool to identify areas of ER and background, and assign them to either Class 1 (ER) or Class 2 (Background). Use a minimum of five selections for each class to train the classifier, and save as a ".model" file.
5. Load the classifier file, and select Create Result to segment the image ("Classified Image"). Convert this image into a binary image (Process> Binary> Make Binary).
6. Duplicate the binary segmented image (A and B). Apply an erode function to ImageA (iterations = 3; count = 1) to remove thinner structures from the image such as tubules (Process> Binary> Erode). Save eroded image file (ER sheets only) (A) and unaltered binary image (total ER network) (B).
7. Color Pixel Counter plugin [9] (*see* **Note** 7). Convert both images to RGB Color (from 8-bit) (Image> Type> RGB Color). Run the Color Pixel Counter to count green pixels in the unaltered binary image (B) and red pixels in the eroded image file (A) (cells = 20; pixels = 0.2100; minimum = 30) (Plugins> Color Pixel Counter) (*see* **Note** 7).
8. The proportion of sheets in the image can be calculated from the ratio of pixels in the eroded image (A) to that in the total binary image (B) (*see* **Note 8**).

## 4 Notes

1. It may help to have to small piece of Whatman filter paper or microscope lens cleaning tissue on the petri dish to absorb excess water and limit movement of the seed during uncoating.
2. Uncoating embryos, and ensuring that they remain intact and undamaged, is technically challenging and likely requires extensive practice. The use of two syringe needles, in addition to forceps, may help with both the manipulation of the seed and the application of pressure to ease the embryo out of the seed coat.

3. The imaging chamber is relatively flexible, and so it is possible to bend it slightly to allow the agar disc to become dislodged.
4. When transferring the uncoated embryos to the imaging chamber, care should be taken to ensure that they don't become damaged. Once in the chamber, it is beneficial to ensure that the embryos are well-spaced and not clumped together to allow them sufficient room to expand over the time course.
5. It is important to ensure that only a minimal amount of water is used to transfer the embryos to the imaging chamber so that when the chamber is inverted, the surface tension of the water keeps the embryos on the coverslip.
6. If the image file is a Z-stack, then it must be converted into single images, by either selecting an appropriate slide (Image> Stacks> Stack to Image) or converting the stack to a projection (Image> Stacks> Z-project....).
7. The Color Pixel Counter plugin is not included in the standard Fiji package and so must be downloaded separately and saved to Program Files\Fiji.app\plugins (or equivalent location; NB the filename must be altered so that each word is capitalized). It is freely available from http://imagejdocu.tudor.lu/doku.php?id=plugin:color:color_pixel_counter:start.

   The choice of red or green pixels for the eroded and binary images, respectively, is arbitrary and simply allows the user to identify which results from which image the results were obtained since the Color Pixel Counter identifies the number of white pixels as colored pixels.
8. Be aware that the zoom level applied in the images will impact on the results obtained. This can be negated by converting the pixel count to an area (microns). First identify the resolution of the image (Image> Show Info...), and then calculate the square root of the number of green or red pixels, divided by the resolution. By squaring this result, the area in microns can be calculated and used to determine the proportion of ER sheets present in the image.

## References


1. Bewley JD (1997) Seed germination and dormancy. Plant Cell 9:1055–1066
2. Finch-Savage WE, Leubner-Metzger G (2006) Seed dormancy and the control of germination. New Phytol 171:501–523
3. Holdsworth MJ, Bentsink L, Soppe WJJ (2008) Molecular networks regulating Arabidopsis seed maturation, after-ripening, dormancy and germination. New Phytol 179:33–54
4. Boevink P, Cruz S, Hawes C, Harris N (1996) Virus-mediated delivery of the green fluorescent protein to the endoplasmic reticulum of plant cells. Plant J 10:935–941
5. Schindelin J, Arganda-Carreras I, Frise E et al (2012) Fiji: an open-source platform for biological-image analysis. Nat Methods 9:676–682

6. Linkert M, Rueden CT, Allan C et al (2010) Metadata matters: access to image data in the real world. J Cell Biol 189:777–782
7. Arganda-Carreras I, Kaynig V, Rueden C et al (2017) Trainable weka segmentation: a machine learning tool for microscopy pixel classification. Bioinformatics. https://doi.org/10.1093/bioinformatics/btx180
8. Arganda-Carreras I, Kaynig V, Rueden C et al Trainable_segmentation: release v3.1.2 [data SET]. Zenodo. https://doi.org/10.5281/zenodo.59290
9. Pichette B. (2012) Color pixel counter. http://imagejdocu.tudor.lu/doku.php?id=plugin:color:color_pixel_counter:start

# Chapter 7

# Dancing with the Stars: Using Image Analysis to Study the Choreography of the Endoplasmic Reticulum and Its Partners and of Movement Within Its Tubules

Lawrence R. Griffing

## Abstract

In this chapter, approaches to the image analysis of the choreography of the plant endoplasmic reticulum (ER) labeled with fluorescent fusion proteins ("stars," if you wish) are presented. The approaches include the analyses of those parts of the ER that are attached through membrane contact sites to moving or non-moving partners (other "stars"). Image analysis is also used to understand the nature of the tubular polygonal network, the hallmark of this organelle, and how the polygons change over time due to tubule sliding or motion. Furthermore, the remodeling polygons of the ER interact with regions of fundamentally different topologies, the ER cisternae, and image analysis can be used to separate the tubules from the cisternae. ER cisternae, like polygons and tubules, can be motile or stationary. To study which parts are attached to non-moving partners, such as domains of the ER that form membrane contact sites with the plasma membrane/cell wall, an image analysis approach called persistency mapping has been used. To study the domains of the ER that move rapidly and stream through the cell, image analysis of optic flow has been used. However, optic flow approaches confuse the movement of the ER itself with the movement of proteins within the ER. As an overall measure of ER dynamics, optic flow approaches are of value, but their limitation as to what exactly is "flowing" needs to be specified. Finally, there are important imaging approaches that directly address the movement of fluorescent proteins within the ER lumen or in the membrane of the ER. Of these, fluorescence recovery after photobleaching (FRAP), inverse FRAP (iFRAP), and single particle tracking approaches are described.

**Key words** Endoplasmic reticulum, Network flow, Network movement, Persistency mapping

## 1 Introduction

The endoplasmic reticulum (ER) is an extremely dynamic organelle that ramifies throughout the cytosol of plant cells [1, 2]. One might expect this, given that it is tethered to other dynamic organelles like the Golgi body, the peroxisome, and the mitochondrion. Remarkably, it is also tethered to less dynamic organelles such as the chloroplast, the nucleus, and the plasma membrane. How image analysis is currently used and could be used in the future to analyze

Verena Kriechbaumer (ed.), *The Plant Endoplasmic Reticulum: Methods and Protocols*, Methods in Molecular Biology, vol. 2772, https://doi.org/10.1007/978-1-0716-3710-4_7,

the choreography of this dance with still, slow, and fast partners is the subject of the first part of this chapter. But there is another dance going on as well: the dance of the proteins, lipids, ions, and secondary metabolites in the endoplasmic reticulum itself. Sometimes, these two dances are confused. As proteins move through the ER, they are sometimes considered as the movement of the ER itself. It is as if the movement of costume of the dancer was taken to be the movement of the dancer herself. Although this is an easy-enough mistake to make, it is important to follow the movement of the dancer, the ER, and the costume, the surface, or luminal markers of the ER, separately. The costume movement could well be dictated by physics, but here instead of the complicated physics of fabric draping, we are concerned with the complicated physics of the small (the ER is about 60 nm in diameter) in a highly viscous, high Reynold's number, environment.

We approach this by deconstructing ER dynamics into two broad categories: *network movement*, the movement and remodeling of the tubules and cisternae, and *network flow*, the movement of protein and lipid (and ions) within the ER lumen and the ER membrane. Approaches to the analysis of network movement are first described. The ER is made of network of tubules and cisternae, and in special cases like brassicaceous plants, e.g., *Arabidopsis* species, the lozenge-shaped ER body. One quantifiable characteristic of the network is the mesh size of the network (Figs. 1 and 2). Other quantifiable features are the tubules and the cisternae, which

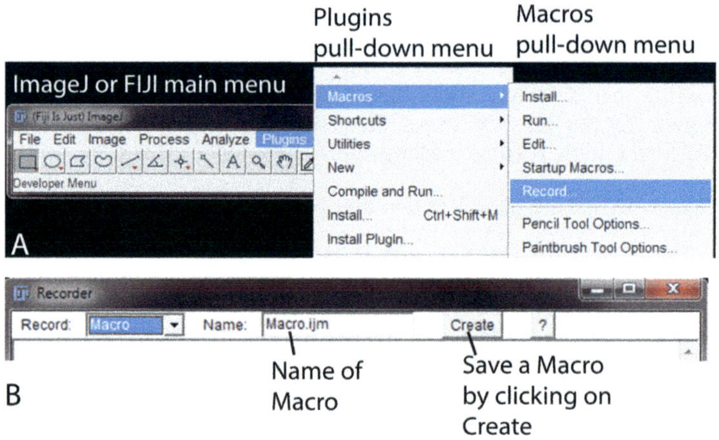


**Fig. 1** ImageJ menu structure and accessing macro commands. (**a**) Top menu bar of ImageJ (or the FIJI version) with the Plugins command highlighted in blue. The Plugins>Macros command is shown in the plugins pulldown menu. The Plugins>Macros>Record command is shown in the macros pulldown menu. (**b**) Macro recorder window. The top of the window is shown. The name of the macro can be edited, but the recorder does not save the recorded actions until the Create button is clicked. Closing the recorder without clicking on create loses the macro

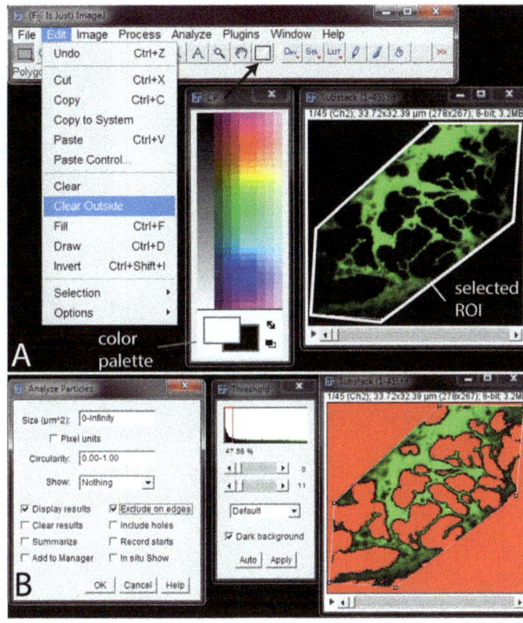


**Fig. 2** ImageJ commands for ER mesh counting. (**a**) ImageJ top menu bar showing the Edit>Clear Outside command. This can be achieved after a region of interest (ROI) is selected with the polygon tool that excludes out-of-focus regions (right-hand panel). If the image has a black background, the out-of-focus regions can be excluded with the Clear Outside command once the background has been set to black using the color palette (center panel) accessed by double-clicking the eye-dropper icon (arrow). (**b**) Analyzing the mesh. The dialog box for the Analyze>Analyze Particles command is shown (left panel), with the "Exclude on edges" option checked. The Threshold dialog box (center panel) from the Image>Adjust>Threshold command is shown selecting the low-intensity regions (mesh openings). During thresholding, the selected region is red (right panel)

can be separated using morphological image processing routines on binary images (Figs. 3 and 5). The ends and junctions of tubules can be quantified (Fig. 4). These operations, when combined with image interval subtraction, can be used to create separate persistency maps for tubules and cisternae that identify the most and least persistent tubules over the time period of observation and interval of analysis (Fig. 6). Using persistency mapping, the movement and form of the ER was shown to be differentially sensitive to dominant-negative tail domain expression of the six (of thirteen in *Arabidopsis*) paralogs of myosin XI that are involved in the motility of spheroid organelles like peroxisomes, mitochondria, and Golgi [3–5]: XI-2 (Myo11B2), XI-C (Myo11C1), XI-E (Myo11C2), XI-K (Myo11E), XI-1 (Myo11F), and XI-I (Myo11G). Note that the new nomenclature for these genes [6, 7] follows their old nomenclature in parentheses. While XI-C (Myo11C1) and XI-E (Myo11C2) have their largest effects on ER

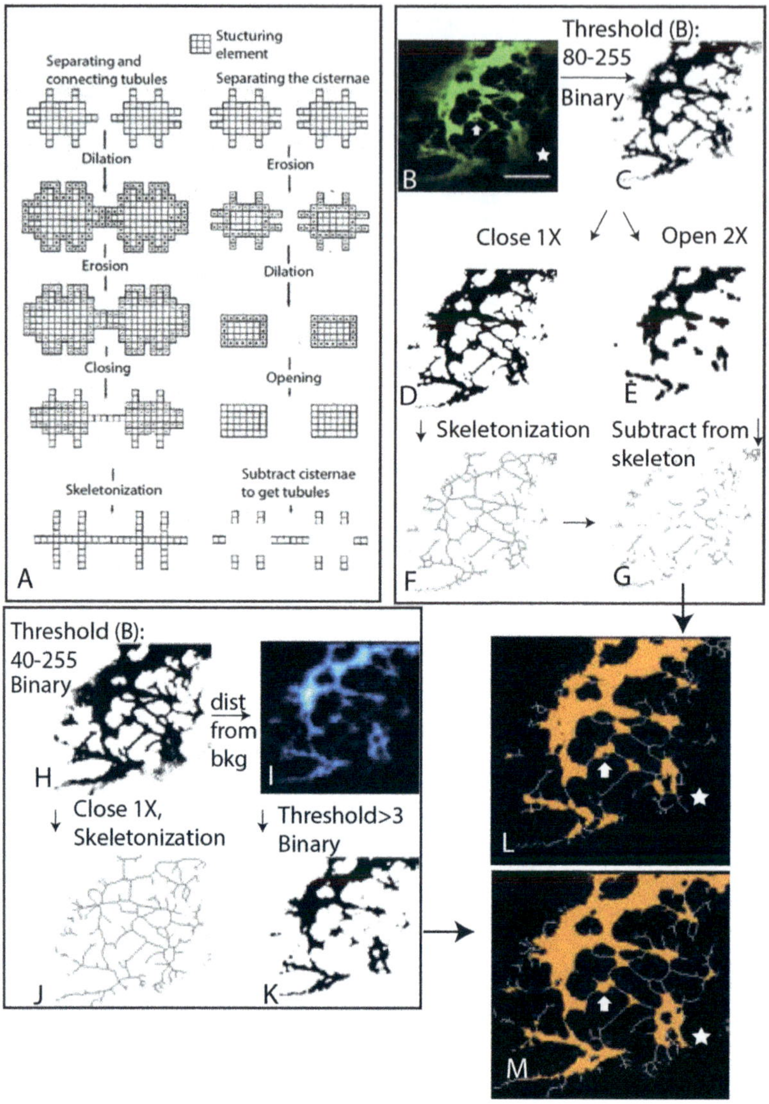


**Fig. 3** Separating tubules from cisternae using 8-bit (Image>Type>8bit) morphological image processing. (**a**) Morphological procedures for binary images summarized using a 3 × 3 pixel structuring element. Squares represent neighboring pixels. Squares with minus signs produced during erosion or skeletonization mean that the pixel is removed. Squares with plus signs produced during dilation means the pixel is added. The center of the structuring element is put over the target pixel, and if any of the structuring element pixels don't cover a neighboring pixel, the pixels underlying the structuring element are added (dilation) or removed (erosion). (**b**) An image of the ER. Scale bar = 10 μm. (**c**) Result of thresholding at a gray level of 80 and using the Binary>Make Binary command. (**d**) One iteration of a Process>Binary>Close command. (**e**) The result of two iterations of the Process>Binary>Open command (number of iterations can be adjusted in the Binary>Options command). (**f**) Skeletonization of (**d**) using the Process>Binary>Skeleton command. (**g**) Subtracting the cisternal regions from the skeletons using the Process>Image Calculator command. (**h**) Binary image resulting from thresholding at a grey value of 40 using the Process>Binary>Make Binary command. (**i**) Euclidean distance map of (**h**) (pseudo-colored with the Cyan>Hot lookup table, LUT) using the Process>Binary>Distance Map command. (**j**) Skeletonization of (**h**) after one closing operation. (**k**) Result of thresholding

tubulation, XI-K (Myo11E), XI-1, (Myo11F) and XI-I (Myo11G) have their largest effects on cisternalization. Of the latter, XI-I (Myo11G) localizes to the nuclear envelope and is involved in forming or maintaining tubules that are coextensive with the cisternal structure of the nuclear envelope [8].

In a separate study examining insertional mutants of XI-K (Myo11E), XI-1 (Myo11F), and XI-2 (Myo11B2) [9], XI-1 (Myo11F) mutation was only found to have an effect on ER dynamics if it was crossed with XI-2 (Myo11B2) or XI-K (Myo11E) mutations. This contrasts with the work [4], where XI-1 (Myo11F) tail domain expression alone caused an increase in cisternal and tubule persistency, while XI-2 (Myo11B2) tail domain expression alone shows no effect on tubule or cisternal persistency [4] but has been shown to slow streaming rates of spheroid organelles. However, the way that dynamics were measured in [9] was different from persistency analysis. LPX flow (originally called Kbi flow, [9]) uses an optic flow approach and is described below (Fig. 7). The reason for the difference between the two analyses is that optic flow approaches do not distinguish network movement from network flow and, using a mask, excludes the dynamics of the fainter, tethered, polygonal ER, where much of the remodeling occurs (Fig. 7).

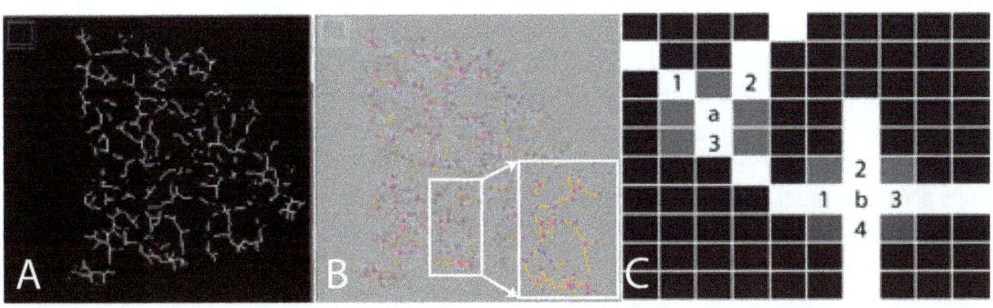


**Fig. 4** Analysis of skeletons representing tubules. (A) Skeletonized region of ER from a time-lapse series with the cisternal component subtracted (see Fig. 6a for original image). (B) Result of the Analyze>Skeleton>Skeleton 2D/3D command in FIJI. Inset is blow-up of region showing the blue single and end pixels, orange-yellow slab pixels, and pink/purple junction pixels. (C) Identification of junctions as triple or quad junctions: (**a**) is a triple junction because it has three neighboring pixels that are occupied (white). (**b**) is a quad junction showing four neighboring pixels that are occupied. Gray pixels are unoccupied neighboring pixels

**Fig. 3** (continued) (**i**) above a gray value of 3. (**l**) Result of combining (**e**) and (**g**), pseudo-coloring (**e**) orange and (**g**) white. (**m**) Result of combining (**k**) and (**j**) (**k** can be subtracted from **j** for analysis), pseudo-coloring (**k**) orange and (**j**) white. Arrows in (**l**) and (**m**) show a differently-processed cisterna. Stars show region of low signal

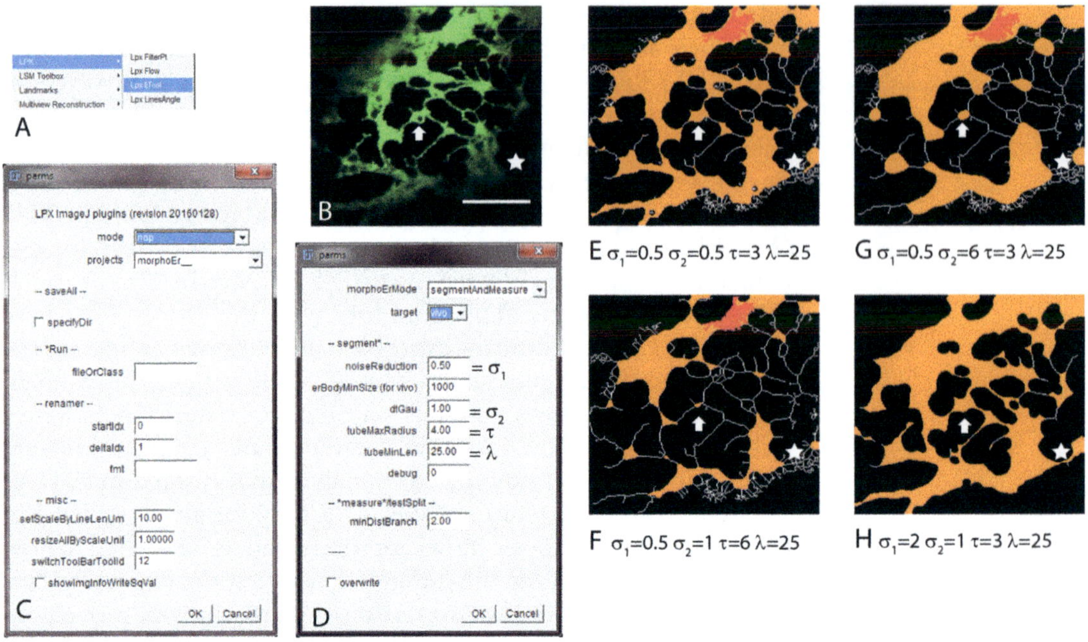


**Fig. 5** Separation ER tubules and cisternae using the morphoER plugin from LPX. (**a**) Pulldown plugin menu showing the LPX option and the LpxljTools option. (**b**) Dialog box brought up by LpxljTools. The nop "mode" should be selected. The morphoEr "projects" should be selected. Other options can be run using the default choices. (**c**) The first image in a time-lapse sequence of ER for analysis. Scale bar = 10 μm. (**d**) After clicking OK in (**c**), this dialog box comes up. The segmentAndMeasure option is shown. The target (in vivo as opposed to in vitro) is shown. In order to segment (select) the tubules vs the cisternae, there are several user variables that can be chosen. (**e**–**h**) show the results of different choices after clicking OK in (**d**). σ1 is the radius of the Gaussian (blur filter) used to reduce noise. σ2 is the radius of the Gaussian used to "fill in" the cisternae eroded by Euclidean distance subtraction. τ is the radius of the tubules in the image. λ is the minimum length of skeletons shown, in order to remove skeleton fragments resulting from noise. Tubules are generated by the same skeletonization as in Figs. 3j and f, but without binary closing. Cisternae are generated by the same Euclidean distance subtraction shown in Fig. 3k. Tubules are white and cisternae are orange, with saturating pixels red. Arrows show a cisternae that changes shape and area with different settings. All of the settings do not show the hole in this cisterna shown in Fig. 3l. The stars show regions of low signal and the consequent large proliferation artefactual skeletonization that occurs in these regions in (**e**–**g**) that is excluded with manual thresholding (Figs. 3l and m)

The implementation of persistency mapping has identified persistent junctions in the ER at which it remains tethered to the plasma membrane (persistent cisternae less than 0.3 square μm in area [2]. Originally called anchor points, they are now considered ER/PM membrane contact sites (MCS) [5]. However, it is important to note that the cortical ER probably has more than one manner of tethering to the plasma membrane, with some tethers more persistent than others. To date, persistency mapping has only been used on cortical ER but could be used for other organelles or features of them. For example, if organelles need to be anchored to the plasma membrane/cell cortex in order to divide, as occurs in

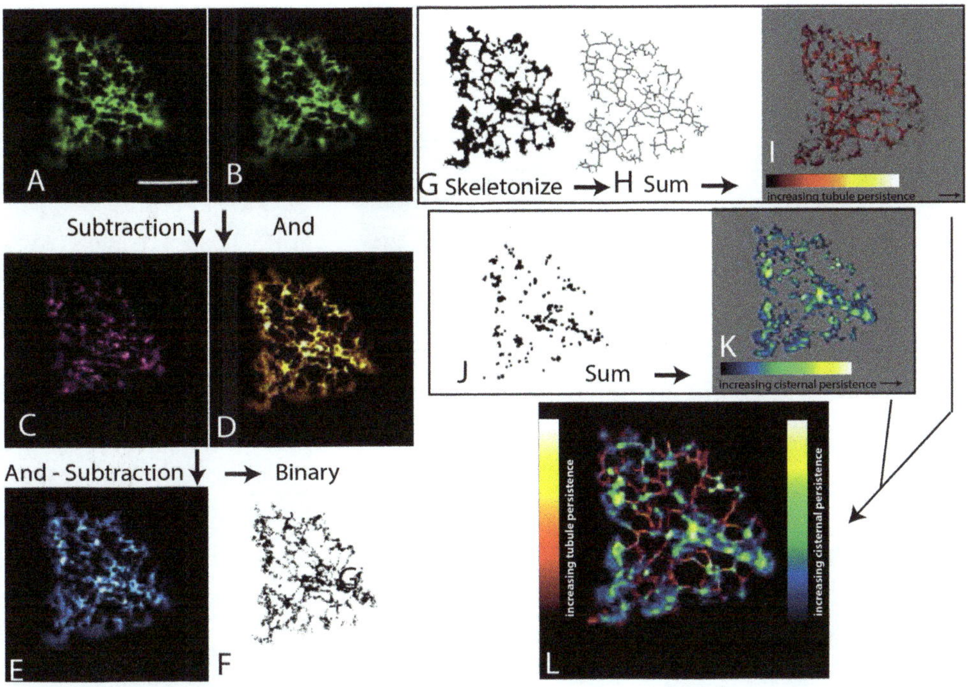


**Fig. 6** Persistency mapping. (**a**) Image of ER taken at 0 s. Scale bar = 10 μm. (**b**) Image taken 8 s later than (**a**). (**c**) Result of subtracting image (**a**) from image (**b**) using the Analyze>Image Calculator command, pseudo-colored magenta hot. This is done on the entire stack of images as a stack. So all of the following images represent stacks of images every 8 s apart (0 and 8, 1 and 9, 2 and 10, etc.) (**d**) Result of the Boolean "And" operation between (**a**) and (**b**) using the Process>Image Calculator command, pseudo-colored orange hot. (**e**) Result of subtracting (**c**) from (**d**) using the Process.Image Calculator command, pseudo-colored cyan hot. (**f**) Result of converting E to a binary image with the Process>Binary>Make Binary command. (**g**) Closed version of (**f**). (**h**) Skeletonized version of (**g**). (**i**) Tubule Persistency Map: The individual images in stack (**h**) are summed using the Images>Stack>Z Project>Sum Images command, following subtraction of the cisternae. Pseudo-colored red hot, with the lighter regions being the more persistent. (**j**) Opened version of (**f**). (**k**) Cisternal Persistency Map: The individual images in the stack (**j**) are summed using the Images>Stack>Z Project>Sum Images command. Pseudo-colored green-fire-blue, with the more persistent cisternae being white-yellow. (**l**) Composite image of (**i**) and (**j**) with the more persistent features being pseudo-colored white-yellow

with mitochondria in yeast [10], the population of those mitochondria becoming attached to the membrane/wall could be quantified in plants and related to those events which accompany division. It could map the choreography that leads to ER associating with the mitochondria and then tethers to the plasma membrane as mitochondrial division is completed.

Just as the movement of the dancer's costume is different from the movement of the dancer, so the flow of the proteins and, probably, lipids in the membrane and lumen of the ER network is different from the movement of the network itself. As mentioned above, optic flow approaches combine these two types of movement into one measurement [9]. However, there is a remarkable,

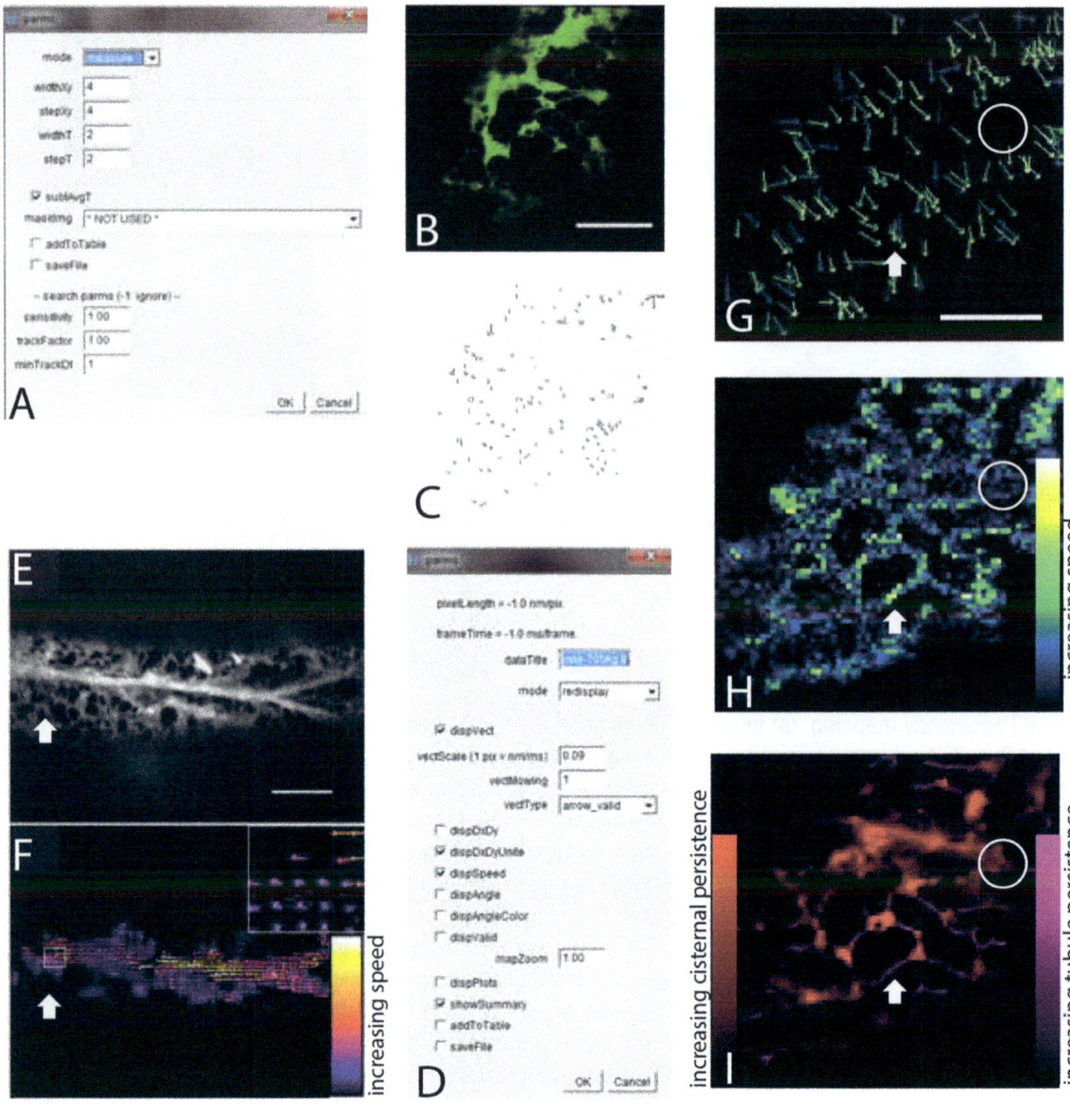


**Fig. 7** Optic flow analysis of movement in ER using LpxFlow. (**a**) The dialog box that comes up with the Plugins>LPX>LPXflow command. The settings are set so that there is 4 × 4 matrix being analyzed using spatiotemporal image correlation over an interval of every two frames. Subtracting the average intensity over time is selected. The mask option is not available. (**b**) Image of first frame of time series with background subtracted. Scale bar = 10 μm. (**c**) First image of image stack resulting from clicking OK in (**a**): positioned of vectors shown in various directions with different colors representing different magnitudes. (Pseudo-colored here for contrast). (**d**) Second dialog box that comes after the Plugins>LPX>LPXflow command is re-entered with the image stack (**c**) open. DispVector (display vector) is checked. The vector scale is derived by dividing pixel size in nm by the frame interval in milliseconds. Vector type is chosen as arrows. (**e**) Image 2S from [9] supplementary data showing fluorescence of the ER network in the petiole tissue from *Arabidopsis*. Scale bar = 10 μm. (**f**) The arrow vectors shown from one frame from a 16 × 16 pixel matrix, pseudo-colored to show speed. Arrows in (**e**) and (**f**) show masked region where polygonal network is excluded. (**g**) Arrow vector display from (**c**). Bar = 10 μm. (**h**) Summed speed values from network from checking the DispSpeed output setting in (**d**). (**i**) Persistency Map of same region showing more persistent tubules (magenta) and more persistent cisternae (red). Note that in areas of high tubule persistence, the optic flow program shows high speed (arrows **g**, **h**, and **i**). Also note that areas with low cisternal persistence are areas of low speed (circles in **g**, **h**, and **i**)

directional network flow of membrane [3, 11] and luminal [12] protein in plants that is not found in animal cells where network flow in the ER is primarily by diffusion [13] and diffusional constraints imposed by glycosylation and the actin cytoskeleton [14]. When considering network flow in plants, the distinction is therefore made between advective flow (movement in a constant velocity field), active flow (movement in changing velocity fields), and flow by diffusion. The directional flow in the ER is dependent on the actomyosin cytoskeleton, but whether it is advective or active is unknown. Directional flow of photoactivated GFP fused to the cytoplasmic domain of truncated calnexin (CNX-paGFP, [3], [11]) can be inhibited with latrunculin b treatment or tail domain expression of XI-K (Myo11E). Directional flow can also be shown with fluorescence recovery after photobleaching (FRAP) approaches (Fig. 8), where the movement of the photobleached GFP-HDEL can be studied. A similar kind of directional flow also occurs in the tubules arising from chloroplasts, the stromules, but it has been shown, using fluorescence correlation spectroscopy (FCS), to be active and dependent of cellular ATP levels [12]. While myosin XI and the actin cytoskeleton have been shown to associate with stromules and are involved in movement of stromules [15], evidence for cytoskeletal involvement in stromule directional flow is lacking.

Even though *flow directionality* in the plant ER is dependent on the actomyosin cytoskeleton, how the cytoskeleton modulates *flow rate* is unknown. In fact, it is difficult to get consistent measurements of flow rates in plant epidermal cells. Half-times of fluorescence spread after photoactivation of CNX-paGFP vary by 50% in the same cell type, at the same developmental stage [3]. Half-times of fluorescence recovery after photobleaching (FRAP) of CNX-GFP vary less, but by 30% [3]. In the face of this high variability, no significant difference in the rate of flow has been detected between control (untreated) cells and those with latrunculin b treatment and XI-K (Myo11E) tail domain expression [3]. A major contributor to this variability may be the different, changing geometries of the 3D structure of the ER [16–20]. In animal cells, during diffusion-limited network flow, very different FRAP half-times of recovery can be completely attributed to different 3D geometries [16, 19]. In the last method in this chapter (Fig. 9), it is shown how particle tracking of ER aggregates in constitutively stressed cells in the mutant *ceh-1* [21] can be used to examine rates of flow in the ER after the cell, thereby overcoming the problems of calculating diffusion constants from fluorescence recovery rates and providing a means to examine rates in specific 3D domains.

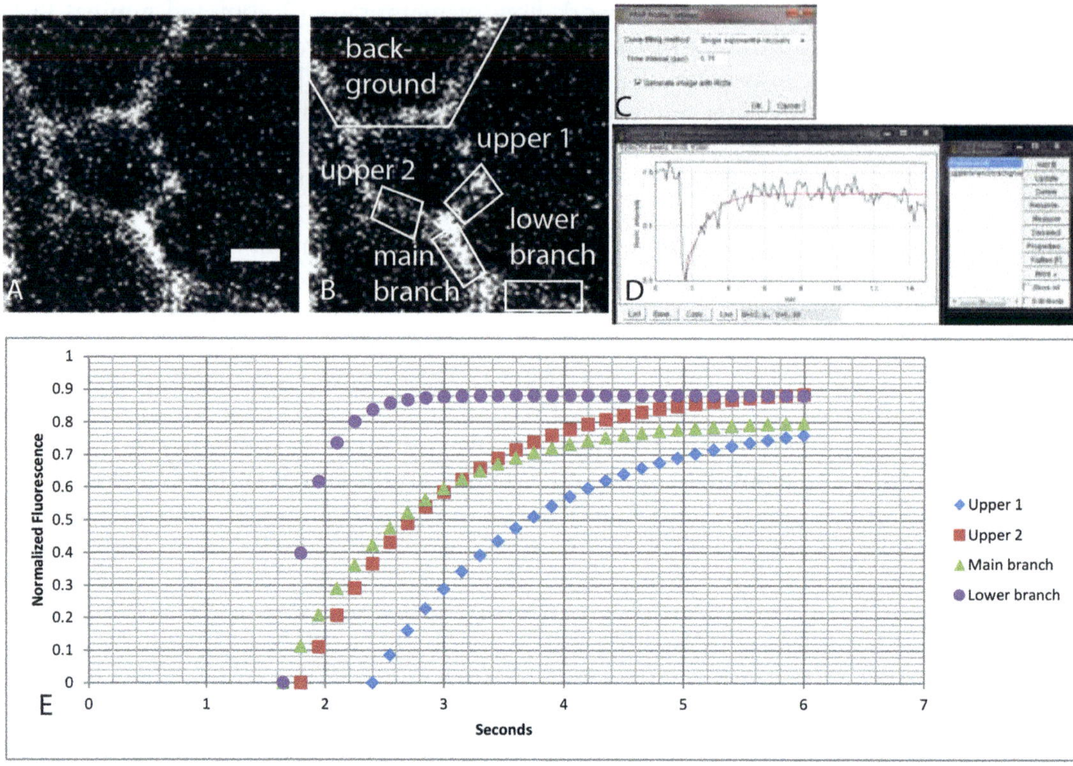


**Fig. 8** Analysis of movement of photobleached GFP in ER tubules using FRAP and iFRAP (inverse FRAP). (**a**) Persistent tubule network in the cortical cytoplasm of a cell. Scale bar = 2 μm. (**b**) Regions used for FRAP and iFRAP. FRAP is conducted on the main branch, which contained the region of photobleaching (smaller than ROI shown). iFRAP is conducted on two upper branches, 1 and 2, and a lower branch. The background ROI is used to normalize the data and account for any overall photobleaching that results from viewing. (**c**) FRAP Profile dialog box from the Plugins>-FRAP profile command, with the frame time interval entered and a single exponential used to model the data. (**d**) Normalized raw data output from main branch, showing the drawn FRAP curve that is replotted in (**e**). Note that only the main branch and background ROIs are used in this analysis, as shown in the ROI manager in the left-hand panel. (**e**) FRAP for main branch and iFRAP for peripheral branches. Note the time lag in the appearance of upper branch 1, indicating slower movement of the photobleached GFP into that tubule, compared with upper branch 2 and lower branch. The very fast recovery in the lower branch is probably due to the small amount of photobleached GFP entering that branch, indicating the GFP is flowing away from this branch into the upper branches via the main branch

## 2 Materials

1. Medium: Murashige & Skoog basal medium (Cat # MSPO1-1LT, Caisson Labs, Smithfield, UT, USA) with or without 0.25 (w/v) sucrose and 0.9% w/v Agar (Plant cell culture tested, Cat # A8678, Sigma-Aldrich, St. Louis, MO, USA).
2. VALAP: a 1:1:1 mixture of Vaseline, lanolin, and beeswax (or Paraffin mix, 22).

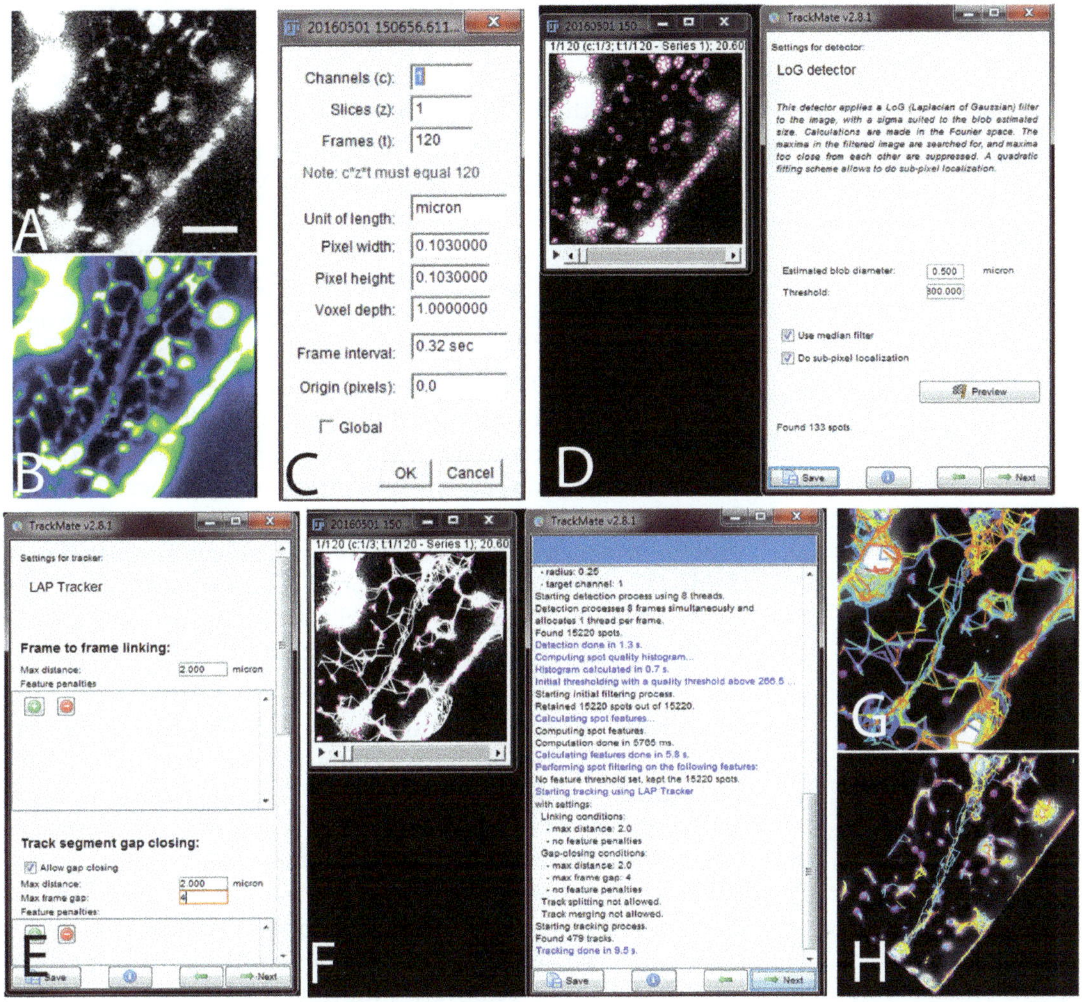


**Fig. 9** Analysis of particle movement inside slightly stressed ER using TrackMate. (**a**) ER network as seen in a single frame of a time series in *ceh-1* mutant of *Arabidopsis*, which has a low level of constitutive ER stress. Scale bar = 5 μm. (**b**) Summed time sequence of ER dynamics of region in (**a**) showing that the tracks of the ER aggregates move along a typical ER tubule network (low intensity, blue color). (**c**) Dialog box from Image>Properties command in ImageJ showing that the sequence is time interval (not a Z-stack), what the time interval is between stacks, and how big the pixels are. It is important to set this up correctly before running TrackMate. (**d**) Dialog box in TrackMate (Plugins>Tracking>TrackMate). Object identification dialog box (right hand side) allowing particle size and minimum intensity to be selected. Identified objects are previewed as small purple circles (right hand image). (**e**) Dialog box for user-entered calculation constraints for the Linear Assignment Problem (LAP) algorithm in TrackMate, showing frame-to-frame linking and gap closing variables in μm and frame to frame gap closing in time. (**f**) Output of track assignment showing calculations and calculation steps (right hand side) and tracks (left hand side). (**g** and **h**) Comparison of tracks using 4 frame, 3 μm particle gaps (**g**), and 2 frame, 2 μm particle gaps (**h**)

3. 100 mM MES (Free Acid, Cat # 475893, Sigma-Aldrich, St. Louis, MO, USA) Buffer, titrated to pH 5.8 with 100 mM KOH—as stock, diluted 1:10 for imaging and for treatment with different fluorescent dyes.
4. Fluorescent dyes (*see* **Note 1**): Cell walls (and nuclei of dead cells): Propidium iodide (Cat # P1470, Sigma, St. Louis, MO, USA) used at a final concentration of 1 μg/ml. DiOC6 stains the ER and mitochondria of the abaxial epidermis of onion scales adequately [23], but there are problems with the generation of reactive oxygen species with exposure to light [24]. DiOC6 (DiOC6 [3] (3,3′-Dihexyloxacarbocyanine Iodide, (Cat# D273, Thermo Fisher Scientific, Molecular Probes) 2 mM stock solution dissolved in 100% ethanol (or acetone) as a stock solution and used at 2 μM concentration (*see* **Note 2**). ER Tracker red and green dyes (Thermo Fisher Scientific) are not recommended for *Arabidopsis*, since the protein to which they bind through the glibenclamide adduct, AtMRD5/AtABCC5, the sulfonyl receptor portion of the ATP-sensitive potassium channel that transports phytate, is on the plasma membrane in guard cells [25] and the vacuole in other plant cell types [26].
5. Transgenic *Arabidopsis* expressing GFP-HDEL, mCherry-HDEL, or YFP-HDEL (plasmids and/or seed is available through the *Arabidopsis* Biological Resource Center, www.arabidopsis.org) and transgenic *Nicotiana benthamiana* expressing GFP-HDEL [28] or *Nicotiana tabacum* leaves transiently expressing the above fluorescent ER-labeling constructs [29] and the fluorescent tail domains of myosin XI [4]. A significant fraction of eGFP could be non-fluorescent and misfolded [30]. The relatively non-interactive mCherry-HDEL and superfolded (sf)GFP-HDEL are good passive markers of the ER lumen. mCherry constructs are reported to be "sticky" when labeling oligomeric or membrane proteins [24], so should probably only be used for studies on luminal proteins. sfGFP [31] has the advantages of being monomeric, lacking N-glycosylation sites, and being resistant to disulfide bond formation [24]. It should be fusion protein of choice for both luminal fusion proteins and membrane fusion proteins.
6. Laser scanning confocal microscope with FRAP software. The ones used for these analyses are the Zeiss LSM510 and the Olympus Fluoview 1000.
7. In all of the following methods, implementation is described using ImageJ (https://imagej.nih.gov/ij/) version 1.45 or higher and as it is used in the ImageJ version containing many useful plugins, FIJI (www.fiji.sc), version 2.0.0. Additional plugins not part of the FIJI package are listed below. For security reasons, it is recommended that Java version 1.8 jre or higher is used (*see* **Note 3**).

## 3 Methods

### 3.1 Microscopy of Living Plant Tissue

1. Select relatively planar surfaces for imaging 25 square mm or less (half a cm on a side or less) sections of mature tobacco leaves in buffer under a 22 × 22 mm coverslip sealed with VALAP.
2. Select relatively planar regions from epidermal cells of roots, shoots, or cotyledons from 6–7-day-old seedlings of *Arabidopsis*, which fit entirely under a single coverslip (*see* **Notes 4** and **5**).

### 3.2 Using ImageJ Macros for Image Processing Pipelines

1. In order to provide the methods of several approaches in this chapter, an example imaging pipeline is included as a Macro file, available for download at http://griffing.tamu.edu.
2. Copying these files or following the sequence given below while recording a Macro, saving them in ImageJ macro format (as an .ijm file in the Macro folder), and editing them so that they use file systems on the host computer of choice should provide an easily accessible way to implement the approaches described below.
3. Recording a Macro. ImageJ provides a useful Macro implementation that can be set up for processing pipelines using the command, Plugins>Macro>Record. Command sequences such as this one are described by the sequence of "pulldown" sub-commands from the menu bar (Fig. 1), which appears when ImageJ or FIJI is started. To record a Macro, click on (select) Plugin in the main menu bar. Select the top option Macro. Select the Record option. Besides being useful for processing image pipelines, the Macro feature can be used as an imaging "lab book" to record the steps by which an image is processed (*see* **Note 6**).
4. After recording the sequence of steps in the imaging pipeline, the Create command on the Macro dialog box (Fig. 1) has to be selected because just closing the Macro dialog box does not save the recorded Macro, and it will be lost.
5. For ImageJ to "see" the Macro, the Macro file (written in the ImageJ macro or .ijm format) is saved or placed in the Macro folder in the ImageJ folder.

### 3.3 Installing, Running, and Editing Macros

1. To Install a Macro, use the Plugins>Macro>Install command, and the Macro comes up as a separate command in the Plugins>Macro bar. This is very useful for shortcuts and for making Macros that include other Macros.
2. To Run a Macro, use the Plugins>Macro>Run command, and select the .ijm file manually.

3. To Edit a Macro, use the Plugins>Macro>Edit command, and select the .ijm file for editing. This can then be text-edited in the Macro dialog screen.
4. Using image stacks for measurement after processing using a Macro:
   (a) The Macro saves a series of images for measurement under a common name for each sample in a group of replicates. For example, each replicate could have an image named AND.tif. These are transferred to a new folder sequentially, and Windows automatically assigns them numbers if they have the same name (e.g., AND.tif, AND [2].tif, AND [3].tif, etc.). Measurements are assigned with the command, Analyze>Set Measurements.
   (b) The separate images can then be stacked in sequence with the command, Image>Stacks>Images to Stack.
   (c) The different replicates of one treatment can be measured in the stack with one Analyze>Analyze Particles command following selection with the Image>Adjust>Threshold command.
5. Network movement: removing the drift.
   (a) Make a time series movie.
   (b) Check to see if reference points are moving, such as cell corners or other external wall features that can be seen after treatment with propidium iodide. This is done by mousing over the precise points in the first and last frame in an imaging sequence and noting if their coordinates have changed.
   (c) If reference points are moving, use the optic-flow-based stabilization plugin by Kang Li using the Lucas-Kanade algorithm with the command Plugins>Image_Stabilizer.
   (d) Another plugin that has more options in terms of template matching and the implementation of cross-correlation is the "Align slices in stack" plugin by Qingzong Tseng (*see* **Note 7**).
6. Network movement: measuring average mesh size (Fig. 2).
   (a) Make a time series movie, usually 80 frames at about a second per frame for standard line-scanning confocal microscopes.
   (b) The in-focus region is selected for the entire time series (movie). This can be done manually (Fig. 2a) with the polygon tool. This region of interest is saved as infocus.roi using the File>SaveAs>Selection command. Those out-of-focus areas are cleared by the Edit>Clear Outside command after the color picker (CP) is set to white on a black

background. The color picker is accessed by clicking twice on the "eye dropper" icon.

(c) Those regions not occupied by membrane, i.e., the mesh spaces (low intensity values), are selected using manual or automatic gray scale thresholding. If the intensity differences between pixels with and without ER are large enough, many different automatic thresholding routines that isolate the low values from the majority of the high values could be used (e.g., Minimum, Otsu, or IsoData segmentation algorithms found in the Image>Adjust>Auto Threshold), but if not, then the image is thresholded manually with the Image>Adjust>Threshold command (Fig. 2b)—they turn red.

(d) The low-value regions in each frame are then counted and measured. The "Area" and the "Stack Position" are selected in the Analyze>Set Measurements dialog box. The area of the mesh spaces for each image in the stack is measured with the Analyze>Analyze Particles command. To avoid counting very small (single pixel) regions, a lower limit of mesh space is put in the Size variable in the Analyze Particles dialog box. Also, make sure to click the exclude edge regions (Fig. 1b) before clicking OK. This will produce a Results table with Area and Slice values—this can be ported to a spreadsheet program, such as Microsoft Excel.

(e) The mesh number (counted or calculated by average area per mesh space/total mesh space area measured) and the number and average area of mesh space per cubic μm (derived from the in-focus area imaged and the depth of field) of cytoplasm can then be calculated in a spreadsheet program, such as Microsoft Excel (using the Averageif command to specify that feature areas are averaged if they are in a certain frame), or with a simple separate script in Python or MATLAB.

7. Network movement: identification of tubules vs. cisternae (Fig. 3).

   (a) Select the fluorescent membrane with a threshold command, Image>Adjust>Threshold, either on 8 or 16 bit images. The original bit depth (intensity range) of the acquired image should be maintained. In the example image (Fig. 3b), an 8 bit image is used.

   (b) The out-of-focus region can be cropped away using the instructions in 6b. This will make the choice in the next step easier.

   (c) A manual threshold value is used that is a compromise between eliminating the out-of-focus low-intensity blur and eliminating continuity between thin tubules. Once a manual threshold level is set, it is generally not changed between samples of procedures. Two different threshold values are shown in Fig. 3. One is chosen to eliminate the out-of-focus low-intensity values (80–255 Fig. 3c), while the lower-intensity values are chosen (40–255 Fig. 3h) to preserve the continuity of tubules. Automatic thresholding based on the histogram can be used as in MorphoER (LPX plugin, [32], Fig. 5), but in the current implementation of MorphoER, the threshold is so low that much the out-of-focus image is included if smoothing is not used (compare Fig. 5b with Fig. 2m). Automatic minimum value thresholding algorithms such as the Minimum, Otsu, and IsoData (from the pulldown menu in the Image>Adjust>Threshold dialog box or from the Image>Adjust>Auto threshold dialog box) and WEKA trainable segmentation (Plugin>Segmentation>WEKA trainable segmentation) could be explored to make the initial threshold values less subjective.

(d) Create a binary image with the Make Binary command, Process>Binary>Make Binary. Morphological image processing (Fig. 3a) takes a binary image (Fig. 3c–d).

(e) Operate on the binary image with the Open command, Process>Binary>Open. Morphological processing can either add (dilation) or subtract (erosion) pixels from a binary object's periphery based on the shape of the structuring element. Opening is erosion followed by dilation. The number of iterations for opening (and closing, step 5) can be set in the Process>Binary>Options dialog box. In Fig. 3a, two side-by-side cisternal structures with emanating tubules are considered. The distance between them is two pixels. Opening eliminates the emanating tubules. After opening, the cisternae are isolated but remain the same area (Fig. 3a and e) [2, 4]. The areas of all of the cisternae in a frame can then be measured using the Analyze>Measure Particles command, described above in mesh measurement. The Results table is saved as an . xls file, opened in Excel, and converted to a worksheet format (.xlsx).

(f) The Euclidean distance of each pixel from a background (0) value can be calculated and mapped as an intensity map or image (Fig. 3i) using the Process>Binary>Distance Map command. The image is adjusted so that that the largest structures have the brightest values in their center using the Image>Adjust>Brightness, and Contrast command (Fig. 3i) also has a Cyan Hot lookup table applied using the command, Image>Look up tables>Cyan Hot. That image is then thresholded using the Image>Adjust>Threshold command to select only the cisternae; in the example in Fig. 3i, the tubules can be eliminated by selecting values that have intensity values of three or greater (i.e., are more than 3 pixels away from background). This operation on the Euclidean distance map removes all pixels a distance of three or less from the periphery of the cisternae, as well as the tubules, unlike the open operation that preserves the area of the cisternae, but creates object edges that are similar in shape to the structuring element. MorphoER addresses this by increasing the cisternal size with a manually applied Gaussian blur function (see below).

(g) Operate separately on the binary image with the Close command, Process>Binary>Close. After closing (dilation followed by erosion), the tubules that are separated by two pixels become connected (Figs. 3a, d and h). This thresholding operation (either automatic or manual) is used to create the binary image that may exclude faint tubule pixels (undesirable) as well as faint noise pixels (desirable). If the condition, set by the size of the structuring element, is met that two disconnected tubules are separated by one or two pixels, then the closing operation connects them, and they are counted as a continuous tubule. Faint noise is mostly eliminated with manual thresholding. The closing operation also fills in pixels between two side-by-side tubules that are separated by one or two pixels; hence, if clusters of tubules are being examined, then the pixel size should be at least twofold smaller than the diffraction limit (the Nyquist criterion).

(h) For tubules, the closed image is skeletonized (Fig. 3f) using the Process>Binary>Skeletonize command. Skeletonize implements a 2D thinning algorithm [33].

(i) Skeletonization is followed by subtraction of the regions that contain cisternae (Figs. 3a and g) using the Process>Image Calculator command. The 8-bit images need to be inverted first, with the Edit>Invert command. This turns the skeletons and cisternae to white on a black background instead of black on a white background. In the Image Calculator dialog box, the upper image selection should be the skeletonized image, and the lower image selection should be the opened cisternal image. When cisternae are skeletonized, they give rise to multiple radiating skeletons, and depending on the skeletonization algorithm, these have diagnostic radiating or angular geometries (Fig. 3f and j). Consequently, to eliminate this artefactual data set from the tubule population, these regions are subtracted from the tubule population.

8. Network movement: measurement of cisternal areas and numbers.
   (a) Set the Analyze>Set Measurement dialog box for area and stack location.
   (b) The opened binary image (Fig. 3e) or the thresholded Euclidean distance map (Fig. 3i) can be analyzed using the Analyze>Analyze Particles command.
   (c) The results table is saves as an .xls file.
   (d) The CountIf formula is used in Excel to count the number of cisternae in each slice of the stack (stack location).
   (e) The AverageIf formula is used to get the average size of cisternae.
   (f) These are normalized using the area of the membrane imaged, which is the area of the original binary thresholded object (Fig. 2c or Fig. 2h).
9. Network movement: measurement of tubule numbers, lengths, and junctions (Fig. 4).
   (a) The skeletonized region can be evaluated using the Analyze>Skeleton>Skeleton 2D/3D command (Fig. 4). This procedure provides a list of the length of all of the separate skeletonized tubules, with their number of junctions and the type of junction.
   (b) To analyze the skeletons, they need to be white on a black background (using the Edit>Invert command as above, if that has not already been done) (Fig. 4b; *see* **Note 8**).
   (c) There are several images and tables that are produced by this algorithm: a color-tagged skeleton image showing blind ends as blue, junctional regions purple-pink, and tubule lengths yellow-orange (Fig. 4b and inset); a grayscale image (labeled skeletons, shown for clarity with a color table in Fig. 4d) has a different gray value for each identified skeleton; a results table that includes a list of all of the independent skeletons with the following: # branches, # junctions, # end-point voxels (blue in Fig. 4b, they have less than two pixel neighbors), # junction voxels (pink in Fig. 4b, they have more than 2 pixel neighbors), # slab voxels (yellow in Fig. 4b, they if they have exactly 2 pixel neighbors), average branch length, # triple points (a in Fig. 4c), # quadruple points (b in Fig. 4c), and maximal branch length (this can be saved as an .xls file); a branch information table that includes the skeleton ID, the branch length, the coordinates of the start, and end of the branches; and the Euclidean (straight line) distance between the start and the end of the branches. Branch length divided by the Euclidean distance is a measure of the complexity of the branch. This can be saved as an .xls file (*see* **Note 9**).

10. Using MorphoER for analysis of tubules and cisternae (Fig. 5).

MorphoER (see LPX plugins above), accessed through LPX plugin set with the command Plugin>LPX>LPX IjTool command (Fig. 5a), can be opened from the IjTool dialog box by selecting morphoER from the projects pulldown menu (Fig. 5c) once an image for analysis is open (Fig. 5b).

MorphoER uses automatic thresholding/segmentation as mentioned above, selecting as a threshold, the value which maximizes the distance between the histogram values to a line defined by the histogram peak value and histogram blackest value [35]. However, it also has several user-defined variables (Fig. 5d): Gaussian blur radius to reduce noise ($\sigma_1$), ER body minimum size (for arabidopsis ER body identification), the threshold value in the Euclidean distance map for the width of tubules ($\tau$), and the Gaussian blur radius ($\sigma_2$) that enlarges the cisternae after the threshold operation on the Euclidean distance map. A tubule minimum length value ($\lambda$) can also be specified that removes small skeletonized isolated tubules from counting. In Figs. 5e–h, the cisternae are colored orange and red, while the tubules are identified as white.

(a) Effect of setting $\sigma_1$: This blurs the initial image with a Gaussian function of the radius set by the user. It has a large influence on the region selected by automatic thresholding; compare Figs. 5e ($\sigma_1 = 0.5$) and h ($\sigma_1 = 2.0$). As shown in Fig. 5h, it will thicken the tubules, changing the value needed for $\tau$.

(b) Effect of setting $\tau$: This is the threshold of the Euclidean distance map. Larger values will threshold away the tubules, but with higher values, more of the cisternal area is lost,(Fig. 5f).

(c) Effect of setting $\sigma_2$: This is the radius of a Gaussian blur of the selected cisternae. Larger values produce cisternae with different areas and shapes (Fig. 5g).

(d) Clicking OK in the second dialog box produces a final image that is a composite of three values, tubules with a value of 1 (white in Fig. 5e–h), cisternae with a value of 2 (orange in Fig. 5e–h), and saturated pixels with a value of 3 (red in Fig. 5e–h). These can be selected for measurement using a thresholding command. Note that the arrows in Figs. 5b and e–h show a cisterna with a fairly large hole or tubule loop. All of the settings remove the hole or loop. The tubule loop is preserved in Fig. 3l (arrow), which uses a higher threshold and opening instead of a Euclidean distance map (Fig. 3m). The stars in Figs. 5b and e–h are in regions of low or out-of-focus signal. These produce an array of tubules in the

MorphoER program, Figs. 5e–f, but they do not exist. These are excluded in both thresholding approaches shown in Fig. 3 (stars in 3 L and M). MorphoER produces a MorphoER results table that has several aspects of tubules and cisternae measured, but the nature of evaluation is not currently published or open source. It is apparently done in pixel number values, but the units are undefined in the software.

11. Network movement: persistency mapping (Fig. 6).
    (a) In order to identify the regions that are moving rapidly, one can subtract frames between defined intervals in an 80 s time series of 50 frames. In the original approach [3], every fifth frame subtracts an image taken 8 s earlier. This is done by making substacks using the Image>Stacks>Tools>Make Substack command twice: first to remove the first five in the stack leaving a Substack of 6–50 and the second to remove the last five, leaving a Substack of 1–45. Substack (6–50, Fig. 6b) subtracts Substack (1–45, Fig. 6a) using the Process>Image Calculator command and specifying the subtract option. Only those regions which are moving have a value, so this is the "moving" or SUB data set (Fig. 6c; *see* **Note 10**).
    (b) The non-moving data, or AND data set, is the result of a Boolean AND operation between every fifth frame of Substack (1–45, Fig. 6a) and Substack (6–50, Fig. 6d) using the Process>Image Calculator command and specifying the AND option. To generate the persistent data set, the moving data set is subtracted from the non-moving data set (Fig. 6e) by using the Process>–Image Calculator command and subtracting the SUB data set from the AND data set. These operations are done on 8-bit images, and ImageJ truncates the results so that negative-valued pixels are set to zero.
    (c) The persistent data set is then thresholded by a value that includes the in-focus membrane (step 7c, Fig. 3c or 3h) using the Image>Adjust>Threshold command.
    (d) Total detected persistent membrane area can be measured at this stage using the Analyze>Analyze Particles command, specifying stack location and area in the Analyze>Set Measurements dialog box. In the Analyze Particles dialog box, if the Masks option is taken, then a binary mask is generated that can then be used instead of the next step (Fig. 6f).
    (e) A binary image is generated from this thresholded image using the Process>Binary>Make Binary command (if a binary image mask has not been generated in the prior

step). Use the default settings in the Make Binary dialog box – do NOT check the option to have the threshold calculated for each image or any of the options.

(f) The binary image stack is separately converted into the persistent tubule stack using the closing (Fig. 6g) and skeletonization (Fig. 6h) commands (steps 7g and h), followed by inverting (Edit>Invert), subtracting the (inverted) persistent cisterna stack, and summing them, using the Image>Stacks>ZProject command, specifying the Sum Slices option, Fig. 6I.

(g) The binary image stack is then converted into the persistent cisternae stack using either the opening operation (step 7d), Fig. 6j, or the Euclidean distance operation followed by thresholding (step 7e), inverted (Edit>Invert), and summed as above (Fig. 6k).

(h) The Tubule Persistency Map (Fig. 6l) is the sum of all of the skeletons, and only those that persist the longest are the brightest. A colored ramp is made as above but for tubules (Image>Lookup Table>Red Hot). Because of the variation in skeletonization that occurs with cisternae as they are changing shape, this image can also be thresholded to eliminate these skeletons, without subtracting the cisternae from tubules.

(i) The Cisternal Persistency Map (Fig. 6k) is the sum of all of the opened binary images and has, as its lowest-value objects, the least persistent cisternae and, as its highest-value objects, the most persistent cisternae (Fig. 6k). A colored ramp is made with the Image>New>Ramp command and given the same color table as the cisternae (Image>Lookup Table>GreenFireBlue).

(j) Both Cisternal Persistency Maps and Tubule Persistency Maps are color-coded to show the most and least persistent values (Fig. 6l). Those that persist (are added together) in the same position for 32 s (the first 8-s difference interval and then 24 s thereafter or 15 of the 45 difference frames) can be thresholded, measured, and counted.

12. Network movement: analysis of network movement using optical flow with LPX flow (Fig. 7).

(a) LPX flow (Plugin>LPX>LPXflow) starts by subtracting the temporally averaged image from the data set. This is done by checking the subtAvgT box in the first dialog box that comes up (Fig. 7a). The temporally averaged image includes several of the persistent features in the data set. This differentiates it from the persistency map above and emphasizes the fact that it is the change between frames that is being measured.

(b) It then implements a cross-correlation function based on spatiotemporal image correlation spectroscopy (STICA, [36, 37]). The motion is examined by the cross-correlation of pixel blocks. The position of peaks produced by the cross-correlation is a measure of how well the two blocks in subsequent frames match. These peaks, therefore, represent the amount and direction (a vector) by which the second image block has to be moved in time and space in order to best match the first image block. The variables defined by the user and the system are:

- The spatial lag variables in the x and y direction, respectively, $\xi$ and $\eta$.
- The temporal lag variable, $\tau$, or the time difference between frames.

(c) The above variables are entered in the interface (Fig. 7a), and the space block and time difference (frame difference) over which the optic flow operates are defined. In the case of the example data set (Fig. 7b, also used in Figs. 3 and 5), the time is 1.6 s per frame, and the pixel size is 121 nm. In the analysis between every second frame is chosen (width T and step T = 2). A pixel block size of 4 pixels is chosen (widthXy and step Xy = 4). The output from clicking OK in the first dialog box is shown in Fig. 7c. This shows the displacement or flow vector calculated on all pairs of correlated time-successive 4x4 blocks.

(d) Once Fig. 7c is generated, other image plots of the data can be produced by using the Plugin>LPX>LPXflow command while this image is open. This second reiteration of the command produces the dialog box in Fig. 7d. Figures 7e and f show part of Supplemental Fig. S2 from [9], which has an image of the cortical cytoplasm of petiolar tissue from *Arabidopsis* labeled with GFP-HDEL (Fig. 7e) and a flow map of the region in Fig. 7f. Figure 7f has, however, been masked using the MaskImg pulldown from the first LPXflow dialog box (Fig. 7a). The mask applied to Fig. 7e is generated to remove the regions containing "Error vectors." These are regions which exclude the relatively static polygonal regions of the ER, leaving areas that include the fast flow rates of ER motion (Fig. 7f and inset). As shown by the large white arrows in Figs. 7e and f, much of the labeled network is removed from the image, but the regions containing bundles of dynamic ER remain have been rendered using the vectType pull-down in Fig. 7d using the arrow_valid choice. Figure 7g shows an arrow vector map of one of the sequences in the dataset starting with image Fig. 7b but with no masking. Note that when the mask is not included in the approach, the flow velocities do not correlate with persistency mapping. Comparing circled regions in Fig. 7h (Sum of speed maps shown when dispSpeed is checked in Fig. 7d with Fig. 7i; persistency map like that shown in Fig. 6i, but with different color tables), it can be seen that areas of low persistency (circled region, Fig. 7i, high degree of remodeling) can show low flow velocities (circled region, Fig. 7h), whereas in the persistent tubule region indicated by the arrow in Fig. 7i, there can be high flow velocities (arrow in Fig. 7h). This lack of correlation is because *network movement* of the ER is not the same as *network flow* within the ER. However, optic flow approaches do not distinguish between these two types of network dynamics. To assess network flow, the two following approaches are described.

13. Network flow: measurement of directional vs. non-directional (diffusive) flow in tubules using FRAP and iFRAP (Fig. 8).
    (a) FRAP investigations on directional network flow can be carried out on tubular branch networks that are persistent for about 30 s using persistency mapping. The example used has two separate junctions, one on each side of the main branch (Fig. 8a). Movement of tubules into or out of the region that is being FRAP'ed. will invalidate the results, so it is difficult to measure the movement in regions of fast flow, where there are many tubules moving as well as movement of materials within the tubules.
    (b) The recovery of the region of interest (FRAP ROI, main branch) exposed to the FRAP (Fig. 8b) is monitored using a time window (in this case 150 msec per frame) that allows the exponential curve to be drawn by the software in the FRAP profile plugin (the plugin source is in methods). This time can be entered into the FRAP plugin dialog box (Fig. 8c).
    (c) To distinguish between directional and non-directional movement of the photobleached GFP, similar ROIs (Fig. 8b) in adjacent branches are monitored. This is a form of inverse FRAP or iFRAP, so these are iFRAP ROIs (lower branch, upper branch 1, and upper branch 2). These can be monitored using the same FRAP profile plugin.
    (d) The FRAP profile plugin can compensate for the general bleaching that occurs in the region outside the region containing the FRAP ROI before and after the photobleaching. Because the iFRAP ROIs are also, in part,

monitoring the drop in fluorescence outside of the FRAP ROI, it is important to use a region well outside of the FRAP influence, the background ROI (Fig. 8b).

(e) The measurement of iFRAP with this technique gives a time of maximal bleaching, followed by options for one- or two-exponent exponential fit to the data.

(f) This approach can identify gated or directional flow, as evidenced by the differential movement of photobleached GFP out of the FRAP ROI (main branch) and into the iFRAP ROIs, with very little going into the lower branch, which recovers very quickly, and with a definite lag in the appearance of maximal photobleached GFP going into upper branch 1 and upper branch 2. The role of the 3D organization of the ER network in this differential flow [16, 18] has yet to be determined (*see* **Note 11**).

14. Network flow: analysis of advective and active flow of aggregated YFP-HDEL in constitutively stressed ER within the lumen of the ER using the particle tracking plugin, TrackMate (Fig. 9).

When the ER is constitutively stressed (*ceh1* mutants, 21), aggregates appear in the tubules (Fig. 9a). However, in regions of the cortex with polygonal networks, these aggregates are usually at the vertices of polygons that can be seen when all of the sequences are summed (Fig. 9b). Others tubule bundles are very dynamic and move by active flow. This can be measured with TrackMate.

(a) TrackMate uses a temporal optimization algorithm based on a mathematical framework called the Linear Assignment Problem (LAP), which calculates the cost matrices for certain linkages of the identified particles in frame 1 to frame 2, etc. (Jaqaman et al. 2008) and the merging and splitting of particles (simple LAP ignores merging and splitting).

(b) The spatial and time characteristics of the image sequence needs to be carefully calibrated and entered. It is often best to do this in the Image>Properties command, which then defines the metadata used for the TrackMate program (Fig. 9c).

(c) The program uses a variety of filters to identify the particles, including particle size, intensity thresholding, Laplacian of Gaussian or Difference of Gaussian filters, and a median filter to reduce noise (Fig. 9d). The spots can be previewed by clicking the Preview button in the spot detector dialog box (Fig. 9d). When the particles or spots are identified, they can be displayed by intensity and confidence level. The confidence level of particle assignment can then be used to further refine the selection of particles.

(d) The maximal distance between particles that are to be linked and the maximal number of frames in which the particle disappears are entered manually as part of the LAP program (Fig. 9e).

(e) The tracks are then calculated and displayed as overlays using either the track index or other track properties (Fig. 9f). The merging or splitting of particles is calculated after the tracks have been assigned. Figure 9g and h compare the tracks in a trimmed and untrimmed sequence. The trimming was done to eliminate the several of the large ER bodies in the cytoplasm. In the trimmed sequence, the distance for connection between particles in subsequent frames was decreased from 4 times the particle diameter (0.5 μm) to 3 times the particle diameter. The track statistics are saved in data tables and can be accessed by spreadsheet programs such as Microsoft Excel. They can also be plotted by track length and time of entry into the field of view or by simply plotting things like track duration vs. track maximum speed, track median speed, or track standard deviation. Track speed standard deviation gives a measure of active flow because its value indicates the degree to which the velocity field varies—high values being more active than low values. By identifying the tracks with higher track speed standard deviation, the regions within the network that have more active flows can be determined.

## 4 Notes

1. Fluorescent dyes are not recommended for examining ER dynamics in tobacco and *Arabidopsis.*
2. The final preparation of the solvent should equal to or less than a 1:1000 dilution; this is particularly true with DMSO, which we do not recommend in *Arabidopsis.*
3. There are MATLAB (https://www.mathworks.com/products/matlab.html) and Python (https://www.python.org/downloads/) implementations of some of these algorithms, but MATLAB is proprietary (Octave is the open source project, https://www.gnu.org/software/octave/), and many of the useful Python implementations use Python as an interpreted (non-compiled) wrapper for imaging libraries written in C++ (https://www.visualstudio.com/vs/cplusplus/), e.g., VTK and ITK (www.kitware.org). The great advantage of

both the Python and MATLAB implementations is that a specific image processing pipeline can be coded for larger data sets from sequential files in sequential directories, but the syntax and file management commands for these languages are beyond the scope of these methods. The additional plugins used are the Image Stabilizer plugin of Kang Li (http://www.cs.cmu.edu/~kangli/code/Image_Stabilizer.html), the Align_slices_in_stack plugin by Qingzong Tseng (https://sites.google.com/site/qingzongtseng/template-matching-ij-plugin), the FRAP plugin of Jeff Hardin (http://worms.zoology.wisc.edu/research/4d/4d.html#frap)—for ImageJ 1.45 s—and LPixel's LPX plugin package (https://lpixel.net/ijp/lpx_iij_plugins_s.jar).

4. Make sure to place the plant grown in 0.9% w/v washed agar in half-strength Murashige-Skoog medium on a slide that already has ~50 μL of buffer, 10 mM MES pH 5.8, on it so that it does not dry out.

5. For young seedlings of *Arabidopsis* and other species, growth and streaming are inhibited by the pressure of the drying coverslip, so elevating the coverslip with two 2 mm × 10 mm single folded strips of Kim Wipe tissues or single strips of filter paper soaked in buffer (about 350 μm thick total) is recommended.

6. Recording Macros is a little tricky for the novice because every click on each window is recorded, but those records can be removed by editing the Macro file (Fig. 1), which appears as the Macro is being recorded in the Macro dialog box.

7. Drift can be avoided (but not always eliminated) by putting VALAP around the coverslip of the object being imaged so that drying doesn't occur [22]. Drift can also be avoided with the use a perfusion vessel (however, there may be movement during flow of the perfusion media). During osmotic shock, where the tissue visibly shrinks during analysis (even prior to plasmolysis) and image stabilization is therefore necessary.

8. This algorithm assumes that an image stack is a 3D stack of z-dimensional optical sections. In order to evaluate the whole stack, they can be simply made into a montage using the command, Image>Stacks>Make montage. However, this approach makes it difficult to get the information for each image in the stack independently.

9. This algorithm implements a different 3D thinning algorithm [34], written by Ignacio Arganda-Carreras, based on an implementation of the 3D thinning algorithm in ITK (Insight Registration and Segmentation Toolkit, http://itk.org).

10. Large moving cisternae may not be detected as moving within the interval because they are large, i.e., a large region that moves over another large region will be "subtracted out" and not revealed as moving. This is a form of the standard aperture problem, whereby some large object that takes up the region of interest may move, but since its moving edges are outside of the region of interest, the direction and magnitude of the movement cannot be determined.
11. FRAP'ing a region repeatedly can produce ER stress, and immediate ER stress is seen following photostimulation or photobleaching of the chloroplast-ER nexus with 405 nm laser [27]. Although photobleaching with laser lines at 456 and 488 nm does not induce a visible stress response, other work has shown that blue light lasers can cause reorientation of the cortical microtubules in hypocotyl cells [37].

## References


1. Griffing LR (2015) Chapter 1: plant and cell architecture. In: Taiz L, Zeiger E, Moeller I, Murphy A (eds) Plant physiology and development, 6th edn. Sinauer, Sunderland
2. Griffing L (2010) Networking in the endoplasmic reticulum. Biochem Soc Trans 38L:747–753
3. Sparkes I, Runions J, Hawes C, Griffing L (2009) Movement and remodeling of the endoplasmic reticulum in nondividing cells of tobacco leaves. Plant Cell 21:3937–3949
4. Griffing L, Gao H, Sparkes I (2014) ER network dynamics are differentially controlled by myosins XI-K, XI-C, XI-E, XI-1, and XI-2. Front Plant Sci 5:218
5. Griffing LR, Lin C, Perico C, White RR, Sparkes I (2016) Plant ER geometry and dynamics: biophysical and cytoskeletal control during growth and biotic response. Protoplasma 254:43. https://doi.org/10.1007/s00709-016-0945-3
6. Madison S, Nebenfuehr A (2013) Understanding myosin functions in plants: are we there yet? Curr Opin Plant Biol 16:710–717
7. Muehlhausen S, Kollmar M (2013) Whole genome duplication events in plant evolution reconstructed and predicted using myosin motor proteins. BMC Evol Biol 13:202
8. Tamura K, Iwabuchi K, Fukao Y, Kondo M, Okamoto K, Ueda H, Nishimura M, Hara-Nishimura I (2013) Myosin XI-I links the nuclear membrane to the cytoskeleton to control nuclear movement and shape in Arabidopsis. Curr Biol 23:1776–1781
9. Ueda H, Yokota E, Kutsuna N, Shimada T, Tamura K, Shimmen T, Hasezawa S, Dolja VV, Hara-Nishimura I (2010) Myosin-dependent endoplasmic reticulum motility and F-actin organization in plant cells. Proc Natl Acad Sci U S A 107:6894–6899
10. Lackner LL, Ping H, Graef M, Murley A, Nunnari J (2013) Endoplasmic reticulum-associated mitochondria-cortex tether functions in the distribution and inheritance of mitochondria. Proc Natl Acad Sci U S A 110: 458–467
11. Runions J, Brach T, Kuehner S, Hawes C (2006) Photoactivation of GFP reveals protein dynamics within the endoplasmic reticulum membrane. J Exp Bot 57:43–50
12. Koehler R, Schwille P, Webb W, Hanson M (2000) Active protein transport through plastid tubules: velocity quantified by fluorescence correlation spectroscopy. J Cell Sci 113:3921–3930
13. Dayel M, Hom E, Verkman A (1999) Diffusion of green fluorescent protein in the aqueous-phase lumen of endoplasmic reticulum. Biophys J 76:2843–2851
14. Nagaya H, Tamura T, Higa-Nishiyama A, Ohashi K, Takeuchi M, Hashimoto H, Hatsuzawa K, Kinjo M, Okada T, Wada I (2008) Regulated motion of glycoproteins revealed by direct visualization of a single cargo in the endoplasmic reticulum. J Cell Biol 180:129–143
15. Natesan S, Sullivan J, Gray J (2009) Myosin XI is required for actin-associated movement of plastid stromules. Mol Plant 2:1262–1272

16. Sbalzarini I, Mezzacasa A, Helenius A, Koumoutsakos P (2005) Effects of organelle shape on fluorescence recovery after photobleaching. Biophys J 89:1482–1492
17. Sbalzarini I, Hayer A, Helenius A, Koumoutsakos P (2006) Simulations of (an)isotropic diffusion on curved biological surfaces. Biophys J 90:878–885
18. Means S, Smith AJ, Shepherd J, Shadid J, Fowler J, Wojcikiewicz R, Mazel T, Smith G, Wilson B (2006) Reaction diffusion modeling of calcium dynamics with realistic ER geometry. Biophys J 91:537–557
19. Sbalzarini I (2013) Modeling and simulation of biological systems from image data. BioEssays 35:482–490
20. Ölveczky B, Verkman A (1998) Monte Carlo analysis of obstructed diffusion in three dimensions: application to molecular diffusion in organelles. Biophys J 74:2722–2730
21. Walley J, Xiao Y, Wang J-Z, Baidoo EE, Keasling JD, Shen Z, Briggs SP, Dehesh K (2015) Plastid-produced interoganellar stress signal MEcPP potentiates induction of the unfolded protein response in endoplasmic reticulum. Proc Natl Acad Sci U S A 112:6212–6217
22. Lutz DA, Inoué S (1986) Techniques for observing living gametes and embryos. Methods Cell Bio 27:89–110
23. Knebel W, Quader H, Schnepf E (1990) Mobile and immobile endoplasmic reticulum in onion bulb epidermis: short and long-term observations with a confocal laser scanning microscope. Eur J Cell Biol 52:238–340
24. Constantini L, Snapp E (2013) Probing endoplasmic reticulum dynamics using fluorescence imaging and photobleaching techniques. Curr Protocols Cell Biol 60:21.7.1–21.7.29
25. Leonhardt N, Marin E, Vavasseur A, Forestier C (1997) Evidence for the existence of a sulfonylurea-receptor-like protein in plants: modulation of stomatal movements and guard cell potassium channels by sulfonylureas and potassium channel openers. Proc Natl Acad Sci U S A 94:156–161
26. Jaquinod M, Villiers F, Kieffer-Jaquinod S, Hugouvieuz V, Bruley C, Garin J, Bourquignon J (2007) A proteomics dissection of Arabidopsis thaliana vacuoles isolated from cell culture. Mol Cell Proteomics 6:394–412
27. Griffing L (2011) Laser stimulation of the chloroplast/endoplasmic reticulum nexus in tobacco transiently produces protein aggregates (boluses) within the endoplasmic reticulum and stimulates local ER remodeling. Mol Plant 4:886–895
28. Sparkes IA, Runions J, Kearns A, Hawes C (2006) Rapid, transient expression of fluorescent fusion proteins in plants and generation of stably transformed plants. Nat Protoc 1:2019–2025
29. Jain R, Joyce P, Molinete M, Halban P, Gorr S (2001) Oligomerization of green fluorescent protein in the secretory pathway of endocrine cells. Biochem J 360:645–649
30. Pedelacq J, Cabantous S, Tran T, Terwilliger T, Waldo G (2006) Engineering and characterization of a superfolder green fluorescent protein. Nat Biotechnol 24:79–88
31. Ueda H, Yokota E, Kuwata K, Kutsuna N, Mano S, Shimada T, Tamura K, Stefano G, Fukao Y, Brandizzi F, Shimmen T, Nishimura M, Hara-Nishimura I (2016) Phosphorylation of the C terminus of RHD3 has a critical role in homotypic ER membrane fusion in Arabidopsis. Plant Physiol 170:867–880
32. Zhang TY, Suen C (1984) A fast parallel algorithm for thinning digital patterns. Commun ACM 27:236–239
33. Arganda-Carreras I, Fernandez-Gonzalez R, Munoz-Barrutia A, Ortiz-De-Solorzano C (2010) 3D reconstruction of histological sections: application to mammary gland tissue. Microsc Res Tech 73:1019–1029
34. Zack GW, Rogers WE, Latt SA (1977) Automatic measurement of sister chromatid exchange frequency. J Histochem Cytochem 25:741–753
35. Hebert B, Costantino S, Wiseman PW (2005) Spatiotemporal image correlation spectroscopy (STICS) theory, verification, and application to protein velocity mapping in living CHO cells. Biophys J 88:3601–3614
36. Ji L, Danuser G (2005) Tracking quasi-stationary flow of weak fluorescent signals by adaptive multi-frame correlation. J Microsc 220:150–167
37. Lindeboom JJ, Nakamura M, Hibbel A, Shundyak K, Gutierrez R, Ketelaar T, Emons AMC, Mulder BM, Kirik V, Ehrhardt DW (2013) A mechanism for reorientation of cortical microtubule arrays driven by microtubule severing. Science. 342:1245533 https://doi.org/10.1126/science.1245533

# Chapter 8

# Preparation of Highly Enriched ER Membranes Using Free-Flow Electrophoresis

**Harriet T. Parsons**

## Summary

Free-flow electrophoresis (FFE) is a technique for separation of proteins, peptides, organelles, and cells. With zone electrophoresis (ZE-FFE), organelles are separated according to surface charge. The ER is the only remaining major cellular compartment in Arabidopsis not to have been isolated using density centrifugation, immune-isolation, or any other method previously applied to purification of plant membranes. By using continuous-flow electrophoresis, ER vesicles of similar surface charge, which may have been fragmented during cell lysis, can be focused. A large portion of these vesicles are of sufficiently different surface charge that separation from the majority of Golgi and other contaminants is possible. Here we adapt an earlier ZE-FFE Golgi isolation protocol for the isolation of highly pure ER vesicles and for tracking the migration of peripheral ER vesicles. Isolating ER vesicles of homogeneous surface charge allows multi-omic analyses to be performed on the ER. This facilitates investigations into structure-function relationships within the ER.

**Key words** Free-flow electrophoresis, Arabidopsis, Endoplasmic reticulum, Proteomics

## 1 Introduction

The endoplasmic reticulum (ER) is a dynamic, network-like, membranous organelle that extends throughout the cell via tubules connected by three-way junctions. It seemingly forms connections with most cellular compartments between where it surrounds the nucleus and extends out to tethering points at the plasma membrane (PM). Functionally, it is an important site of protein, lipid, and hormone synthesis, as well as protein quality control and modification, calcium storage, and stress responses. Despite its essential biochemical roles, the ER remains one of the few locations within the plant cell that has not been isolated. Proteins have been localized to the Arabidopsis ER by LOPIT [1] and fluorescent tagging, rather than the typical organelle-purification route by which most Arabidopsis organelles proteomes have been obtained [2, 3]. The ER is the least structurally discrete of all cellular

Verena Kriechbaumer (ed.), *The Plant Endoplasmic Reticulum: Methods and Protocols*, Methods in Molecular Biology, vol. 2772, https://doi.org/10.1007/978-1-0716-3710-4_8,

compartments, as shown by recent advances in imaging [4, 5], which no doubt contributes to the difficulty in capturing the ER within a single location on a density gradient. In an earlier "Methods in Molecular Biology" chapter, the isolation of high purity vesicles from the Golgi by free-flow electrophoresis was described [6]. Dynamic and varied structural features make both these organelles difficult to purify using density centrifugation, but domains within each organelle that have uniform surface charge can be focused on an electrophoretic gradient, after an initial enrichment. This chapter builds on methods described previously for the Golgi [6] but with specific adaptations for the ER. Recognizing the increased diversity of tubular and cisternal structures within the ER, the initial enrichment gradient is simplified. Membrane washes are included before and after electrophoresis, as ribosomes and abundant ER-lumen proteins released from any ruptured vesicles may dominate proteomic analyses. The majority of ER proteins are focused within approximately 10 fractions, but like the Golgi, the ER appears to show evidence of sub-domains with variable surface charge. This is not likely to arise from migration differences of rough vs. smooth ER, as ribosome migration resembles neither group. Additionally, a sub-group of ER proteins has been observed to co-migrate with the Golgi. Therefore, particular emphasis is placed on suitable ER markers and tracking ER migration after electrophoretic separation and estimating purity. Instead of using Western blotting to monitor marker proteins in FFE generated fractions, selected reaction monitoring (SRM) offers an excellent way of rapidly monitoring membrane migration [7, 8]; otherwise data-dependent acquisition and spectral counting can be used [9].

## 2 Materials

Prepare all solutions using ultrapure water and analytical-grade reagents. Prepare all reagents at room temperature. Perform all centrifugation steps at 4 °C. Unless otherwise stated, prepare all buffers the day before and store at 4 °C.

### *2.1 Arabidopsis Protoplast Preparation*

1. Arabidopsis cell-suspension culture, temperature controlled shaking incubator (22 °C, 120 rpm) with constant light (*see* **Note 1**).
2. Arabidopsis cell culture medium: 2% (w/v) glucose, α-naphthalenacetic acid (0.5 mg/L), kinetin (0.05 mg/L), 1× Murashige and Skoog basal salt mixture [10]. Prepare media, and adjust to pH 5.7–5.8 with potassium hydroxide (KOH), autoclave for 20 min at 121 °C, and store at 4 °C.
3. Variable speed bench top orbital shaker with wide orbital (*see* **Note 2**).

4. Protoplasting buffer: 500 mM Sorbitol, 20 mM 2-(N-morpholino) ethanesulfonic acid (MES); adjust to pH 5.7–5.8 with KOH. Store at 4 °C. Just prior to use, add 1% (w/v) Cellulase "Onozuka" R-10 and 0.05% (w/v) pectolyase Y-23 (Duchefa Biochemie, Netherlands) by vortexing in 30 mL of protoplasting buffer.
5. Large-capacity preparative centrifuge with 2 × 250 mL tube capacity, e.g., Sorvall SLA-1500.

### 2.2 Cell Rupture and Crude Membrane Enrichment

1. Homogenization buffer: 1% (w/v) dextran (Mw 200,000), 0.4 M sucrose, 20 mM disodium hydrogen phosphate ($Na_2HPO_4$), 3 mM ethylenediaminetetraacetic acid (EDTA), 0.1% (w/v) bovine serum albumin (BSA), 5 mM dithiothreitol (DTT), and one cOmplete EDTA-free protease inhibitor tablet (Roche) (*see* **Note 3**), pH to 7.1 with sodium hydroxide (NaOH).
2. Glass-Teflon Dounce homogenizers (30–50 mL capacity), cooled on ice.
3. Preparative centrifuge with 4 × 50 mL tube capacity capable of 5000 × *g*, e.g., Sorvall SS-34.
4. Ultracentrifuge and swing-out rotor with 40 mL tube capacity capable of 100,000 × *g* for gradients, e.g., Beckman SW-32.
5. Gradient buffer 1: 25% iodixanol, 20 mM $Na_2HPO_4$, 3 mM EDTA, pH 7.1 with NaOH (*see* **Note 4**).
6. Gradient buffer 2: 1.0 M sucrose, 20 mM $Na_2HPO_4$, 3 mM EDTA, dextran Mw 200,000 (1% w/v), 5 mM DTT (*see* **Note 3**), pH 7.1 with NaOH; can store at −20 °C.
7. Gradient buffer 3: 0.2 M sucrose, 20 mM $Na_2HPO_4$, 3 mM EDTA, dextran Mw 200,000 (1% w/v), 5 mM DTT (*see* **Note 3**), pH 7.1 with NaOH; can store at −20 °C.

### 2.3 Focusing of ER Membranes Using FFE

1. Free-Flow Electrophoresis System: BD™ FFE System (BD Diagnostics, NJ, USA for model years 2006–2010) or FFE System (FFE Service GmbH, Germany, http://www.ffeservice.com/ for model years 2011–present).
2. Spacers and filters for ZE-FFE including 0.5 mm spacer and 0.8 mm electrode filter papers for ZE-FFE (FFE Service GmbH, Germany).
3. 96-well deep well plates (2 mL).
4. UV-transparent 96-well plates, e.g., UV-Star (Greiner Bio One, NC, USA).
5. Microplate reader capable of reading absorbance at 280 nm, e.g., Single-Mode Microplate Readers (Molecular Devices, CA, USA).

6. FFE buffer 1: 280 mM sucrose, 10 mM acetic acid, 10 mM triethanolamine, 1 mM EDTA, pH to 7.0 with NaOH (*see* **Note 5**).
7. FFE buffer 2: 200 mM sucrose, 100 mM acetic acid, 100 mM triethanolamine, 10 mM EDTA, pH to 6.5 with NaOH (*see* **Note 5**).
8. FFE buffer 3: 100 mM acetic acid, 100 mM triethanolamine, 10 mM EDTA, pH to 6.5 with NaOH (*see* **Note 5**).
9. Ultracentrifuge and fixed angle rotor with 10–15 mL tube capacity, capable of 100,000 × *g* for sample concentration.
10. 50 mM tris(hydroxymethyl)aminomethane (Tris–HCl), pH 8.0, stored at 4 °C.

### 2.4 Post-FFE Sample Analysis

1. High-grade trypsin, e.g., Trypsin, from Porcine pancreas (Sigma-Aldrich, MO, USA).
2. SpeedVac concentrator.
3. Ultra-micro Spin Columns with $C_{18}$ (Harvard Apparatus, MA, USA).
4. ACN1 solution: 80% acetonitrile (v/v) with 0.1% trifluoroacetic acid (v/v).
5. ACN2 solution: 2% acetonitrile (v/v) with 0.1% trifluoroacetic acid (v/v).
6. LC-MS/MS buffer A (98% water, 2% acetonitrile, 0.1% formic acid).
7. LC-MS/MS buffer B (98% acetonitrile, 2% water, 0.1% formic acid).
8. Optional: Agilent Technologies (Santa Clara, CA) 6460QQQ mass spectrometer operating in SRM mode, with a Sigma (St. Louis, MO) Ascentis Peptide Express C-18 column (2.1 mm × 50 mm) or similar (*see* **Note 21**).
9. Tandem mass spectrometer (MS/MS) with online liquid chromatography (LC) capabilities (nanoflow or capillary flow rates) capable of data dependent acquisition. Data in Fig. 1 was generated on a nano-ESI-Q-TOF (TripleTOF 5600 System, AB SCIEX) coupled to an Eksigent nano LC system (AB Sciex) and an Acclaim Pepmap100 C18 column (75 μm × 150 mm, Dionex-LC Packings) at 300 nL/min flow rate.
10. Search engine for analyzing mass spectrometry data to identify proteins, e.g., Mascot (Matrix Science, UK).
11. Optional: Targeted proteomics software, e.g., Skyline (freely available online at http://skyline.maccosslab.org) and Scaffold 4 (Proteome Software, OR, USA).

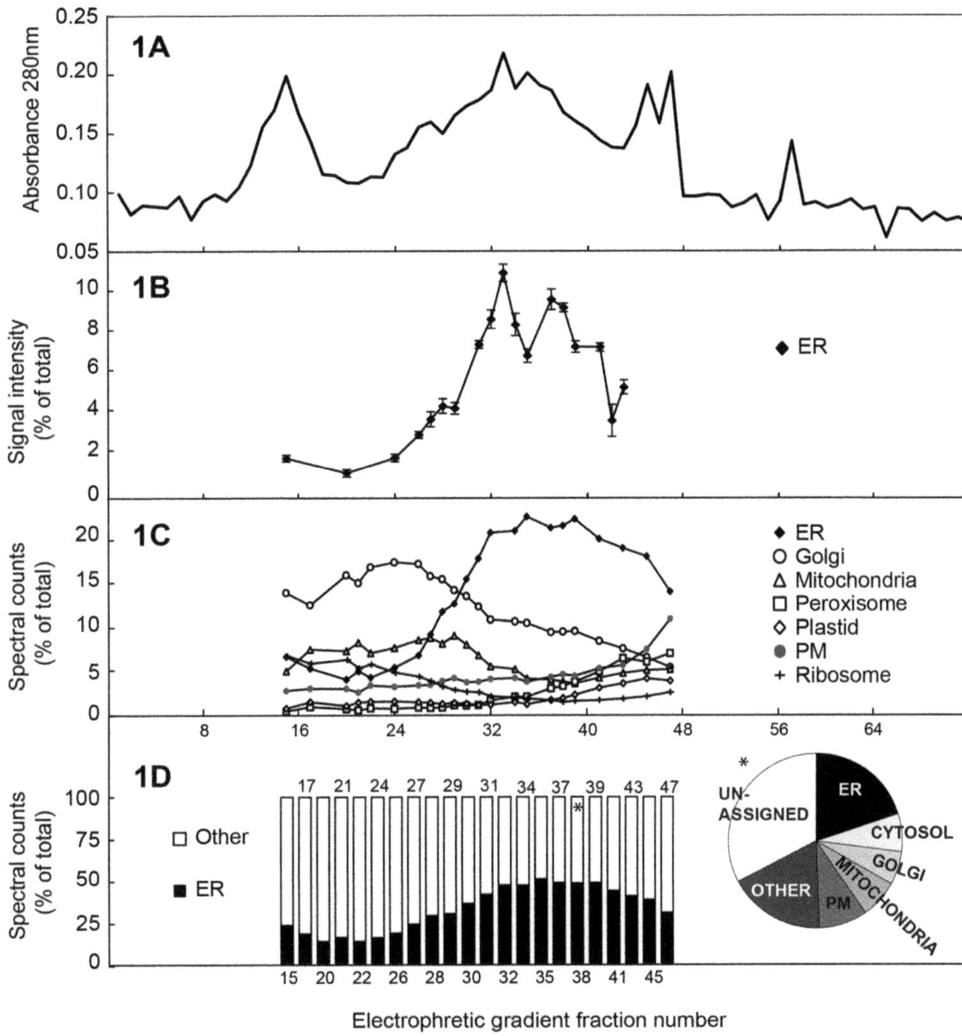


**Fig. 1** Distribution of proteins along electrophoretic gradient, after ZE-FFE. (**a**) Total protein distribution was estimated by measuring the absorbance at 280 nm of 250 μL aliquots of fractions after electrophoresis. (**b**) Using the transitions and collision energies given in Table 1, the distribution of the ER was estimated using SRM measurements averaged from 6 proteins, each protein being averaged from 2 peptides. Peptide signal intensities were calculated by summing the 2 highest transition intensities, as described in [12]. (**c**) Spectra were summed for the proteins in each fraction that could be unambiguously assigned to subcellular compartments (based on experimental data housed at SUBA) and presented as a proportion of the total spectra for each fraction to give an estimation of ER migration relative to other organelles. (**d**) Spectral counts from all other organelles were summed and compared to summed spectral counts for ER-localized proteins, to give an estimate of the purity of the most ER-rich fractions. A pie chart gives an overview of protein numbers in ER-enriched fractions, showing that ER abundance is calculated from over 100 proteins in fraction 38*, from a total of 500 proteins with >5 spectra

**Table 1**
**Transitions used for identification of ER-rich fractions using rapid gradient profiling by SRM. Retention times and collisions energies refer to those used in this protocol using the Agilent 6460QQQ**

| Protein identifier | Gene description | Peptide | Precursor m/z | Fragment m/z | Retention time (min) | Collision energy (eV) | Precursor ion charge | Fragment type | Fragment number | Fragment charge |
|---|---|---|---|---|---|---|---|---|---|---|
| AT5G61790.1 | CNX | NSDYEGVWK | 549.2 | 896.4 | 4.9 | 12.4 | 2.0 | y | 7 | 1 |
| | | | | 781.4 | | | 2.0 | y | 6 | 1 |
| AT5G61790.1 | CNX | VVFDLLNK | 474.3 | 848.5 | 11.0 | 8.6 | 2.0 | y | 7 | 1 |
| | | | | 749.4 | | | 2.0 | y | 6 | 1 |
| AT1G56340.1 | CRT1 | TLVFQFSVK | 534.8 | 854.5 | 12.1 | 11.7 | 2.0 | y | 7 | 1 |
| | | | | 755.4 | | | 2.0 | y | 6 | 1 |
| AT1G56340.1 | CRT1 | YVGVELWQVK | 534.8 | 755.4 | 11.6 | 15.6 | 2.0 | y | 7 | 1 |
| | | | | 608.3 | | | 2.0 | y | 5 | 1 |
| AT1G52260.1 | PDIL1-5 | FAEAATALK | 461.3 | 703.4 | 4.4 | 8.0 | 2.0 | y | 7 | 1 |
| | | | | 574.4 | | | 2.0 | y | 6 | 1 |
| AT1G52260.1 | PDIL1-5 | YTVYDGSYK | 548.3 | 732.3 | 2.5 | 12.4 | 2.0 | y | 6 | 1 |
| | | | | 569.3 | | | 2.0 | y | 5 | 1 |
| AT1G67490.1 | GCS1 | LPFTIDIPFI SGIK | 781.0 | 989.6 | 19.3 | 24.3 | 2.0 | y | 9 | 1 |
| | | | | 761.5 | | | 2.0 | y | 7 | 1 |
| AT1G67490.1 | GCS1 | TLSSGLDDYPR | 612.3 | 1009.5 | 6.2 | 15.7 | 2.0 | y | 9 | 1 |
| | | | | 835.4 | | | 2.0 | y | 8 | 1 |
| AT5G13640.1 | PDAT1 | ITGPLP DPPGVK | 595.8 | 709.4 | 8.7 | 14.8 | 2.0 | y | 7 | 1 |
| | | | | 497.3 | | | 2.0 | y | 5 | 1 |
| AT5G13640.1 | PDAT1 | LWGGTFGE VYK | 628.8 | 1143.5 | 10.6 | 16.5 | 2.0 | y | 10 | 1 |
| | | | | 957.5 | | | 2.0 | y | 9 | 1 |

## 3 Methods

### 3.1 Protoplast Preparation of Arabidopsis Cell Suspension Cultures

1. Maintain growing Arabidopsis cells in 100 mL aliquots, and subculture weekly at 1:10 ratio (*see* **Note 1**). Harvest cells from 4–6 flasks by centrifugation 800 x *g*, 20 °C for 5 min, using a 250–500 mL capacity rotor and weigh pellet (*see* **Note 6**).
2. For up to 50 g cells (fresh weight), make 400 mL of protoplasting buffer. To 150 mL of digestion buffer, add 1.0% (w/v) cellulase and 0.05% (w/v) pectolyase in a 4 L wide-bottomed, conical flask. Add cells, ensuring they are adequately suspended in the buffer. Place on an orbital shaker, and rotate slowly in the dark for 2.5 h (*see* **Note 7**).
3. Protoplasts are harvested by centrifugation at 800 x *g* for 5 min, the supernatant discarded, and cells gently resuspended in the remaining 250 mL of protoplasting buffer without enzymes (*see* **Note 8**).
4. Centrifugation at 800 x *g* is repeated, the supernatant discarded, and cells resuspended in homogenization buffer using a ratio of 1:1 (w/v) using the original fresh weight of cells to buffer volume.

### 3.2 Homogenization of Protoplasts and Pre-FFE Enrichment

1. Working at 4 °C, homogenize the protoplasts in homogenization buffer using four to five strokes at even pressure with a Dounce homogenizer (*see* **Note 9**).
2. Compare pre- and post-homogenized protoplasts under a light microscope to ensure adequate disruption (*see* **Note 9**).
3. Centrifuge homogenate at 5000 x *g* for 15 min (rotor with capacity for 50 mL samples, e.g., Sorvall SS-34).
4. Transfer the supernatant to thin-wall polypropylene tubes suitable for use in an SW-32 rotor, such that tubes are two-thirds full and underlay with 6 mL of Gradient Buffer 1 (*see* **Note 10**).
5. Ultracentrifuge at 100,000 x *g* for 1.5 h (*see* **Note 11**).
6. Remove the supernatant such that the cushion surface is disturbed as little as possible, then remove membranes from the cushion, gently mix in the same volume of homogenization buffer as previously, and repeat **steps 4** and **5** (*see* **Note 12**).
7. After removing the supernatant for a second time, gently layer 15–20 mL of gradient buffer 2 on to the membrane cushion, then 12–15 mL of gradient buffer 3, or until tubes are at least two-thirds full.
8. Ultracentrifuge samples at 100,000 x *g* for 1.5 h.
9. A single band should be present at the boundary between gradient buffer 2 and 3 (*see* **Note 13**).

10. Measure the protein concentration of the collected sample, and dilute so the sucrose concentration is as close to FFE Buffer 1 as possible, although a total protein concentration less than 0.75 μg/μL is not recommended. Keep the desired band in a 15 mL tube on ice (*see* **Note 14**).

#### 3.3 Separation of ER Membranes Using ZE-FFE

1. Ensure the FFE system is correctly set up for ZE-FFE, as detailed in Fig. 1 in [6], the temperature set to 8 °C, the media pump calibrated, and all necessary quality control tests undertaken (*see* **Note 15**).
2. Load **FFE Buffer 1** (inlet tubes 2–6), **FFE Buffer 2** (inlets 1 and 7), **FFE Buffer 3** (electrodes), and slowly fill chamber with buffers avoiding air bubbles. Set the media flow rate to 250 mL/h. Switch on electrode circuits; set the voltage, current, and power to 700 V, 150 mA, and 100 W, respectively; and switch on the voltage (*see* **Note 16**). Allow the system to equilibrate and stabilize for 20–30 min prior to use.
3. Load the sample at a flow rate of 2500–3000 μL/h. The sample should appear homogeneous and be visible in the lower half of the separation chamber but is usually too dilute to be seen in the upper half (*see* **Note 17**).
4. After about 10 min, start collecting samples in the precooled 2 mL deep-well plates (*see* **Note 18**).
5. Throughout the ZE-FEE process, monitor protein distribution by removing 150–250 μL from plates and measuring at 280 nm using UV-transparent 96-well plates (Fig. 1a). Protein peak positions and profiles should remain constant throughout separation.
6. Assuming successful enrichment of membrane by density centrifugation, the main peak at 280 nm will constitute a mixture of cis-Golgi and ER membranes (Fig. 1c). The proportion of ER proteins will increase at the cathodic edge of this main peak and the cis-Golgi to the anodic edge (Fig. 1b, c).
7. After electrophoretic separation, pool fractions from multiple plates, and centrifuge at 100,000 x *g* for 60 min (*see* **Note 19**).
8. Remove the supernatant, and rinse pellets gently in chilled water (*see* **Note 20**). Resuspend in 50 mM Tris–HCl (pH 8.0), and then sonicate for 15 min at 30 s pulse intervals. Measure protein concentration using the Bradford's assay, and store resuspended pellets at −80 °C.

#### 3.4 Tryptic Digestion of Samples and Analysis by Mass Spectrometry

1. Reduce, alkylate, and digest at least 10 μg of protein overnight (37 °C) at a 1:40 trypsin/protein ratio in 50% (v/v) ACB.
2. Remove ACN in a SpeedVac concentrator, and then resuspend in 50 μl ACN2.

3. Clean and concentrate samples in Ultra-micro SpinColumns (10–25 μL capacity), after washing columns once in ACN1 and priming twice in ACN2. Elute peptides in 50 μL ACN1.
4. Remove ACN in a SpeedVac concentrator, and then resuspend in ACN2 to give a concentration of 1 μg/μL, assuming complete protein digestion.
5. Optional analysis using SRM: This method was developed using a 6460QQQ mass spectrometer operating in SRM mode (*see* Subheading 2.4, **item 8**), although other possibilities exist (*see* **Note 21**). Five micrograms of peptide were injected and separated at a flow rate of 400 μL/min over the following gradient: 95% buffer, 5% buffer B for 2 min, increased to 40% Buffer B over 15 min, rapidly increased to 90% B over 5 min, held at 90% B for 5 min, decreased rapidly to 5% B, and held for 2 min before the next run. Eluting peptides were ionized by using an Agilent Jet Stream source (sheath gas flow, 11 L/min; sheath gas temperature, 350 c; nozzle voltage, 1000 v; nebulizing pressure, 30 psi; chamber voltage, 4500 V) operating in positive-ion mode. ER peptide transitions are given in Table 1. These parameters delivered the ER peptide-profiles shown in Fig. 1b.
6. If targeted proteomics analysis is not possible, then peptide samples can be analyzed by liquid chromatography tandem mass spectrometry (LC-MS/MS) using software such as Mascot (Matrix Science, UK). Organelle profiles in Fig. 1c, d were produced using instrumentation as detailed (*see* Subheading 2.4, **item 9**), with a gradient starting at 5% buffer B, increasing to 35% B over 60 min. Over 3 min, buffer B was increased to 90% over 3 min, held for 15 min, and then decreased to 5% Buffer B over 3 min and held for 15 min to re-equilibrate the column (*see* **Note 22**).
7. With sufficient protein of known localization to the ER and other cellular compartments, it is possible to monitor the migration of organelles (*see* **Note 22**). Information on all predicted and previous experimental localizations for proteins can be found at SUBA3 (suba3.plantenergy.uwa.edu.au). By assessing the proportion of proteins within a fraction that have a majority consensus to a given cellular location, the migration of the ER and other organelles can be mapped along the gradient, as in Fig. 1c, d. This proportion can be assessed by simply counting identifications, emPAI values, or the number of spectra associated with each protein [9] (*see* **Note 23**).

### 3.5 Summary and Future Uses of This Technique

In summary, this method describes how ER proteins can be enriched to approximately 50% of an endomembrane fraction. Some proteins identified in ER-rich fractions may have been substrates transiting the ER, so ER vesicles may comprise more than 50% of some fractions. Although this technique does not yield completely pure ER fractions, it is an appreciable increase on the purity achieved by density gradient centrifugation. As ER vesicles can be pelted from enriched fractions, this technique offers much interesting potential for lipidomic and metabolomic, as well as proteomic investigation.

## 4 Notes

1. This protocol uses a *L. erecta* line similar to MMd2 cells [11] and tolerates a range of light levels. An essential characteristic of whichever cell line is used is that it is fine, suspension culture, which readily forms protoplasts.
2. A variable speed bench top orbital shaker with a large orbital throw is optimal (at least 2 cm). This allows slow rotation while maintaining the cells in solution. This permits efficient enzymatic digestion of the cell wall.
3. DTT and protease inhibitors are added immediately prior to use and can be quickly solubilized by vortexing in 1.0 mL of buffer.
4. 25% iodixanol is prepared by diluting the stock solution twofold with 2× homogenization buffer (without BSA) to give a 30% solution and then diluting to 25% with homogenization buffer. This results in less osmolality extremes the previous 1.6 M cushion [6].
5. All FFE buffers must be used within 24 h and can be stored at 4 °C. Triethanolamine should be weighed. A generous volume of FFE Buffer 3, e.g., 800 mL for a 1–2 h run, is recommended so the electrode buffers are not exhausted. For FFE Buffer 1, about 1.5 L is required for a run using the conditions described.
6. Start with a minimum of 50 g fresh pellet weight. Before protoplasting, the cell pellet can be spooned out and weighed without damaging cells.
7. Rotate at the lowest possible speed at which cells remain in suspension. Enzymes should be added to the protoplasting buffer immediately before use. Enzymes are easily solubilized by vigorous shaking in a 30 mL aliquot of protoplasting buffer prior to their addition to the main solution.
8. The pellet/protoplasts are delicate; decant the supernatant gently to avoid breaking cells.
9. The idea is to rupture protoplasts while minimizing organelle rupture. This is dependent on the number of strokes, the force

applied to each stroke, and the fit between the Teflon plunger and glass wall. It is recommended that this step be carefully optimized, by examining the number of intact protoplasts under a light microscope and the efficiency with which chloroplasts can be pelleted at low speeds to leave a pale and beige, but not green, supernatant.

10. Ideally, use about 15 g fresh weight starting material per gradient.
11. This step should result in a yellow-white-colored cushion about 2 mm thick (assuming starting with 60 g fresh weight cells). Green coloration implies organelle rupture during protoplast homogenization. Elimination of green should be possible if starting from cell suspension culture but otherwise may not be possible if, e.g., liquid-grown plantlets are used as the starting material.
12. Complete removal of the supernatant is a compromise between quality of step gradient formation and disturbance of the cushion, with ensuing loss of yield.
13. A distinctly formed yellow-white band 1–3 mm thick at the 1.0 M/0.2 M interface will give optimize results obtained after FFE.
14. Ideally, **steps 1** and **2** (Subheading 3.2) should be completed as fast as practicably possible. Tubes containing Gradient Buffer 1 can be prepared in advance. Setup of the FFE and stabilization of the current should have been completed so as to coincide with **step 9**. The current normally takes not more than 20–30 min to stabilize. The conductivity may increase as the sample buffer enters the chamber. This can be rectified by lowering the voltage by, e.g., 20–30 V and waiting approximately 5 min before starting to collect samples.
15. This is a nontrivial task, and some previous experience or instructions in the setup of the system for ZE-FFE has been assumed. Use of fresh filters, membranes, and electrode gaskets can improve separation performance considerably. Great care should be taken that the plastic spacer is exactly centered. Tubing sections under the peristaltic pumps should be checked thoroughly for hairline cracks prior to setup and changed if necessary. The sample inlet tubing should be exchanged if any kinks are present. Excessive indentation from the spacer or inlet tube on the lower-temperature cooled plate indicates a change of plastic cover sheeting is required for optimal separation. All FFE buffers should be made up precisely to the stated pH; otherwise, the conductivity will be too high, and the current will not stabilize as desired. Much useful material, including references, is available at http://www.ffeservice.com.

16. Media flow rate and voltage are the two principal parameters that affect sample migration. For different types of sample preparation, a ratio between the two should be optimized; however, below 200 MFR diffusion between streams of subcellular compartments may occur. The effect of changing parameters on protein distribution can be instantaneously verified by measuring protein content at 280 nm in UV-transparent 96-well plates.
17. If particulates are present in the sample, pass the sample once through a glass Pasteur pipette.
18. At a flow rate of 250 mL/h, plates will fill in approximately 17 min, although shorter fill times will prevent sample warming if a plate cooling device is not available. Store plates at 4 °C, and work at 4 °C for **steps 6** and 7 (Subheading 3.3).
19. Assuming 8 plates at approximately 1.5 mL per well have been collected, pellets will be visible in all fractions above 0.15 $A_{280}$ (measuring 250 μL at $A_{280}$); otherwise, it may be necessary to pool fractions.
20. If only proteomic analysis of membrane proteins is required, pellets can be resuspended in 100 mM $NaCO_3$, shaken for 30 min at 4 °C, and re-pelleted at 100,000*g* for 1 h. This will remove soluble proteins sticking to membranes but may result in vesicle rupture.
21. If analyzing ER membrane migration over many fractions, then analysis by a targeted proteomics technique such as SRM is recommended, using the transitions in Table 1. Using a QQQ mass spectrometer using microflow rates is ideal, as short gradients of 30 min or less can be used, dramatically reducing analysis time. Targeted analysis is also possible using a Triple-TOF running at microflow rates. Use of a Q-Exactive operating in PRM mode is also possible, but as this must be conducted at nanoflow rates, there is little gain in analysis time.
22. When separating organelles by free-flow electrophoresis using the described methods, variations in instrument setup, buffer conductivity, and cell growth may combine to give a shift in absolute migration distances of about five fractions between experiments. This shift can be up to ten fractions when different instruments are used. However, the relative distribution of organelles should not change. It is therefore recommended that users carefully monitor organelle migration and relate the to the total protein profile for several experiments, as in Fig. 1, until users are familiar with their experimental setup.
23. Software such as Scaffold 4.0 facilitates spectral counting but is not essential.

## References


1. Dunkley TP et al (2006) Mapping the Arabidopsis organelle proteome. Proc Natl Acad Sci U S A 103:6518–6523
2. Agrawal GK et al (2011) Plant organelle proteomics: collaborating for optimal cell function. Mass Spectrom Rev 30:772–853
3. Drakakaki G et al (2012) Isolation and proteomic analysis of the SYP61 compartment reveal its role in exocytic trafficking in Arabidopsis. Cell Res 22:413–424
4. Kittelmann M, Hawes C, Hughes L (2016) Serial block face scanning electron microscopy and the reconstruction of plant cell membrane systems. J Microsc 263:200–211
5. Nixon-Abell J et al (2016) Increased spatiotemporal resolution reveals highly dynamic dense tubular matrices in the peripheral ER. Science 354(6311):aaf3928
6. Parsons HT, Fernández-Niño SM, Heazlewood JL (2014) Separation of the plant Golgi apparatus and endoplasmic reticulum by free-flow electrophoresis in plant proteomics. In: Methods and protocols. Humana Press, Totowa, pp 527–539
7. Picotti P, Bodenmiller B, Aebersold R (2013) Proteomics meets the scientific method. Nat Methods 10:24–27
8. Aebersold R, Burlingame AL, Bradshaw RA (2013) Western blots versus selected reaction monitoring assays: time to turn the tables? Mol Cell Protes 12:2381–2382
9. Arike L, Peil L (2014) Spectral counting label-free proteomics. In: Martins-de-Souza D (ed) Shotgun proteomics: methods and protocols. Springer New York, New York, pp 213–222
10. Murashige T, Skoog F (1962) A revised medium for rapid growth and bio assays with tobacco tissue cultures. Physiol Plant 15:473–497
11. Menges M, Murray JA (2002) Synchronous Arabidopsis suspension cultures for analysis of cell-cycle gene activity. Plant J 30:203–212
12. Dahl RH et al (2013) Engineering dynamic pathway regulation using stress-response promoters. Nat Biotech 31:1039–1046

# Chapter 9

# Preparation of ER Microsomes from *Arabidopsis thaliana*


## Verena Kriechbaumer



## Abstract

Microsomes are vesicles derived from the endoplasmic reticulum (ER) when cells are broken down in the lab. These microsomes are a valuable tool to study a variety of ER functions such as protein and lipid synthesis in vitro.

Here we describe a protocol to isolate ER-derived microsomes *Arabidopsis thaliana* seedlings and exemplify the use of these purified microsomes in enzyme assays with the auxin precursors tryptophan (Trp) or indole-3-pyruvic acid (IPyA) to quantify auxin synthetic capacity in microsomal and cytosolic fractions.



**Key words** Endoplasmic reticulum, Microsomal fraction, Enzyme assays, Arabidopsis


## 1 Introduction

Microsomes are generally described as vesicle-like structures with a diameter of 100–200 nm that are formed from pieces of the endoplasmic reticulum (ER) after eukaryotic cells are broken down experimentally (Fig. 1); hence, microsomes are not found in healthy living cells. Microsomes are still capable of ER functions such as protein synthesis, protein glycosylation, $Ca^{2+}$ uptake, and lipid synthesis and can be used to study all these functions in a test tube [1]. Therefore microsomes have found use in a plethora of experimental approaches and studies.

Microsomes are purified from other cell compartments and membranes by differential centrifugation at 100,0000*g*. Whole cells, nuclei, and mitochondria pellet at 10,000*g* [2] and chloroplasts already at 1000*g* [3, 4].

In his publication on the *Constitution of Protoplasm*, Albert Claude named microsomes in 1943 [5] to distinguish these "small granules of undefined nature." He then used this fractionation technique to determine the localization of nucleic acid in leukemic cells [6]. Current applications and use of microsomes


Verena Kriechbaumer (ed.), *The Plant Endoplasmic Reticulum: Methods and Protocols*, Methods in Molecular Biology, vol. 2772, https://doi.org/10.1007/978-1-0716-3710-4_9,

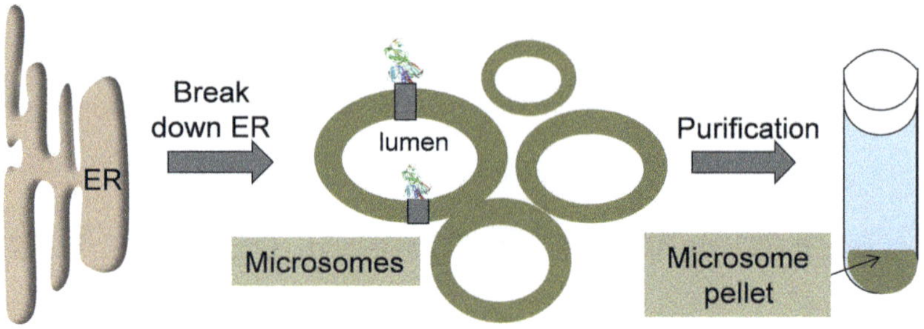


**Fig. 1** Schematic representation of microsome purification

for research are for example—among various others—the areas of lipid secretion [7], P450 assays [8], or auxin biosynthesis [9–11].

A common source for microsomes is dog pancreas cells [12] or rabbit reticulocyte lysate [13]. In plants, microsomes wheat germ is a good source [14], or they can be purified from soybeans [15].

Here, I describe the isolation of microsomes from *Arabidopsis thaliana* seedlings (adapted from [9–11]).

## 2 Materials

### 2.1 Stock Solutions (see Note 1)

#### 2.1.1 For Microsomal Preparation

1. M TEA-HOAc, pH 7.5.
2. 2 M KOAc, pH 7.5.
3. 0.1 M $Mg(OAc)_2$.
4. 1 M Sucrose.
5. 1 M DTT.
6. 0.5 M EDTA.

#### 2.1.2 For IAA Quantification

1. 1 M Tris–HCl, pH 8.0.
2. 1 M $Na_2CO_3$.
3. 0.1 M NADPH.
4. 0.1 M FAD.
5. 0.1 M indole-3-pyruvic acid (IPyA).
6. 1 M Tryptophan (Trp).

### 2.2 Buffers

1. Buffer A: 25 mM TEA-HOAc (pH 7.5), 50 mM KOAc, (pH 7.5), 5 mM $Mg(OAc)_2$, 0.25 M Sucrose, 4 mM DTT.
2. Buffer B: 100 mM TEA-HOAc (pH 7.5), 20 mM EDTA.
3. Buffer C: 25 mM TEA-HOAc (pH 7.5), 25 mM KOAc (pH 7.5), 2 mM $Mg(OAc)_2$, 0.5 M Sucrose, 4 mM DTT.
4. Buffer D: 25 mM TEA-HOAc (pH 7.5), 0.25 M Sucrose, 1 mM DTT.

#### 2.3 Equipment

1. Porcelain mortar and pestle.
2. Glass bottles for buffers.
3. Cheese cloth.
4. Refrigerated table centrifuge.
5. Ultracentrifuge with swing-out rotor (e.g., SW41, Beckman Coulter).
6. Ultracentrifuge Corex tubes.
7. 2 mL Potter–Elvehjem homogenizer with glass rod.
8. water bath (37 °C).
9. SpeedVac.
10. NanoDrop spectrophotometer, or equivalent, to determine protein concentration.

## 3 Methods

#### 3.1 Endoplasmic Reticulum (ER) Microsome Preparation

(This part of the procedure will take between 3 and 4 h depending on sample size.)

1. All steps are carried out on ice or at 4 °C unless indicated otherwise.
2. All buffers, tubes, etc. used in the procedure should be precooled (*see* **Note 2**).
3. Five grams of Arabidopsis seedling tissue (7 days after germination; *see* **Notes 3**, **4**, and **5**) are ground to fine powder in liquid nitrogen using a precooled mortar and pestle.
4. The powder is then homogenized in 4 mL of ice-cold buffer A and transferred into a 50 mL falcon tube.
5. Four milligrams of ice-cold buffer B are added to the tube, and the suspension is incubated on ice for 10 min. Then the homogenate is centrifuged at 1000 × *g* for 10 min at 4 °C. The resulting supernatant is poured over four layers of cheese cloth into a fresh falcon tube. At this step, the resulting extract is considered total plant extract for later enzymatic assays. The extract is centrifuged again at 4500 × *g* for 25 min at 4 °C.
6. A 4 mL sucrose cushion (buffer C) is layered on the bottom of ultracentrifuge Corex tubes.
7. The 8 mL of plant suspension is layered on top of this sucrose cushion by slightly angling the tube and carefully and slowly pipetting the suspension at the side of the tube. The tube is centrifuged for 90 min at 93,000 × *g* (ultracentrifuge with swing-out rotor, e.g., SW41).

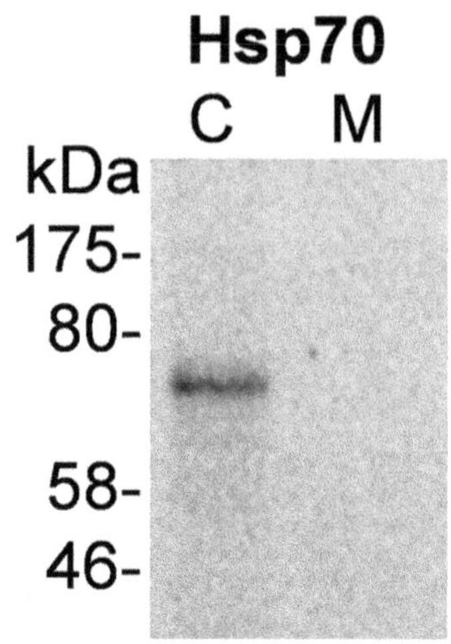


**Fig. 2** Immunoblot analysis for purity of microsomal fractions. Microsomal (M) and cytosolic (C) fractions were tested for heat shock protein 70 (HSP70) proteins using immunoblot analysis. 100 μg of protein from each fraction were probed with anti-HSP70 antibodies (1:1000) recognizing the cytosolic HSP70

8. The resulting pellet is removed from the ultracentrifuge tube if necessary with 20 μL of buffer D and transferred to a 2 mL Potter–Elvehjem homogenizer. The supernatant is kept and used as cytosolic extract in later IAA quantifications.
9. The final pellet was resuspended in 200 μL of buffer D using a glass rod and a 2 mL Potter-Elvehjem homogenizer. Protein content is measured using a NanoDrop spectrophotomoter. Freshly prepared microsomes should be used for enzymatic assays straight away (*see* **Note 6**).

### 3.2 Purity of Microsomes

To check for the purity of the microsomal fraction, both the microsomal and the cytosolic fraction can be probed with an anti-HSP70 antibody detecting cytosolic heat shock protein 70 (Fig. 2). The HSP70 band is only detected in the cytosolic fraction indicating a rather pure microsomal fraction.

Further purity tests using immunoblot analysis could include testing for the presence of plasma membrane or mitochondrial proteins in the microsomal fraction.

### 3.3 Example for Microsomal Enzymatic Tests

The YUC-route of auxin biosynthesis is a two-step process (Fig. 3): In Arabidopsis TAA-proteins (Tryptophan Aminotransferase of Arabidopsis), TAA1, TAR1, and TAR2 convert tryptophan (Trp) to indole-3-pyruvic acid (IPyA) [16], which is then converted by YUC proteins to IAA (YUC1-11) [17]. It was shown that TAR2, YUC4.2, 5, 7, 8, and 9 are located in the ER, whereas TAA1 and YUC1, 2, 3, 6, and 11 are located to the cytosol [10].

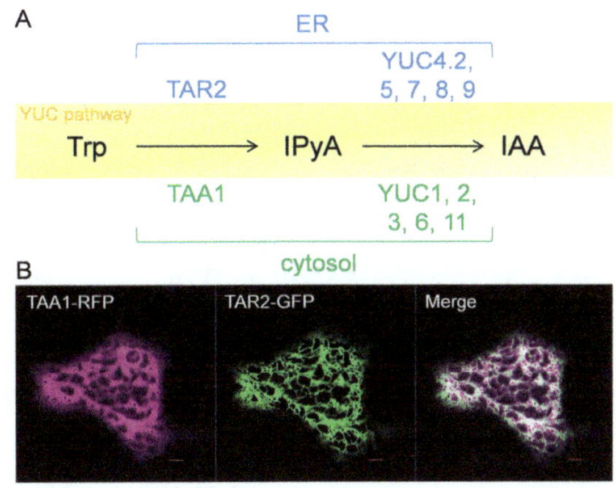


**Fig. 3** Subcellular compartmentation of auxin biosynthesis in *Arabidopsis thaliana*. (**a**) YUC-route of auxin biosynthesis in Arabidopsis. ER-located enzymes are labeled in blue, cytosolic enzymes in green. (**b**) Confocal localization for two enzymes in the first step of IAA biosynthesis. TAA1 is present in the cytosol, whereas TAR2 shows ER-location

Enzymatic activity tests (100 μL total volume) with microsomal and cytosolic fractions were carried out in the following manner:

1. In 2 mL Eppendorf tubes, mix carefully 20 μL of microsomal or cytosolic extract, 1 mM NADPH, 100 μM FAD, 100 μM IPyA or Trp (depending on experimental interest).
2. 100 mM Tris–HCl (pH 8.0) up to a total volume of 100 μL.
3. The assays are incubated for 1 h in a 37 °C water bath and snap-frozen in liquid nitrogen straight after the incubation time.
4. At this stage, the assays can be stored at −20 °C before the IAA analysis is carried out.

### 3.4 Auxin Biosynthetic Capacity in Arabidopsis Microsomes

Representative data for the conversion of tryptophan and IPyA to IAA in Arabidopsis seedlings shown in Fig. 4 [10].

## 4 Notes

1. Filter buffers 1–4 through 0.45 μm syringe filters. Add DTT from 1 M stock fresh prior to use.
2. Pre-chill tubes, glassware, buffers, centrifuges, etc., and keep your samples on ice as much as possible.

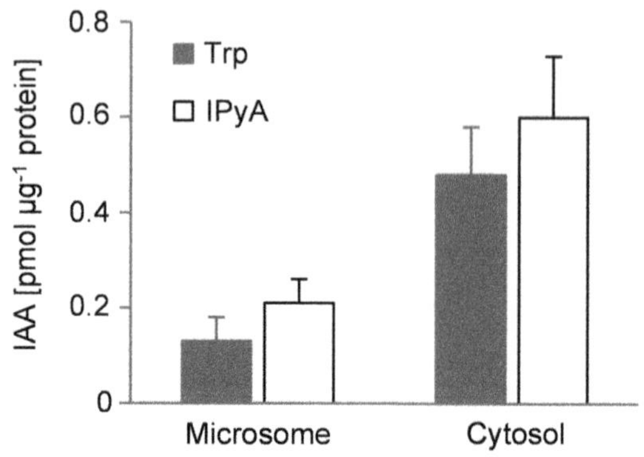


**Fig. 4** Auxin biosynthetic capacity in Arabidopsis seedlings. Enzymatic conversion of tryptophan to IAA by microsomal (Microsome) fractions or cytosolic (Cytosol) fractions from Arabidopsis seedlings 7 d after germination. Standard errors are indicated. $n = 3$

3. Tissue: depending on the biological question, instead of using whole seedlings for the microsomal preparation, it can, e.g., be distinguished between root and shoot, or other ages and developmental stages can be used.
4. This preparation has also been successfully used for maize [9] and tobacco tissues.
5. Collect plant material straight into a chilled falcon tube that is kept on ice, and try to collect the material as quickly as possible.
6. Where possible, timewise freshly prepared microsomes should be used for enzyme assays. Freezing in 20% glycerol for later assays is possible but not recommended.

## Acknowledgments

This work was supported by a research scholarship from the Korean Federation of Science and Technology Societies (KOFST) awarded to Dr. Verena Kriechbaumer and the British Biotechnology and Biological Sciences Research Council (grant no. BB/J004987/1 research grant) awarded to CH.

### References

1. Alberts B, Johnson A, Lewis J, Morgan D, Raff M, Roberts K, Walter P (2014) Molecular biology of the cell, 6th edn. Garland Science, New York
2. Nagahashi J, Hiraike K (1982) Effects of centrifugal force and centrifugation time on the sedimentation of plant organelles. Plant Physiol 69:546–548
3. Tobin AK, Bowsher CG (2004) Subcellular fractionation of plant tissues: isolation of plastids and mitochondria. Methods Mol Biol 244: 53–63
4. Kriechbaumer V, Shaw R, Mukherjee J, Bowsher CG, Harrison AM, Abell BM (2009) Subcellular distribution of tail-anchored proteins in Arabidopsis. Traffic 10:1753–1764

5. Claude A (1943) The constitution of protoplasm. Science 97:451–456
6. Claude A (1944) The constitution of mitochondria and microsomes, and the distribution of nucleic acid in the cytoplasm of a leukemic cell. J Exp Med 144(80):19–29
7. Yao Z, Zhou H, Figeys D, Wang Y, Sundaram M (2013) Microsome-associated lumenal lipid droplets in the regulation of lipoprotein secretion. Curr Opin Lipidol 24:160–170
8. Clausen M, Kannangara RM, Olsen CE, Blomstedt CK, Gleadow RM, Jørgensen K, Bak S, Motawie MS, Møller BL (2015) The bifurcation of the cyanogenic glucoside and glucosinolate biosynthetic pathways. Plant J 84:558–573
9. Kriechbaumer V, Seo H, Park WJ, Hawes C (2015) Endoplasmic reticulum localization and activity of maize auxin biosynthetic enzymes. J Exp Bot 66:6009–6020
10. Kriechbaumer V, Botchway SW, Hawes C (2016a) Localization and interactions between Arabidopsis auxin biosynthetic enzymes in the TAA/YUC-dependent pathway. J Exp Bot 67: 4195–4207
11. Kriechbaumer V (2016b) ER microsome preparation and subsequent IAA quantification in maize coleoptile and primary root tissue. Bio-protocol 6(9):e1805
12. Walter P, Blobel G (1983) Preparation of microsomal membranes for cotranslational protein translocation. Methods Enzymol 96: 84–89
13. Jackson R, Hunt T (1983) Preparation and use of nuclease-treated rabbit reticulocyte lysates for the translation of eukaryotic messenger RNA. Methods Enzymol 96:50–74
14. Erickson AH, Blobel G (1983) Cell-free translation of messenger RNA in a wheat germ system. Methods Enzymol 96:38–50
15. Abell BM, Holbrook LA, Abenes M, Murphy DJ, Hills MJ, Moloney MM (1997) Role of the proline knot motif in oleosin endoplasmic reticulum topology and oil body targeting. Plant Cell 9:1481–1493
16. ZhaoY CSK, Fankhauser C, Cashman JR, Cohen JD, Weigel D, Chory J (2001) A role for flavin monooxygenase-like enzymes in auxin biosynthesis. Science 291:306–309
17. Stepanova AN, Robertson-Hoyt J, Yun J, Benavente LM, Xie DY, Dolezal K, Schlereth A, Jürgens G, Alonso JM (2008) TAA1-mediated auxin biosynthesis is essential for hormone crosstalk and plant development. Cell 133:177–191

# Chapter 10

# ER Membrane Lipid Composition and Metabolism: Lipidomic Analysis


**Laetitia Fouillen, Lilly Maneta-Peyret, and Patrick Moreau**



## Abstract

Plant ER membranes are the major site of biosynthesis of several lipid families (phospholipids, sphingolipids, neutral lipids such as sterols and triacylglycerols). The structural diversity of lipids presents considerable challenges to comprehensive lipid analysis. This chapter will briefly review the various biosynthetic pathways and will detail several aspects of the lipid analysis: lipid extraction, handling, separation, detection, identification, and data presentation. The different tools/approaches used for lipid analysis will also be discussed in relation to the studies to be carried out on lipid metabolism and function.



**Key words** Ceramides, GC-FID, GC-MS, Glucosylceramides, HPTLC, LC-MS, Long-chain bases (LCB), Phospholipids, Phytosterols, Triacylglycerols


## 1 Introduction

ER membranes are known to be involved in many key cellular functions, and understanding more precisely their role, being able to analyze their lipid composition and metabolism/homeostasis is as crucial as it is for ER proteins.

Of course, to be able to perform such lipid analysis, we must handle highly purified ER membranes devoid of other membrane contaminants, something which is not so obvious as it will depend on the plant material considered. According to the starting material, different methodologies with various efficiencies in isolating highly purified ER membranes (differential ultracentrifugation, free-flow electrophoresis, two phase partition systems, affinity purification) alone or in combination have to be used ([1–3] and references therein).

Table 1 details the lipid composition of the ER membranes from different plant species and tissues that have been combined from the literature to give a simplified view of the different lipid families/species and their relative abundance. The corresponding


Verena Kriechbaumer (ed.), *The Plant Endoplasmic Reticulum: Methods and Protocols*, Methods in Molecular Biology, vol. 2772, https://doi.org/10.1007/978-1-0716-3710-4_10,

**Table 1**
**Lipid composition of plant ER membranes in mol%[a]**

| PA | lPC | PC | PE | PG | PI | PS | GlcCer | Free Sterols | Conjugated Sterols |
|---|---|---|---|---|---|---|---|---|---|
| 0.5–7 | 0.8–1.7 | 36–65 | 11–23 | 3.5–5 | 2.5–17.5 | 1–7 | 2–8 | 3–14 | 1–4 |

Generally, PA is considered and found to be low/very low in the ER membranes and in most plant cell membranes because it is an intermediary metabolite and a signaling phospholipid. So most of the time, high amounts of PA are probably due to phospholipase D activation during material handling and/or lipid extraction
*PA* phosphatidic acid, *lPC* lysoPhosphatidylCholine, *PC* phosphatidylCholine, *PE* phosphatidylEthanolamine, *PG* phosphatidylGlycerol, *PI* phosphatidylInositol, *PS* phosphatidylSerine, *GlcCer* glucosylceramides. Conjugated sterols: sum of sterol glycosides + steryl esters
[a]Data combined from several plant species and tissues ([13–15] and references therein, and unpublished results)

biosynthetic pathways of the different lipid classes of the ER membranes are presented in Fig. 1 for phospholipids, in Fig. 2 for glucosylceramides, and in Fig. 3 for sterols.

The pretty large variations in lipid amounts which can be observed (Table 1) can be due of course to the differences in the plant species/tissues but also to some extent to the purity of the ER membranes according to the isolation procedures used. Anyway, Table 1 indicates that the major lipid families/species are represented by the phospholipids, followed by the sterols (free and conjugated) and the glucosylceramides (GluCer). Other lipid species such as phosphoinositides (PIP, PIP2), acyl-CoAs, and ceramides (precursors of GluCer) are also present but in lower amounts. The levels of diacylglycerol in ER membranes (precursors of phospholipids and also of triacylglycerols for lipid bodies) are not mentioned because their levels can even be more variable.

The purpose of this chapter is not to give an extensive overview of the various lipid methodologies because this has already been done by well-recognized lipid experts in exhaustive set of publications such as in the AOCS Lipid Library (http://lipidlibrary.aocs.org) or in the *Arabidopsis* book [4] but to select some lipid technologies that were, are, and will be relevant for studying several aspects of plant ER membranes lipid metabolism.

## 2 Materials

### *2.1 Technologies/Equipment*

1. Lipid extraction-preparation/Bench centrifuge, 110 °C dry bath incubator, 110 °C resistant screwed glass tubes, bench vortex.
2. Densitometric analysis/TLC sampler Linomat 5 (Camag, Muttenz, Switzerland), TLC scanner 3 (CAMAG, Muttenz, Switzerland).

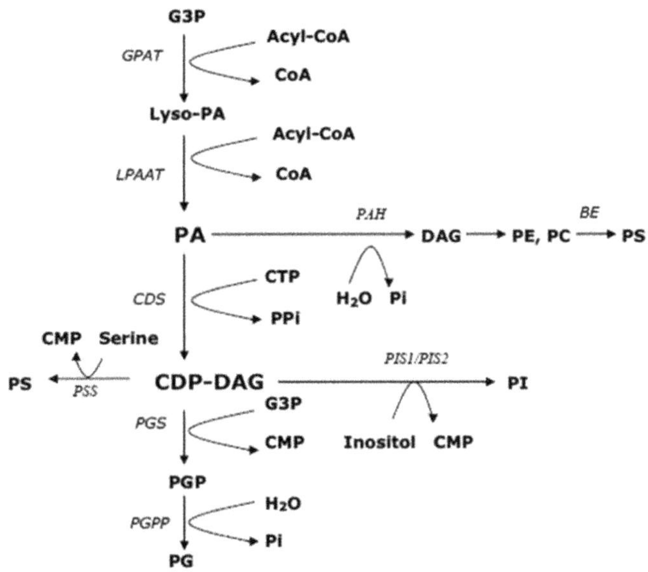


**Fig. 1** Simplified phospholipid biosynthetic pathways in the ER. CDP-DAG, cytidine diphosphate diacylglycerol; CoA, coenzyme A; CMP, cytidine monophosphate; CTP, cytidine triphosphate; DAG, diacylglycerol; G3P, glycerol-3-phosphate; Lyso-PA, lysophosphatidic acid; PA, phosphatidic acid; PC, phosphatidylcholine; PE, phosphatidylethanolamine; PG, phosphatidylglycerol; PGP, phosphatidylglycerol phosphate; PI, phosphatidylinositol; PS, phosphatidylserine. *Enzymes*: CDS, cytidine-diphosphate diacylglycerol synthase; GPAT, glycerol-3-phosphate acyltransferase; LPAAT, lysophosphatidic acid acyltransferase; PAH, phosphatidic acid phosphohydrolase; PGPP, phosphatidylglycerol phosphate phosphatase; PGS, phosphatidylglycerol phosphate synthase; PIS, phosphatidylinositol synthase; PSS, phosphatidylserine synthase

3. HPTLC (High-Performance Thin Layer Chromatography)/ ADC2 automatic developing chamber (Camag, Muttenz, Switzerland), HPTLC plates F254 (Merck, Darmstadt, Germany).
4. GC-FID (Gas Chromatography-Flame Ionization Detector)/ Agilent 7890 gas chromatograph equipped with flame ionization detection (Santa Clara, CA, USA), Agilent GC Colum DB-Wax 15 m × 0.53 mm × 1.0 μm.
5. GC-MS (Gas Chromatography-Mass Spectrometry)/Agilent 6850 gas chromatograph coupled with a MS detector MSD 5975-EI (Santa Clara, CA, USA), HP-5MS capillary column (5% phenyl-methyl-siloxane), 30 m, 250 mm, and 0.25 mm film thickness (Agilent, Santa Clara, CA, USA).
6. LC-MS (Liquid Chromatography-Mass Spectrometry)/ QTRAP 5500 (Sciex) mass spectrometer coupled to a liquid chromatography system (Ultimate 3000, Dionex). Luna C8 150x1 mm column with 100 Å pore size and 5 mm particles (Phenomenex) for phospholipids and neutral lipids analyses.

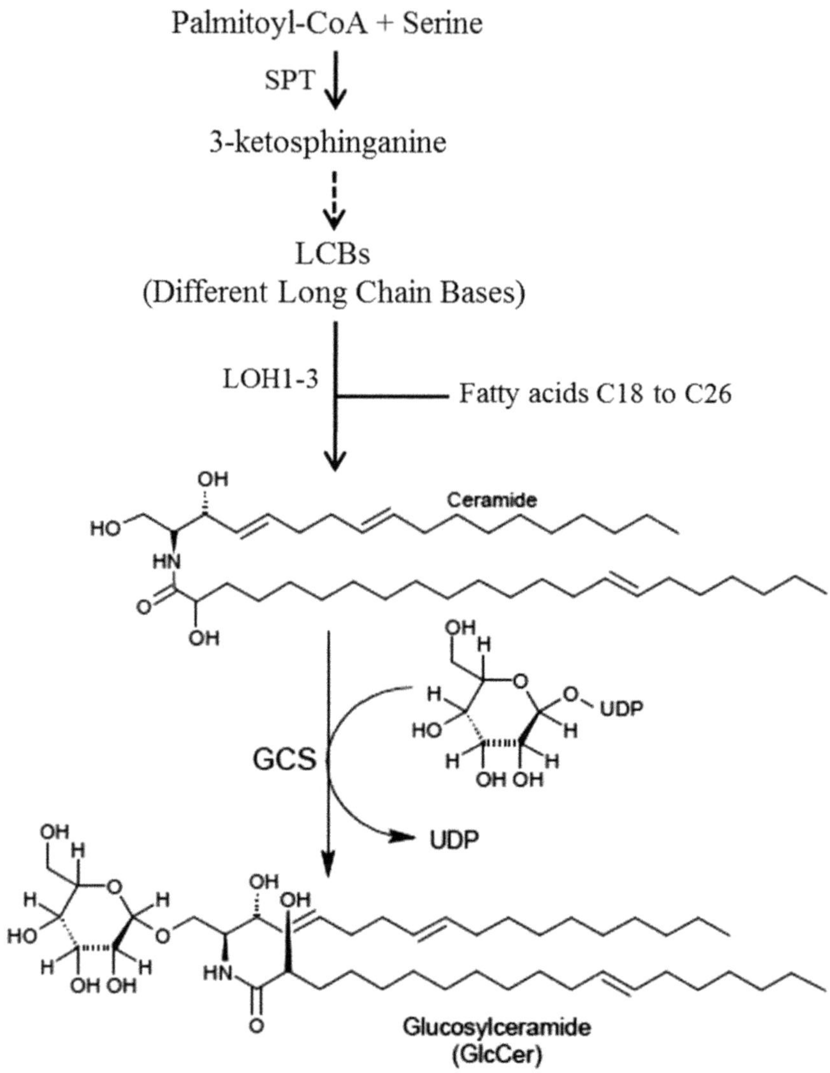


**Fig. 2** Simplified GlcCer biosynthetic pathway in the ER. GCS, glucosylceramide synthase; LOH1-3, ceramide synthases 1, 2 and 3; SPT, serine palmitoyl transferase. The dotted arrow indicates several enzymatic steps. All the enzymes involved in the biosynthesis of glucosylceramides and its precursors are located in the ER

7. LC-MS (Liquid Chromatography-Mass Spectrometry)/ QTRAP 6500 (Sciex) mass spectrometer coupled to a liquid chromatography system (1290 Infinity II, Agilent, Supercolsil ABZ+, 100 × 2.1 mm column and 5 mm particles (Supelco) for sphingolipid analyses.
8. Radiolabeling analysis/PhosphorImager (such as STORM or TYPHOON from GE Healthcare).

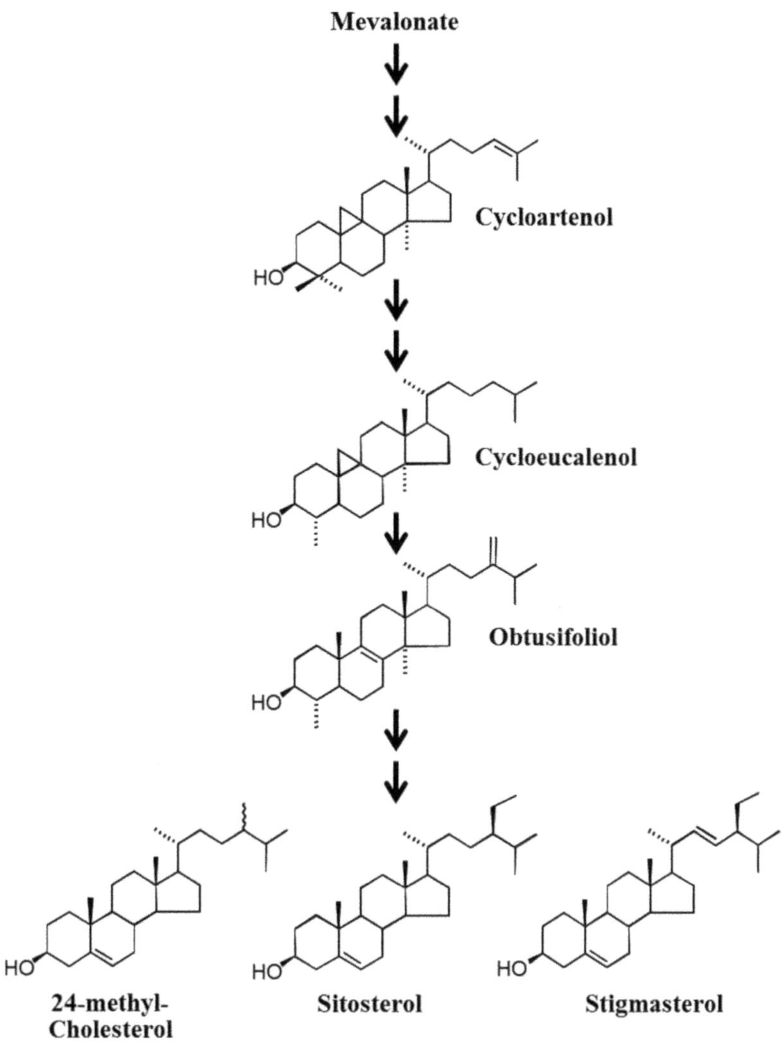


**Fig. 3** Simplified Sterol biosynthetic pathway in the ER. The two successive arrows indicate two or more enzymatic steps. The biosynthesis of the precursor mevalonate by the HMG-CoA reductase is the first step of the sterol biosynthetic pathway, which is located in the ER

### 2.2 Lipid Standards

1. Lipid standards are purchased from Avanti Polar Lipids (Alabaster, AL, USA), Nu-Chek-Prep (Elysian, MN, USA), Matreya (State College, PA, USA), and/or Sigma (St. Louis, MO, USA).
2. For HPTLC analysis, all the phospholipid standards (PC, PE, PI, PG and PS) are from soybean extracts except PA, which can be from any other source, DAG (e.g., di-18:1), TAG (e.g., tri-18:1), ceramides (bovine extract), glucosylceramides (soybean extract), or sterol (plant extract) (*see* **Note 2**).

3. For GC-FID or GC-MS analysis, C17:0, h14:0,d20:0, t17:0, d14:1$^{4(E)}$ and cholestanol.
4. For LC-MS analysis, PC, 17:0/17:0; PI, 17:0/14:1; PE, 17:0/17:0; PG, 17:0/17:0;PS, 17:0/17:0, TAG 17:0/17:0/17:0, DAG 17:0/17:0, Cer d18:1/C17:0 and GlcCer d18:1/C12:0.

### 2.3 Solutions

1. Eluent A: isopropanol/methanol/$H_2O$ 5/1/4, v/v containing 0.2% formic acid and 0.028% $NH_3$.
2. Eluent B: isopropanol containing 0.2% formic acid and 0.028% $NH_3$.
3. Eluent C: THF/ACN/5 mM ammonium formate (3/2/5 v/v/v) with 0.1% formic acid.
4. Eluent D: THF/ACN/5 mM ammonium formate (7/2/1 v/v/v) with 0.1% formic acid.
5. Eluent E: acetonitrile/methanol/$H_2O$ (19/19/2, v/v) containing 0.2% formic acid and 0.028% $NH_3$).
6. Eluent F: isopropanol containing 0.2% formic acid and 0.028% $NH_3$.

## 3 Methods

### 3.1 Lipid Extraction

Whatever the isolation procedure of the purified ER membranes (differential ultracentrifugation, free-flow electrophoresis, two-phase partition systems, affinity purification), the ER fraction will be generally obtained as a pellet after centrifugation in order to concentrate the membranes.

Analyzing lipids requires extraction using organic solvents. Extraction should be rapid in order to avoid lipid modifications during treatment. In the case of ER membranes, extraction should be carried out immediately after recovering the ER membranes; otherwise the membranes can be stored at −80 °C (but there is a risk of changes to the lipids).

1. The pellet should be suspended in a small volume (100–300 μL) of distilled $H_2O$ (if protein activity is not necessary to be measured) or in buffer, by gentle grinding rather than vortexing, and transferred into glass tubes (with Teflon protected screw caps) for lipid extraction (plastic apparatus and containers (other than Teflon) should be avoided). One part (5%) of the aliquot should be conserved for protein measurement.
2. Then extract the lipids with chloroform/methanol 2/1, v/v [5] (*see* **Note 3** for acyl-CoAs), using a ratio of chloroform, methanol, and aqueous phase in the final mixture close to 8/4/3, v/v to be efficient.

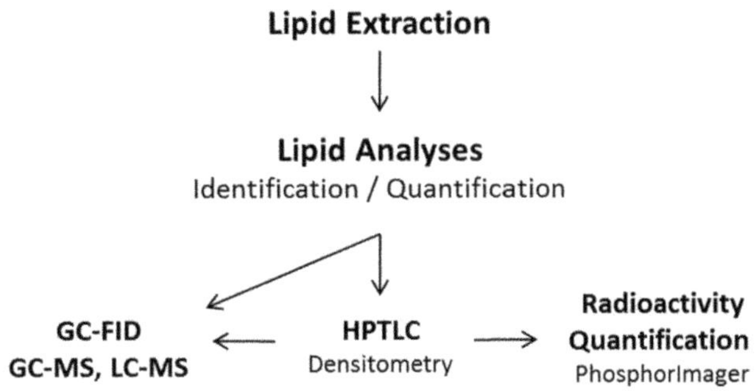


**Fig. 4** Different steps of lipid analysis processes. The principal methods for lipid class analysis are HPTLC and LC-MS and for their fatty acid content are GC-FID or GC-MS. Long-chain bases of sphingolipids, and sterols, can be analyzed by GC-MS. GC-FID, gas chromatography-flame ionization detector; GC-MS, gas chromatography-mass spectrometry; SPE, solid-phase extraction; HPTLC, high-performance thin layer chromatography

3. The solution is vortexed or shaken for 5 min, at room temperature. Centrifuge the solution (10 min 3000 rpm), and discard the upper phase (as well as interphase).
4. This organic phase is washed with ¼ of volume of $H_2O$ or 0.9% NaCl; after vortexing for 1 min, the solution is centrifuged (10 min, 3000 rpm) and the lower phase recovered in a new glass tube.
5. The organic solvent is evaporated under air (or nitrogen if particular protection of unsaturated fatty acids is needed). Resuspend the lipids into appropriate solvents required for the subsequent analysis.

The principal methods for lipid class analysis are HPTLC and LC-MS/MS and for their fatty acid content are GC-FID or GC-MS. Long-chain bases of sphingolipids and sterols can be analyzed by GC-MS (a general scheme of lipid analysis is given in Fig. 4).

### 3.2 HPTLC Analysis of Phospholipids and Acyl-CoAs

1. For analysis of phospholipids with HPTLC plates, the lipids should be resuspended into 50–200 μL of chloroform/methanol 2/1, v/v. Close the cap until use to prevent solvent loss and changes in lipid concentration.
2. Phospholipids may be analyzed by loading the extract, using a CAMAG Linomat IV, onto 10 × 10 or 10 × 20 cm HPTLC plates, which are developed in methyl acetate/n-propanol/chloroform/methanol/0.25% aqueous KCl (25/25/25/10/9, v/v) according to Heape et al. [6].
3. External phospholipid standards (10 μg per standard; *see* **Note 2**) are loaded on the same plates as the samples to be analyzed.

4. The migration may be carried out into a simple developing chamber (20 min saturation/45 min migration/45 min drying; developing solvent volume 20 mL) or into an ADC2 automatic developing chamber (developing solvent volume 10 mL; program, migration distance 85 mm; drying time 5 min; saturation time 0 min).
5. Lipids can be detected by dipping the plates into a 3% copper acetate (w/v) in 8% aqueous phosphoric acid solution [7, 8], followed by heating at 110 °C during 30 min.
6. Densitometric analysis of phospholipids must then be performed in the following hour (stability of coloration), using, for example, a TLC scanner 3 at 366 nm. Lipids are quantified (μg of lipids) by comparing lipids and standard lipids peak areas (*see* **Note 2**).
7. If analysis of the fatty acid content is needed, the lipids from the HPTLC plates are only labeled by iodine vapors. Place the chromatogram 5–15 min into a strong iodine atmosphere (into a covered beaker in which some iodine resublimed crystals have been placed); double bonds containing lipids will be stained in yellow over a few minutes. The desired lipid bands are scraped off and transferred into glass tubes (with Teflon covered screw caps) for further analysis by GC-FID or GC-MS.
8. For HPTLC analysis of acyl-CoAs, the lipids should be resuspended into 50–200 μL of chloroform/methanol/$H_2O$, 3/3/1, v/v.

### 3.3 GC-FID Analysis of Fatty Acids from Phospholipids

1. Fatty acid methyl esters (FAMEs) are prepared by resuspending the scraped lipid bands in 1 mL of 2.5% $H_2SO_4$ (v/v) in methanol using heptadecanoic acid 2 μg/mL as internal standard.
2. Tubes are heated at 80 °C for 1 h and cooled to room temperature.
3. Then 400 μL hexane and 1.5 mL $H_2O$ are added to extract fatty acid methyl esters.
4. The tubes are vortexed vigorously for at least 1 min and centrifuged 10 min at 2000 rpm with a bench centrifuge, and the organic (upper) phase is transferred (preferentially with a syringe) to injection GC vials.
5. GC-FID can be performed using an Agilent 7890 gas chromatograph equipped with an Agilent GC Colum DB-Wax 15 m × 0.53 mm × 1.0 μm and flame ionization detection, with helium as carrier gas at a flow rate of 3 mL/min (4 psi).
6. The temperature gradient conditions can be 160 °C for 1 min, increased to 190 °C at 20 °C/min, increased to 210 °C at 5 °C/min, and then remaining at 210 °C for 5 min.

7. FAMES are identified by comparing their retention times with commercial fatty acid standards and quantified using ChemStation to calculate the peak areas and then comparing them with the response of an internal standard (C17:0).

### 3.4 Phosolipidomic Analysis by LC-MS

1. For the analysis of ceramides and glucosylceramides by LC-MS, lipid extracts were incubated 1 h at 50 °C in 2 mL of methylamine solution (7 mL methylamine 33% [w/v] in EtOH combined with 3 mL of methylamine 40% [w/v] in water) (Sigma-Aldrich) in order to hydrolyze phospholipids. After incubation, methylamine solutions were dried at 40 °C under a stream of air and, finally, were resuspended into 100 μL of THF/MeOH/H2O (40:20:40, v/v) with 0.1% formic acid containing synthetic internal lipid standards (Cer d18:1/C17:0 and GluCer d18:1/C12:0, Avanti Polar Lipids) added, thoroughly vortexed, incubated at 60 °C for 20 min, sonicated for 2 min, and transferred into LC vials.
2. LC-MS/MS (multiple-reaction monitoring mode) analyses were performed with a model QTRAP 6500 (ABSciex) mass spectrometer coupled to a liquid chromatography system (1290 Infinity II, Agilent). Analyses were performed in the positive mode. Nitrogen was used for the curtain gas (set to 30), gas 1 (set to 30), and gas 2 (set to 10). Needle voltage was at +5500 V with needle heating at 400 °C. The multiple-reaction monitoring mode is based on the loss of the LCB part. The de-clustering potential was adjusted at 10 for Cer and GluCer; collision energy is set to 35 eV for Cer and 40 eV for GluCer.

Reverse-phase separations were performed at 40 °C on a Supercolsil ABZ+, 100 × 2.1 mm column and 5 mm particles (Supelco). During the run, the proportions of eluent C and eluent D vary: 0–1 min, 1% D; 40 min, 80% D; 40–42 min, 80% D. The flow rate is set up at 0.2 mL/min, and 5 μL sample volumes are injected. The areas of LC peaks are determined using the MultiQuant software for sphingolipid quantification [11].

### 3.5 Neutral Lipid Analysis by HPTLC and GC-FID

To isolate and quantify neutral lipids (including phytosterols, DAG, FFA, TAG, and steryl esters), the lipids, resuspended into 50–200 μL of chloroform/methanol 2/1, v/v, are loaded onto 10 × 10 or 10 × 20 cm HPTLC plates developed with hexane/ethylether/acetic acid (90/15/2, v/v) as described [8, 12]. External neutral lipid standards (10 μg per standard; *see* **Note 2**) are loaded on the same plates than the samples to be analyzed. The migration may be carried out into a simple developing chamber (10 min saturation/20 min migration/30 min drying; developing solvent volume 20 mL) or into an ADC2 automatic developing chamber (developing solvent volume 10 mL; program: migration distance 85 mm; drying time 5 min; saturation time 0 min). Quantification by densitometry can be performed as explained for phospholipids in Subheading 3.2.

For GC-FID analysis of fatty acids from fatty acid-containing neutral lipids, FAMEs are analyzed and quantified as described in Subheading 3.2.

### 3.6 Phytosterol Analysis by GC-MS

Phytosterols are prepared by resuspending the corresponding scraped lipid bands in 1 mL of ethanol containing the internal standard α-cholestanol (25 mg/mL) in screwed glass tubes. For saponification, add 100 μL of 11 N KOH (56.11 g/mol, 12.3 g/ 20 mL $H_2O$), close the tube(s) under nitrogen, incubate in a dry bath incubator at 80 °C for 1 h, add 1 mL of hexane, vortex for 15 s, add 2 mL of distilled $H_2O$, vortex for 15 s, centrifuge 5 min at 2000 rpm in a bench centrifuge, and recover the phytosterol-containing upper phase in a new screwed glass tube, and evaporate the solvent under an N2 gas stream. For silylation of phytosterols, add 200 μL of BSTFA/TMCS, incubate 15 min at 110 °C in a dry bath incubator, and evaporate the solution to dryness under nitrogen.

For GC-MS analysis: add 200 μL of hexane before injection on GC-MS, and use the same program as described in Subheading 3.3.

### 3.7 LC-MS Analysis of DAG and TAG

For the analysis of DAG and TAG by LC-MS, phospholipid extracts were dissolved in 100 μL of eluent C; synthetic internal lipid standards (DAG 17:0/17:0; TAG, 17:0/17:0/17:0) were added at different amounts adapted for the detection.

LC-MS/MS (multiple-reaction monitoring mode) analysis is performed with a model QTRAP 5500 mass spectrometer coupled to a liquid chromatography system Ultimate 3000. The multiple-reaction monitoring mode is based on the loss of the fatty acid part during the fragmentation.

Analyses were achieved in positive mode, and nitrogen was used as curtain gas (set to 30), gas1 (set to 25), and gas2 (set to 0). Needle voltage was at +5500 V without needle heating. The collision gas was also nitrogen. The multiple-reaction monitoring mode is based on the loss of the fatty acid part of the TAG or DAG molecules. The de-clustering potential was set at +40 V for TAG and +86 V for DAG; collision energy was adjusted at +34 V for TAG and DAG. The dwell time was set to 3 ms. Reversed-phase chromatography was carried out at 30 °C using a Luna 3u C8 150 × 1 mm column, with 100 Å pore size, 3 μm particles. Eluent E and eluent F were then used as follows. The gradient elution program was 0–5 min, 15% F; 5–35 min, 15–40% F; 35–50 min, 55% F; 51–58 min, 80% F; 59–70 min, 15%F. The flow rate was 40 μL/min, and injection volume was 3 μL. The relative levels of DAG and TAG species were determined using MultiQuant and normalizing to the area of the DAG or TAG internal standards.

### 3.8 Radiolabeling In Vivo

Radiolabeled lipid precursors can be used for incubation with plant cells/tissues to investigate ER membrane lipid metabolism in vivo. According to the starting material, ER membranes can be isolated through different methodologies (see Introduction). Subsequently, ER membrane lipids are then extracted and washed as detailed in Subheading 3.1. Polar and neutral lipids can then be separated onto HPTLC plates as described in the corresponding paragraphs. Their radiolabeling can then be determined using a PhosphorImager equipment. HPTLC plates are generally exposed to the adapted screens for 24–48 h, and then the screens are scanned with the PhosphorImager. The ImageQuant software translates the radioactivity on HPTLC plates into peaks and gives a surface area for each. Then, from a calibrating curve (made with any radioactive standard) defining the correspondence between peak surface area and labelling amount, we can calculate for each lipid the level of labeling, and from the specific radioactivity of the lipid precursor used, we can determine the amount of each lipid in moles.

### 3.9 Radiolabeling In Vitro

Radiolabeled lipid precursors can also be used for incubation with ER membranes to investigate in vitro the enzymatic activity of ER membrane enzymes involved in lipid metabolism. According to the starting material, ER membranes can be isolated through different methodologies (see Introduction). Subsequently, ER membrane lipids are then extracted and washed as detailed in Subheading 3.1. Polar and neutral lipids can then be separated onto HPTLC plates as described in Subheading 3.2. Their radiolabeling can then be determined using a PhosphorImager equipment. The procedure determining the formed products amounts in moles is described above.

## 4 Notes

1. All chemicals and solvents were either of analytical or mass spectrometric grades. Prepare all solutions using deionized $H_2O$ (18.2 M Ω cm at 25 °C), for example, produced using a Synergy UV Millipore System. Follow all disposal regulations when disposing waste material.
2. For HPTLC analysis, each of the lipid classes should be calibrated with appropriate standards, i.e., with a similar fatty acid degree of unsaturation to the samples because the revelation with both cupric acetate/phosphoric acid or iodine vapor solutions are sensitive to the degree of fatty acid unsaturation.
3. If acyl-CoAs are required, lipids should be extracted with 6 times the sample's volume of chloroform/methanol 1/1 in order to obtain an homogeneous phase permitting to conserve the more polar molecules such as the acyl-CoAs.
4. Be extremely careful that all tube caps contain a Teflon disk.

5. Standards (e.g., d20:0, t17:0 and d14:$1^{4(E)}$) should be added in the dioxane. Each standard initial concentration (1 g/L in methanol) should be added at 5 μg/mL final, i.e., 10 μL in the 2 mL of barium hydroxide + dioxane.
6. The $NaIO_4$ stock concentration should be prepared only freshly.

Equivalent instrumentation can be used.

## Acknowledgments

Bordeaux Metabolome Facility, MetaboHUB (ANR-11-INBS-0010).

## References

1. Morré DJ, Panel C, Morré DM, Sandelius AS, Moreau P, Andersson B (1991) Cell-free transfer and sorting of membrane lipids in spinach. Donor and acceptor specificity. Protoplasma 160:49–64
2. Parsons HT, Drakakaki G, Heazlewood JL (2013) Proteomic dissection of the Arabidopsis Golgi and trans-Golgi network. Front Plant Sci 3:298
3. Parsons HT, Fernández-Niño SM, Heazlewood JL (2014) Separation of the plant Golgi apparatus and endoplasmic reticulum by free-flow electrophoresis. Methods Mol Biol 1072: 527–539
4. Li-Beisson Y, Shorrosh B, Beisson F, Andersson MX, Arondel V, Bates PD, Baud S, Bird D, DeBono A, Durrett TP, Franke RB, Graham IA, Katayama K, Kelly AA, Larson T, Markham JE, Miquel M, Molina I, Nishida I, Rowland O, Samuels L, Schmid KM, Wada H, Welti R, Changcheng X, Zallot R, Ohlrogge J (2013) Acyl-lipid metabolism. In: The Arabidopsis book, vol 11, p e0161
5. Folch J, Lees M, Stanley GHS (1957) A simple method for the isolation and purification of total lipides from animal tissues. J Biol Chem 226:497–509
6. Heape AM, Juguelin H, Boiron F, Cassagne C (1985) Improved one dimensional thin layer chromatographic technique for polar lipids. J Chromatogr 332:391–395
7. Macala LJ, Yo RK, Ando S (1983) Analysis of brain lipids by high performance TLC and densitometry. J Lipid Res 24:1243–1250
8. Laloi M, Perret AM, Chatre L, Melser S, Cantrel C, Vaultier MN, Zachowski A, Bathany K, Schmitter JM, Vallet M, Lessire R, Hartmann MA, Moreau P (2007) Insights into the role of specific lipids in the formation and delivery of lipid microdomains to the plasma membrane of plant cells. Plant Physiol 143: 461–472
9. Cacas JL, Melser S, Domergue F, Joubès J, Bourdenx B, Schmitter JM, Mongrand S (2012) Rapid nanoscale quantitative analysis of plant sphingolipid long-chain bases by GC-MS. Anal Bioanal Chem 403:2745–2755
10. Hillig I, Leipelt M, Ott C, Zahringer U, Warnecke D, Heinz E (2003) Formation of glucosylceramides and sterol glucoside by a UDP-glucose-dependent glucosylceramide synthase from cotton expressed in Pichia pastoris. FEBS Lett 553:365–369
11. Ito Y, Esnay N, Platre MP, Wattelet-Boyer V, Noack LC, Fougère L, Menzel W, Claverol S, Fouillen L, Moreau P, Jaillais Y, Boutté Y (2021) Sphingolipids mediate polar sorting of PIN2 through phosphoinositide consumption at the trans-Golgi network. Nat Commun 12(1):4267. https://doi.org/10.1038/s41467-021-24548-0
12. Juguelin H, Heape MA, Boiron F, Cassagne C (1986) A quantitative developmental study of neutral lipids during myelinogenesis in the peripheral nervous system of normal and trembler mice. Brain Res 390:249–252
13. Brown DJ, Dupont FM (1989) Lipid composition of plasma membranes and endomembranes prepared from roots of barley *(Hordeum vulgare* L.): effects of salt. Plant Physiol 90:955–961
14. Grindstaff KK, Fielding LA, Brodl MR (1996) Effect of gibberellin and heat shock on the lipid composition of endoplasmic reticulum in barley aleurone layers. Plant Physiol 110:571–581
15. Moreau P, Bessoule JJ, Mongrand S, Testet E, Vincent P, Cassagne C (1998) Lipid trafficking in plant cells. Progr Lipid Res 37:371–339

# Chapter 11

# 2 in 1 Vectors Improve in Planta BiFC and FRET Analysis

**Dietmar Mehlhorn, Niklas Wallmeroth, Kenneth W. Berendzen, and Christopher Grefen**

## Abstract

Protein-protein interactions (PPIs) play vital roles in all subcellular processes, and a number of tools have been developed for their detection and analysis. Each method has its unique set of benefits and drawbacks that need to be considered prior application. In fact, researchers are spoilt for choice when it comes to deciding which method to use for the initial detection of a PPI and which to corroborate the findings. With constant improvements in microscope development, the possibilities of techniques to study PPIs in vivo, and in real time, are continuously enhanced and expanded. Here, we describe three common approaches, their recent improvements incorporating a 2-in-1 cloning approach, and their application in plant cell biology: ratiometric bimolecular fluorescence complementation (rBiFC), FRET acceptor photobleaching (FRET-AB), and fluorescent lifetime imaging (FRET-FLIM), using *Nicotiana benthamiana* leaves and *Arabidopsis thaliana* cell culture protoplasts as transient expression systems.

**Key words** Protein-protein interaction, rBiFC, FRET, Acceptor photobleaching, FLIM, Gateway, SEC61, 2 in 1

## 1 Introduction

Core cellular processes such as signal perception and transduction, vesicle trafficking, transport activities, and metabolic pathways rely on formation of complex protein networks. Analysis of the biochemical mechanisms at the molecular level requires a fundamental understanding of the protein-protein interactions (PPIs) involved. Over the past decades, a number of techniques such as Co-immunoprecipitation (CoIP), the Yeast-Two Hybrid (Y2H), the Split Ubiquitin System (SUS), Förster Resonance Energy Transfer (FRET), and related methods, as well as Bimolecular Fluorescence Complementation (BiFC), have become routine laboratory techniques to dissect PPIs (reviewed in (1)). Technological advancements in microscope development have enhanced possibilities further by facilitating real/life-time imaging of PPIs. Even techniques that

Verena Kriechbaumer (ed.), *The Plant Endoplasmic Reticulum: Methods and Protocols*, Methods in Molecular Biology, vol. 2772,
https://doi.org/10.1007/978-1-0716-3710-4_11,

were thought to be predominantly in vitro techniques such as CoIP are now applicable for microscopy analysis at the single-molecule level (2).

Recently, our lab has introduced 2-in-1 cloning, a Gateway-compatible approach to simultaneously clone two genes-of-interest into two independent expression cassettes on the same T-DNA/plasmid (3) (Fig. 1a). This cloning system is well suited for PPI techniques that rely on transient transfection systems, as a mixture of two (or more) independent agrobacteria, each carrying their own plasmid, leading to unequal gene dosage, resulting in high variability of co-expression (4). We have combined the 2-in-1 cloning system with bimolecular fluorescence complementation (3) as well as with a range of suitable fluorophores for enhanced FRET/FLIM analysis (4).

This chapter describes the application of the optimized 2 in 1 vector sets that can be used for ratiometric bimolecular fluorescence complementation (rBiFC), FRET-acceptor photobleaching (FRET-AB), and fluorescent lifetime imaging (FRET-FLIM), using either *Nicotiana benthamiana* leaves or *Arabidopsis thaliana* cell culture protoplasts as expression system (3, 4).

### 1.1 Ratiometric Bimolecular Fluorescence Complementation (rBiFC)

Bimolecular fluorescence complementation (BiFC) is a protein fragment complementation analysis whose first application was in bacteria (5). The principle of BiFC is straightforward: a previously split fluorescent protein (FP) reconstitutes due to an interaction of two proteins of interest that are fused to the complementary non-fluorescent fragments (Fig. 1b). One major advantage of BiFC is the applicability of the system to almost all types of proteins (cytosolic, nuclear, organellar, or membrane bound) and the detection of interaction in an in vivo context. It also allows visualization of PPIs at their site of first interaction—which should not be confused with the subcellular position where these proteins normally reside or are trafficked to, a misconception that is often inferred. Nevertheless, the flaws of the technique almost match the advantages. Overexpression of fusion proteins as well as their FP fragment tag can lead to mis-localization and alter the likelihood of a positive readout. Reassembly of the FP is irreversible, promoting and stabilizing weak or transient interactions but thereby also causing artefactual results even in the absence of a true interaction (1, 6). However, the biggest disadvantage in conventional BiFC systems used in the plant field is the use of two individual plasmids and their co-infiltration in transient transformation. This leads to very high variability in gene dosage and only about 70–80% of co-expression ratios (4). In other words, out of four or five cells analyzed, one does not contain both fusion proteins present, making a meaningful interpretation of the results very hard. In addition, classical BiFC constructs do not contain concurrent

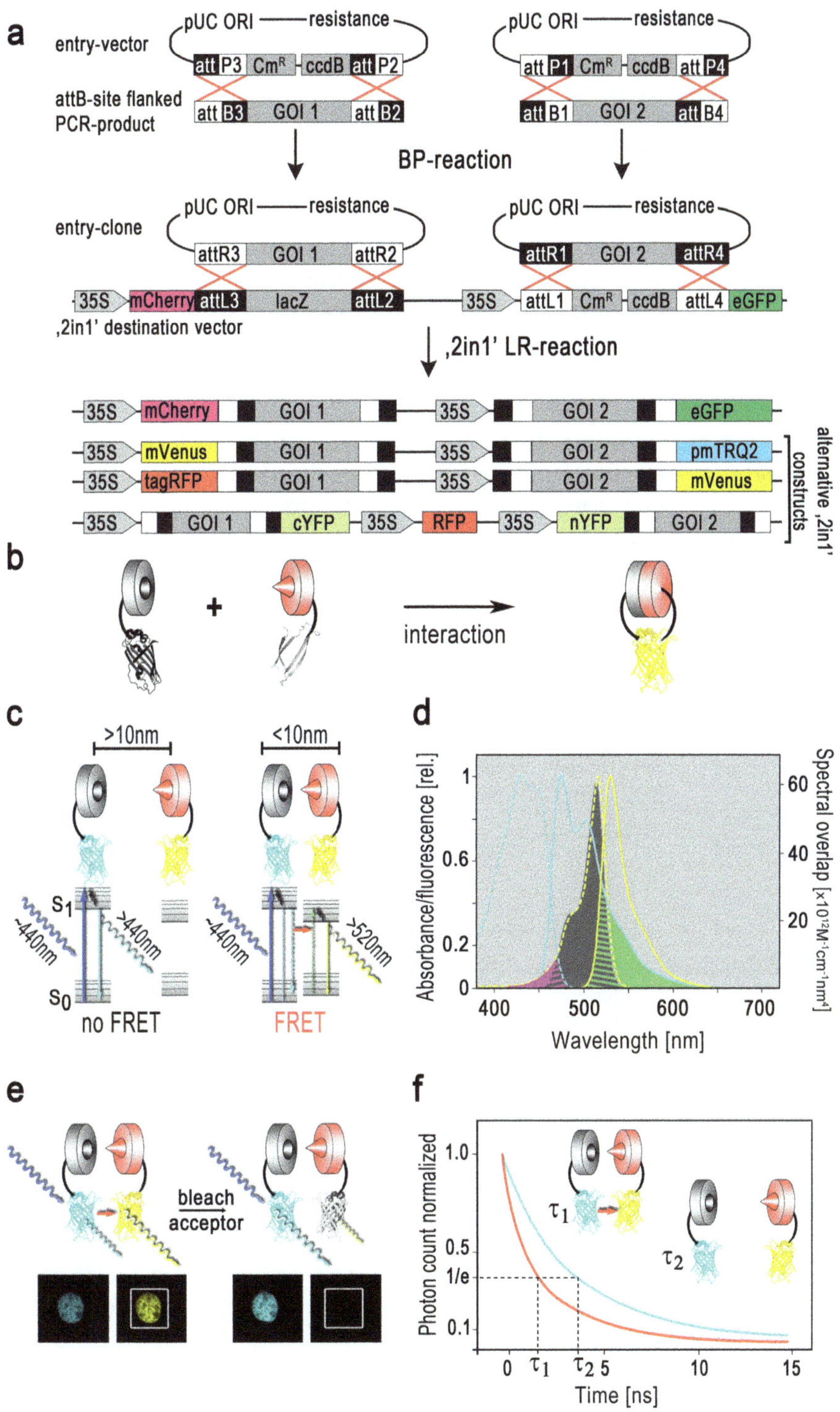


**Fig. 1** Concepts of 2-in-1 cloning, bimolecular fluorescence complementation, and Förster resonance energy transfer (**a**) Site-specific BP recombination of PCR products and Entry vectors generates Entry clones. A subsequent one-step LR reaction recombines the two genes of interest (GOI) simultaneously in a 2-in-1 destination vector. Depicted are the 2-in-1 vectors used in this chapter. See Table 1 for details. 35S,

reference markers, making quantification against presumed negative controls (which could be non- or unequally transformed cells) a cherry-picking exercise.

Contrary to classical BiFC, the ratiometric BiFC (rBiFC) approach addresses most of these issues and provides means for internal, ratiometric quantification of results (3). Placing both FP fragments on the same T-DNA guarantees equal gene dosage of both non-fluorescent fusion proteins, and the inclusion of a soluble RFP marker provides a readout for ratiometric analysis and transformation control of the cell under study (3) (Fig. 2). The technique has since been used in a number of studies for the detection of PPIs or their structure-function analysis (7–13) (*see* **Note 1**).

### *1.2 Förster Resonance Energy Transfer-Acceptor Bleaching (FRET-AB)*

An alternative, more reliable, yet sophisticated technique exploits the physical phenomenon of resonance energy transfer, which was first postulated by Theodor Förster in 1946 and has been named after him as commemoration (14) (Fig. 1c–f). The principle of FRET is a non-radiative energy transfer from an excited donor fluorophore to an acceptor molecule—often, but not necessarily, fluorescent itself. Prerequisite is an overlap of the donor emission with the excitation spectrum of the acceptor. FRET only occurs if donor and acceptor are in close enough proximity of 10 nm or less (15, 16). Such resolution of molecular distances is more than a magnitude lower than the diffraction limit of light microscopy at 200 nm, allowing distinction of protein co-localization from interaction.

**Fig. 1** (continued) Cauliflower mosaic virus 35S promoter and omega translational enhancer sequence; pUC ORI, origin of replication; $Cm^R$, chloramphenicol acetyltransferase resistance gene; ccdB, gyrase inhibitor gene; lacZ, lacZ expression cassette. (**b**) Cartoon depicting the concept of bimolecular fluorescence complementation (BiFC). Two proteins (gray and red) are tagged with non-fluorescent YFP fragments, nYFP and cYFP, respectively. Interaction of both proteins leads to reconstitution of fluorescent YFP. (**c**) Simplified Jablonski diagram depicting the excitation of an electron in a cyan-fluorescing protein (CFP). After internal conversion (black wavy arrow) to the S1 ground state (cyan solid arrow), a crossing of the energy barrier between S1 and S0 can cause the emission of a photon at a longer wavelength (red shifted, Stokes shift, i.e., fluorescence). If an acceptor molecule is in close enough proximity (here, YFP attached to a red-colored interacting protein), the energy can be transferred non-radiatively to the acceptor molecule. (**d**) Absorbance/fluorescence spectrum of mTurquoise2 and mVenus. The dark gray surface depicts the $\lambda^4$ weighted overlap integral. Acceptor and donor spectral bleed through are shown in magenta and green, respectively. (**e**) Example of a FRET acceptor photobleaching experiment. A nucleus expressing two interacting proteins attached to a donor and an acceptor fluorophore. After bleaching (right), the acceptor fluorescence is almost completely lost, but an increase in donor fluorescence can be detected. (**f**) Exemplary fluorescence lifetime decay curve of a putative donor molecule when FRET is occurring (red solid curve; t1) or in the absence of FRET (cyan solid line; t2). Fluorescence lifetime t is the average time that a fluorescent protein resides in the excited state and at which fluorescence intensity decreased to 1/e of its initial value. (**b**–**f**) Modified from (1, 4)

**Fig. 2** Method handling. (**a**) Infiltration of 3–4-week-old *N. benthamiana* leaves using a syringe as described in Subheading 3.2. (**b**) White arrows point at areas of infiltration at the abaxial side of the leaves. (**c**) The first two youngest (top) leaves of a 4–6-week-old *N. benthamiana* plant should *not* be used for infiltration. Instead, slightly older leaves (white arrows) are ideal with respect to suitability for injection and transgene expression (24). (**d**) "Tapping method" for gentle mixing of protoplasts with DNA. (**e**) Preparation of a *N. benthamiana* leaf disc for microscopic analysis. The syringe is filled with water, which, by application of a vacuum, replaces the air in the intercellular space of the leaf epidermis enhancing image quality

**Table 1**
**Available 2-in-1 expression vector sets for rBiFC and FRET experiments[a]**

| | Origin of replication | | Resistance marker | | | | |
|---|---|---|---|---|---|---|---|
| Name | *E. coli* | *A. tumefaciens* | Bacteria | Plant | Promoter | FPs[a] | Reference |
| pBiFC-2in1 | ColE1 | pVS1 | Spectinomycin | None | CaMV35S | nYFP, cYFP, RFP[b] | (3) |
| pFRETcg-2in1 | ColE1 | pVS1 | Spectinomycin | Basta | CaMV35S | mCherry, mEGFP | (4) |
| pFRETvt-2in1 | ColE1 | pVS1 | Spectinomycin | Basta | CaMV35S | mVenus, mTRQ2 | (4) |

[a]Plasmids exist in all possible tag combinations of fluorescent proteins (FPs)
[b]RFP fluorescence serves as transformation control and ratiometric marker

In fluorescence microscopy, the FRET phenomenon can be detected using different approaches. Classical FRET approaches simply measure fluorescence of the acceptor when the donor is excited. However, this can lead to artefactual results due to spectral bleed through, a phenomenon deriving from an overlap of excitation spectrum of the donor with the emission spectrum of the acceptor, which—unfortunately—is a common property with most current FRET FP couples.

An alternative, more credible method based on FRET is "acceptor photobleaching" (AB). Here, the intensities of the donor fluorophore before and after photobleaching of the acceptor are measured in a specified region of interest (ROI). Bleaching of the acceptor leads to an increase in donor fluorescence in the ROI as the energy that would have been transferred to the acceptor remains within the donor, leading to increased fluorescence of the latter.

AB does not require high-end microscopes and can be carried out at any fluorescence microscope, which is an advantage compared to fluorescence lifetime imaging microscopy (FLIM) measurements (15) (see below). However, photobleaching in AB can potentially photo-damage the samples and is therefore not suitable in prolonged time-lapse experiments. Other pitfalls are concomitant bleaching of acceptor and donor, as well as movement of the cytoplasm, both of which may lead to unfeasible results.

### 1.3 Fluorescence Lifetime Imaging Microscopy (FLIM)

Classical FRET or FRET-AB are both intensity-based methods and—as such—can be compromised by suboptimal fluorophore concentration ratios. Measurement of fluorescence lifetime, however, might overcome this issue (16, 17). Briefly, irradiation leads to raising of the FP to an excited—$S_1$—state before relaxation to the ground, $S_0$, state. The average time the fluorophore resides in the excited state represents its fluorescence lifetime and is unique for each FP (18). When an acceptor molecule is in close enough range ($<10$ nm), it can serve as energy sink, resulting in a lifetime "decay" of the donor molecule (i.e., the donor population fluoresces for a shorter time when the acceptor is present). Such reduction in lifetime can be compared to negative controls or "donor only" samples, which inform on the original lifetime of an FP—the difference between the two informs on PPI (Fig. 3).

Still, while small imbalances in concentration of donor and acceptor do not affect FLIM measurements, the lifetime of fluorophores is affected by other factors such as changes in redox state, pH, or temperature. Yet, FRET-FLIM delivers more credible results as it is almost unaffected by concentration differences and spectral properties compared to FRET AB, but it requires sophisticated and expensive equipment as well as considerable training and experience of the scientist who handles it.

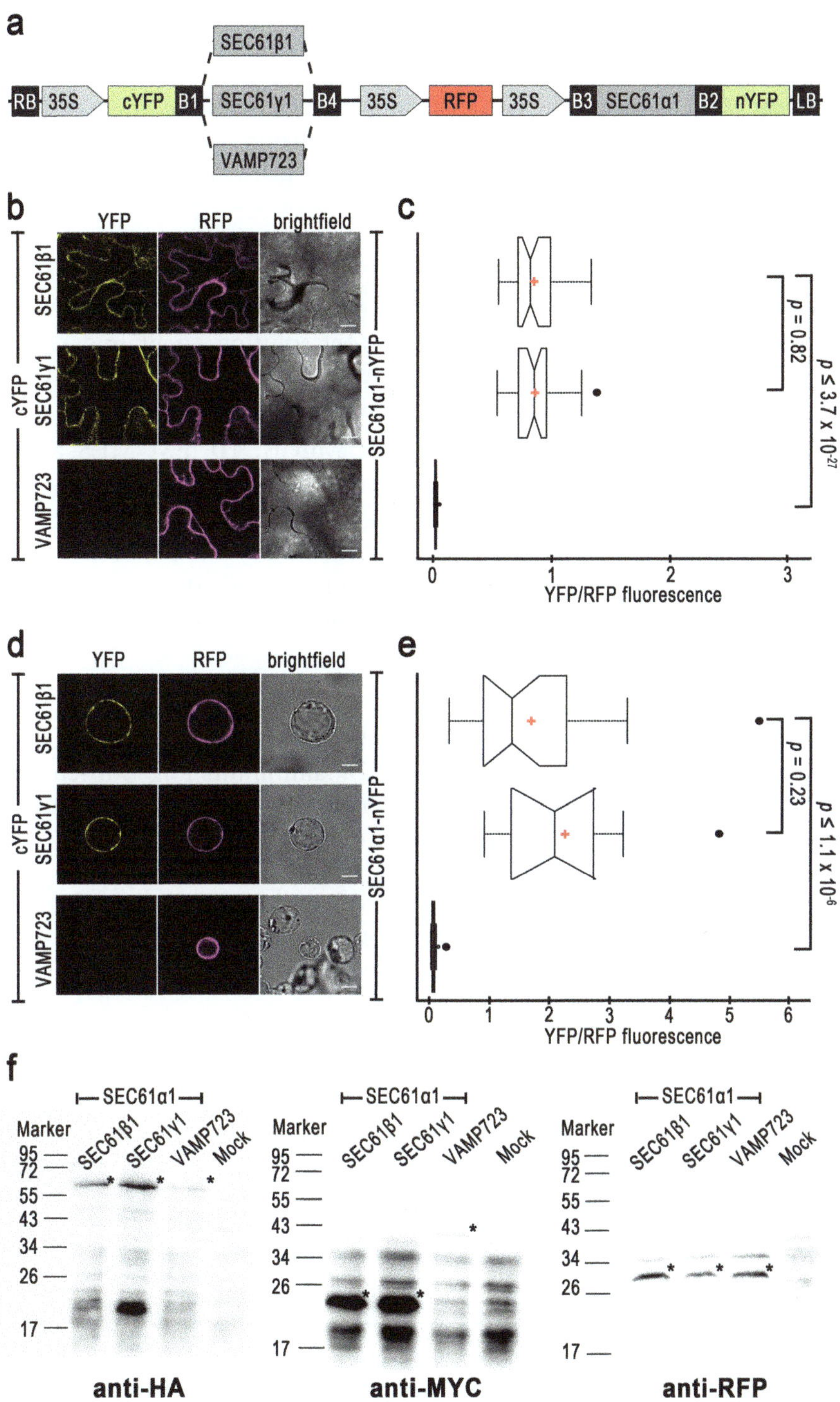


**Fig. 3** rBiFC analysis of putative *Arabidopsis* SEC61 subunits in *N. benthamiana* leaves and *A. thaliana* protoplasts. (**a**) Cartoon of the "2-in-1" rBiFC destination constructs used in this experiment (3). (**b**) Representative images of *N. benthamiana* leaves expressing SEC61α-nYFP, RFP, and either cYFP-SEC61β, cYFP-SEC61γ, or, cYFP-VAMP723, respectively. (**c**) At least 40 images were recorded per construct, and YFP/RFP mean fluorescence intensity was calculated. Box plot depicts the median with outer limits at the 25th

## 2 Materials

### 2.1 Bacteria and Plants

For transient transformation of *Nicotiana benthamiana* leaves via infiltration, *Agrobacterium tumefaciens* strain GV3101 pMP90 carrying the selection markers for growth on rifampicin and gentamicin was chosen (19, 20).

### 2.2 Media and Reagents (Tobacco Leaf Infiltration)

1. Luria-Bertani (LB) medium: 1% tryptone, 0.5% yeast extract, 0.5% NaCl, pH 7.0–7.5 (for solid media add 2% agarose).
2. Antibiotics (1000 × stock solution): gentamicin (20 mg/mL in distilled water), spectinomycin (100 mg/mL in distilled water), and rifampicin (50 mg/mL in DMSO).
3. AS medium: 10 mM $MgCl_2$, 10 mM MES-KOH, pH 5.6, 150 μM acetosyringone.
4. Sterile distilled water.

### 2.3 Media and Reagents (Protoplasts)

#### 2.3.1 Sterile Filtration

The most economic sterile filtration method is to acquire a peristaltic pump and use membrane filters with a Swinnex filter holder (Millipore).

#### 2.3.2 Sterile Erlenmeyer Flasks for Cell Culture

250 mL Erlenmeyer flasks are stoppered with Rotilabo®-culture plugs and covered well by aluminum foil. Sub-culturing is performed under highly sterile conditions, and the flasks are flamed when opened to prevent contamination.

#### 2.3.3 Pipetting of Protoplasts

We use 200 μL cut tips to handle protoplasts and try to aim for a diameter of 2–3 mm. Protoplasts are extremely sensitive to shearing stress, and larger-tip pore sizes circumvents this.

#### 2.3.4 Cell Culture Maintenance Media

1. MS Col Medium (for 1 L): 4.3 g/L Murashige and Skoog basal salt mixture (MS), 5 mL of NPT-Vitamin Stock, 100 mg/mL *myo*-inositol, 30 g/L sucrose, pH 5.8 with KOH. After autoclaving and before use, add 2,4-D to 1 mg/L. Store at 4 °C in the dark.

**Fig. 3** (continued) and 75th percentile, respectively. Notches indicate the 95% confidence intervals; Tukey whiskers extend to the 1.5× IQR, and outliers are depicted as black dots. Red crosses mark sample means of each dataset. Exemplary *p*-values are indicated in graph. (**d**) Representative images of *Arabidopsis* protoplasts expressing the same constructs. (**e**) At least ten protoplasts were recorded per construct and YFP/RFP mean fluorescence intensity calculated. Boxplot as before. (**f**) Immunoblot analysis of HA-tagged SEC61α1, MYC-tagged proteins of interest, and soluble RFP expressed in *Arabidopsis* protoplasts used for rBiFC analysis. Asterisks indicate expected protein size. Scale bar in (**b**, **d**) = 10 μm

2. NPT-Vitamin Stock: 1 mg/mL nicotinic acid, 1 mg/mL pyridoxine-HCl, 10 mg/mL thiamine-HCl in $dH_2O$. Sterilize by filtration (0.22 μm), and store at −20 °C.
3. 2,4-D Stock (for 100 mL): Dissolve 100 mg in 10 mL absolute ethanol, and bring to 100 mL with $dH_2O$. Concentration is 1 mg/mL. Sterilize by filtration (0.22 μm), and store at −20 °C.

### 2.4 Media for Protoplast Generation

1. Wall digestion solution without enzymes: 8 mM $CaCl_2 \times 2H_2O$, 0.4 M mannitol, pH 5.5 with KOH. Sterilize by filtration (0.22 μm).
2. Wall digestion solution: 1% cellulase, 0.25% macerozyme, 8 mM $CaCl_2 \times 2H_2O$, 0.4 M mannitol, pH 5.5 with KOH. Sterilize by filtration (0.22 μm), use immediately, or store at −20 °C.
3. W5: 154 mM NaCl, 125 mM $CaCl_2$, 5 mM KCl, 5 mM glucose, pH 5.8–6.0 with KOH. Sterilize by filtration (0.22 μm), or autoclave.

### 2.5 Media for Protoplast Transfection

1. 40% PEG-1500: 10 g w/v PEG (1500), 1.275 g mannitol, 0.413 g $Ca(NO_3)_2 \times 4H_2O$, 4 mM MES, pH 6 with KOH. Dissolve mannitol and calcium nitrate in 17.5 mL $ddH_2O$ first, and then add the PEG. Thereafter, adjust the pH with KOH. Sterilize by filtration (0.45 μm), and prepare aliquots, and store at −20 °C. After defrosting, make sure that all salts are dissolved; if not, shake vigorously until they are; slightly warming of the solution helps.
2. MM: 0.4 M mannitol, 5 mM MES. pH 6.0 with KOH. Sterilize by filtration (0.22 μm), or autoclave.
3. K3 (for 100 mL): 10 mL macro stock, 0.1 mL micro stock, 0.1 mL vitamin stock, 0.5 mL EDTA stock, 1 mL Ca-Phosphate stock, 10 mg *myo*-Inositol, 25 mg D(+)-Xylose, 13.7 g Sucrose, pH 5.6 with KOH. Sterilize by filtration (0.22 μm), prepare aliquots (15–40 mL), and store at −20 °C.
4. macro stock (for 1 L): 1.5 g $NaH_2PO_4 \times H_2O$, 9 g $CaCl_2 \times 2H_2O$, 25 g $KNO_3$, 2.5 g $NH_4NO_3$, 1.34 g $(NH_4)_2SO_4$, 2.5 g $MgSO_4 \times 7H_2O$; add $H_2O$ up to 1 L, and autoclave.
5. micro stock (for 100 mL): 75 mg KI, 300 mg $H_3BO_3$, 1 g $MnSO_4 \times 7H_2O$ (or 0.6 g $MnSO_4 \times H_2O$), 200 mg $ZnSO_4 \times 7H_2O$, 25 mg $Na_2MoO_4 \times 2H_2O$, 2.5 mg $CuSO_4 \times 5H_2O$, 2.5 mg $CoCl_2 \times 6H_2O$; add $H_2O$ up to 100 mL. Sterilize by filtration (0.22 μm), and store at −20 °C.

6. Vitamin stock (for 100 mL): 100 mg nicotinic acid, 100 mg pyridoxine-HCl, 1 g thiamine-HCl; add $H_2O$ up to 100 mL. Sterilize by filtration (0.22 μm), and store at −20 °C.
7. EDTA stock (for 1 L): dissolve 7.46 g EDTA in 300 mL pre-warmed $H_2O$ (approx. 30 °C), separately; dissolve 5.56 g $Fe(II)SO_4 \times 7H_2O$ in 300 mL pre-warmed $H_2O$ (approx. 30 °C). Combine, and add $H_2O$ up to 1 L. Autoclave, aliquot, and store in the dark at 4 °C. Protect from light.
8. Ca-phosphate stock (for 200 mL): 1.26 g $CaHPO_4 \times 2H_2O$; add $H_2O$ up to 200 mL, pH 3 with 25% HCl. Autoclave, and keep in the dark at 4 °C. Protect from light.

## 3 Methods

### *3.1* Agrobacteria *Preparation*

1. Select a single colony of transformed agrobacteria that contains the 2-in-1 vector plasmid or any other plasmid of choice.
2. Inoculate 5 mL LB medium containing the appropriate antibiotics (rifampicin 50 μg/mL, gentamicin 20 μg/mL, and the plasmid specific antibiotic) with the agrobacterium colony, and grow it overnight (28 °C, 200–230 rpm) (*see* **Note 2**).
3. Transfer 500 μL of the overnight culture into 4.5 mL of fresh LB medium supplied with rifampicin, gentamycin, and specific antibiotic, and grow for another 3–4 h (28 °C, 200 rpm) to an $OD_{595}$ of approximately 0.2–0.8.
4. Determine the $OD_{595}$.
5. Centrifuge at 4000 rpm for 15 min at 4 °C, and wash the pellet once or twice with 5 mL of 4 °C cold water.
6. Adjust to a final $OD_{595}$ of 0.5 with 4 °C cold AS medium. (Calculation: $OD_{595}/0.5 \times 5$ mL = volume (in mL) of AS-medium needed.)
7. Before proceeding with transient transformation of *Nicotiana benthamiana* leaves, incubate samples at least 1 h on ice.

### *3.2 Transient Transformation of* Nicotiana benthamiana *Leaves*

1. Water 3–4-week-old (8th–10th leaf stage) *Nicotiana benthamiana* plants 4–6 h prior to infiltration.
2. Start preparation of *Agrobacteria* as mentioned in Subheading 3.1.
3. Inject with a 1 mL syringe (without cannula) *Agrobacterium* suspension into the abaxial side of the third to fifth youngest leaves. For this purpose, position the syringe between leaf and veins, and support injection by gentle counterpressure with a finger on the adaxial side (Fig. 4a–c).

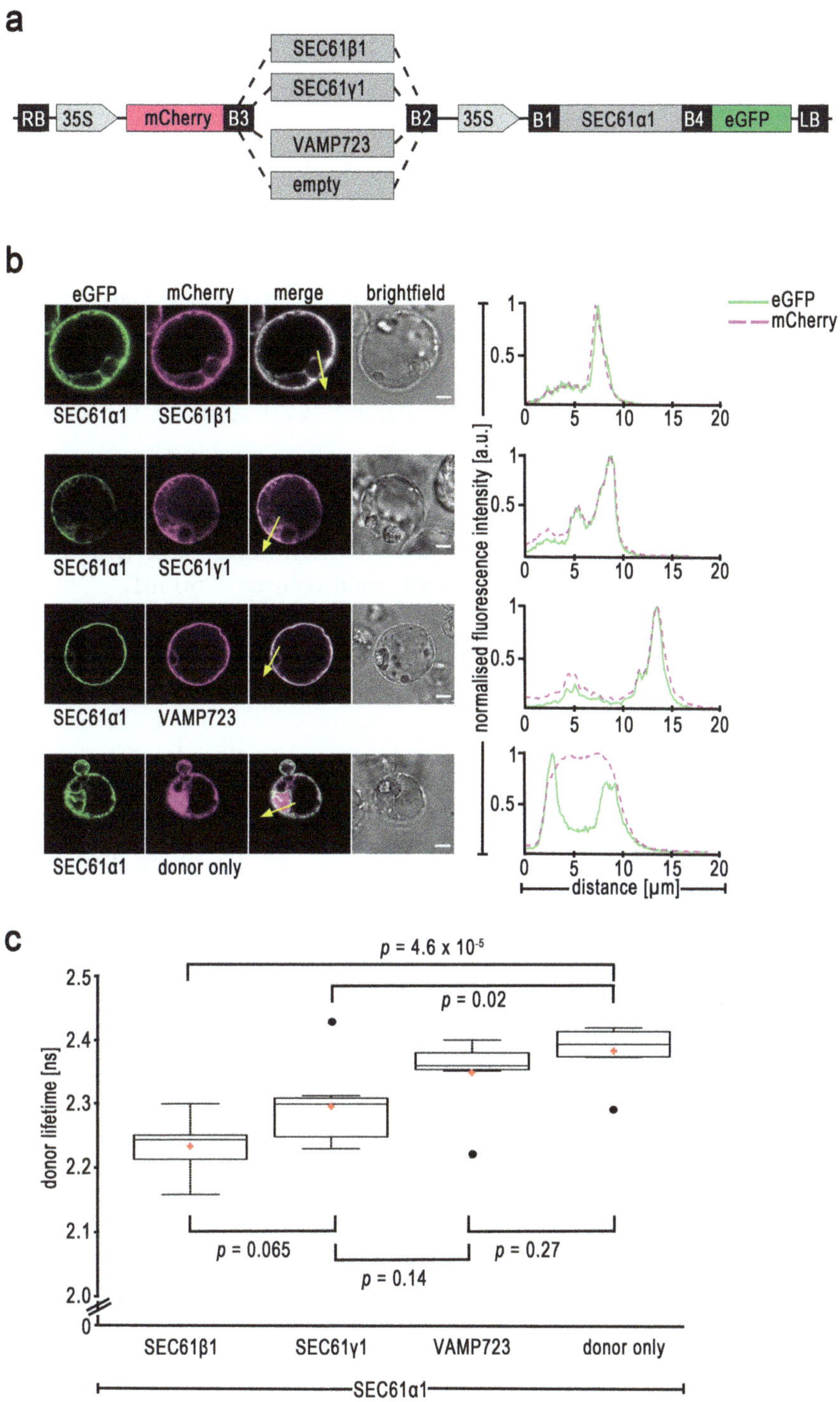


**Fig. 4** FRET-FLIM analysis of putative *Arabidopsis* SEC61 subunits in *A. thaliana* protoplasts. (**a**) Cartoon of the "2-in-1" FRET Destination constructs used in this experiment (4). (**b**) Representative images of *A. thaliana* protoplasts expressing SEC61α-eGFP with either mCherry-SEC61β, mCherry-SEC61γ, mCherry-VAMP723, or, mCherry, respectively. Line histograms to the right show normalized intensities of either eGFP or mCherry fluorescence along yellow arrows in merged images. a.u., arbitrary units. Scale-bar = 10 μm. (**c**) Box plot of SEC61α1-eGFP donor lifetime in ns of at least seven independent measurements for each of the various acceptor samples. Centerlines of boxes represent the median with outer limits at the 25th and 75th percentile, respectively. Notches indicate the 95% confidence intervals; Tukey whiskers extend to the 1.5× IQR, and outliers are depicted as black dots. Red crosses mark sample means of each dataset. Exemplary *p*-values are indicated in graph

4. After infiltration, cover, and return plants to the growth chamber.
5. After approximately 36–72 h post-infiltration, proceed to Subheading 3.6.

### 3.3 Cell Culture Maintenance (Protoplasts)

The particular cell culture in this protocol (*Arabidopsis thaliana*, Col-0, root derived) was generated and donated by Mathur and Koncz (21).

1. Cell cultures are maintained in 250 mL Erlenmeyer flasks under constant shaking (120 rpm) in the dark at 23.5 °C ± 1 °C.
2. The cell culture is propagated by re-inoculation every 7 days by transferring 10 mL of a 1-week-old (7-day-old) culture to a fresh, sterile 250 mL Erlenmeyer flask with 50 mL MSCol medium. The final, total volume is 60 mL.

### 3.4 Generation of Protoplasts

1. Protoplasts should be generated from cells 3 days after sub-cultivation when they are in their peak growth phase (*see* **Note 3**).
2. Aliquot 10 mL of cells into 12 mL poly-propylene (PP) round-bottom tubes, and collect them at 400*g* for 5 min (4–25 °C okay).
3. Wash cells by resuspension with 10 mL "wall digestion solution without enzymes," and spin down at 400*g* for 5 min (4–25 °C okay).
4. Re-suspend pellet in 7 mL "wall digestion solution," and dispense into a 90 mm diameter Petri dish.
5. Incubate by shaking at 50 rpm for 6 h at 23.5 °C ± 1 °C in the dark.
6. Collect protoplasts from one Petri dish in a 12 mL PP tube. Centrifuge at 100*g* for 5 min at 4 °C. Discard the supernatant.
7. Wash protoplasts by resuspension in up to 10 mL "wall digestion solution without enzymes," and spin down at 100*g* for 5 min at 4 °C. This is Wash #1.
8. Remove supernatant, and dissolve pellet in remaining solution. Add up to 10 mL with W5 solution slowly, and mix slowly. This is Wash #2. Spin down the cells at 100*g* for 5 min at 4 °C.
9. Re-suspend cells in 10 mL W5 solution as in **step 8**. This counts as Wash #3. Take 20 μL (use cut 200 μL tips to handle protoplasts) to determine cell concentration with a hemocytometer.
10. Adjust the cells to $1.5 \times 10^6$ cells/mL$^{-1}$. Leave cells in the refrigerator (4 °C) for at least 20 min; overnight is also okay.

11. Determine the protoplast concentration you will use. Transfection will occur in ranges of $1.5 \times 10^6$ cells/mL$^{-1}$ up to $6.7 \times 10^6$ cells/mL$^{-1}$. Our standard concentrations are $6.6 \times 10^6$ cells/mL$^{-1}$ and $3.5 \times 10^6$ cells/mL$^{-1}$. Use higher cell concentrations when working with multiple plasmid species (≥3) and the lower concentration for ≤2 plasmids. It is recommended that you test this for each experimental condition.
12. Collect the protoplasts under 50 g at room temperature (RT).
13. Decant the supernatant (some W5 will remain, ≤500 μL), and adjust to the determined cell concentration with MM and incubate cells for 10–30 min (but not longer) at RT.
14. The cells are ready for transfection.

### 3.5 Protoplast Transfection

This protocol was designed to transfect protoplasts in 96-well plate format (2.2 mL, round-bottom, deep well) using the "Liquidator™ 96 Manual Pipetting System" (Mettler-Toledo), but it can also be executed using 8- or 12-channel pipettes. It is recommended to mix after each row as soon as you add PEG. A small set of reactions can be performed in 2 mL round-bottom Eppendorfs; mix as soon as you add PEG.

1. Prepare up to 16 μg of high-quality DNA in 20 μL $H_2O$ in each reaction-well.
2. Add 60 μL of cell suspension to the DNA, and mix the plate from every side by tapping the plate against your forearm (Fig. 4d) until no cell clumps are visible. Make sure that the DNA is evenly mixed in the total volume.
3. Add 60 μL of PEG, and start a timer to a 5 min countdown. Mix the plate from every side by tapping the plate against your forearm until the PEG is evenly dispensed. This will take approximately 1 min. You can see the solution mixing if you hold the plates or tubes against the lighting in your laboratory so that you see light diffraction through the liquid. After 3 min, mix again.
4. Add 60 μL of MM solution, and mix as before.
5. Add 500–700 μL of K3 medium, mix gently, and cover the plate with a film that allows gas exchange (e.g., AeraSeal™ film, SIGMA) (*see* **Note 4**).
6. Incubate the cells in the dark at 25 °C from 3 to 24 h depending on your assay (*see* **Note 5**).

### 3.6 Mounting for Microscopy (Tobacco Leaf Infiltration)

1. Cut a leaf section of approximately 1 cm$^2$, and transfer it to a 20 mL syringe half-filled with water. Use your finger to stop water loss at the syringe's exit, insert the plunger, and carefully remove the remaining air in the syringe (Fig. 4e).

2. Close the tip of the syringe again, and pull the plunger back. This will produce a partial vacuum in the syringe.
3. Releasing the plunger gently will replace air space within the leaf slice with water.
4. Repeat **steps 3–4** several times until the tissue appears translucent.
5. Take out the slice, and mount it upside down on a slide. Drop a bit of water on the slide, and place a coverslip over the tissue.
6. Tap coverslips gently to remove air bubbles.

### 3.7 Mounting for Microscopy (Protoplasts)

Tightly wrap two stripes of tape around a microscope slide with a gap of around 5 mm between them. These will serve as spacers. Put approx. 50–100 μL of cell suspension between the stripes, and place a coverslip on top (*see* **Note 6**).

### 3.8 Confocal Imaging

Our studies are done using the Leica TCS SP8 confocal microscope with laser lines diode 405, pulsed 440, pulsed 470, Argon 488, 496, 514, DPSS 561, HeNe 594, and HeNe 633, respectively. The microscope is equipped with a fast 8 kHz resonant scanner, HyD detectors, and FLIM unit (PicoQuant). Leica Application Suite (LAS) X software was used for image acquisition.

#### 3.8.1 rBiFC

1. Set up the confocal microscope for YFP fluorescence with 514 nm excitation and 520–560 nm emission range and for RFP fluorescence with 561 nm excitation and 570–630 nm emission range. To avoid spectral overlap, sequential scanning should be applied. Include a bright-field channel for guidance from transmission of either laser line.
2. Use a 40×/0.75 NA water-immersion objective to focus on epidermal cells. A zoom factor between 3 and 4 is recommended.
3. Start with a positive control, and collect several images while adjusting gain and—if necessary—offset, for an optimal dynamic range. Settings should be adjusted to yield a mean fluorescence intensity ratio between YFP and RFP of one (*see* **Note 7**). Once the parameters are adjusted, select, and save for quantification 20–40 images of randomly selected areas (*see* **Note 8**).

#### 3.8.2 Image Analysis

For image analysis, we use the ImageJ software (freely available at http://imagej.net/Downloads) with the Bio-Formats plugin, which must be installed separately before using it for the first time.

1. Open the *.lif file with the ImageJ software. Using the Bio-Format tool the first time, choose "view stack with: Hyperstack," and set a mark in the "split channels" box. Select all series.

2. Close bright-field images.
3. Select the first RFP image, and press button M or STRG-M (measure). A table with different values (Area, Mean, Min, Max) will appear.
4. Select the corresponding YFP image, and measure again.
5. Proceed for all images as described in steps 2–4. Afterward, you can copy the data table into Microsoft Excel.
6. Calculate the ratio between the mean fluorescence intensity values of YFP and RFP for each image, and use for graphical display after/including statistical analysis.

*3.8.3 FRET-AB*

1. Set up the confocal microscope for eGFP fluorescence with 488 nm excitation and 490–530 nm emission range and for mCherry fluorescence with 561 nm excitation and 565–610 nm emission range. To avoid spectral overlap, sequential scanning should be applied. Include a bright-field channel for guidance from transmission of either laser line.
2. Use a 40×/0.75 NA water-immersion objective.
3. Adjust gain and offset settings for your sample exploiting the full dynamic range.
4. Switch to the FRET-AB mode from the dropdown menu in the LAS X software, and adjust donor and acceptor settings as before. Set acceptor-excitation-laser intensity in the "bleach" tab to 100% and frames between 300 and 500.
5. Search an area where both proteins of interest are sufficiently expressed, set up an ROI, and start the bleaching procedure. LAS X software automatically captures images before and after bleaching and calculates the FRET-efficiency (*see* **Note 9**) according to the following formula:

$$\text{FRET}_{\text{eff}} = \left(D_{\text{post}} - D_{\text{pre}}\right) / \left(D_{\text{post}}\right)$$

with $D_{\text{pre}}$ and $D_{\text{post}}$ as mean fluorescence intensities prior and after bleach, respectively.

*3.8.4 Image Analysis*

Note the values provided by the LAS X software under the "Quantification" tab for further statistical evaluation and graphical display.

*3.8.5 FLIM*

For FLIM applications, LAS X software and SymPhoTime 64 (from PicoQuant) software are used.

1. Switch the microscope to FLIM-mode. Start SymPhoTime, and create a new workspace.
2. Start lasers, and set the laser combining unit to 0.

3. Click on the tab "Setup Imaging" in the Leica software, and choose your appropriate laser and detectors.
4. Go to the tab "Setup FLIM," and choose the detector that is connected to the time-correlated single photon counting (TCSPC) module. Choose the wavelength to excite the donor.
5. Under "Acquisition" set resolution to 256 × 256 and scanning speed to 20–50 MHz while adjusting pixel dwell time to approximately 20 μs.
6. Set up the pulsed laser.
7. Choose the number of photons per pixel that should be counted. We recommend 500–1000. Pick a region of interest (ROI), and run a FLIM test. The kilo-counts/s should not exceed 10% of the excitation frequency (e.g., 2 kilo counts/ s for 20 MHz). If necessary, close the shutter of the laser combining unit to get less counts/s.
8. Run FLIM measurements at 7–10 different positions/cells/ protoplasts. Zoom in, and define ROIs for measurement. We recommend performing at least three independent biological replicates for FLIM measurements.
9. Measure donor-only samples, as well as a meaningful non-interacting candidate as negative controls.

#### *3.8.6 Measure Instrument Response Function (IRF)*

Since the microscope, objective, coverslip, and the pulse type of the laser do have an influence on the detection, the IRF must be measured for later evaluation.

1. Put only a coverslip on the objective without sample.
2. Set the mode from xyz to xzy, the laser to 5%, and the required wavelength of the donor (e.g., for GFP, 488 nm). Choose one PMT detector for emission at the same wavelength as the excitation (e.g., GFP: 488 nm).
3. Set the donor fluorescence to reflexion in the tab "AOBS."
4. Search the upper edge of the coverslip (Fig. 5a).
5. Setup FLIM as before in 3.8.5, but use the PMT to detect exactly at the same wavelength as the pulsed laser.
6. In the tab "fluorifier disc setting," deselect the autoselect, and choose an empty position for the filters.
7. Close the laser combining unit.
8. Run a FLIM test with kilo-counts per second at approximately 2000.
9. Measure IRF once, and save it.

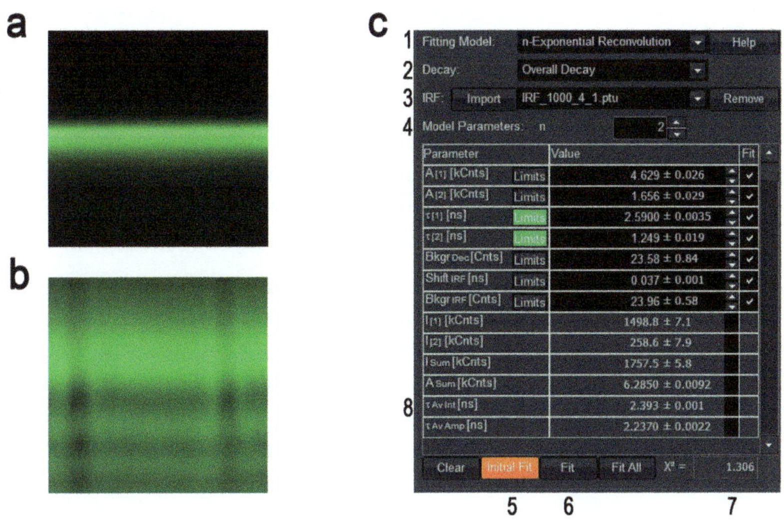


**Fig. 5** Comparison between the edge of (**a**) the upper coverslip, which is needed for measuring the IRF, and (**b**) the edge of the microscope slide, which should not be measured. (**c**) Section of SymPhoTime 64 software for evaluation of FRET-FLIM data (1). The "Fitting Model" should be "n-Exponential Reconvolution," while the (2) "Decay" should be "Overall Decay" (3). The experimentally measured IRF (Subheading 3.8.6) has to be imported here (4). "Model Parameters – n" have to be set up according to the decay variant of the donor. Calculation is started with (5) "Initial Fit" and (6) Fit (7). If $\chi^2$ is in the range of approximately 1–2, the fluorescence lifetime $\tau$ in (8) can be copied for further evaluation

### 3.8.7 Evaluation

1. Open SymPhoTime, and load your data. Choose one ".ptu" file, and click Analysis → Imaging → FLIM → Start.
2. Set up "n-Exponential Reconvolution" and overall decay, and load your measured IRF sample in the designated tab. Choose the "Model Parameter – n" according to the decay variant of your donor fluorophore (e.g., mono- or bi-exponential decay) (Fig. 5).
3. We recommend decreasing the front border width (e.g., 0.55 ns).
4. Press "Initial FIT" and afterward "FIT." The $\chi^2$ should be approximately between 1 and 2. Higher values indicate that the measured decay does not correlate to the calculated ideal decay (Fig. 5).
5. Lifetime values for further statistical evaluations are "τ av int (ns)" (Fig. 5).
6. Note these values in a .txt file or Excel, and use them for statistical evaluation and graphical display.

## 4 Notes

1. For structure-function analysis using alanine-scanning mutagenesis, we routinely design primers using our SDM-assist program that allows—in addition to the desired mutation—introduction of a silent restriction site to distinguish different point mutants via endonuclease digest (22, 23).
2. We recommend construct verification of *Agrobacteria* clones used for transfection through plasmid rescued in *E. coli* followed by restriction analysis (20). Even antibiotic-resistant *Agrobacteria* colonies sometimes do not carry the recombinant plasmid, and we had cases, where plasmid restriction digests revealed significant alterations in plasmids. Verified *Agrobacteria* clones should be stored at −80 °C in 7% DMSO.
3. To gain more *Arabidopsis* protoplasts, it is possible to subculture 10 mL of a 3-day-old culture into 50 mL fresh MSCol medium. Cells for protoplast generation are taken 3 days after this second subculture. These cells are nicknamed "turbo cells" and produce finer and smaller protoplasts. You must not use these turbo cells for your main culture though; otherwise, the culture will outgrow itself too fast.
4. It is not necessary to wash out the PEG (1500) as its effective concentration is diluted in subsequent steps; alternatively, addition of a washing step with MM or W5 can be included.
5. Reporter assays benefit from shorter incubation times as they have better signal-to-noise ratios when applying treatments, whereas an overnight incubation is sufficient for localization or interaction assays.
6. Living cells will float on the top of the solution when they are in K3. Alternative incubation buffers found in the literature can also be used: for example, WI or TEX.
7. The ratio needs to be set up for a positive control and should be examined on several different epidermal cells. A ratio of 1 is preferred but not always feasible due to overall low expression. It is important that once the parameters are adjusted, there will be no change in settings over the complete set of samples to assure comparability.
8. Even though RFP acts as an expression and transformation/transfection control, verification of protein expression via immunoblot is recommended. The 2-in-1 plasmid used contains a 3xHA tag as well as a MYC tag fused to the proteins of interest for antibody detection (Fig. 2f).
9. Due to temporal delay in recording caused by the bleaching step, cellular movement leads to a shift of the region of interest out of the focal plane, making measurements meaningless. To

avoid this issue, we fixed samples in 4% (w/v) paraformaldehyde (PFA) in microtubule stabilization buffer (MTSB) (50 mM PIPES, 5 mM EGTA, 5 mM $MgSO_4 \times 7H_2O$, pH ~ 7 (KOH)).

## Acknowledgments

We are grateful to Eva Schwörzer and Laure Grefen for excellent technical support. Work in our lab is supported through seed funding of the SFB1101 and an Emmy Noether fellowship of the German Research Foundation (DFG) to C.G. (GR 4251/1-1).

### References

1. Xing S, Wallmeroth N, Berendzen KW, Grefen C (2016) Techniques for the analysis of protein-protein interactions in vivo. Plant Physiol 171:727–758
2. Husbands AY, Aggarwal V, Ha T, Timmermans MC (2016) In planta single-molecule pull-down reveals tetrameric stoichiometry of HD-ZIPIII:LITTLE ZIPPER complexes. Plant Cell 28:1783–1794
3. Grefen C, Blatt MR (2012) A 2in1 cloning system enables ratiometric bimolecular fluorescence complementation (rBiFC). Biotechniques 53:311–314
4. Hecker A, Wallmeroth N, Peter S, Blatt MR, Harter K, Grefen C (2015) Binary 2in1 vectors improve in planta (co-) localisation and dynamic protein interaction studies. Plant Physiol 168:776–787
5. Ghosh I, Hamilton AD, Regan L (2000) Antiparallel leucine zipper-directed protein reassembly: application to the green fluorescent protein. J Am Chem Soc 122:5658–5659
6. Horstman A, Tonaco IA, Boutilier K, Immink RG (2014) A cautionary note on the use of split-YFP/BiFC in plant protein-protein interaction studies. Int J Mol Sci 15:9628–9643
7. Grefen C, Karnik R, Larson E, Lefoulon C, Wang YZ, Waghmare S, Zhang B, Hills A, Blatt MR (2015) A vesicle-trafficking protein commandeers Kv channel voltage sensors for voltage-dependent secretion. Nat Plants 1: 15108
8. Lipka E, Gadeyne A, Stockle D, Zimmermann S, De Jaeger G, Ehrhardt DW, Kirik V, Van Damme D, Muller S (2014) The phragmoplast-orienting kinesin-12 class proteins translate the positional information of the preprophase band to establish the cortical division zone in *Arabidopsis thaliana*. Plant Cell 26:2617–2632
9. Albert I, Bohm H, Albert M, Feiler CE, Imkampe J, Wallmeroth N, Brancato C, Raaymakers TM, Oome S, Zhang HQ, Krol E, Grefen C, Gust AA, Chai JJ, Hedrich R, Van den Ackerveken G, Nurnberger T (2015) An RLP23-SOBIR1-BAK1 complex mediates NLP-triggered immunity. Nat Plants 1:15140
10. Karnik R, Zhang B, Waghmare S, Aderhold C, Grefen C, Blatt MR (2015) Binding of SEC11 indicates its role in SNARE recycling after vesicle fusion and identifies two pathways for vesicular traffic to the plasma membrane. Plant Cell 27:675–694
11. Zhang B, Karnik R, Wang Y, Wallmeroth N, Blatt MR, Grefen C (2015) The arabidopsis R-SNARE VAMP721 interacts with KAT1 and KC1 K+ channels to moderate K+ current at the plasma membrane. Plant Cell 27:1697–1717
12. Kumar MN, Hsieh YF, Verslues PE (2015) At14a-Like1 participates in membrane-associated mechanisms promoting growth during drought in *Arabidopsis thaliana*. Proc Natl Acad Sci U S A 112:10545–10550
13. Le CT, Brumbarova T, Ivanov R, Stoof C, Weber E, Mohrbacher J, Fink-Straube C, Bauer P (2016) Zinc finger of Arabidopsis Thaliana12 (ZAT12) interacts with FER-like iron deficiency-induced transcription factor (FIT) linking iron deficiency and oxidative stress responses. Plant Physiol 170:540–557
14. Forster T (1946) Energiewanderung und Fluoreszenz. Naturwissenschaften 33:166–175
15. Ishikawa-Ankerhold HC, Ankerhold R, Drummen GP (2012) Advanced fluorescence

microscopy techniques--FRAP, FLIP, FLAP, FRET and FLIM. Molecules 17:4047–4132

16. Bucherl CA, Bader A, Westphal AH, Laptenok SP, Borst JW (2014) FRET-FLIM applications in plant systems. Protoplasma 251:383–394
17. Wallrabe H, Periasamy A (2005) Imaging protein molecules using FRET and FLIM microscopy. Curr Opin Biotechnol 16:19–27
18. De Los SC, Chang CW, Mycek MA, Cardullo RA (2015) FRAP, FLIM, and FRET: detection and analysis of cellular dynamics on a molecular scale using fluorescence microscopy. Mol Reprod Dev 82:587–604
19. Koncz C, Schell J (1986) The promoter of Tl-DNA Gene5 controls the tissue-specific expression of chimeric genes carried by a novel type of agrobacterium binary vector. Mol Gen Genet 204:383–396
20. Blatt MR, Grefen C (2014) Applications of fluorescent marker proteins in plant cell biology. Methods Mol Biol 1062:487–507
21. Mathur J, Koncz C (1997) Method for preparation of epidermal imprints using agarose. Biotechniques 22:280–282
22. Karnik A, Karnik R, Grefen C (2013) SDM-Assist software to design site-directed mutagenesis primers introducing "silent" restriction sites. BMC Bioinformatics 14:105
23. Grefen C, Chen Z, Honsbein A, Donald N, Hills A, Blatt MR (2010) A novel motif essential for SNARE interaction with the K(+) channel KC1 and channel gating in Arabidopsis. Plant Cell 22:3076–3092
24. Sparkes IA, Runions J, Kearns A, Hawes C (2006) Rapid, transient expression of fluorescent fusion proteins in tobacco plants and generation of stably transformed plants. Nat Protoc 1:2019–2025

# Chapter 12

# Immunoprecipitation and FRET-FLIM to Determine Metabolons on the Plant ER

**Verena Kriechbaumer and Stanley W. Botchway**


## Abstract

Metabolons are protein complexes that contain all the enzymes necessary for a metabolic pathway but also scaffolding proteins. Such a structure allows efficient channeling of intermediate metabolites form one active site to the next and is highly advantageous for labile or toxic intermediates. Here we describe two methods currently used to identify metabolons via protein-protein interaction methodology: immunoprecipitations using GFP-Trap®_A beads to find novel interaction partners and potential metabolon components and FRET-FLIM to test for and quantify protein-protein interactions in planta.


**Key words** Endoplasmic reticulum, Metabolon, Enzymes, Protein complex, Protein-protein interaction, Co-immunoprecipitation, GFP-trap, FRET-FLIM

## 1 Introduction

The plant ER has been shown to be dynamic and constantly remodelling [1]. Specific ER-microdomains with unique lipid composition are suggested to direct proper anchoring of proteins and to facilitate and stabilize the subsequent assembly of components required to form a specific protein complex capable of carrying out a complete metabolic pathway [2]. These so-called metabolons enable metabolic processes to take place in the necessary highly coordinated and regulated processes in an efficient manner. Metabolons contain sequential enzymes plus scaffolding proteins allowing for efficient channeling of metabolic intermediates from one active site directly to the next [2]. The concept of these so-called metabolons was introduced in 1985 by Paul Srere in the context of glycolytic and Krebs cycle enzymes [3]. Sequential enzymes of a specialized biosynthetic pathway are transiently linked together by noncovalent binding typically stabilized by membrane or cytoskeletal anchoring [4]. Such a mechanism increases substrate concentration and turnover rates, prevents unwanted diffusion and

Verena Kriechbaumer (ed.), *The Plant Endoplasmic Reticulum: Methods and Protocols*, Methods in Molecular Biology, vol. 2772,
https://doi.org/10.1007/978-1-0716-3710-4_12,

metabolic interference, and is beneficial for containment of labile or toxic intermediates. Such complex organization has been shown for protein translocases in the endoplasmic reticulum, in chloroplasts and mitochondria [5–7], allowing coordinated and efficient transport potentially also preventing back-flux.

Metabolons will typically be comprised of sequential enzymes in the pathway together with scaffolding proteins allowing for efficient channeling of metabolic intermediates from one active site to the next [2]. Candidate proteins for such scaffolding are chaperones as well as membrane-anchored cytochrome P450 enzymes that can serve as nucleation points and platforms for metabolon formation [2]. In addition to this, TA proteins were shown to have the potential to localize metabolons to specific organelle surfaces, e.g., in the TOM complex where the tail-anchors operate as assembly signal [8]. This can also stabilize proteins compared to their cytosolic versions as shown for tobacco expression of the HIV gene product Nef, which has been researched in the context of HIV vaccination [9].

Also membrane-structural proteins have been suggested to play a role in metabolon formation, e.g., P450 enzymes involved in a lignin biosynthetic metabolon were co-purified with reticulon proteins (RTN) [10], which contribute to ER tubule shaping [11].

In plants, metabolons have been shown for several enzymatic pathways in secondary and primary metabolism [12]. Most metabolons involved in secondary metabolism involve less stable interactions than described for metabolons in primary metabolism demanding refined microscopy methods and computational analysis to prove their interactions [13].

Metabolons have been shown, e.g., for the production of flavonoids [14] and sporopollenin [15] in *Arabidopsis* and the glucoside dhurrin [16] in sorghum. Recently, a soybean isoflavonoid metabolon tethered to the ER has been reported [17]. An ER-localized metabolon for the biosynthesis of the auxin indole-3-acetic acid (IAA) has been suggested [18].

The metabolon complexes producing phenylpropanoids [19] and flavonoids [14, 20] seem to cater for specific metabolites made from shared intermediate products. The enzyme composition in the flavonoid pathway complex can change resulting in varying end products [12, 14]. Membrane-anchored cytochrome P450 enzymes are reported to serve as nucleation points and platforms for the metabolon formation. Most P450 monooxygenases are membrane-bound to the ER via an N-terminal tether enhancing metabolon formation on this membrane system [2, 21].

Metabolons for the glucoside dhurrin in sorghum are formed in specific ER domains [22], and metabolon formation between the three enzymes involved in dhurrin biosynthesis is suggested to be highly advantageous as the dhurrin precursor p-hydroxymandelonitrile is very unstable. Another ER-loccalized metabolon was suggested for the biosynthesis of sporopollenin in tapetal cells. Colocalization and protein-protein interaction of these enzymes indicate the existence of a sporopollenin metabolon on the ER [15]. This benefits a coordinated and quick production of pollen cell wall components allowing for rapid pollen development.

Molecular distances (1–10 nm) can be measured using energy transfer processes. Förster or fluorescence resonance energy transfer (FRET), first described by Theodor Förster over 50 years ago [23], relies on the non-radiative energy transfer from an excited fluorescent donor molecule to a different non-excited fluorescent acceptor molecule in its vicinity. The very short distances required for this process to occur (<10 nm) means the two molecules, in this case two proteins, need to be physically close to one another. For FRET to occur, the donor emission spectrum must overlap sufficiently with the acceptor absorption spectrum, and the donor and acceptor dipoles display a mutual molecular orientation. The rate of the energy transfer $k_T$ is described and calculated using Eq. 1:

$$k_T = \left(\frac{1}{\tau_D}\right)\left(\frac{R_0}{R}\right)^6 \tag{1}$$

where $\tau_D$ is the donor excited-state lifetime in the absence of the acceptor, $R$ is the distance between $D$ and $A$, and $R_0$ is the Förster radius. At the Förster radius, 50% of the donor molecules will emit fluorescence, while the rest will undergo energy transfer. Since the energy transfer process is strongly distance dependent with $1/R^6$, FRET can be used to measure distance and examine molecular interactions on a nm spatial scale. During FRET, the rate of decay is reduced by a quenching process that depletes the excited state of the donor fluorophore, i.e., the donor fluorescence lifetime is shortened. By measuring changes in the excited state lifetime of the donor at each pixel making up an image, steady-state FRET is enhanced. This is described as fluorescence lifetime imaging microscopy or FRET-FLIM. Generally, the two proteins under investigation are tagged with GFP and its variants. We have used GFP and RFP as the donor and acceptor, respectively, in our work. We describe here two methods that have been used in metabolon work: immunoprecipitations using GFP-Trap®_A beads to find novel interaction partners and potential metabolon components and FRET-FLIM to test for and quantify protein-protein interactions in planta.

## 2 Materials

### 2.1 Buffers and Media for Immunoprecipitations

1. Murashige and Skoog (MS) medium: 4.4 g/L Murashige and Skoog basal salts, pH 5.7, 10 g/L agar.
2. Lysis buffer: 10 mM Tris-HCl pH 7.5, 150 mM NaCl, 0.5 mM EDTA, 0.5% NP-40, 1 mM PMSF, protease inhibitor according to manufacturers' instructions.
3. Equilibration buffer: 10 mM Tris-HCl pH 7.5, 150 mM NaCl, 0.5 mM EDTA.

### 2.2 Buffers and Media for FRET-FLIM

For plant infiltration and expression in tobacco, see protocols in chapter "Labelling the ER for Light and Fluorescence Microscopy."

### 2.3 Equipment for Immunoprecipitations

1. Petri dishes (large, with lid L × W × H: 245 mm × 245 mm × 25 mm).
2. 50 mL Falcon tubes.
3. Glass bottles for buffers.
4. Cheese cloth.
5. Refrigerated table centrifuge.

### 2.4 Equipment for FRET-FLIM

1. At least two-channel confocal and one-channel FLIM set up.
2. Becker and Hickl SPC 830 or SPC150 time-correlated single-photon counting card.
3. SPCImage analysis software version 5.1 (Becker and Hickl).

## 3 Methods

### 3.1 Immuno-precipitation Using GFP-Trap®_A Beads

(This part of the procedure will take between 4 and 5 h depending on sample size.) This protocol uses GFP-Trap®_A beads (Chromotek, Martinsried, Germany), and the procedure is carried according to the company's protocol with slight modifications (*see* **Note 1**):

*Arabidopsis thaliana* seedlings expressing the protein of interest fused to a fluorescent tag (see Protocol "Labelling the ER for light and fluorescence microscopy" for preparation of stable arabidopsis transformants) are grown for 2 weeks on MS plates (*see* **Note 2**).

1. Approximately 5 g of whole seedling plant material are ground first in liquid nitrogen to a fine powder and then in 5 mL lysis buffer until liquid. The extract is then distributed into three 2-mL-Eppendorf tubes.

2. The extracts are incubated on ice for 30 min and then centrifuged at 10,000*g* for 10 min at 4 °C.
3. The supernatant (about 4 mL) is poured into fresh 50 mL Falcon tubes via two layers of cheese cloth.
4. 100 μL of GFP-Trap®_A beads per sample are equilibrated in 500 μL equilibration buffer and centrifuged at 2500*g* for 2 min. The supernatant is being discarded, and this wash is repeated twice.
5. 100 μL of the washed beads are added to the plant extract, and the mixture is shaken on ice for 2 h with the tubes being placed horizontally in the ice.
6. After this, tubes are centrifuged at 2500*g* for 2 min at 4 °C, and the supernatant is carefully discarded without disturbing the agarose pellet.
7. The resulting agarose pellet is then washed twice with dilution buffer.
8. This pellet can be used straight away for tryptic peptide digest and mass spectrometry analysis for containing proteins [24].

### 3.2 Real-Time Imaging of Metabolon Protein-Protein Interactions Using Two-Photon FRET-FLIM

Potential interaction candidates resulting from the immunoprecipitation are tested for interactions in planta with the bait protein. This procedure is done using agrobacterium-mediated transient expression in tobacco leaf epidermal cells (see Protocol "Labelling the ER for light and fluorescence microscopy" for transient expression in tobacco).

1. Epidermal samples of infiltrated tobacco leaves are excised, and both confocal/multiphoton FRET-FLIM data capture is performed by a two-photon microscope [25] (*see* **Note 3**).
2. The two-photon microscope is built around a Nikon TE2000-U inverted microscope and is used with a modified Nikon EC2 confocal scanning system to allow for near infrared laser wavelength for FLIM [26].
3. Laser light at a wavelength of 920 nm is produced by a mode-locked titanium sapphire laser (Mira; Coherent Lasers), with 200 fs pulses at 76 MHz, pumped by a solid-state continuous wave 532 nm laser (Verdi V18; Coherent Laser).
4. The laser beam is focused to a diffraction limited spot using a water-immersion objective (Nikon VC; 360, numerical aperture of 1.2) to illuminate specimens on the microscope stage.
5. Fluorescence emission is collected without de-scanning (no pin-hole), bypassing the scanning system, and passed through a BG39 (Comar) filter to block the near-infrared laser light. Line, frame, and pixel clock signals are generated and synchronized with an external detector in the form of a fast

microchannel plate photomultiplier tube (Hamamatsu R3809U).

6. Linking these via a time-correlated single-photon counting PC module SPC830 (Becker and Hickl) generated the raw FLIM data. Prior to FLIM data collection, the GFP and mRFP expression levels in the plant samples within the region of interest are confirmed using a Nikon EC2 confocal microscope with excitation at 488 and 543 nm, respectively.
7. A 633 nm interference filter is used to significantly minimize the contaminating effect of chlorophyll autofluorescence emission that would otherwise obscure the mRFP as well as GFP emission.
8. Data are analyzed by obtaining excited-state lifetime values first on a pixel-by-pixel basis then of a region of interest on the nucleus, and calculations are made using SPCImage analysis software version 5.1.
9. The distribution of lifetime values within the region of interest is generated and displayed as a curve. Only values with a $\chi^2$ between 0.9 and 1.4 are taken.
10. The median lifetime, minimum, and maximum values for one-quarter of the median lifetime values from the curve are taken to generate the range of lifetimes per sample.
11. The ER associated with at least three nuclei from a minimum of three independent biological samples per protein-protein combination are analyzed, and the average of the ranges is taken.
12. The degree or efficiency of energy transfer ($E$), from one protein to the other may be determined using Eq. 2:

$$E\% = \left[1 - \left(\frac{\tau_{\mathrm{DA}}}{\tau_{\mathrm{D}}}\right)\right] \times 100 \quad (2)$$

where $\tau_{\mathrm{D}}$ and $\tau_{\mathrm{DA}}$ are the measured excited state lifetime of the donor and acceptor respectively.
Example data is shown in Fig. 1.

## 4 Notes

1. Gloves have to be worn for the whole procedure to reduce contamination.
2. This protocol can also be carried out with tobacco transiently expressing the protein of interest but will result in reduced yield of interacting proteins and will of course only show interactors in tobacco.
3. Best measurement results for the ER have been shown for the nuclear envelope and associated ER. Latrunculin B can be

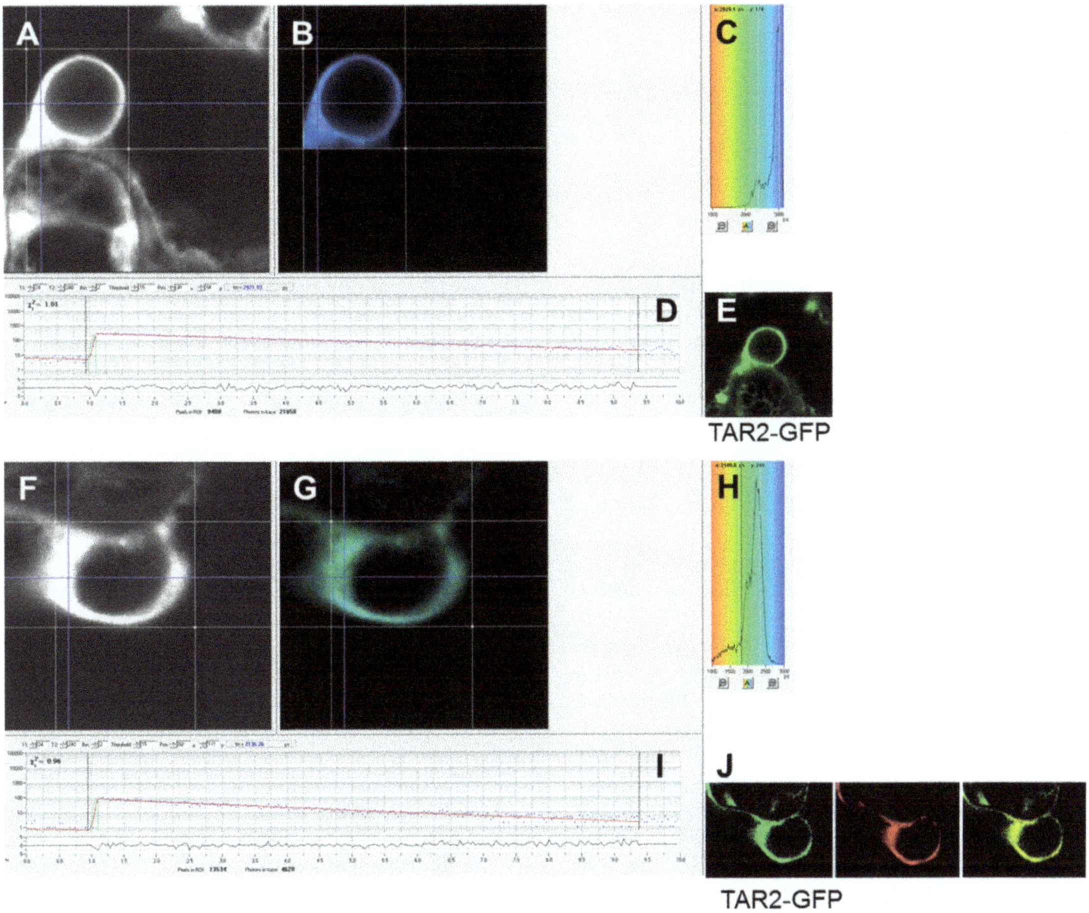


**Fig. 1** FRET-FLIM analysis of TAR2 without an interaction partner (**a**–**e**) or with YUC8 (**f**–**j**). (**a**, **f**) Raw FRET-FLIM data; (**b**, **g**) pseudocolored lifetime maps showing the lifetime values for each point within the region of interest; (**c**, **h**) distribution of lifetimes across the image. Blue shades represent longer GFP fluorescence lifetimes than green ones. (**d**, **i**) Representative decay curves of a single point with an optimal single exponential fit, where $\chi 2$ values from 0.9 to 1.2 were considered an excellent fit to the data points (a binning factor of 2 was applied). The confocal images for the analysis in (**e**, **j**) show the GFP construct in green and the mCherry construct in red. This example of FRET-FLIM analysis shows TAR2–GFP alone as a negative control and YUC8 for protein-protein interaction. The fluorescence lifetime values for TAR2–GFP+YUC8–mCherry are $2.8 \pm 0.02$ ns and therefore statistically lower than the lifetime values for the TAR2–GFP fusion alone ($3.04 \pm 0.03$ ns)

applied to the leaf discs to depolymerize the actin cytoskeleton and therefore inhibit any movement of the ER. This may not be necessary with high expression levels of the donor protein, which result in shorter time for data capture, and therefore, movement is less critical.

## Acknowledgments

This work was supported by the Science and Technology Facilities Council Programme (grant no. 14230008) awarded to CH and VK.

### References


1. Hawes C, Kiviniemi P, Kriechbaumer V (2015) The endoplasmic reticulum: a dynamic and well-connected organelle. J Integr Plant Biol 57:50–62
2. Jørgensen K, Rasmussen AV, Morant M, Nielsen AH, Bjarnholt N, Zagrobelny M, Bak S, Møller BL (2005) Metabolon formation and metabolic channelling in the biosynthesis of plant natural products. Curr Opin Plant Biol 8:280–291
3. Srere PA (1987) Complexes of sequential metabolic enzymes. Annu Rev Biochem 56:89–124
4. Hoppert M, Mayer F (1999) Principles of macromolecular organization and cell function in bacteria and archaea. Cell Biochem Biophys 31: 247–284
5. Jarvis P, Chen LJ, Li H, Peto CA, Fankhauser C, Chory J (1998) An Arabidopsis mutant defective in the plastid general protein import apparatus. Science 28:100–103
6. Werhahn W, Niemeyer A, Jansch L, Kruft V, Schmitz UK, Braun H (2001) Purification and characterization of the preprotein translocase of the outer mitochondrial membrane from Arabidopsis. Identification of multiple forms of TOM20. Plant Physiol 125:943–954
7. Van den Berg B, Clemons WM, Collinson I, Modis Y, Hartmann E, Harrison SC, Rapoport TA (2004) X-ray structure of a protein conducting channel. Nature 427:36–44
8. Horie C, Suzuki H, Sakaguchi M, Mihara K (2003) Targeting and assembly of mitochondrial tail-anchored protein Tom5 to the TOM complex depend on a signal distinct from that of tail-anchored proteins dispersed in the membrane. J Biol Chem 278:41462–41471
9. Marusic C, Nuttall J, Buriani G, Lico C, Lombardi R, Baschieri S, Benvenuto E, Frigerio L (2007) Expression, intracellular targeting and purification of HIV Nef variants in tobacco cells. BMC Biotechnol 26(7):12
10. Bassard JE, Richert L, Geerinck J, Renault H, Duval F, Ullmann P, Schmitt M, Meyer E, Mutterer J, Boerjan W, De Jaeger G, Mely Y, Goossens A, Werck-Reichhart D (2012) Protein-protein and protein-membrane associations in the lignin pathway. Plant Cell 24:4465–4482
11. Sparkes I, Tolley N, Aller I, Svozil J, Osterrieder A, Botchway S, Mueller C, Frigerio L, Hawes C (2010) Five plant reticulon isoforms share ER localisation, topology, ER membrane shaping properties. Plant Cell 22:1333–1343
12. Winkel-Shirley B (2002) Evidence for enzyme complexes in the phenylpropanoid and flavonoid pathways. Plant Physiol 107:142–149
13. Bassard JE, Mutterer J, Duval F, Werck-Reichhart D (2012) A novel method for monitoring the localization of cytochromes P450 and other endoplasmic reticulum membrane associated proteins: a tool for investigating the formation of metabolons. FEBS J 279:1576–1583
14. Hrazdina G, Wagner GJ (1985) Metabolic pathways as enzyme complexes: evidence for the synthesis of phenylpropanoids and flavonoids on membrane associated enzyme complexes. Arch Biochem Biophys 237:88–100
15. Lallemand B, Erhardt M, Heitz T, Legrand M (2013) Sporopollenin biosynthetic enzymes interact and constitute a metabolon localized to the endoplasmic reticulum at tapetum cells. Plant Physiol 162:616–625
16. Nielsen KA, Tattersall DB, Jones PR, Møller BL (2008) Metabolon formation in dhurrin biosynthesis. Phytochemistry 69:88–98
17. Dastmalchi M, Bernards MA, Dhaubhadel S (2016) Twin anchors of the soybean isoflavonoid metabolon: evidence for tethering of the complex to the endoplasmic reticulum by IFS and C4H. Plant J 85:689–706
18. Kriechbaumer V, Botchway SW, Hawes C (2016) Localization and interactions between Arabidopsis auxin biosynthetic enzymes in the TAA/YUC-dependent pathway. J Exp Bot 67: 4195–4207
19. Stafford HA (1974) Possible multienzyme complexes regulating the formation of C6-C3 phenolic compounds and lignins in higher plants. Recent Adv Phytochem 8:53–79
20. Winkel-Shirley B (2001) Flavonoid biosynthesis: a colorful model for genetics, biochemistry,

cell biology and biotechnology. Plant Physiol 126:485–493
21. Ralston L, Yu O (2006) Metabolons involving plant cytochrome P450. Phytochem Rev 5: 459–472
22. Winkel BS (2004) Metabolic channelling in plants. Annu Rev Plant Biol 55:85–107
23. Förster T (1948) Zwischenmolekulare Energiewanderung und Fluoreszenz. Ann Phys 437:55–75
24. Kriechbaumer V, Botchway SW, Slade SE, Knox K, Frigerio L, Oparka K, Hawes C (2015) Reticulomics: protein-protein interaction studies with two plasmodesmata-localized reticulon family proteins identify binding partners enriched at plasmodesmata, endoplasmic reticulum, and the plasma membrane. Plant Physiol 169:1933–1945
25. Schoberer J, Botchway SW (2014) Investigating protein-protein interactions in the plant endomembrane system using multiphoton-induced FRET-FLIM. Methods Mol Biol 1209:81–95
26. Botchway SW, Scherer KM, Hook S, Stubbs CD, Weston E, Bisby RH, Parker AW (2015) A series of flexible design adaptations to the Nikon E-C1 and E-C2 confocal microscope systems for UV, multiphoton and FLIM imaging. J Microsc 258:68–78

# Chapter 13

# Using Optical Tweezers Combined with Total Internal Reflection Microscopy to Study Interactions Between the ER and Golgi in Plant Cells

**Imogen Sparkes, Rhiannon R. White, Benji Bateman, Stanley Botchway, and Andy Ward**

## Abstract

Optical tweezers have been used to trap and micro-manipulate several biological specimens ranging from DNA, macromolecules, organelles, to single-celled organisms. Using a combination of the refraction and scattering of laser light from a focused laser beam, refractile objects are physically captured and can be moved within the surrounding media. The technique is routinely used to determine biophysical properties such as the forces exerted by motor proteins. Here, we describe how optical tweezers combined with total internal reflection fluorescence microscopy (TIRF) can be used to assess physical interactions between organelles, more specifically the ER and Golgi bodies in plant cells.

**Key words** Optical trap, ER, Golgi, Tweezers, GFP

## 1 Introduction

Trapping and micro-manipulation for biological samples using optical tweezers has a wide range of applications from DNA, macromolecules, organelles, to single-celled organisms (reviewed in [1]). The basic premise of optical traps is that only microscopic particles that display a significantly different refractive index to the surrounding medium will be trapped in the focused infrared laser beam (Fig. 1). Infrared laser beams are routinely used as they generally result in the lowest sample absorbance and therefore cause the least photon damage. Optical traps can either be fixed in position and therefore require lateral stage movement to move the sample relative to the trapped object or the system can use mirrors, acousto-optic deflectors, or spatial light modulation to move the position of the trap and therefore the object [2]. Here, we present our system that utilizes a single-beam gradient trap that

Verena Kriechbaumer (ed.), *The Plant Endoplasmic Reticulum: Methods and Protocols*, Methods in Molecular Biology, vol. 2772,
https://doi.org/10.1007/978-1-0716-3710-4_13,

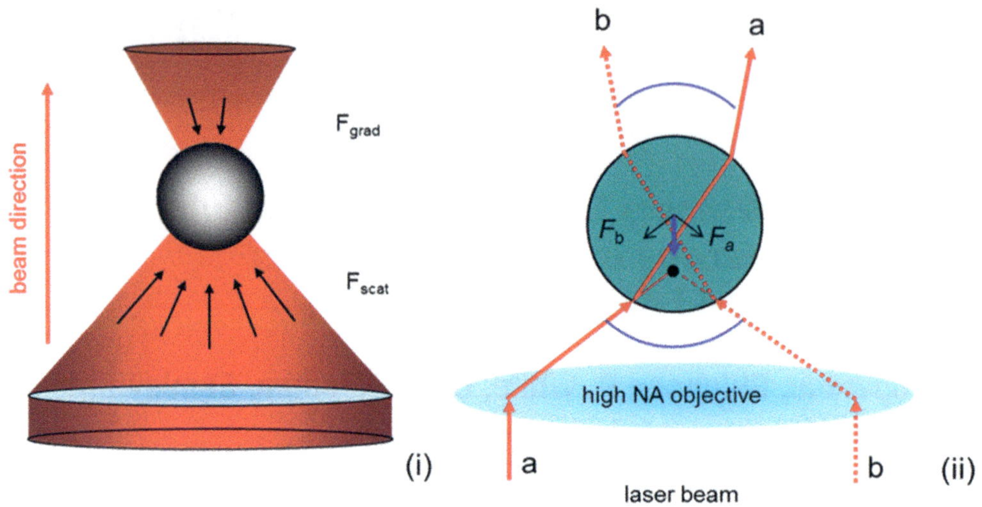


**Fig. 1** Mechanism of action of a gradient optical trap. (i) Scattering and gradient forces are balanced in a tightly focused laser beam using a high numerical aperture objective lens to form a stable three-dimensional trap. (ii) Gradient restoring forces (Fa, Fb) result from the refraction of the laser beam as it passes through the trapped object, with rays of light (a and b) depicting the paths taken by the edges of the beam

is fixed in position combined with TIRF microscopy. Readers are directed to Ketelaar et al. [3] for discussion of integrating optical traps with confocal microscope systems.

Optical trapping has been used to probe several aspects of subcellular architecture in plant cells (reviewed in [3]), for example, the effects of nuclear positioning on root hair growth [4]; interactions between several organelle pairings including peroxisome-chloroplasts [5], ER-Golgi [6], and ER-chloroplasts [7]; and cytoskeletal organization [8]. We have used optical tweezers to characterize interactions between several organelles in plants (ER-Golgi and peroxisome-chloroplast), where the measurement of such interactions is only possible if one of the organelles is preferentially trapped. The efficiency with which organelles are trapped is related to a number of factors including refractive index, size, shape, and laser wavelength. For example, unlike the endoplasmic reticulum (ER), Golgi bodies in leaf epidermal cells are amenable to being trapped and moved within the confines of the cell. Since the ER and Golgi are functionally linked through the secretory pathway, this raised the question as to whether they were also physically linked. Using optical tweezers, we were able to show that lateral movement of trapped Golgi bodies resulted in the remodeling of the ER indicative of a physical association between the two organelles [6]. These studies provided a qualitative analysis of the interaction between the two structures. Here, we provide an overview of our current optical trapping platform, which enables the user to quantify interactions between organelles. We have used this system successfully to characterize the physical interaction between peroxisomes and chloroplasts, which are functionally linked through the photorespiratory pathway [5]. We will discuss important considerations relating to system calibration, sample generation, developing a quantifiable platform for measuring interactions, and post-acquisition analysis. We also include caveats to force measurement calculations in plant cells. Of course, variations will apply subject to the optical trap system available to the user, but the general premise and pitfalls herein are universal.

## 2 Materials

### 2.1 System Setup and LabVIEW Interface

An optical trap with a two-channel TIRF microscope (TIRF-M) was constructed around a Nikon Ti-U inverted microscope. The near-infrared (NIR) trapping laser at 1090 nm was delivered to the trapping objective (100×, oil immersion, NA 1.49 with both temperature and cover glass correction ring) using a commercial Elliot Scientific optical trapping platform. Fluorescence excitation for the TIRF-M imaging was obtained from an Omicron laser hub fiber coupled to a manual Nikon TIRF-M unit. For GFP and RFP chromophores fused to the proteins of interest, 488 nm and 561 nm were used, respectively. Fluorescence emissions were detected using two electron multiplying charge-coupled device (EMCCD, iXon, Andor) cameras. These are connected via a twin-cam (Cairn) unit (Fig. 2). This allowed the flexibility of full field of view per imaging channel as well as the ability to further magnify the image without restricting the observation area. A single camera may be used for a dual-channel imaging (requiring technologies such as the Cairn OptoSplit II). These lead to restricted field of view but has the advantage of reduced cost and no need for synchronization of two separate cameras.

### 2.2 Microscopy Consumables

1. Standard glass slides (76 mm × 26 mm, 1 mm thick) to fit the stage.
2. Borosilicate No. 1 coverslips (*see* **Note 1**).
3. Electrical tape.
4. Polystyrene beads (1 micron diameter) are commercially available.

## 3 Methods

Carry out all procedures at room temperature unless otherwise stated.

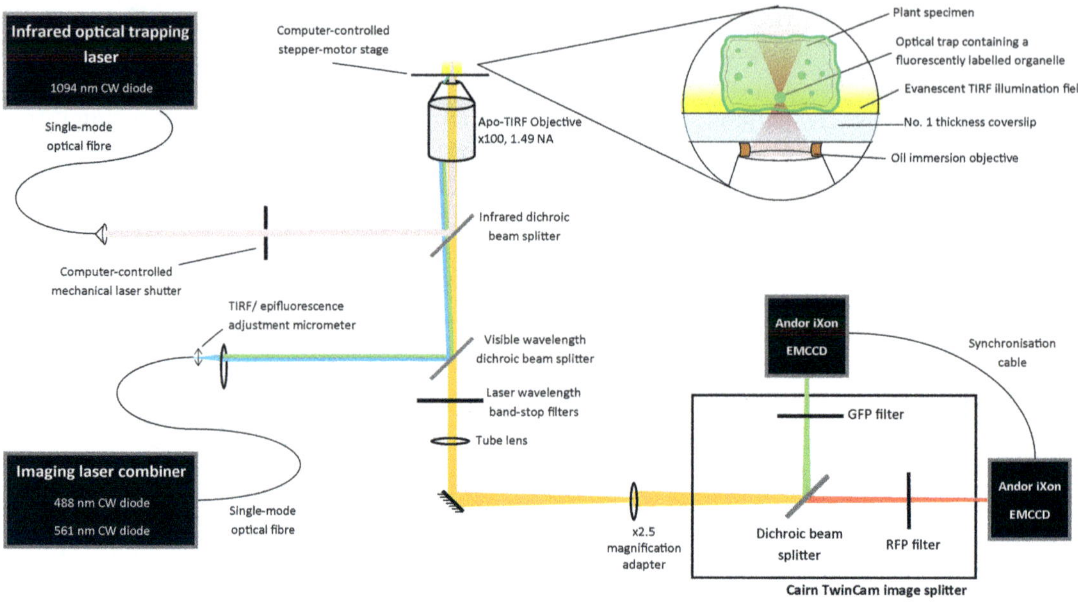


**Fig. 2** Schematic of the microscope and optical trap system setup. The infrared trapping laser and visible-wavelength imaging lasers are fiber-coupled to independent microscope ports. The TIRF angle is tuned by adjusting the position of the fiber aperture in relation to the collimating lens using a micrometer. A computer-controlled mechanical shutter is placed in the trapping laser beam path, which provides fast shuttering of the optical trap. The sample is mounted on a computer-controlled variable speed (Märzhäuser) stepper motor stage. The fluorescence is imaged through a (Nikon) ×100 1.49 NA objective and ×2.5 magnification adapter onto two (Andor) EMCCD cameras, filtered for GFP and RFP fluorescent dyes, respectively. The associated computer controlled hardware is interfaced using National Instruments LabVIEW, which provides full automation for each trapping routine

### 3.1 Calibrating the Optical Trap Against Known Standards

The trap position, optical trapping force, and laser power transmission at the microscope objective are measured daily and compared for consistency to ensure the laser trapping setup is performing to specification. You are referred to **Note 2** for a critical evaluation of objective lenses for tweezers and TIRF microscopy.

### 3.2 Nominal Laser Power Transmission at the Sample

Using a power meter set to the wavelength of the trapping laser (1090 nm), a range of power readings of the optical trap laser transmission (0–130 mW) at the objective are measured on a daily basis. Large fluctuations of these readings (i.e., >5% change) between daily experimental runs could be indicative of changes in laser performance and stability or changes in laser alignment affecting the percentage transmission at the objective.

### 3.3 Trap Position and Force

Trap position is determined by trapping a 1 micron diameter polystyrene bead dispersed in water and imaging the position under bright field illumination or directly observing a significantly attenuated laser using the EMCCDs with the infrared filter removed.

To assess the trap strength, the escape force of spherical polystyrene beads is calculated. This is a relationship between the power of the optical trap that is required to maintain beads in the trap at a given stage speed and is carried out as follows:

1. Place a drop of polystyrene beads on a coverslip, and leave to settle for ~30 min.
2. Trap a polystyrene bead at the upper end of the optical trap power spectrum (*see* **Note 3**).
3. Move the stage back and forth with a pre-set velocity of 100 μm/s.
4. Repeat **step 3** after decreasing the power incrementally.
5. Note down the trap laser power at which the bead escapes the trap at 100 μm/s stage oscillation.
6. Repeat **steps 2–5** using 200, 300, 400, and 500 μm/s stage speeds.
7. Using the equation for Stokes viscous drag, calculate the escape force at which the bead escapes the trap (i.e., is no longer trapped) for each of the five different stage speeds.

$$F_{\text{drag}} = 6\pi\mu r v$$

where $F_{\text{drag}}$ is the viscous drag force acting on the bead, $\mu$ is the fluid viscosity (water, 0.00089 Pa s), $r$ is the object radius (bead), and $v$ is the velocity measured.
8. Plot laser power against escape force for each stage speed (i.e., calculate the $F_{\text{drag}}$ for the given velocity at which the bead escapes the trap). For beads, the relationship is linear as there are no constraints on moving the bead other than viscosity (Fig. 3).

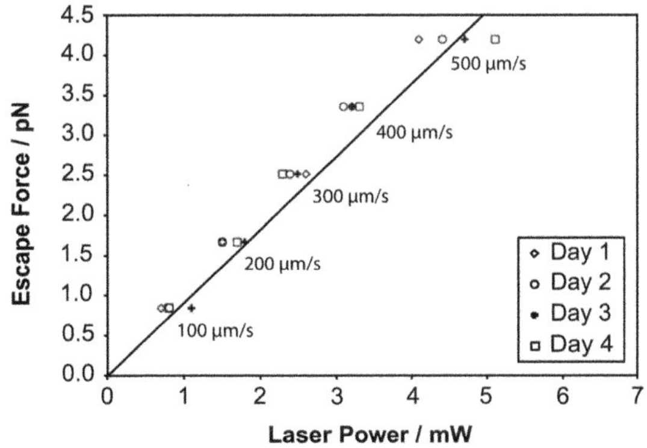


**Fig. 3** Calculation of the optical trap escape force of a 1 μm polystyrene bead. Example plot of laser power (mW) against bead escape force (pN) for a range of indicated stage speeds (100–500 μm/s). Data displayed was collected on four consecutive days

Comparison between escape force profiles between different experimental runs enables the user to determine whether the trap "strength" has altered, i.e., is it "easier" or "harder" to maintain beads in the trap at a given stage speed? Similar bead trapping profiles ensures that the system is behaving in a consistent manner and therefore allows results from biological samples to be directly compared.

### 3.4 Sample Preparation

To be able to specifically trap Golgi bodies, fluorophores need to be introduced, which are specifically targeted to and highlight the Golgi. Similarly, fluorophores for ER, or other organelles, are required if monitoring the effects of interaction between two organelles. Genetically encoded fusions to targeting sequences or functional proteins can be introduced into the plant material through agrobacterium-mediated transformation. Integration and expression can either be stable or transient. Readers are referred to standard protocols for these transformation procedures [9, 10]. Note: a bank of fluorescent organelle markers generated by the Nebenführ lab [11] are available either as plasmids or stable transgenic *Arabidopsis* lines through public repositories (ABRC or NASC).

Plant leaf material expressing the relevant fluorescent markers to be subjected to optical trapping is prepared in the following manner:

1. Pieces of tissue (~5 mm$^2$) are cut from the plant leaves and immersed in latrunculin b (25 μM for 1 h) to depolymerize actin and prevent organelle movement (see below).
2. The leaf samples are carefully mounted on the microscope slide and a No 1 coverslip secured over the top using thin strips of electrical tape at either end of the slide.
3. Slides are mounted coverslip down onto the microscope stage and secured with spring clips. We are aiming to minimize the working distance between the coverslip and leaf to enable efficient trapping and uniform TIRF illumination. If the samples are incorrectly mounted, then trapping efficiency and imaging can be compromised.
4. To this end, a metal plate with a hole to allow imaging of the sample is placed under the slide and the spring clips of the slide holder placed over the top. The metal plate allows equal pressure to be applied evenly over the slide to help minimize the distance between sample and coverslip (i.e., plane of optimal trap force). Note: the level of pressure does not affect cell viability as evidenced through cytoplasmic streaming in samples, which have not been treated with latrunculin b.
5. Regions of the sample are imaged, ensuring the level of excitation lasers for the fluorophores is minimal so as to reduce any effects from fluorophore bleaching.

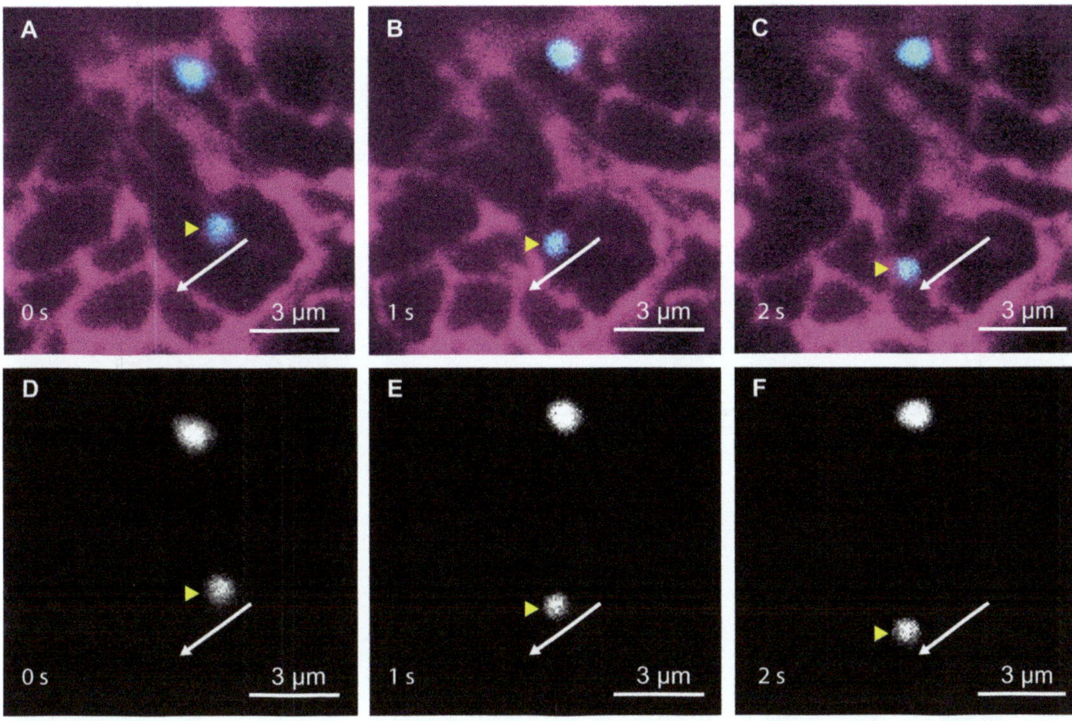


**Fig. 4** Optical trapping and movement of an ER-associated Golgi body in tobacco leaf epidermal cells. Micrographs showing sequential frames of a movie of a Golgi (**d–f**) that is trapped under the force of the optical trap as the microscope stage is moved a distance of 6 μm at 6 μm/s. This movement is referred to as the stage translation. Panels **a–c** show the same frames of the trapped Golgi (cyan) in relation to the position of the surrounding ER (magenta). The ER is observed to be attached to and to move with (behind) the trapped Golgi. The yellow arrow indicates the position of the trapped Golgi, and the white arrow indicates the direction of movement of the stage. Scale bar, 3 μm

We have been able to trap Golgi bodies in both *Arabidopsis thaliana* [6] and *Nicotiana tabacum* leaf epidermal cells (Figs. 4 and 5). Both ER and Golgi are highly motile, and so to be able to carefully dissect interaction, the movement of both organelles needs to be inhibited [6]. Any subsequent motion is therefore due to the trapping and micromanipulation of the trapped object. Figure 4 highlights trapping of Golgi in a motile system. Comparisons between image frames highlight dynamic rearrangement of the ER, making it difficult to observe the resulting changes in the ER as a result of trapping and moving a Golgi body, which is physically associated with the ER. Figure 5 clearly shows trapping and movement of Golgi, which are not attached to the ER.

### 3.5 Developing an Automated Trapping Routine

Qualitative analysis of interaction between ER and Golgi can be carried out by trapping and manually moving the Golgi and observing the effects on the ER [6]. To quantify interactions, the relationship between trap strength and organelle trapping efficiency and behavior is required (*see* **Note 3**).

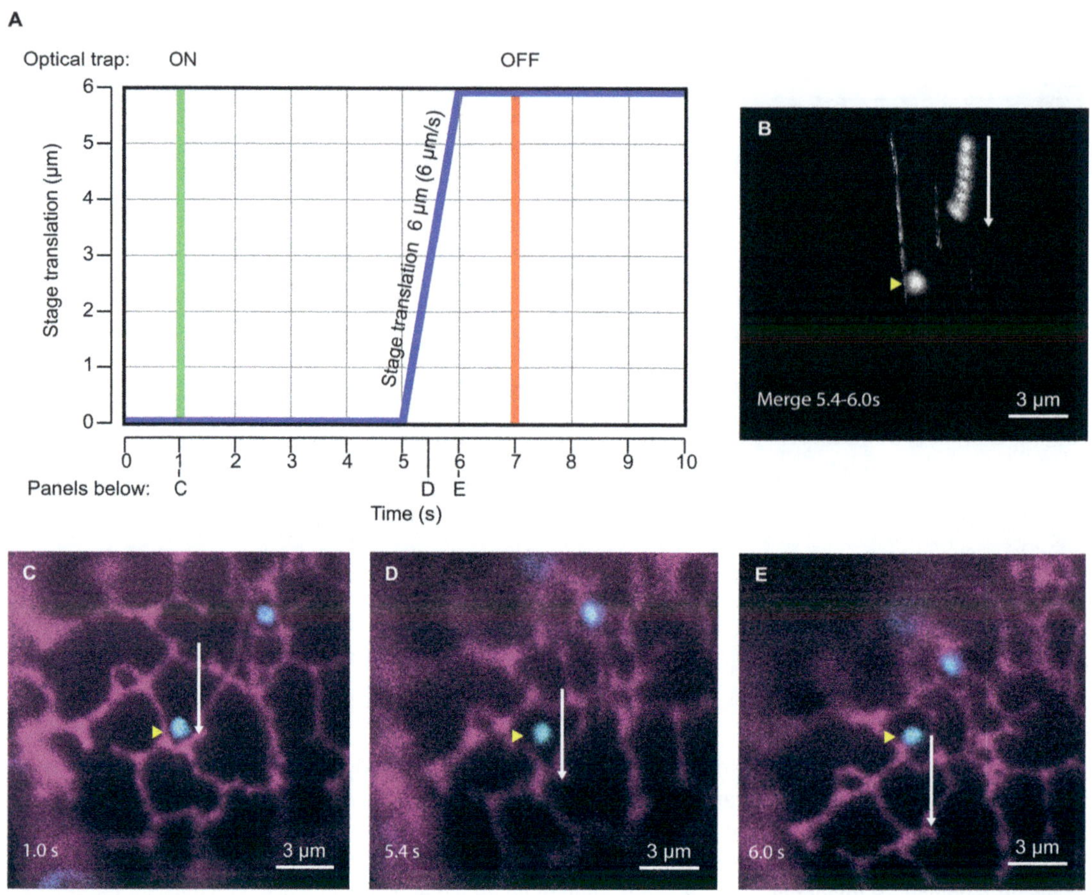


**Fig. 5** Automated trapping routine applied to a Golgi body in tobacco leaf epidermal cells. Schematic representation of the position of the optical trap as considered radially from the starting point of the trapping routine (**a**). The TIRF image is recorded from 0 s, the trap is turned on after 1 s, the stage movement (translation) of 6 μm at 6 μm/s begins at 5 s and ends at 6 s, the trap is turned off after 7 s and the image recording ends at 10 s. Images taken from a representative movie where the trapping routine (**a**) was used to trap a Golgi (cyan) and record the event in relation to the surrounding ER (magenta) are depicted; the moment the trap is turned on (**c**), the midpoint of the translation (**d**) and the end of the translation before the trap are turned off (**e**). The yellow arrow indicates the position of the trapped Golgi, and the white arrow indicates the direction of movement of the stage. In this case, the Golgi is not attached to the ER, and as it is moved, the ER does not follow (**d**). Merged frames of the Golgi signal that correspond to 5.4–6.0 s of the routine are provided (**b**). The trapped Golgi (yellow arrow) does not change position relative to the field of view as the stage moves relative to the trapped object. Merged image therefore shows the trapped organelle as a single point (yellow arrow), whereas an organelle that was not trapped appears to be moving in the merged composite (White arrow) as its position changes relative to the field of view as the stage moves. Scale bar, 3 μm

1. Prior to data acquisition, a live image is displayed on the computer monitor, and a potential organelle of interest is located manually using a joystick and aligned with the optical trapping position.

2. The organelle then undergoes a recorded trapping routine such as that highlighted in Fig. 5. Here, the trap is turned on, stage moved laterally at a fixed speed over a fixed distance (referred to as stage translation), trap turned off, and the movement of the organelle post trapping monitored.
3. Organelles are scored as either being stably trapped (i.e., remaining in the trap during stage translation) or escaping the trap during the stage translation or are not trapped during the procedure.
4. Distribution of trapping outcomes amongst these three categories can serve to determine the relative differences in trapping efficiency between different conditions and can be compared when optical trapping power is changed [5]; tethered organelles will tend to require greater force to trap and move them a fixed distance than organelles that are not tethered. This procedure was followed for characterizing interactions between peroxisomes and chloroplasts [5].
5. The major considerations are the laser power used to trap the organelle and the distance and speed at which to move the surrounding environment relative to the trapped organelle. If the trapping force is too strong, it will simultaneously trap several refractile objects; if it is too weak, then the object will not be trapped. Similarly, if the object is moved too quickly or too far (i.e., beyond the cell boundary), it will escape the trap.

Through development of the experiment, we have built a custom system to allow the interface between imaging and optical trapping to be integrated. The major benefit of an integrated system is that customized scripts can be written in LabVIEW® to control the stage lateral motion (i.e., speed and distances traversed), the timing of trap activation, and the subsequent data capture [5]. The essential components are therefore:

1. An optical trap
2. A means of turning the trap on and off (we use a shutter that can be automated)
3. A microscope stage that can be moved at the sub-micron resolution scale at controlled velocity
4. A means of imaging the organelles of interest in real time (TIRF)

### 3.6 Setting the Optical Trap Laser Power

If the optical trap is too strong, it will likely trap the majority of organelles it is centered over (which includes organelles that do not contain fluorophores). If the goal is to determine whether trapping efficiency is affected by molecular perturbation, then it is important that not all organelles can be trapped under control conditions. Therefore, trapping 20 organelles and assessment of whether they remain in the trap during the micromanipulation procedure (see above) are deemed necessary. In practice, we classify organelles that have undergone the trapping routine into three categories: stably trapped (i.e., stay in the trap over the entire movement), escape the trap (i.e., "fall out" of the trap during the stage movement), or are not trapped at all. Once this distribution has been ascertained, a full characterization can be carried out with at least five samples from each condition where at least ten organelles are trapped per sample.

### 3.7 Optimizing Distance and Speed at Which to Move the Trapped Organelle

Determining the distance to move the trapped organelle is an important consideration: too far and organelles may escape the trap due to opposing force components encountered (e.g., moving it into cytoplasmic streams if working in a motile system or encountering other subcellular structures); too short and it will be difficult to assess whether the trapped organelle is still attached to the ER.

### 3.8 Post-acquisition Data Analysis

Analysis of data generated from the integrated TIRF-M optical tweezer setup requires certain META data to be assigned to each data set. The META data enables the user to determine the coordinates of the optical trap, to overlay trails of trapping events for ease of initial analysis, and indicates the crucial time points during the trapping routine (e.g., trap on/off, etc.).

### 3.9 META Data Output for Post-acquisition Analysis

For each recorded trapping event, five data files are generated based on post-acquisition analysis requirements:

1–2. Raw images acquired from the synchronized EMCCDs as 16-bit TIFF stacks, one for each fluorescent channel.

3. An 8-bit false-color (RGB) TIFF stack containing overlaid channels for quick reference.

4. A txt file containing the lateral ($x/y$) trap (shutter) position at the start of the routine and any additional META data (e.g., optional notes).

5. A csv file containing lateral ($x/y$) positions as measured in μm relative to the top-left stage limit. These positions are measured at the end of each frame but could be measured more often if required.

### 3.10 Data Analysis

Data can be analyzed based on the distribution of organelles into the three categories (stable, escaped, not trapped classification) and how this changes with optical trap power or under molecular perturbation. In addition, the distance and rate at which stably trapped organelles move back toward their original pre-trap position after the trap has been turned off can be measured. These measurements enable the user to understand the nature of physical interaction or the surrounding environment of the organelle. Assuming constant viscosity, differences in recoil rate can be used to model the physical tethering process [5]. Statistical analysis undertaken is dependent on the hypothesis under test and the nature of the material. Keeping sample sizes constant over experimental runs allows for less complicated downstream statistical analysis. Commercial or free software such as ImageJ can be used to analyze the TIFF stacks post-acquisition, and using python or similar programming language the lateral stage positions can be used to map and subtract the movement of the stage for automatic/batch analysis of the relative movement (e.g., speed, distance, recoil) of the trapped organelle. Determination of precise force components in vivo is difficult, owing to variability in the biological specimen, slight variations in organelle size and morphology, changes in microenvironments within the cell such as viscosity, cytoplasmic streaming events, and potential shear forces from boundary layers generated in the constrained cytoplasmic environment. Due to these inherent complications generating variation of trapping efficiencies across the cell environment, a large number of interaction events need to be probed to obtain statistically significant data; we routinely trap at least 50 organelles in total, taken from several samples from at least three independent plants from independent transformations.

## 4 Notes

1. Some batches or individual coverslips can have aberrations in the glass, making trapping difficult. A manifestation of these issues is a significant change in Z focus upon trapping. If this occurs, move to another area on the slide, or make a fresh sample.
2. Using a high numerical aperture TIRF objective lens to create a single-beam gradient trap (laser tweezers) requires some care. The TIRF objectives with oil immersion have aberration when used in aqueous environments such as plant cells, which reduces the effective numerical aperture for trapping from 1.49 to a value between 1.0 and 1.1. This both reduces the efficiency of trapping from a focusing perspective but also means that not all the laser power entering the objective is used for trapping. While we have used the TIRF objective in a conventional manner by slightly overfilling the input pupil, other researchers have found more efficient trapping by underfilling the objective. The transmission of laser light at 1090 nm of our TIRF objective was determined to be 70% using the double-objective method suggested by Mahamdeh et al. [12].
3. Beads which have settled and adhered to the coverslip cannot be trapped and moved laterally. Beads which are “free-floating” can undergo the trapping routine.

## Acknowledgments

This work was supported by The Leverhulme Trust (RPG-2015-106) and STFC (PM-1216 and PAG 2201).

### References

1. Hawes C, Osterrieder A, Sparkes IA, Ketelaar T (2010) Optical tweezers for the micromanipulation of plant cytoplasm and organelles. Curr Opin Plant Biol 13:731–735
2. Osterrieder A, Sparkes IA, Botchway SW, Ward A, Ketelaar T, de Ruijter N, Hawes C (2017) Stacks off tracks: a role for the golgin AtCasp in plant endoplasmic reticulum – Golgi tethering. J Exp Bot 68(13):3339–3350
3. Ketelaar T, de Ruijter N, Niehren S (2014) Optical trapping in plant cells. Methods Mol Biol 1080:259–265
4. Ketelaar T, Faivre-Moskalenko C, Esseling JJ, de Ruijter NCA, Grierson CS, Dogterom M, Emons AMC (2002) Positioning of nuclei in Arabidopsis root hairs: an actin-regulated process of tip growth. Plant Cell 14:2941–2955
5. Gao H, Metz J, Teanby NA, Ward AD, Botchway SW, Coles B et al (2016) In vivo quantification of peroxisome tethering to chloroplasts in tobacco epidermal cells using optical tweezers. Plant Physiol 170:263–272
6. Sparkes IA, Ketelaar T, de Ruitjer NCA, Hawes C (2009) Grab a Golgi: laser trapping of Golgi bodies reveals in vivo interactions with the endoplasmic reticulum. Traffic 10:567–571
7. Andersson M, Goksor M, Sandelius AS (2007) Optical manipulation reveals strong attracting forces at membrane contact sites between endoplasmic reticulum and chloroplasts. J Biol Chem 282:1170–1174
8. Van der Honing HS, de Ruijter NCA, Emons AMC, Ketelaar T (2009) Actin and myosin regulate stiffness in plant cells: a study using optical tweezers. New Phytol 185:90–102
9. Clough SJ, Bent AF (1998) Floral dip: a simplified method for Agrobacterium-mediated transformation of Arabidopsis thaliana. Plant J 16:735–743
10. Sparkes IA, Runions J, Kearns A, Hawes C (2006) Rapid, transient expression of fluorescent fusion proteins in tobacco plants and generation of stably transformed plants. Nat Protoc 1:2019–2025
11. Nelson BK, Cai X, Nebenführ A (2007) A multicolored set of in vivo organelle markers for co-localization studies in Arabidopsis and other plants. Plant J 51:1126–1136
12. Mahamdeh M, Campos CP, Schäffer E (2011) Under-filling trapping objectives optimizes the use of the available laser power in optical tweezers. Opt Express 19:11759–11768

# Chapter 14

# Protein Biosynthesis and Maturation in the ER

## Emanuela Pedrazzini and Alessandro Vitale

### Abstract

The endoplasmic reticulum takes care of the folding, assembly, and quality control of thousands of proteins destined to the different compartments of the endomembrane system or to be secreted in the apoplast. Here we describe how these early events in the life of all these proteins can be followed biochemically by using velocity or isopycnic ultracentrifugation, metabolic labelling with radioactive amino acids, drug treatments, and immunoselection in various conditions and, in certain cases, predicted in silico by algorithms.

**Key words** Protein synthesis, Endoplasmic reticulum, Chaperones, Immunoglobulin binding protein (BiP), Protein folding, Protein oligomerization, Protein-protein interactions, Pulse-chase, Immunoselection, Ultracentrifugation

## 1 Introduction

In a given plant cell, thousands of different polypeptides are synthesized by ribosomes bound to the endoplasmic reticulum membrane (ER), to be introduced into the ER lumen or integrated in its limiting membrane. Most of these proteins will then proceed along the secretory pathway, to reach the correct destination within the endomembrane system or to be secreted. Before starting their journey, all these polypeptides undergo folding and in many cases assembly into oligomers, with the assistance of ER resident folding helpers, and if these events fail, the same helping machinery directs the defective polypeptides to degradation [1–3]. The major ER folding helper is the immunoglobulin binding protein (BiP), member of the heat shock protein 70 family. The extent of interactions with the folding machinery is often predictive of whether folding defects can explain observed effects in activity, intracellular sorting, or turnover, resulting from natural mutations or protein engineering [4–8].

The unfolded protein response senses, mainly through the BiP activity status, developmental or stress changes in the levels of ER

Verena Kriechbaumer (ed.), *The Plant Endoplasmic Reticulum: Methods and Protocols*, Methods in Molecular Biology, vol. 2772, https://doi.org/10.1007/978-1-0716-3710-4_14,

workload and consistently regulates the levels of ER folding helpers and the activity of the secretory pathway, at transcriptional, translational, and post-translational steps [2, 3]. Moreover, newly synthesized polypeptides can undergo co- or post-translational modification, such as asparagine N-glycosylation, proline hydroxylation, or serine/threonine/hydroxyproline O-glycosylation.

Here, we describe biochemical techniques to determine protein structural maturation in the ER, interactions with molecular chaperones of this compartment and in silico prediction of BiP binding sites, and protein intracellular localization (Fig. 1). These techniques can be used to define the steady-state situation or, when coupled with pulse-chase radioactive labelling, to analyze the time-course dynamics of events. Protein radioactive pulse-chase was developed several decades ago and has been used in plant biology since the end of the 1950s [9], but it still remains an approach of choice for subsequent fine biochemical analysis of a given protein regarding its in vivo time course of assembly, interactions with folding helpers, and traffic.

## 2 Materials

### 2.1 Protein Extraction from Leaves or Protoplasts (See Note 1)

#### 2.1.1 Leaf Homogenization

1. Leaf homogenization buffer (HB): 200 mM NaCl, 1 mM EDTA, 50 mM Tris–Cl, 0.2% Triton X-100, pH 7.8. Store at −20 °C in small aliquots, and supplement with protease inhibitors cocktail immediately before use.
2. Reducing HB: same as HB but supplemented with 4% (V/V) 2-mercaptoethanol (*see* **Note 2**).
3. Ice-cold mortar.

#### 2.1.2 Protoplast Homogenization

1. 15 mL or 1.5 mL polypropylene tubes.
2. Protoplast homogenization buffer (ptsHB): 150 mM NaCl, 1.5 mM EDTA, 1.5% Triton X-100, 150 mM Tris–Cl pH 7.5. This is a 1.5× stock. Store at −20 °C in small aliquots, and supplement with protease inhibitors cocktail immediately before use.
3. Reducing protoplast homogenization buffer: same as ptsHB but supplemented with 4% (V/V) 2-mercaptoethanol (*see* **Note 2**).

### 2.2 Subcellular Fractionation on Isopycnic Sucrose Gradients

1. Buffer A: 10 mM KCl containing either 2 mM $MgCl_2$ or 1 mM EDTA (*see* **Note 3**), 100 mM Tris–Cl, pH 7.8. Store at −20 if not autoclaved.
2. 12%, 16%, 55%, 70% (W/W) sucrose buffer: dissolve sucrose in buffer A (with 2 mM $MgCl_2$ or 1 mM EDTA) to 12%, 16%, 55%, and 65% (w/w). To dissolve 55% or 65% sucrose, heat the

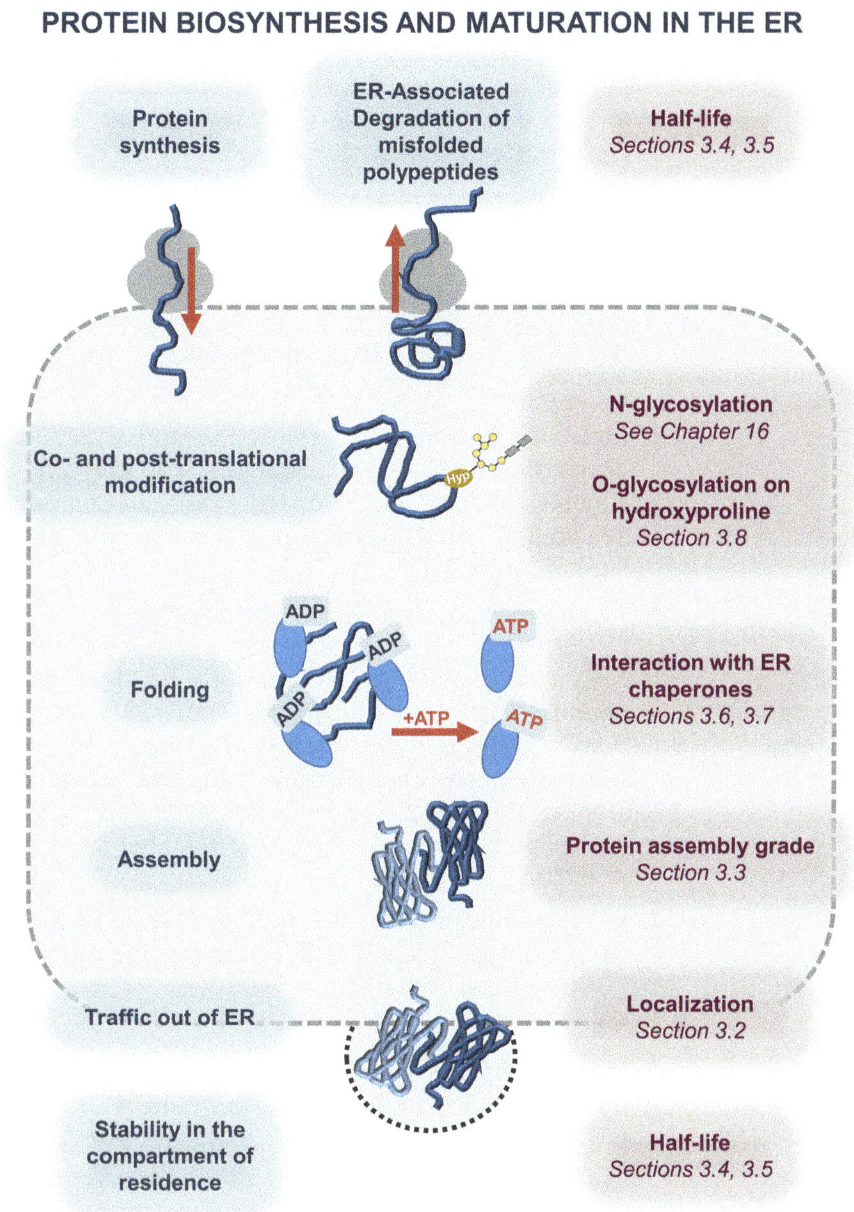


**Fig. 1** Schematic representation of secretory protein synthesis, structural maturation, and sorting in the ER. For each event, the relevant sections of this chapter are indicated. N-glycosylation, a modification occurring on many but not all secretory polypeptides, is treated in Chapter 16. A putative soluble protein is represented, but integral membrane proteins are subjected to very similar processing steps. It should also be kept in mind that certain highly polymerized secretory proteins, such as the prolamins that accumulate in cereal seeds as protein bodies, do not traffic out of the ER even when their structural maturation is completed [2]

solution to 40–50 °C, because sucrose is at the limit of solubility. Once dissolved, it will not precipitate even upon freezing. Do not autoclave (the solution will turn yellowish-brown, indicating sucrose alterations). Store at −20 °C; otherwise, it gets easily contaminated: sucrose is very good for bacteria and fungi.

3. 13 mL polypropylene ultracentrifuge tubes.

### 2.3 Velocity Sucrose Gradient Centrifugation

1. Sucrose gradient buffers: 5% sucrose (W/V) in 150 mM NaCl, 0.1% Triton X-100, 25 mM Tris–Cl, pH 7.5; 25% sucrose (W/V) in 150 mM NaCl, 0.1% Triton X-100, 25 mM Tris-Cl, pH 7.5.
2. Molecular mass markers: prepare a mixture with 200 µg each of cytochrome *c* (12.4 kDa), ovalbumin (43 kDa), bovine serum albumin (67 kDa), aldolase (161 kDa), and catalase (232 kDa) in a volume of HB equal to that of the samples to analyze (*see* Subheading 3.3.1).
3. 13 mL polypropylene ultracentrifuge tubes.

### 2.4 Metabolic Labelling with Radioactive Amino Acids

1. K3 medium: 3.78 g/L Gamborg's B5 basal medium with minimal organics, 750 mg/L $CaCl_2$ $2H_2O$, 250 mg/L $NH_4NO3$, 136.2 g/L sucrose, 250 mg/L xylose, 1 mg/L 6-benzylaminopurine (make a 5 mg/mL stock of 6-BAP in 1 N NaOH), 1 mg/L 1-naphtalenacetic acid (NAA). Bring to pH 5.5 with a few drops of 1 M KOH, filter-sterilize, and store at −20 °C.
2. W5 medium: 9 g/L NaCl, 0.37 g/L KCl, 18.37 g/L $CaCl_2$ $2H_2O$, 0.9 g/L glucose. Filter/sterilize, aliquot, and store at −20 °C.
3. $^{35}S$ Met/Cys mix.
4. Unlabelled Met/Cys (10× stock): 15 mg/mL Met, 6.05 mg/mL Cys in K3. Store at −20 °C (*see* **Note 4**).
5. Bovine serum albumin, 4 mg/mL stock in K3. Store at −20 °C in aliquots.

### 2.5 Protein Immunoselection

1. Gelatin stock: 2% gelatin from porcine skin in distilled water. Autoclave, and when at around 30–40 °C but before it starts forming the gel, add $NaN_3$ to 0.02% to avoid microorganism growth, and store at 4 °C. Melt in microwave oven before use. It can undergo many melting cycles without deteriorating.
2. Immunoselection gelatin buffer (Nonidet-EDTA-Tris-gel buffer, NET-gel): 50 mM Tris–HCl, pH 7.5, 150 mM NaCl, 1 mM EDTA, 0.1% octylphenoxy poly(ethyleneoxy)ethanol, branched (Nonidet P-40 or IGEPAL® CA-630), 0.25% gelatin, 0.02% sodium azide.

3. Washing buffer (NET buffer): same as NET-gel buffer, but without gelatin.
4. Protein A-Sepharose 10% suspension: work at room temperature. Swell Protein A-Sepharose in NET buffer for at least 3 h with occasional agitation, in a graduated 50 or 15 mL polypropylene tubes), and wait until it sediments. Discard the supernatant using a Pasteur pipette connected to a water pump. Add 1 M Tris–HCl pH 7.5. Allow to stand for 1 h with occasional agitation. Spin down the bead at 5000*g*, 5 min at room temperature; discard supernatant. Wash the bead 1× with NET buffer. Spin down, and note volume occupied by the bead (1 g of dry powder should give about 3.5 mL of bead). Discard supernatant, and add 9 bead volumes of NET buffer. Store at 4 °C for up to several months.

#### 2.6 In Vivo Chaperone Activity Assay

1. Tunicamycin stock: dissolve tunicamycin at 5 mg/mL in 10 mM NaOH, and store in small aliquots at 20 °C.

#### 2.7 Drug Treatment to Inhibit Prolyl-4-hydroxylase Activity and Subsequent Hydroxyproline O-glycosylation

1. 2,2′-dipyridyl stock: dissolve 2,2′-dipyridyl at 25 mM in 100% ethanol, and use immediately.

## 3 Methods

### 3.1 Total Protein Extraction from Leaves or Protoplasts (See Note 5)

#### 3.1.1 Leaf Homogenization

1. Collect Arabidopsis rosette leaves or small (4–7-cm-long) leaves from tobacco grown in axenic conditions or in soil, weight, and transfer them (*see* **Note 6**) to an ice-cold mortar kept on ice containing 5–7 V/W of ice-cold HB supplemented, immediately before use, with protease inhibitor cocktail. Rapidly homogenize the leaves keeping the mortar on ice. Alternatively, use reducing HB, and follow the same procedure (*see* **Note 7**).
2. Transfer the homogenate to an appropriate tube (depending on the volume), and clarify the homogenate by centrifugation at 1500*g*, 10 min at 4 °C. Collect the supernatant, and discard the pellet (*see* **Note 8**). The clarified homogenate can be stored frozen at −20 °C for several months.

#### 3.1.2 Protoplast Homogenization

1. Prepare and collect protoplasts as described [5].
2. Add to protoplast pellet (usually around 50 µl) two volumes of ice-cold ptsHB, by pipetting 4–5 times through a 200 µL tip, and freeze-thaw the suspension in liquid nitrogen.

Alternatively, use reducing ptsHB, and follow the same procedure (*see* **Note 7**).

3. Clarify the homogenate by centrifugation at 1500*g*, 10 min at 4 °C (*see* **Note 9**). Protoplast homogenate can be stored at −20 °C for several months.

### *3.2 Subcellular Fractionation to Determine Protein Localization*

#### *3.2.1 Subcellular Fractionation on Isopycnic Sucrose Gradients*

1. To prepare linear 16–55% or 16–65% (W/W) (*see* **Note 10**) sucrose gradients, gently stratify with a micropipette 6 mL of 16% sucrose buffer (with magnesium or EDTA) onto 6 mL of 55% (or 65%) sucrose buffer (with magnesium or EDTA) in a 13 mL ultracentrifuge tube. Seal the tube with Parafilm™, and gently turn it to a perfectly horizontal position. Keep it in this position for 4 h at room temperature, then gently turn it back to a vertical position, and store for at least 10 h at 4 °C before using it.
2. For leaves: homogenize small leaves (3–8-cm-long) in an ice-cold mortar with ice-cold 12% (W/W) sucrose buffer (with magnesium or EDTA) supplemented with protease inhibitor cocktail, using 4–7 mL of buffer per gram of fresh leaf tissue.
3. For protoplasts: homogenize pelleted protoplasts in four volumes of ice-cold 12% sucrose buffer supplemented with protease inhibitor cocktail. With a 200 μL micropipette tip, pipette the suspension up and down about 20 times, to break protoplasts. Do not let the material warm up.
4. With a micropipette, load the homogenate (up to 600 μL) on top of the gradient. Keep everything in ice (*see* **Note 11**).
5. Ultracentrifuge at 150,000*g* ($r_{av}$), in a swinging rotor at 4 °C for 2 h.
6. Collect fractions (usually of about 500–600 μL each) (*see* **Note 12**). A pellet will be visible at the bottom of the tube: do not discard it. Fractions and the tube with its pellet can be stored at −20 °C.
7. Take an equal aliquot of each fraction, and add 0.5 volumes of 3× SDS-PAGE denaturation buffer. Solubilize the pellet at the bottom of the centrifuge tube with a volume of 1× SDS-PAGE corresponding to the same volume of each original fraction.
8. Denature the aliquots from the gradient and the solubilized pellet at 90 °C, 5 min. Analyze by SDS-PAGE and protein blot or fluorography each fraction and an equal aliquot of the solubilized pellet. Remember to load an aliquot of denatured total 12% sucrose homogenate, as a control (*see* **Note 13**).

*3.2.2 Rapid Isolation of Microsomes*

1. Homogenize leaves or protoplast in 12% sucrose buffer, and clarify as described above.
2. Load the clarified homogenate on a cushion made of 17% sucrose buffer. Ultracentrifuge in a swinging rotor 90 min at 100,000*g* ($r_{av}$), 4 °C (*see* **Note 14**).
3. Collect the supernatant, resuspend the pellet, and analyze both fractions by SDS-PAGE and protein blot (or fluorography).

### 3.3 Determination of the Assembly Grade of Proteins

*3.3.1 Velocity Centrifugation on Sucrose Gradient*

1. Prepare linear 5–25% (W/V) sucrose gradients following the procedure described in Subheading 3.2.1, **step 1**, but using the buffers for velocity sucrose gradient centrifugation.
2. Homogenize leaves or protoplast as described in Subheadings 3.1.1 or 3.1.2.
3. Load the clarified lysate (up to 600 μl) on the top of the 12 mL linear 5–25% (W/V) sucrose gradient.
4. In parallel, load an additional gradient with the molecular mass marker mixture, prepared in the same volume of the lysate samples.
5. Centrifuge at 200,000 × $g$ ($r_{av}$) for 25 h at 4 °C in a ultracentrifuge swinging rotor.
6. Fractionate the gradient from the top into 18–22 fractions of about 500–600 μL each (*see* **Note 12**).
7. Solubilize equal volumes of each fraction with 0.5 volumes of 3× SDS-PAGE loading buffer. The pellet at the bottom of the centrifuge tube is solubilized with 500–600 μL of 1× SDS-PAGE loading buffer.

Take equal aliquots of fractions and pellet; denature at 90 °C, 5 min; and analyze by SDS-PAGE and protein blot (or fluorography). Remember to load an aliquot of denatured total homogenate, as a control.

### 3.4 Protein Metabolic Labelling with Radioactive Amino Acids

*3.4.1 Pulse-Chase Labelling of Protoplasts*

1. Pulse: Label protoplasts in K3 medium using 100 μCi/mL of $^{35}$S-label Met/Cys mixture. Incubate for 1 h at 25 °C (tobacco) or 20 °C (*Arabidopsis*) in the dark. A workable concentration of protoplasts for the subsequent steps is 1–2 millions/mL.
2. Stop pulse labelling by diluting protoplasts with four volumes of ice-cold W5 solution; centrifuge at 100*g* for 5 min at 4 °C.
3. Chase: Resuspend the pelleted protoplasts in K3 medium supplemented with unlabelled Met/Cys to a final concentration of 10 mM and 5 mM, respectively, using Met/Cys 10× stocks (*see* **Note 15**). Incubate at 25 °C (tobacco) or 20 °C (Arabidopsis) in the dark for the desired chase time points.

4. Collect an aliquot of protoplasts (usually 100–250 μL) at each chase time point, add three volumes of ice-cold W5, and centrifuge 10 min at 60 × *g*. Remove the supernatant (and save it, if secretion needs to be studied), leaving about 50 μL covering the protoplast pellet.
5. Quickly freeze in liquid nitrogen and store at −80 °C until homogenization. Alternatively, homogenize immediately using the buffer required according to subsequent analysis that will be performed.

#### 3.4.2 Pulse-Chase Labelling of Intact Leaves

1. Pulse: put 150 μL of MS medium supplemented with 150 μCi/mL $^{35}$S-label Met/Cys mixture in a conical 10 or 15 mL plastic tube (*see* **Note 16**).
2. Cut one leaf (about 5 cm long) from axenic plant, and insert it into the tube so that the petiole touches the bottom of the tube.
3. Incubate under vacuum for 1 min, to favor entrance of the labelling medium into the vascular system.
4. Incubate for 1–3 h or longer at 25 °C (tobacco) or 20 °C (Arabidopsis) in the dark, covering the tube with its unscrewed cap.
5. Carefully remove the incubation medium with a micropipette, and replace it with 150 μL 1/2 MS. Incubate 1 h as above, to favor as much as possible incorporation of the radioactive amino acids before starting the chase. You can homogenize leaves now (as described above), or continue with the chase.
6. Chase: remove the medium, and replace it with 1/2MS supplemented with unlabelled Met/Cys to 10 mM and 5 mM, respectively, using the Met/Cys 10× stock.
7. At the desired time points, homogenize leaves as described above.

### 3.5 Immunoselection

1. Add NET-gel buffer to protoplast or leaf homogenates prepared as in Subheadings 3.1.1 and 3.1.2, or to aliquots from isopycnic or velocity gradient fractions (no more than 400 μL of sample), up to a final volume of 1 mL. Mix well by vortexing. Use 1.5 mL polypropylene tubes.
2. Spin twice (4 min. at 4 °C at maximum speed in minifuge) to pellet any insoluble material. Transfer supernatant to a new tube each time.
3. Add the appropriate antibody or antiserum. The final concentration of immunoglobulins depends on the affinity of each antibody/antiserum and must be previously determined, but the volume should be limited to a few microliters. Incubate 2 h on ice (*see* **Note 17**).

4. Add 100–150 μL of a 10% suspension of Protein A-Sepharose. Incubate for 1 h at 4 °C, under gentle agitation (the resin must be in continuous movement).
5. Pellet the resin (30 s to 2 min at maximum speed in minifuge), and discard supernatant.
6. Wash the resin, two times with 1 mL of NET-gel buffer and then once with NET buffer. Spin as above, and discard supernatant after each wash, leaving each time about 1 mm of buffer over the resin.
7. Add to the bead equal volume of 2× SDS-PAGE buffer (bead is usually 30–40 μL).
8. Mix well, denature 5 min at 90 °C, spin, and load the solution on SDS-PAGE. If you use a Hamilton syringe, this is easy because the bead will not enter the needle.
9. Radioactive polypeptides can be revealed by fluorography or autoradiography of the dried gel.

### 3.6 Protein Interaction with BiP

#### 3.6.1 In Silico Prediction of BiP Binding Sites

Information of whether domains of a given polypeptide contain specific putative binding sites for BiP can complement the information provided by protein interaction experiments, which are described in Subheadings 3.6.2 and 3.7. The potential BIP binding sites in a protein can be predicted using the BIPPRED algorithm (https://www.bioinformatics.wzw.tum.de/bippred/submit) [10, 11]. BIPPRED operates by scanning a given polypeptide sequence using a 7mer window and assigning scores from 0 to 1 (low to high probable BiP binding site), based on experimental observations indicating that BiP recognizes peptides enriched in hydrophobic amino acids [12–14]. These are likely exposed on the surface of folding intermediates or structurally defective polypeptides, but they are masked in the interior of correctly folded and oligomerized proteins, thus becoming unavailable for BiP binding.

#### 3.6.2 Analysis of Protein Interaction with BiP

1. Protoplast or leaf homogenate is immunoselected as describe above with anti-BiP antiserum (commercially available or prepared in the investigator's laboratory).
2. After washing three times with NET-gel buffer, split the beads into three aliquots: one aliquot is washed again twice for 30 min with 500 μL of NET-gel buffer only, one with NET-gel buffer containing 6 mM MgCI2, and one with NET-gel buffer containing 6 mM MgCI2 and 3 mM ATP (*see* **Note 18**). Treatments are performed on ice.
3. The material still bound to the beads after these treatments is directly analyzed by SDS-PAGE.

4. The ATP-released material is re-immunoselected using antiserum or antibody against the protein of interest, following the procedure described above, Subheading 3.5, **steps 3–9**.
5. As a further control, repeat the entire procedure from **step 1** in reverse order, that is, using first the antiserum or antibody against the protein of interest and then selecting the ATP-released material with anti-BiP antiserum.

### *3.7 In Vivo Chaperone Activity Assay*

1. Prepare protoplasts from 4- to 7-cm-long leaves of wt tobacco plants cultivated in axenic condition as described [5].
2. Transfect protoplasts with a total 60 μg of plasmid mixture containing: 20 μg of plasmid encoding the marker secretory protein whose synthesis will be measured (*see* **Note 19**); 20 μg each of the plasmid encoding the chosen chaperones to be tested; the amount of a second secretory protein needed to reach the total 60 μg of plasmid mixture (*see* **Note 20**). Let protoplasts recover in K3 medium at 25 °C for 4 h.
3. Incubate protoplasts in the presence or absence of 20 μg/mL tunicamycin for 14–16 h at 25 °C in K3 medium (at $1*10^6$ protoplasts/mL).
4. Pulse-label protoplasts for 1 h, and collect as described at Subheading 3.4.1. You can then store then at −80 °C, as described in the same section.
5. Homogenize protoplasts (fresh or frozen) with ptsHB, and immunoselect the homogenate (*see* Subheading 3.5) with the appropriate antiserum against the marker secretory protein.
6. Analyze the immunoselected material by SDS-PAGE, fluorography, and quantification of band intensities using an appropriate software.

### *3.8 Drug Treatment to Inhibit Prolyl-4-hydroxylase Activity and Subsequent Hydroxyproline O-glycosylation on a Protein*

1. Prepare and transfect protoplasts with plasmid encoding the protein of interest as described in Subheading 3.7.
2. Incubate protoplasts for 24 h at 25 °C in K3 medium in the absence or presence of 0.25 mM 2,2′-dipyridyl, a ferrous chelating agent that inhibits in vivo activity of prolyl-4-hydroxylase, which contains the ferrous ion as a cofactor [15, 16].
3. Collect and homogenize protoplasts (fresh or frozen) with ptsHB.
4. Analyze aliquots corresponding to equal number of protoplasts by SDS-PAGE and protein blot with the appropriate antiserum against the putative O-glycosylated protein.
5. To determine if the protein is actually O-glycosylated, compare its electrophoretic mobility in samples treated or not with dipyridyl (*see* **Note 21**).

## 4 Notes

1. The described methods can be used both to analyze samples from *Nicotiana tobacum* or from *Arabidopsis thaliana*, indistinctly [5, 8, 17].
2. 2-mercaptoethanol is volatile and highly toxic. Avoid breathing vapors or skin contact. Always work under a ventilated hood when working with the stock chemical or any solution containing it. Discard solutions, tubes, and pipet tips in appropriate containers, separated from other disposed material.
3. Magnesium will preserve ribosomes, and therefore the ones attached to the ER will remain there. EDTA dissociates the ribosome subunits and releases them from the ER membrane. As a result, the density of the ER is higher in the presence of magnesium (because the ratio protein/lipid is higher) [6, 17, 18].
4. When melting the unlabelled amino acid mix, check it for precipitates: after several freezing and melting, the solution may become cloudy. If this occurs, discard and prepare a new stock.
5. The methods in this section are used to prepare proteins for all subsequent analysis described in this chapter, with the exception of the analysis of protein subcellular localization: non-ionic detergents dissolve membranes and therefore destroy the integrity of subcellular compartments. As described in Subheading 2.2, the homogenization buffers to analyze subcellular localization therefore do not contain ionic detergents.
6. If you start from frozen leaves work fast and do not allow them to warm up before homogenization, because freezing breaks membranes and thus vacuolar proteases will have access to all proteins as the tissue is defrosted.
7. HB contains salts and ionic detergent sufficient to bring into solution the vast majority of soluble and integral membrane proteins. However, certain proteins are not solubilized unless their disulfide bonds are reduced. The most striking examples are the prolamins of cereal seeds [19]. Therefore, if their recombinant forms are expressed in leaves of transgenic plants or transiently in protoplasts, the use of reducing HB or reducing ptsHB is mandatory for their solubilization [20, 21]. Other proteins that fold in the ER oxidizing environment may also require reducing conditions to be solubilized. When working on a new protein introduced in the secretory pathway, it is thus recommended to first compare its solubility in reducing and non-reducing homogenization buffers.

8. The pellet mainly contains cell wall polysaccharides and starch from plastids. Unbroken cells may also be present, as well as insoluble proteins that may require reducing conditions to enter into solution (*see* **Note** 7), or even strong denaturing agents such as SDS or urea. However, the use of strong denaturants will preclude further analysis of assembly or interactions with chaperones.
9. The pellet of protoplast homogenates may be not easy to see, especially if you are homogenizing less than 30,000 protoplasts. Proceed anyway to collect the supernatant, taking care of avoiding touching the bottom of the tube.
10. 16–65% sucrose is recommended to fractionate highly dense organelles, such as ER protein bodies [17, 21].
11. Remember to load each gradient with homogenate made in the *same* buffer (magnesium or EDTA)!
12. Use a 1 mL micropipette. Slowly and carefully collect each fraction by remaining as close as possible to the top of the gradient. Unavoidably, a very low amount of the top of the gradient will remain in the tube as you remove the different fractions. Due to the increasing difference in density, this will not contaminate the other fractions, until it will remain at the bottom of the tube where it will slightly contaminate the last fraction.
13. In the presence of $MgCl_2$, the plant ER has a density corresponding to 1.17–1.18 g/ml sucrose. When $Mg^{2+}$ is chelated by EDTA, ribosomes are detached from the ER membrane, which thus becomes less dense, with a consequent shift in migration along the gradient [6, 17, 18].
14. Microsomes derived from Golgi and ER, tonoplast and plasma membrane fragments, vesicles, mitochondria, and chloroplasts will precipitate at the bottom of the tube. Cytosolic proteins as well as soluble proteins of vacuoles (vacuoles break during homogenization) will remain in the 12% supernatant [6].
15. If proteins secreted by protoplasts into the incubation medium are the object of study, 150 μg/ml of bovine serum albumin (from the 4 mg/mL stock) should be added to the K3 medium. This will constitute a very abundant substrate for secreted proteases, thus strongly lowering the probability of degradation of the newly secreted polypeptides.
16. Upon the addition of non-radioactive amino acids to protoplast samples, the incorporation of radioactive ones into protein rapidly drops to almost undetectable levels. This occurs within a few minutes from the starting of chase. Do not expect the same to occur when leaves are labelled, because of the much slower incorporation in intact tissue and the variability

between different leaves. This makes very difficult to analyze the time course of events occurring within relatively short times (1–2 h) when leaves are labelled. Keep this in mind when you interpret your results.

17. Immunoprecipitation works also if homogenization is performed in reducing conditions, such as in the presence of 4% 2-mercaptoethanol [21, 22]. The immunoglobulin tetramer contains inter-chain disulfide bonds, but its assembly depends mainly on other types of interactions, and therefore it is not disrupted by 2-mercaptoethanol in the absence of strong denaturants such as SDS.
18. Like the other members of the heat shock 70 protein family, BiP is an ATPase and releases its bound ligands upon ATP hydrolysis. Because ATPase activity requires magnesium, the presence of EDTA during homogenization and immunoselection ensures that ligands are not released from the chaperone. Subsequent treatment with buffer containing ATP and magnesium promotes the release of polypeptides that have been immunoselected because BiP is acting on them as a molecular chaperone, rather than because of unspecific interactions [5, 6, 21, 23].
19. This assay is based on the observation that treatment with the N-glycosylation inhibitor tunicamycin, which negatively affects the folding of many secretory glycoproteins, increasing the workload of ER folding helpers to levels that are compensated by the unfolded protein response through many activities [2, 3]. These include a general reduction in the synthesis of secretory proteins, which can also be detected by ectopic expression of a model marker secretory protein and can be rescued by concomitant overexpression of the ER folding helper BiP [24]. The assay can therefore be used to test the rescuing activity of other putative ER folding helpers, on model or specific secretory proteins [25, 26].
20. It is important to maintain constant the total amount of secretory protein that is ectopically expressed in each sample, to make reliable comparisons when testing putative ER folding helpers (which are also secretory proteins). This is because the unfolded protein response quantitatively responds to the ER workload.
21. O-glycosylation decreases the electrophoretic mobility of a polypeptide to extents that depend on the number of O-glycans and their post-translational modifications. An increase in electrophoretic mobility in the presence of the inhibitor thus indicates O-glycosylation [16].

## Acknowledgments

We thank the present and past members of our laboratories for their fundamental contributions in developing and continuously improving the methods described here.

### References

1. Vitale A, Ceriotti A, Denecke J (1993) The role of the endoplasmic reticulum in protein synthesis modification and intracellular transport. J Exp Bot 44:1417–1444
2. Vitale A, Boston R (2008) Endoplasmic reticulum quality control and the unfolded protein response: insights from plants. Traffic 9:1581–1588
3. Liu XJ, Howell SH (2016) Managing the protein folding demands in the endoplasmic reticulum of plants. New Phytol 211:418–428
4. Zhang F, Boston RS (1992) Increases in binding protein (BiP) accompany changes in protein body morphology in three high lysine mutants of maize. Protoplasma 171:142–152
5. Pedrazzini E, Giovinazzo G, Bollini R, Ceriotti A, Vitale A (1994) Binding of BiP to an assembly-defective protein in plant cells. Plant J 5:103–110
6. Pedrazzini E, Giovinazzo G, Bielli A, de Virgilio M, Frigerio L, Pesca M, Faoro F, Bollini R, Ceriotti A, Vitale A (1997) Protein quality control along the route to the plant vacuole. Plant Cell 9:1869–1880
7. Hong Z, Jin H, Tzfira T, Li J (2008) Multiple mechanism-mediated retention of a defective brassinosteroid receptor in the endoplasmic reticulum of Arabidopsis. Plant Cell 20:3418–3429
8. Maîtrejean M, Wudick MM, Voelker C, Prinsi B, Mueller-Roeber B, Czempinski K, Pedrazzini E, Vitale A (2011) Assembly and sorting of the tonoplast potassium channel AtTPK1 and its turnover by internalization into the vacuole. Plant Physiol 156:1783–1796
9. Pollard JK, Steward FC (1959) The use of proline-C14 by growing cells: its conversion to protein and hydroxyproline. J Exp Bot 10: 17–32
10. Schneider M, Rosam M, Glaser M, Patronov A, Shah H, Back KC, Daake MA, Buchner J, Antes I (2016) BIPPred: combined sequence- and structure-based prediction of peptide binding to the Hsp70 chaperone BIP. Proteins Struct Funct Bioinf 84:1390–1407
11. Brocca L, Zuccaro M, Frugis G, Mainieri D, Marrano C, Ragni L, Klein EM, Vitale A, Pedrazzini E (2021) Two γ-zeins induce the unfolded protein response. Plant Physiol 187: 1428–1444
12. Flynn GC, Pohl J, Flocco MT, Rothman JE (1991) Peptide-binding specificity of the molecular chaperone BiP. Nature 353:726–730
13. Blond-Elguindi S, Cwirla SE, Dower WJ, Lipshutz RJ, Sprang SR, Sambroo JF, Gething M-JH (1993) Affinity panning of a library of peptides displayed on bacteriophages reveals the binding specificity of BiP. Cell 75:717–728
14. Nordquist EB, Clerico EM, Chen J, Gierasch LM (2022) Computationally-aided modeling of Hsp70–client interactions: past, present, and future. J Phys Chem B 126:6780–6791
15. Moriguchi R, Matsuoka C, Suyama A, Matsuoka K (2011) Reduction of plant-specific arabinogalactan-type O-glycosylation by treating tobacco plants with ferrous chelator 2,2′--dipyridyl. Biosci Biotechnol Biochem 75:994–996
16. Mainieri D, Morandini F, Maitrejean M, Saccani A, Pedrazzini E, Vitale A (2014) Protein body formation in the endoplasmic reticulum as an evolution of storage protein sorting to vacuoles: insights from maize γ-zein. Front Plant Sci 5:331
17. Ceresoli V, Mainieri D, Del Fabbro M, Weinstein R, Pedrazzini E (2016) A fusion between domains of the human bone morphogenetic protein-2 and maize 27 kD γZein accumulates to high levels in the endoplasmic reticulum without forming protein bodies in transgenic tobacco. Front Plant Sci 7:358
18. Barbante A, Irons S, Hawes C, Frigerio L, Vitale A, Pedrazzini E (2008) Anchorage to the cytosolic face of the endoplasmic reticulum membrane: a new strategy to stabilize a cytosolic recombinant antigen in plants. Plant Biotechnol J 6:560–575
19. Pedrazzini E, Mainieri D, Marrano CA, Vitale A (2016) Where do protein bodies of cereal seeds come from? Front Plant Sci 7:1139
20. Shani N, Rosenberg N, Kasarda DD, Galili G (1994) Mechanisms of assembly of wheat high molecular weight glutenins inferred from

expression of wild-type and mutant subunits in transgenic tobacco. J Biol Chem 269:8924–8930

21. Mainieri D, Rossi M, Archinti M, Bellucci M, De Marchis F, Vavassori S, Pompa A, Arcioni S, Vitale A (2004) Zeolin, a new recombinant storage protein constructed using maize gamma-zein and bean phaseolin. Plant Physiol 136:3447–3456
22. Pompa A, Vitale A (2006) Retention of a bean phaseolin/maize γ-zein fusion in the endoplasmic reticulum depends on disulfide bond formation. Plant Cell 18:2608–2621
23. Foresti O, Frigerio L, Holkeri H, de Virgilio M, Vavassori S, Vitale A (2003) A phaseolin domain directly involved in trimer assembly is a BiP binding determinant. Plant Cell 15:2464–2475
24. Leborgne-Castel N, Jelitto-Van Dooren EPWM, Crofts AJ, Denecke J (1999) Overexpression of BiP in tobacco alleviates endoplasmic reticulum stress. Plant Cell 11:459–469
25. Klein EM, Mascheroni L, Pompa A, Ragni L, Weimar T, Lilley KS, Dupree P, Vitale A (2006) Plant endoplasmin supports the protein secretory pathway and has a role in proliferating tissues. Plant J 48:657–673
26. De Marchis F, Colanero S, Klein EM, Mainieri D, Prota VM, Bellucci M, Pagliuca G, Zironi E, Gazzotti T, Vitale A, Pompa A (2018) Expression of CLAVATA3 fusions indicates rapid intracellular processing and a role of ERAD. Plant Sci 271:67–80

# Chapter 15

# ER Membrane Protein Interactions Using the Split-Ubiquitin System (SUS)

**Lisa Yasmin Asseck, Niklas Wallmeroth, and Christopher Grefen**

## Abstract

Protein-protein interactions (PPIs) play fundamental roles in all cellular processes. Especially membrane proteins facilitate a range of important biological functions in stimuli perception, signalling, and transport. Here we describe a detailed protocol for the yeast mating-based Split-Ubiquitin System (mbSUS) to study PPIs of ER membrane proteins in vivo. In contrast to the prominent yeast two hybrid, mbSUS enables analysis of full-length membrane proteins in their native cellular context. The system is based on the ubiquitin proteasome pathway leading to the release of an artificial transcription factor followed by activation of reporter genes to visualize PPIs. The mating-based approach is suitable for both small- and large-scale interaction studies. Additionally, we describe protocols to apply the recently established SUS Bridge assay (SUB), which is optimized for the detection of ternary protein interactions.

**Key words** Protein-protein interaction, Ternary interaction, Yeast, Split ubiquitin, mbSUS, SUB, Membrane proteins, Gateway

## 1 Introduction

Protein-protein interactions (PPIs) are prerequisite for a wide range of cellular functions. A large number of biophysical and biochemical techniques have been developed to unravel PPI networks for a better understanding of cellular complexities (for review, see [1]). For example, a technique such as the yeast two hybrid is often applied for their ease of use and applicability to large-scale approaches. One downside of this particular technique, however, is the obligatory nuclear localization of the interaction partners, which imposes truncation of membrane or large proteins to facilitate their nuclear (mis-)localization. Truncation and artificial nuclear localization in turn might cause artefacts and misinform on interactions among—for example—membrane proteins. Yet, many key processes in cell signalling and vesicle trafficking are mainly coordinated by interactions between or with membrane

Verena Kriechbaumer (ed.), *The Plant Endoplasmic Reticulum: Methods and Protocols*, Methods in Molecular Biology, vol. 2772, https://doi.org/10.1007/978-1-0716-3710-4_15,

proteins, which due to their natural environment and hydrophobic composition has remained experimentally difficult.

The mating-based Split-Ubiquitin System (mbSUS, [2]) described in this chapter is an approach well suited to elucidate full-length membrane protein interactions in vivo. The system relies on reassembly of the N- (Nub) and C-terminal (Cub) ubiquitin moieties that reconstitute a functional molecule when brought into close proximity by interacting proteins fused to these fragments [3]. To prevent spontaneous reassembly, Nub is mutated at residue Ile13 to either glycine or alanine that reduces its affinity for Cub [4] (*see* **Note 1**). Upon reconstitution, ubiquitin is recognized by cytosolic ubiquitin-specific proteases (USPs), which cleave off the artificial transcription factor PLV (ProteinA-LexA-VP16) fused to the C-terminus of Cub [5]. The released transcription factor then translocates into the nucleus to mediate activation of reporter genes allowing auxotrophy selection *(ADE2, HIS3)* and quantification of the relative interaction strength (*lacZ*). Thus, split ubiquitin as interaction sensor circumvents the use of mis-localized or truncated versions of fusion proteins significantly reducing artefacts associated with such protein modifications.

The mating-based strategy can be used not only to assay the interaction between two known proteins but is particularly adapted for large-scale interaction tests to screen a multitude of potential binding partners (*see* **Note 2**). In such screens, the mating approach simplifies and accelerates the experimental procedure compared to sequential transformations or co-transformations. Experimentally verified, binary interactions represent an idealized condition of the real-world biological complexity of multi-protein complexes. However, only a few methods have thus far been established to study multimeric interactions, an increasingly important task for our understanding of the cellular interactome. The SUS Bridge assay (SUB) is adapted from the classical Split-Ubiquitin System with the addition of a third (tagged or untagged) protein being expressed that can either facilitate, enhance, or inhibit a non-existent, weak, or previous interaction between bait and prey, respectively [6, 7].

In this book chapter, we detail the application of both the mbSUS and SUB technique, respectively, to detect interactions between membrane proteins. As textbook example of an ER-membrane complex, we choose the subunits of the heterotrimeric Sec61 translocon complex of *Arabidopsis.* Via Blastp analysis, we identified the *Arabidopsis* orthologues of the mammalian SEC61 translocon subunits. Assembly of the three α, β, and γ SEC61 subunits yields the ER-membrane localized channel, which mediates translocation of secretory proteins across and integration of membrane proteins into the lipid bilayer [8]. Here we show that the Sec61 subunit isoform α1 of *Arabidopsis thaliana* (At2g34250) directly interacts with SEC61β1 (At2g45070) and γ1 (At5g50460), respectively (Fig. 1).

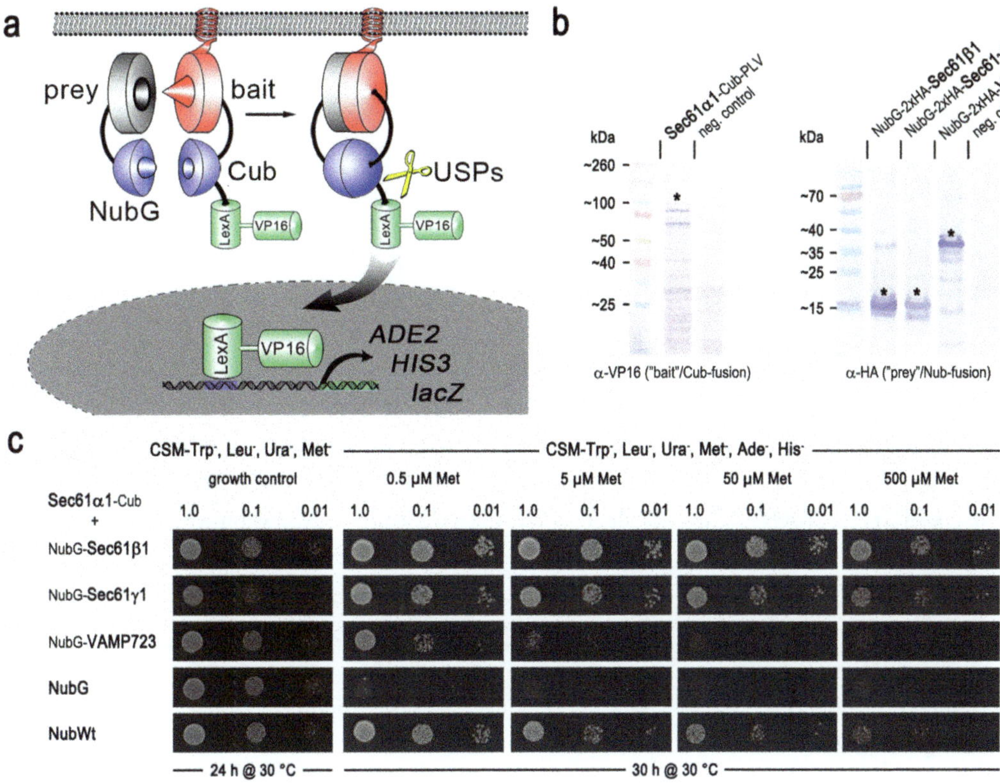

**Fig. 1** mbSUS analysis of the *Arabidopsis* Sec61 subunit α1 with Sec61β1 and Sec61γ1. (**a**) Schematic depiction of the SUS principle. Upon interaction of bait (red) and prey (gray) NubG and Cub reassemble to ubiquitin, which is recognized by ubiquitin-specific proteases (USPs) that cleave off the LexA-VP16 transcript activator. Diffusion of this construct into the nucleus activates reporter genes ADE2, HIS3, and lacZ, which can be monitored by growth or colorimetric assays. (**b**) Western blot analysis of haploid yeast prior mating and used in (**c**) with antibodies detecting the VP16 domain of the chimeric PLV and the HA-tag fused between NubG and GOI, respectively. (**c**) Growth assay of diploid yeast expressing the indicated fusion proteins. Yeast was dropped in serial dilutions (of OD600 from 1.0 to 0.01) on vector-selective (CSM-Leu⁻, Trp⁻, Ura⁻) and interaction-selective (CSM-Leu⁻, Trp⁻, Ura⁻, Ade⁻, His⁻) media with increasing methionine concentrations. NubWt was used as positive control and NubG as well as VAMP723 as negative controls (*see* **Note 3**). Sec61α1-Cub-PLV (~106 kDa), NubG-2xHA-Sec61β1 (~16 kDa), NubG-2xHA-Sec61ɣ1 (~15 kDa), NubG-2xHA-VAMP723 (~32 kDa)

## 2 Materials

### *2.1 Vectors and Strains*

The Gateway-compatible (exception: pNubWt-Xgate) mbSUS and SUB vectors, which are necessary for interaction analysis, are listed in Table 1. Maps and sequences of the vectors can be downloaded from http://www.zmbp.uni-tuebingen.de/dev-genetics/grefen/resources/yeast-vectors.html. The yeast strains that are used in this book chapter are listed in Table 2.

**Table 1**
**Destination vectors for mbSUS and SUB analysis**

| Plasmid name | Promoter | Origin | | Selection | | Function | Reference |
|---|---|---|---|---|---|---|---|
| | | *E. coli* | Yeast | *E. coli* | Yeast | | |
| pMetYC-Dest | Met25 | pUC | ARS/CEN | Amp, Cm | LEU2 | Met-repressible fusion protein with C-terminal Cub-PLV | [9] |
| pNX35-Dest | ADH1 | pUC | 2μ | Amp, Cm | TRP1 | Constitutive expression with N-terminal NubG-2xHA | [7] |
| pXNubA22-Dest | ADH1 | pUC | 2μ | Amp, Cm | TRP1 | Constitutive expression with C-terminal NubA-3xHA | [10] |
| pNubWt-Xgate | ADH1 | pUC | 2μ | Amp, Cm | TRP1 | Positive control vector, NubWt peptide; not a Gateway vector | [2] |
| pMZU-Dest | ADH1 | pUC | 2μ | Spec, Cm | URA3 | Constitutive expression with N-terminal myc | [7] |
| pZMU-Dest | ADH1 | pUC | 2μ | Spec, Cm | URA3 | Constitutive expression with C-terminal myc | [6] |

**Table 2**
**Yeast strain genotypes**

| Name | Organism | Genotype | Function | Reference |
|---|---|---|---|---|
| THY.AP4 | *S. cerevisiae* | MATa; *ade2*$^-$, *his3*$^-$, *leu2*$^-$, *trp1*$^-$, *ura3*$^-$; lexA::ADE2, lexA::HIS3, lexA::lacZ | Reporter strain, used for transformation of Cub-clones, used for SUB assay | [2] |
| THY.AP5 | *S. cerevisiae* | MATα; *ade2*$^-$, *his3*$^-$, *leu2*$^-$, *trp1*$^-$ | Used for transformation of Nub-clones; mate with THY.AP4 for binary interactions | [2] |

### 2.2 Media and Solutions for Growth and Transformation of Yeast

1. YPD media: 2% peptone, 2% glucose, 1% yeast extract; adjust pH to 6–6.3 with KOH before adding 2% oxoid agar.
2. Sterile deionized water (dd$H_2O$).
3. 1 M lithium acetate (LiAc): dissolve LiAc in dd$H_2O$. Adjust pH to 7.5 with acetic acid; sterilize by filtration.
4. 50% polyethylene-glycol 3350 (PEG 3350): dissolve PEG 3350 in deionized water to a final concentration of 50% (w/v); sterilize by filtration. Avoid water loss through autoclaving or during storage as this significantly decreases the transformation efficiency.

5. Single-stranded carrier DNA (ssDNA): dissolve 10 mg/mL ssDNA in deionized water, sonicate, and/or boil for 10 min following cooling on ice before use.
6. CSM-Ade$^-$, His$^-$, Leu$^-$, Met$^-$, Trp$^-$, Ura$^-$ as dropout.
7. Chemicals for auxotrophy selection, each dissolved in 100 mL water and sterilized by filtration; store in darkness at 4 °C:

   ADE: 0.4 g of adenine sulfate (add 5 mL per liter media).

   HIS: 0.4 g of l-histidine HCl (add 5 mL per liter media).

   LEU: 2.0 g of l-leucine (add 5 mL per liter media).

   TRP: 1.0 g of l-tryptophane (add 5 mL per liter media).

   URA: 0.4 g of uracile (add 5 mL per liter media).

   MET: 1.5 g of methionine (equals a 100 mM stock; add appropriate amount to obtain 0.5, 5, 50, and 500 μM final concentrations).
8. Selection media (1 L): 0.17% YNB (without amino acids), 0.5% ammonium sulfate, 2% glucose, 0.056% CSM-dropout mix; adjust pH to 6–6.3 with KOH before adding 2% oxoid agar; add appropriate auxotrophy selection chemical before or after autoclaving, e.g., ADE, HIS, TRP, and URA for transformation of THY.AP4 in the mbSUS assay (*see* **Note 4**).

### 2.3 Solutions for Western Blot Analysis

1. Lyse and load (LL-) buffer: 50 mM Tris (pH 6.8), 4% SDS, 4 M urea, 30% glycerol, 0.1 M DTT, 0.005% bromophenol blue; store at −20 °C.
2. Acid-washed glass beads (0.25–0.5 mm).
3. 20 mL SDS-PAGE resolving gel (12.5%): 6.2 mL $H_2O$, 8.4 mL acrylamide mix (30%), 5.0 mL 1.5 M Tris (pH 8.8), 0.2 mL SDS (10%), 0.2 mL $(NH_4)_2S_2O_8$ (10%), 0.008 mL TEMED.
4. 5 mL SDS-PAGE stacking gel (5%): 3.4 mL $H_2O$, 0.83 mL acrylamide mix (30%), 0.63 mL 1 M Tris (pH 8.8), 0.05 mL SDS (10%), 0.05 mL $(NH_4)_2S_2O_8$ (10%), 0.005 mL TEMED.
5. 10× SDS running buffer: 250 mM Tris, 1.9 M glycine, 0.15% SDS.
6. 100% methanol.
7. PVDF membrane.
8. 10× TBS: 250 mM Tris, 1.5 M NaCl, 20 mM KCl; adjust pH 7.4 (HCl).
9. Washing buffer, 1× TBS-Tween: 100 mL 10× TBS, 900 mL $H_2O$, 0.1% Tween20.
10. Blocking buffer: 1× TBS-Tween, 5% milk powder.
11. Antibodies (dilute 1:1000 in 1× TBS-Tween, and add 0.1% $NaN_3$).

   (a) Primaries: α-VP16, α-HA, α-myc.
   (b) Secondaries: anti-Rabbit IgG alkaline phosphatase, anti-Mouse IgG alkaline phosphatase.
12. Staining buffer: 100 mM Tris (pH 9.5), 100 mM NaCl, 5 mM $MgCl_2$.
13. NBT-solution: 50 mg/mL Nitro blue tetrazolium chloride in 70% DMF; aliquot and store at −20 °C.
14. BCIP solution: 50 mg/mL 5-Bromo-4-chloro3-indolylphosphate-p-Toluidin. in 100% DMF; aliquot and store at −20 °C.
15. Staining solution: 66 μL NBT-solution, 33 μL BCIP solution in 10 mL staining buffer; always prepare freshly.

## 3 Methods

The mbSUS technique allows fast screening of PPIs in vivo. The flowchart in Fig. 2 should provide a detailed overview of the steps necessary for execution of the methods and includes their approximate duration to aid in planning of the experiment.

### 3.1 Yeast Transformation

#### 3.1.1 mbSUS Transformation

1. Streak THY.AP4 and THY.AP5 yeast strains out on YPD plates, and incubate for 2 days at 30 °C.
2. Pick one colony, and inoculate into 5 mL YPD liquid media each. Grow overnight at 30 °C while shaking (240 rpm).

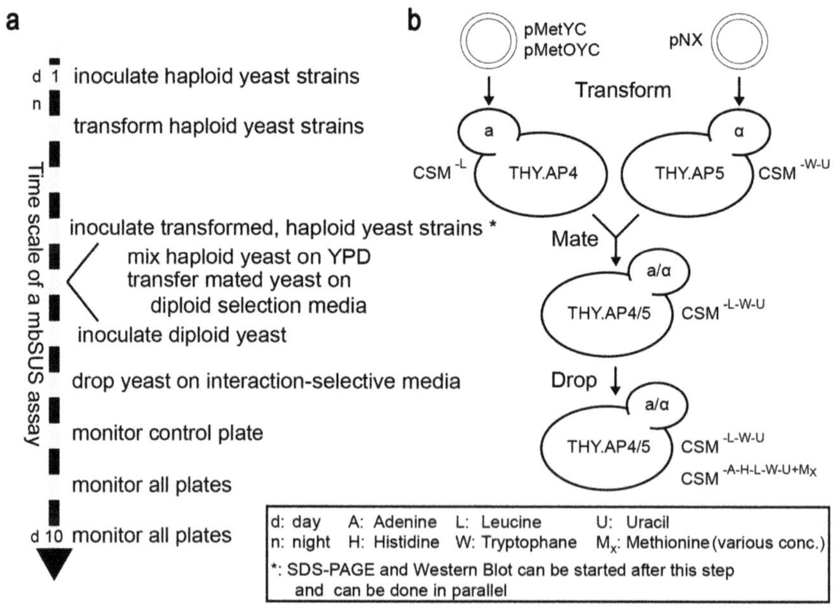


**Fig. 2** Flowchart of mbSUS analysis. (**a**) Timeline listing the necessary steps of an mbSUS analysis in consecutive order. (**b**) Cartoon depicting the different ploidy states of the yeast at the different stages of an mbSUS protocol

3. Add 2 mL of the overnight culture to 100 mL of fresh YPD, and incubate for 3–5 h at 30 °C while shaking (240 rpm) until the $OD_{600}$ reaches 0.5–0.8.
4. Harvest cells by centrifugation (5 min at 2000*g*) in sterile 50 mL tubes. Discard the supernatant.
5. Wash with 20 mL of sterile $ddH_2O$ and centrifuge again (5 min at 2000*g*). Discard the supernatant.
6. Resuspend the pellet in 1 mL of 0.1 M LiAc, and transfer to a 2 mL tube. Spin down (2 min at 1000*g*), and discard the supernatant.
7. Resuspend the cells in an appropriate amount of 0.1 M LiAc (20 μL per transformation), and incubate at room temperature for 30 min.
8. Meanwhile, prepare sterile PCR stripes with 10 μL ssDNA and 5 μL of plasmid DNA for each transformation (*see* **Note 5**).
9. Prepare a master mix by mixing 70 μL of 50% PEG, 10 μL 1 M LiAc, and 20 μL of competent yeast (**step** 7) for each transformation.
10. Distribute 100 μL of the master mix into each PCR tube, and mix carefully with the prepared DNA mixture.
11. Incubate for 20 min at 30 °C using a PCR cycler. Mix the reactions by gently pipetting up and down several times with a multi-channel pipette.
12. Incubate an additional 10 min at 30 °C.
13. Heat shock at 43 °C for 15 min.
14. Spin down briefly. Carefully remove the supernatant using a pipette.
15. *Optional:* Wash the pellet with 100 μL of sterile $ddH_2O$.
16. Resuspend the cells in 100 μL of sterile $ddH_2O$.
17. Plate the entire volume on appropriate selective minimal media (THY.AP4: CSM-$Leu^-$, THY.AP5: CSM-$Trp^-$, $Ura^-$).
18. Seal the plates, and incubate for 3 days at 30 °C.

#### 3.1.2 Bridge Assay Transformation

The transformation protocol is the same as that in Subheading 3.1.1 with the following exceptions:

1. Use haploid THY.AP4 cells exclusively.
2. Prepare DNA-mix with only 9 μL ssDNA and 2 μL each of the three different plasmid DNAs. For efficient transformation use more highly concentrated plasmid DNA (*see* **Note 5**).
3. Plate on CSM-$Leu^-$, $Trp^-$, $Ura^-$ media. Incubate for 3 days at 30 °C.

### 3.2 Mating

1. Pool several colonies per transformation and grow shaking overnight in 5 mL of appropriate selective media at 30 °C.
2. Harvest 2 mL each for Western blot analysis. Store the pellet at −20 °C.
3. Harvest 2 mL each by centrifugation (5 min at 1000*g*), and resuspend the pellet in 200 μL YPD. Scale up the volume for a higher number of crossings (20 μL per mating).
4. Mix 20 μL each, bait (THY.AP4) and prey (THY.AP5) of any desired combination in sterile PCR stripes.
5. Drop 5 μL of each mating onto a YPD plate.
6. Incubate (right side up) for 6–8 h at 30 °C.
7. Transfer mated yeast on CSM-Leu$^-$, Trp$^-$, Ura$^-$ plates using sterile pipette tips or a replicator stamp. Be careful NOT to transfer YPD medium with the cells.
8. Incubate overnight at 30 °C (*see* **Note 6**).

### 3.3 Screening Approach

1. Transform bait construct in THY.AP4 as described in Subheading 3.1.1 (*see* **Notes 7** and **8**).
2. Inoculate 5–10 colonies in 50 mL selective media (CSM-Leu$^-$) and incubate shaking overnight at 30 °C until an OD of ~$10^8$ is reached. Determine the total number of cells using a Neubauer Chamber.
3. Harvest cells by centrifugation (5 min at 1000*g*). Discard the supernatant.
4. Resuspend the cells in 1 mL of YPD (pH 4.5).
5. Simultaneously, transform prey library in THY.AP5 in 10 small-scale approaches.
6. Collect cells from the plates with selective media (CSM-Trp$^-$, Ura$^-$).
7. Harvest cells by centrifugation (5 min at 1000*g*). Discard the supernatant.
8. Resuspend the pellet in 10 mL of CSM-Trp$^-$, Ura$^-$. Determine the total cell number.
9. Mix both transformations 1:1. Fill up to 2 mL with YPD (pH 4.5).
10. Incubate 2 h at 30 °C, shaking at 35 rpm.
11. Plate in 300 μL aliquots on complete medium (YPD pH 4.5) on square plates (12 × 12 cm).
12. Incubate 5 h at 30 °C.
13. Wash cells with 15 mL dd$H_2O$.
14. Harvest cells by centrifugation (10 min at 1000*g*). Discard the supernatant.

15. Resuspend the cells in 5 mL $ddH_2O$.
16. Plate in 100 μL aliquots on CSM including 500 μM methionine and incubate for 3 days at 30 °C.
17. A total of ~$10^6$ colonies should be obtained.
18. Transfer colonies using sterile pipette tips on fresh CSM including 500 μM methionine, and incubate for 3 days at 30 °C. Be careful *not* to transfer too much yeast material to avoid growth of false positives.
19. Analyze by colony PCR (3.6) and sequencing.

### 3.4 Detection Assay

1. Inoculate in 2 mL selective media (CSM-$Leu^-$, $Trp^-$, $Ura^-$), and grow shaking overnight at 30 °C.
2. Pipette 100 μL in a 1.5 mL tube and 100 μL into a cuvette containing 900 μL $H_2O$.
3. Determine 1:10 diluted $OD_{600}$. Note the values.
4. Harvest cells in the tubes by centrifugation (2 min at 2000*g*). Carefully remove the supernatant by pipetting.
5. Add appropriate volume of sterile $ddH_2O$ to reach a final $OD_{600}$ of 1.0 (e.g., 1:10 dilution has an $OD_{600}$ value of 0.450; resuspend yeast pellet in 450 μL sterile $H_2O$).
6. Make tenfold serial dilutions (1:100, 1:1000): Prepare tubes with 900 μL of sterile $ddH_2O$. Add 100 μL of the appropriate 1:10 dilution, and mix well by shaking by hand. For the 1:100 dilution, transfer 100 μL of the 1:100 dilution into another tube containing 900 μL $ddH_2O$.
7. Drop 7 μL yeast of each dilution on selective plates containing increasing methionine concentrations as well as on vector-selective media (CSM-$Leu^-$, $Trp^-$, $Ura^-$) as growth control. Allow the drops to dry (*see* **Notes 9** and **10**).
8. Seal the plates, and incubate for 1–3 days at 30 °C. Remove the vector-selective growth control after 24 h of incubation to see the gradient of the yeast ODs.
9. Document results by scanning or photography using a black background.

### 3.5 Western Blot

1. Harvest 2 mL of overnight yeast culture or use deep-frozen aliquots.
2. Add ~50 μL glass beads.
3. Resuspend cells in 200 μL LL-buffer by vortexing for approx. 2 min.
4. Incubate shaking for 10 min at 65 °C.
5. Centrifuge at 16500*g* for 10 min.

6. Transfer the supernatant to a fresh tube.
7. Load 10–20 μL on a SDS-PAGE gel. *Optional:* Store at −80 °C.
8. Run gel with appropriate conditions.
9. Run Western blot (e.g., semidry transfer on PVDF membrane).
10. Block membrane in blocking buffer on a shaker for 1 h at room temperature. *Optional:* overnight at 4 °C.
11. Wash 3 times, 10 min each, with 1× TBST.
12. Transfer membrane into primary antibody solution. Incubate at least 1 h at room temperature. *Optional:* overnight at 4 °C.
13. Wash 3 times, 10 min each, with 1× TBST.
14. Transfer membrane into secondary antibody solution. Incubate at least 1 h at room temperature. *Optional:* overnight at 4 °C.
15. Wash 3 times, 10 min each, with 1× TBST.
16. Equilibrate membrane in staining buffer for 1 min.
17. Transfer membrane into staining solution. Incubate until clear signals can be monitored. Stop substrate reaction by washing with $H_2O$.

### 3.6 Yeast Colony PCR and Sequencing

1. Prepare sterile PCR tubes, strips, or plates with 50 μL dd$H_2O$ (*see* **Note 11**).
2. Add small amount of yeast material using a sterile pipette tip.
3. Boil the samples at 100 °C for 10 min using a PCR cycler.
4. Cool and spin down briefly to avoid cross-contaminations when opening.
5. Prepare fresh tubes with 20 μL PCR master mix.
6. Use 10 μL of supernatant as sample and add into each well.
7. Run the following PCR program:
   (a) 95 °C—2.00 min.
   (b) 95 °C—0.30 min.
   (c) 58 °C—0.30 min.
   (d) 72 °C—1.30 min.
   (e) 72 °C—3.00 min.
   (f) Repeat **steps (b)–(d)** 37 times.
8. Run 5 μL on a 1.5% agarose/TAE gel to determine successful amplification, and insert size. Use the remainder for sequencing.

## 4 Notes

1. Depending on the presence of signal peptides or transmembrane domains, prey proteins can be tagged with either N-terminal NubG or C-terminal NubA. As mentioned in the introduction, mutating the isoleucine at position 13 prevents spontaneous reassembly with Cub. Replacement by alanine shows a slightly increased affinity toward Cub [4] and can hence be used in the C-terminal Nub vectors that usually perform weaker compared with the N-terminal Nubs [1, 10].
2. The mbSUS is ideally suited for structure function analysis of proteins [11, 12] since the mating-based approach allows biased screening of many different mutants against the same interaction partner. Generation of mutants can best be done by including a point mutation together with a silent mutation that introduces a restriction site as identifier for each mutation [13].
3. In order to correctly interpret your data, it is crucial to always include the whole set of controls, comprising both NubI as positive control and "empty" vectors as negative controls. Moreover, biological negative controls might provide additional evidence for the specificity of an interaction.
4. For solid media, add a magnetic stir bar before autoclaving. To give uniform growth conditions, measure 80 mL medium per square plate (12 cm × 12 cm) or 30 mL into round plates (diam. 9 cm), respectively. Try to avoid air bubbles.
5. Efficient (co-)transformation of yeast depends on the cell concentration as well as on the amount of carrier and plasmid DNA. Therefore, we recommend to use at least 1 μg of plasmid DNA. However, it is important to note that transforming three plasmids at once results generally in lower transformation efficiency, which can be counterbalanced through upscaling of the transformation mixture.
6. Red/white selection of diploid yeasts can be used as a preliminary tool to estimate positive/negative interactions. The effect is based on the *ADE2* reporter gene, which encodes an enzyme involved in the de novo synthesis of adenine. If *ADE2* is not activated due to the lack of interaction between bait and prey peptides, cells accumulate a red-colored intermediate, whereas cells expressing interacting proteins become white.
7. The native structure and topology of an integral membrane protein are important parameters that have to be considered to determine tag orientation of bait proteins. It should always be taken into account that the Cub-PLV tag has to reside in the cytosol (not in the lumen of the ER or other compartments), as the USPs are cytosolic enzymes and the cleaved PLV needs to

be able to diffuse into the nucleus. Also, the bait protein should be large and attached or integrated into a membrane to prevent leakage of the fusion protein into the nucleus. Tools for the prediction of membrane protein topology and signal peptides such as "TMHMM" (http://www.cbs.dtu.dk/services/TMHMM/) and "SignalP" (http://www.cbs.dtu.dk/services/ SignalP/) might be helpful to decide which bait should be used.

To determine the interaction with soluble baits, we recommend "cytoSUS" using a vector with an N-terminal OST4 transmembrane domain to artificially anchor the protein to the ER membrane [14]. Otherwise, the construct itself might diffuse into the nucleus and activate reporter genes due to its PLV fusion, leading to false-positive results.

8. Before starting an experiment, yeast expressing the bait/Cub-fusion proteins should be mated with soluble NubG and NubI peptides, respectively, to verify functionality/stability and excluding self-activation (*see* **Note 12**). This includes growth assays as well as immunoblots of the bait proteins (Subheadings 3.4 and 3.5).
9. Increased assay stringency is achieved by expressing bait fusions in low-copy ARS/CEN vectors under the control of the methionine-repressible Met25 promoter. To identify both strong and weak interactions, we usually apply methionine concentrations of 0.5 μM–500 μM. However, methionine above 500 μM should be avoided as yeast growth becomes negatively affected by such concentrations.
10. It is helpful to print a dropping scheme and stick it under the media plates to guide pipetting. To avoid satellite colonies, the yeast drops should be placed carefully, and under no circumstances should the second pressure point of the pipette be used. The surface of the media should be relatively dry to prevent dispersion of the spots. Also, high (500 μM) levels of methionine seem to break the surface tension of the solid media causing carefully placed yeast drops to disperse on the surface. For this reason, we recommend not to place more than eight by eight yeast drops on a 12 cm square plate.
11. Contrary to our previous protocol [6], we repeatedly found in the evaluation of colony PCR results since that DMSO seems to inhibit the PCR reaction and should hence be omitted from the PCR mix. Also, addition of 0.2 M NaOH, which should aid in lysis of yeast cells, seems to negatively affect PCR fidelity. Instead, boiling the yeast colonies in water should suffice to extract an appropriate amount of template DNA for subsequent PCR amplification.

12. If one of your fusion proteins is not expressing well, it might be related to protein instability due to the tagging and thereby masking of signal peptides or other recognition/processing sites, especially in the N-terminal region. Also codon bias has a crucial role in heterologous gene expression and non-optimal codon usage can limit the expression. Prior to codon optimization, we recommend to verify single colonies by Western blot (Subheading 3.5).

## Acknowledgments

We are grateful to Eva Schwörzer and Laure Grefen for excellent technical support. Work in our lab is supported through seed funding of the SFB1101 and an Emmy Noether fellowship of the German Research Foundation (DFG) to C.G. (GR 4251/1-1).

### References

1. Xing S, Wallmeroth N, Berendzen KW, Grefen C (2016) Techniques for the analysis of protein-protein interactions in vivo. Plant Physiol 171:727–758
2. Obrdlik P et al (2004) K+ channel interactions detected by a genetic system optimized for systematic studies of membrane protein interactions. Proc Natl Acad Sci U S A 101:12242–12247
3. Johnsson N, Varshavsky A (1994) Split ubiquitin as a sensor of protein interactions in vivo. Proc Natl Acad Sci U S A 91:10340–10344
4. Raquet X, Eckert JH, Muller S, Johnsson N (2001) Detection of altered protein conformations in living cells. J Mol Biol 305:927–938
5. Stagljar I, Korostensky C, Johnsson N, Heesen S (1998) A genetic system based on split-ubiquitin for the analysis of interactions between membrane proteins in vivo. Proc Natl Acad Sci U S A 95:5187–5192
6. Grefen C (2014) The split-ubiquitin system for the analysis of three-component interactions. Methods Mol Biol 1062:659–678
7. Grefen C, Blatt MR (2012) Do calcineurin B-like proteins interact independently of the serine threonine kinase CIPK23 with the K+ channel AKT1? Lessons learned from a menage a trois. Plant Physiol 159:915–919
8. Park E, Rapoport TA (2012) Mechanisms of Sec61/SecY-mediated protein translocation across membranes. Annu Rev Biophys 41:21–40
9. Grefen C, Lalonde S, Obrdlik P (2007) Split-ubiquitin system for identifying protein-protein interactions in membrane and full-length proteins. Curr Protoc Neurosc Chapter 5:Unit 5.27
10. Grefen C, Obrdlik P, Harter K (2009) The determination of protein-protein interactions by the mating-based split-ubiquitin system (mbSUS). Methods Mol Biol 479:217–233
11. Zhang B et al (2015) The Arabidopsis R-SNARE VAMP721 interacts with KAT1 and KC1 $K^+$ channels to moderate $K^+$ current at the plasma membrane. Plant Cell 27:1697–1717
12. Grefen C et al (2015) A vesicle-trafficking protein commandeers Kv channel voltage sensors for voltage-dependent secretion. Nat Plants 3: 15108
13. Karnik A, Karnik R, Grefen C (2013) SDM-Assist software to design site-directed mutagenesis primers introducing "silent" restriction sites. BMC Bioinf 14:105
14. Karnik R et al (2015) Binding of SEC11 indicates r role in SNARE recycling after vesicle fusion and identifies two pathways for vesicular traffic to the plasma membrane. Plant Cell 27: 675–694

# Chapter 16

# Analysis of Protein Glycosylation in the ER

**Jennifer Schoberer, Yun-Ji Shin, Ulrike Vavra, Christiane Veit, and Richard Strasser**

## Abstract

Protein N-glycosylation is an essential posttranslational modification which is initiated in the endoplasmic reticulum (ER). In plants, the N-glycans play a pivotal role in protein folding and quality control. Through the interaction of glycan processing and binding reactions mediated by ER-resident glycosidases and specific carbohydrate-binding proteins, the N-glycans contribute to the adoption of a native protein conformation. Properly folded glycoproteins are released from these processes and allowed to continue their transit to the Golgi where further processing and maturation of N-glycans leads to the formation of more complex structures with different functions. Incompletely folded glycoproteins are removed from the ER by a highly conserved degradation process to prevent the accumulation or secretion of misfolded proteins and maintain ER homeostasis. Here, we describe methods to analyze the N-glycosylation status and the glycan-dependent ER-associated degradation process in plants.

**Key words** N-glycosylation, Glycoprotein, Oligosaccharyltransferase, Quality control, ERAD

## 1 Introduction

Glycosylation is an ubiquitous modification of newly synthesized proteins in the endoplasmic reticulum (ER). Dependent on the linkage of the oligosaccharide to the amino acid side chain of the protein, there are two major types of glycosylation: N- and O-glycosylation. O-Glycosylation is quite diverse and various types of O-glycans have been described for secretory proteins in eukaryotes including mucin-type O-linked N-acetylgalactosamine (GalNAc) or O-mannosylation [1, 2]. These O-glycosylation reactions are initiated by the transfer of a single monosaccharide to serine or threonine residues of proteins. While the O-glycan biosynthesis can already be initiated in the ER, the majority of the O-glycan modifications take place in the Golgi. In plants, it has been proposed that the plant-specific serine O-galactosylation of extensins or arabinogalactan proteins is initiated in the ER, while further maturation of glycans takes place in the Golgi [3].

Verena Kriechbaumer (ed.), *The Plant Endoplasmic Reticulum: Methods and Protocols*, Methods in Molecular Biology, vol. 2772, https://doi.org/10.1007/978-1-0716-3710-4_16,

In contrast to O-glycosylation, N-glycosylation and early N-glycan processing steps are highly conserved in all eukaryotes [4]. A hallmark of N-glycosylation is the en bloc transfer of a common preassembled oligosaccharide ($Glc_3Man_9GlcNAc_2$) from the lipid carrier dolichol pyrophosphate to selected asparagine residues in the sequence Asn-X-Ser/Thr ($X \neq P$) within nascent polypeptides [5] (*see* **Note 1**). The transfer of the oligosaccharide takes place at the luminal side of the ER and is catalyzed by the oligosaccharyltransferase (OST) complex. In the ER, the attached N-glycans are crucial for protein folding and for quality control processes. Notably, by exposing defined motifs, the N-glycans orchestrate the fate of glycoproteins in the ER [6]. In contrast to the conserved role of ER-type oligomannosidic structures, complex N-glycans are generated in the Golgi and serve different protein-specific roles that are still not very well understood in plants (Fig. 1a) [7–9].

### 1.1 N-Glycosylation

The biosynthesis of the lipid-linked oligosaccharide starts on the cytosolic side of the ER membrane by a number of different glycosyltransferases that have been initially characterized in yeast (asparagine linked glycosylation proteins (ALGs)) [10]. The synthesized $Man_5GlcNAc_2$-PP-Dol is flipped through the ER membrane (Fig. 1b) and further elongated in the lumen of the ER by three α-mannosyltransferases (ALG3, ALG9, and ALG12) and three α-glucosyltransferases (ALG6, ALG8, and ALG10) (Fig. 1b). Putative orthologs of these enzymes are present in the plant genome, and mutants with defects in the assembly of the oligosaccharide precursor have been described [4]. For example, knockout of the *Arabidopsis thaliana* ALG10 gene results in an incomplete synthesis of the oligosaccharide precursor leading to reduced glycosylation efficiency of proteins [11]. The observed underglycosylation defect affects different glycoproteins and results in a leaf growth phenotype.

The fully assembled oligosaccharide precursor is transferred by OST to asparagine residues within the consensus sequence. In yeast, OST is a heteromeric membrane-bound protein complex [12] consisting of one catalytically active subunit (STT3) and several different non-catalytic subunits that contribute to N-glycosylation by regulation of the substrate specificity, stability, or assembly of the complex [12]. Humans harbor two different catalytic isoforms (STT3A and STT3B) that are present in distinct OST complexes and differ in their catalytic activity and acceptor substrate selectivity [13]. While STT3A is predominately involved in co-translational glycosylation, STT3B displays a preference for posttranslational glycosylation. The transfer of the preassembled oligosaccharide in plants involves a similar OST multi-subunit complex, which is still poorly characterized [14]. In *A. thaliana*, two proteins (STT3A and STT3B) with homology to yeast and

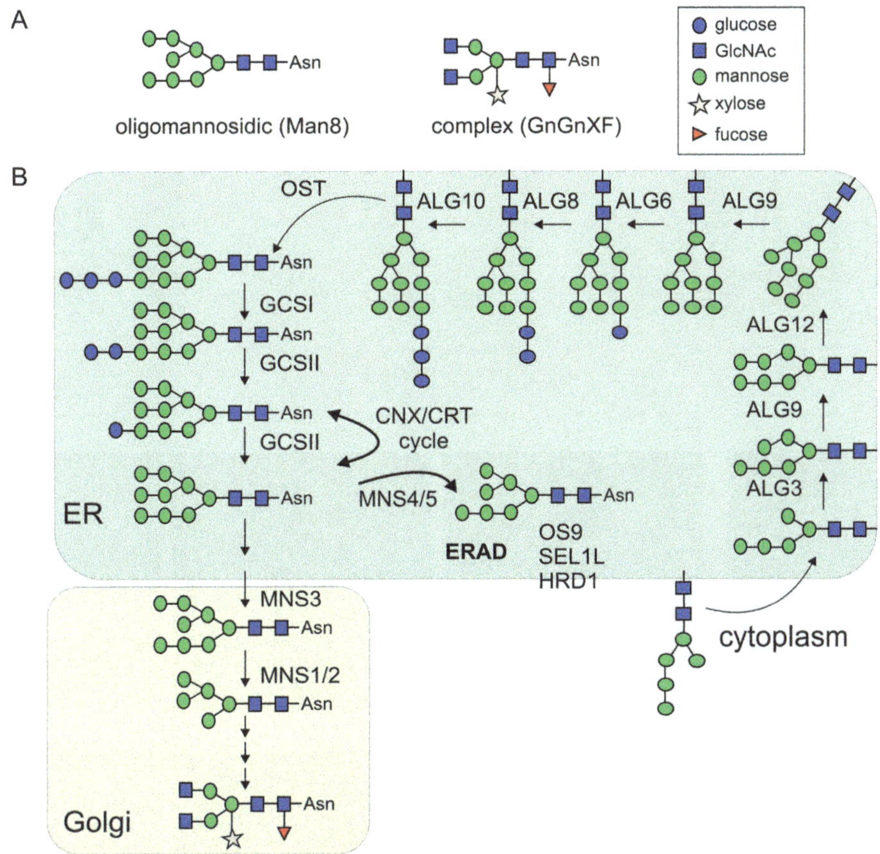


**Fig. 1** (**a**) Schematic representation of two major N-glycan structures in plants: ER-type oligomannosidic and Golgi-processed complex N-glycans. (**b**) N-glycosylation and N-glycan processing in the ER of plants. The dolichol-linked $Man_5GlcNAc_2$ intermediate is translocated from the cytosol into the lumen of the ER (the preceding cytosolic biosynthesis steps are not shown). In the lumen of the ER, mannose residues are sequentially added by the α-mannosyltransferases ALG3, ALG9, and ALG12. Further elongation of the oligosaccharide is carried out by the α-glucosyltransferases ALG6, ALG8, and ALG10. The fully assembled $Glc_3Man_9GlcNAc_2$ oligosaccharide is transferred to selected asparagine residues by the oligosaccharyltransferase (OST) complex. Upon transfer, terminal glucose residues are trimmed by α-glucosidase I (GCSI) and α-glucosidase II (GCSII). A terminal glucose residue is required for the interaction with calnexin (CNX) and calreticulin (CRT) that support folding of the glycoproteins. Properly folded proteins may be further processed in the ER by the α-mannosidase MNS3 [22] and are allowed to exit to the Golgi which is the site for complex N-glycan formation. In the ER, terminally misfolded proteins are subjected to processing by the α-mannosidases MNS4 and MNS5 which results in the degradation by the OS9–SEL1L–HRD1 complex [24]. The symbols for representation of the glycan structures follow the style of the Consortium for Functional Glycomics (www.functionalglycomics.org)

mammalian catalytic subunits have been identified [15]. STT3A-deficient plants are viable, but display a protein underglycosylation defect that impairs the biogenesis of different heavily glycosylated proteins, such as the pattern recognition receptor EF-TU RECEPTOR (EFR) or the endo-β1,4-glucanase KORRIGAN1 [7, 16, 17]. By contrast, much less is known about STT3B function and

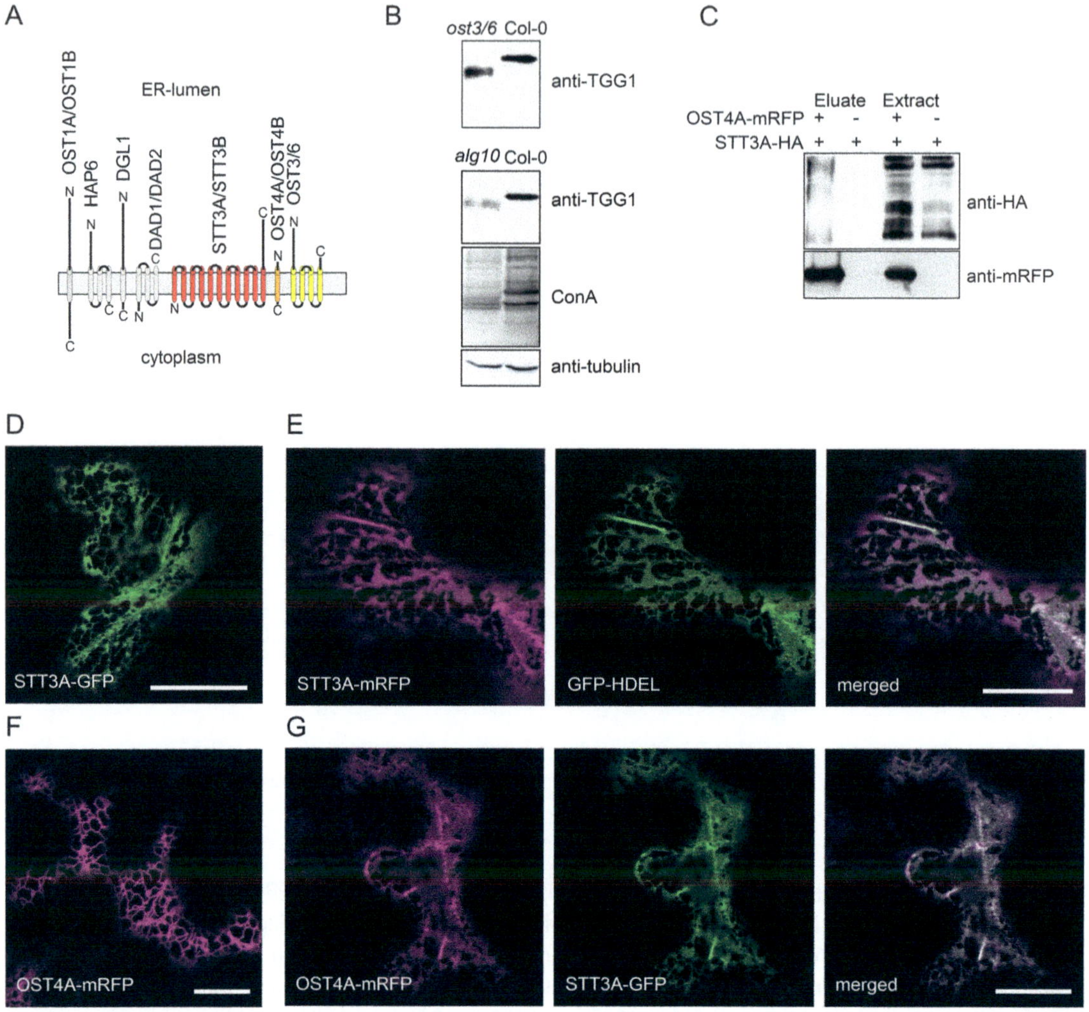


**Fig. 2** (**a**) Schematic representation of the OST subunits and their predicted topology [4]. Subunits which likely interact (based on co-IP experiments) [19] are depicted in red, orange, and yellow. Other subunits are shown in gray. (**b**) Protein extracts from *A. thaliana* leaves were analyzed by immunoblotting with anti-TGG1 or anti-tubulin (non-glycosylated control) antibody. The shift in mobility indicates underglycosylation of TGG1 in the *ost3/6* and *alg10* mutants. The ConA lectin blot shows an overall weaker signal in the *alg10* mutant, and some bands display an altered migration which is consistent with the underglycosylation defect. (**c**) OST4A-mRFP interacts with STT3A-HA. Proteins were transiently co-expressed in *N. benthamiana* leaves and extracts were analyzed 2 dpi. OST4A-mRFP was purified by RFP-Trap®_A beads and co-isolated STT3A-HA was monitored by immunoblotting with anti-HA antibodies. Note: due to the presence of numerous transmembrane helices, STT3A-HA displays anomalous gel mobility. (**d–g**) Analysis of fluorescent proteins was done by confocal microscopy. Bars = 20 μm. The indicated constructs were transiently expressed in *N. benthamiana* leaves and analyzed 2 dpi

its substrates. Notably, *A. thaliana stt3a stt3b* double knockouts are gametophytic lethal [15], highlighting the importance of the catalytic OST subunit for protein N-glycosylation in plants. Other subunits that have been characterized include DGL1, OST3/6,

OST1A/B, and DAD1/2 [14, 18, 19] (Fig. 2a). While DGL1 and OST1A/B appeared essential for plants, the *dad1 dad2* double mutant is viable but displayed a plant growth and pollen production defect [14]. OST3/6-deficient plants do not display any growth phenotype but are compromised in their immunity [19], and OST3/6 is required for the interspecific gametophyte recognition in *A. thaliana* [20].

### 1.2 N-Glycan Processing in the ER

Specific processing of N-glycans in the ER generates distinct oligosaccharide structures that expose signals for quality control processes or degradation. Upon transfer of the preassembled oligosaccharide to the polypeptide chain, glucosidase I (GCSI) and glucosidase II (GCSII) remove the two outermost glucose residues. The resulting monoglucosylated N-glycan can interact with the lectins calnexin (CNX) and calreticulin (CRT) to promote folding. In the current models, release from the CNX/CRT interaction requires the trimming of the remaining glucose by GCSII. Re-addition of the single glucose by the folding sensor UDP-glucose:glycoprotein glucosyltransferase (UGGT) facilitates prolonged interaction with CNX/CRT [21]. Several rounds of glucose trimming, re-glucosylation, and interaction with CNX/CRT are possible until proteins attain their final conformation and are released from this glycan-dependent quality control process.

The N-glycans from properly folded proteins or from folding intermediates may be subjected to further processing in the ER which is catalyzed by the α-mannosidase MNS3. Analysis of the substrate specificity revealed that MNS3 displays the typical ER-α-mannosidase I activity that has been described in mammals and yeast. In contrast to yeast, however, MNS3 is mainly located in small punctate structures resembling Golgi bodies [22]. This specific subcellular localization is a means to spatially separate MNS3 from ER-localized mannose trimming steps that generate the glycan signal required for flagging terminally misfolded glycoproteins for ER-associated degradation (ERAD). Further processing takes place in the Golgi and is initiated by Golgi α-mannosidase I (MNS1/MNS2)-mediated trimming of mannose residues resulting in the formation of $Man_5GlcNAc_2$.

### 1.3 N-Glycan-Dependent Quality Control and ER-Associated Degradation

To maintain protein homeostasis in the ER, all eukaryotic cells have a conserved machinery that selects aberrant or incompletely assembled proteins for degradation. In mammals and yeast, this ERAD pathway involves recognition of the misfolded protein, retrotranslocation from the ER to the cytoplasm, ubiquitination of the protein, and its subsequent disposal by the 26S proteasome [23]. While the latter steps have not been characterized in detail in plants, the selection and delivery reactions that take place in the lumen of the ER appear highly conserved. If folding attempts are unsuccessful, the α-mannosidases MNS4 and MNS5 cleave a terminal mannose residue from the C-branch of the oligomannosidic N-glycan [24]. The exposed α1,6-linked mannose residue is recognized by the lectin-like protein OS9 and delivered to the SEL1L–HRD1 ERAD complex resulting in the degradation of the aberrant glycoprotein [25] (Fig. 1b).

Here, we provide protocols to address different aspects of protein N-glycosylation, N-glycan-processing, and glycan-dependent ERAD that take place in the lumen of the ER.

## 2 Materials

1. *A. thaliana* seeds with characterized underglycosylation defects can be purchased from the European Arabidopsis Stock Center (*alg10-1*: NASC ID: N874475) [11, 15]. Seeds with defects in glycan-dependent ERAD are available from NASC (*os9-1*: N529413; *mns4 mns5*: double mutant can be obtained by crossing of N656907 and N162962) [24, 26]. The *bri1-9* seeds were kindly provided by Frans E. Tax. The *ost3/6-1* seeds can be requested from the Strasser group [19].
2. Mixer mill with steel beads.
3. Container with liquid nitrogen.
4. Refrigerated benchtop centrifuge with a relative centrifugal force of 9.400 or higher and a rotor for 1.5/2 mL microcentrifuge tubes.
5. Thermo block.
6. Orbital shaker.
7. Vertical electrophoresis system.
8. Tank transfer system.
9. Power supply with at least 100 V and 400 mA.
10. SDS–PAGE (10 to 12%) gels.
11. Tris/glycine/SDS–PAGE running buffer: (25 mM Tris–HCl, 192 mM glycine, 0.1% (w/v) SDS.
12. 3× Laemmli sample buffer: 187.5 mM Tris–HCl pH 6.8, 30% (w/v) glycerol, 6% (w/v) SDS, 15% (v/v) β-mercaptoethanol, 0.15% (w/v) bromophenol blue.
13. Prestained protein standard.
14. Tris/glycine/methanol transfer buffer: 25 mM Tris–HCl, 192 mM glycine, 20% (v/v) methanol.
15. Blotting paper.
16. Nitrocellulose blotting membrane.

17. 1× PBS: phosphate-buffered saline: 137 mM NaCl, 2.7 mM KCl, 10 mM $Na_2HPO_4$, 2 mM $KH_2PO_4$, pH 7.4.
18. PBST: 1× PBS + 0.1% (v/v) Tween 20.
19. 1× TBS: Tris buffered saline (25 mM Tris–HCl, pH 7.5, 150 mM NaCl, 2 mM KCl).
20. TBST: 1× TBS + 0.1% (v/v) Tween 20.
21. Blocking solution for immunoblots: TBST +5% (w/v) skimmed milk powder or PBST +3% (w/v) BSA.
22. Western blotting detection reagent: as an example, we use SuperSignal West PICO Chemiluminescent Substrate or ECL Prime Western Blotting Detection Reagent.
23. Sensitive films for detection: for example, we use Amersham Hyperfilm ECL (GE Healthcare).
24. Film developer and fixer solutions.
25. Rat monoclonal HA antibody (Roche 11867423001): 1:2000 diluted in TBST + 1% (w/v) BSA.
26. Polyclonal TGG1 antibody (Agrisera: AS20 4416): 1:1000 diluted in TBST.
27. Polyclonal BRI1 antibody (Agrisera: AS12 1859): 1:5000 diluted in TBST + 5% (w/v) skimmed milk powder.
28. Rat monoclonal RFP antibody (ChromoTek: 5F8). 1:1000 diluted in PBST + 1% (w/v) BSA.
29. Mouse monoclonal GFP antibody (Roche: 11814460001). 1: 2000 diluted in PBST + 3% (w/v) BSA.
30. Mouse monoclonal α-tubulin antibody (Sigma-Aldrich: T6074). 1:5000 diluted in PBST + 1% (w/v) BSA.
31. Concanavalin A-peroxidase (ConA-HRP, Sigma-Aldrich: L6397): ConA solution: 0.5 μg/mL Con A-HRP in PBST containing 1 mM $MnCl_2$, 1 mM $CaCl_2$, 1 mM $MgCl_2$.
32. Anti-rabbit IgG–peroxidase (such as Sigma A0545): 1:5000 diluted in TBST +5% (w/v) skimmed milk powder for BRI1 detection and 1:50000 diluted in TBST for TGG1 detection.
33. Anti-mouse IgG–peroxidase (such as Sigma A9044): 1:10,000 diluted in TBST for detection of GFP and α-tubulin.
34. Anti-rat IgG–peroxidase (Jackson): 1:10,000 diluted in PBST + 0.5% (w/v) BSA for detection of RFP and HA-tagged proteins.
35. Endoglycosidase H (Endo H; we use 500,000 units/mL, New England Biolabs) + GlycoBuffer 3 (New England Biolabs) + 10× Glycoprotein Denaturing Buffer (New England Biolabs).
36. Empty micro spin chromatography columns.

37. Nonidet P-40 (10% solution).
38. Kifunensine (Santa Cruz Biotechnology): class I α-mannosidase inhibitor, dissolved in ultrapure water.
39. RIPA buffer: 150 mM NaCl, 1.0% (v/v) Nonidet P-40, 0.5% (v/v) sodium deoxycholate, 0.1% (w/v) SDS, 50 mM Tris–HCl, pH 8.0.
40. RFP-Trap®_A beads (ChromoTek) for purification of RFP-tagged proteins.
41. Protein A Sepharose (such as Protein A Sepharose Fast Flow from GE Healthcare).
42. Coomassie Brilliant Blue staining solution: 0.25% (w/v) Coomassie Brilliant Blue R-250, 10% (v/v) acetic acid, 50% (v/v) methanol.
43. Infiltration buffer: 0.5% (w/v) D-glucose, 50 mM 2-(N-morpholino)ethanesulfonic acid (MES), 2 mM $Na_3PO_4$, 0.1 mM acetosyringone.
44. *Agrobacterium tumefaciens* strain UIA143 pMP90 [27].
45. LB broth (Lennox): 10 g/L tryptone, 5 g/L NaCl, 5 g/L yeast extract. Sterilize by autoclaving, supplemented with 50 mg/L kanamycin 25 mg/L gentamicin.
46. MS medium: 0.5× Murashige & Skoog (MS) medium including MES buffer supplemented with 0.8% (w/v) agar and 1% (w/v) sucrose.

## 3 Methods

### 3.1 Monitoring of Protein Underglycosylation

N-Glycosylation efficiency can be monitored by comparison of differences in mobility upon SDS–PAGE and immunoblotting of glycoproteins (*see* **Note 2**). Reduced N-glycosylation efficiency causes underglycosylation of glycoproteins which typically results in a faster migrating protein due to the reduced molecular weight (*see* **Note 3**). As a result, a single faster migrating band representing a partially or non-glycosylated variant can be detectable (Fig. 2b) or several bands might appear representing a mixture of differentially glycosylated proteins (fully glycosylated, partially glycosylated, and unglycosylated variants) (*see* **Note 4**). Protein underglycosylation has been described in several *A. thaliana* mutants which can be used as controls (e.g., mutants with defects in one of the OST subunits such as *stt3a-2* [15] or *ost3/6-1* [19] or mutants with defects in the assembly of the oligosaccharide precursor like *alg10* [11]) (*see* **Note 5**). *A. thaliana* beta-thioglucoside glucohydrolase 1 (TGG1) is a heavily glycosylated protein that is expressed in high amounts in leaves [28] and has been used in several studies to show a reduction in protein glycosylation efficiency [11, 15]. In addition, protein underglycosylation can also be detected by an overall reduced signal with the lectin ConA that recognizes terminal α-linked mannose residues from oligomannosidic N-glycans (*see* **Notes 6** and 7).

1. Harvest 10–100 mg of plant material (e.g., rosette leaves) from mutants and Columbia (Col-0) wild-type and transfer to 2 mL Eppendorf Safe-lock microcentrifuge tubes containing two steel beads (5 mm diameter) per tube.
2. Submerge 2 mL tubes with plant material in a container with liquid nitrogen.
3. Mount 2 mL tubes in a mixer mill and run mixing for 2 min at 50–60 amplitude.
4. Add 4 μL extraction buffer such as PBST per mg of plant material, vortex shortly, and transfer mixture into a 1.5 mL tube.
5. Incubate on ice for 15 min, and invert tube every 3 min.
6. Centrifuge two times for 15 min, 9600 × *g* at 4 °C, and transfer supernatant each time to a new tube.
7. Mix samples with 3× Laemmli sample buffer and heat to 95 °C for 5 min.
8. Load SDS–PAGE gel with 20 μL and run protein separation for 1.5 h at 100 V.
9. Soak nitrocellulose membrane, sponges, and blotting paper in a transfer buffer.
10. Perform the gel–membrane assembly according to the user manual.
11. Blot for 1 h at 100 V, disassemble gel–membrane sandwich, and carefully rinse the membrane with ultrapure water.
12. Incubate in a blocking solution for 1 h at room temperature.
13. Rinse briefly with PBST and incubate the membrane on a shaker for 1.5 h at room temperature in the antibody solution (in our example, use a TGG1 antibody to detect changes in N-glycosylation and an antibody against a non-glycosylated protein like α-tubulin as a control).
14. After incubation, wash four times for 5 min with TBST and add secondary antibody solution to the membrane, and incubate for 1.5 h at room temperature on a shaker. Omit this step for ConA detection, because the used ConA protein is covalently linked to horseradish peroxidase.
15. Wash four times for 5 min in TBST, perform detection using the chemiluminescent substrate, and develop the film.

### 3.2 Co-IP and Subcellular Localization of OST Subunits

Agroinfiltration of *Nicotiana benthamiana* leaves is a fast procedure for protein expression and subsequent co-localization studies by confocal laser scanning microscopy or for monitoring of complex formation by co-immunoprecipitation (co-IP). These protocols have, for example, been used to characterize unknown subunits of the *A. thaliana* OST complex [19] (Fig. 2a, c).

1. Grow *A. tumefaciens* in LB liquid culture supplemented with the appropriate antibiotics for 16 h at 29 °C.
2. Dilute the overnight culture containing the construct for expression of the glycan-modifying enzyme with an infiltration buffer to an OD600 (optical density at 600 nm) of 0.1.
3. For co-localization of GFP and mRFP fusion proteins, mix two or more agrobacteria suspensions and co-infiltrate the whole mixture into fully expanded leaves of 5–6-week-old *N. benthamiana* plants using a 1 mL syringe without a needle. The analysis of subcellular localization by confocal microscopy has been described in detail previously [22, 29].
4. Briefly, sections of infiltrated leaves are analyzed 1–3 days post-infiltration (dpi) on a Leica TCS SP5 confocal microscope equipped with ×63 and ×100 oil immersion objectives using appropriate spectral selections. Post-acquisition image processing was performed in ImageJ and Adobe Photoshop CS (Fig. 2d–g).
5. For co-IP, harvest 500 mg of leaves and extract the proteins with RIPA buffer following the grinding steps as described in Subheading 3.1.
6. Per sample equilibrate 50 μL RFP-Trap®_A bead slurry in 500 μL of RIPA buffer.
7. Centrifuge at 2500 × *g* for 2 min, discard the supernatant, and repeat wash twice.
8. Add 2 mL of the clear protein extract to equilibrated RFP-Trap®_A beads and incubate for 1 h at 4 °C.
9. The RFP-Trap®_A beads with the bound proteins are washed four times with RIPA buffer, and proteins are eluted by incubation at 95 °C for 10 min in 50 μL 1× Laemmli buffer.
10. To collect the bound proteins, perform centrifugation at 2500 × *g* for 2 min using micro spin columns.
11. The protein extracts and the eluate are subjected to SDS–PAGE and immunoblotting with anti-HA and anti-mRFP antibodies as described above for monitoring of protein underglycosylation (*see* **Note 8**) (Fig. 2c).

### 3.3 Characterization of ER-Type Oligomannosidic N-Glycans

ER-resident proteins typically carry oligomannosidic N-glycans because these proteins are not in contact with the Golgi-located machinery for complex N-glycan processing (Fig. 1b). Such oligomannosidic N-glycans are sensitive to digestion by endoglycosidase H (Endo H) which cleaves between the chitobiose cores to release the majority of the oligosaccharide (Fig. 3a). As a consequence, Endo H digestion of oligomannosidic N-glycans will result in a reduced molecular weight that can be monitored by SDS–PAGE and immunoblotting. Proteins with complex N-glycans are resistant to Endo H digestion and do not cause a similar change in the molecular weight (Fig. 3a). Endo H treatment has been used to show that a misfolded variant of the brassinosteroid receptor BRI1 accumulates in the ER in the *bri1-9* mutant [30]. By contrast, correctly folded BRI1 is secreted and carries primarily Golgi-processed complex N-glycans (*see* **Note 9**).

1. Incubate 22.5 μL protein extract with 2.5 μL 10× Glycoprotein Denaturing Buffer for 10 min at 50 °C, transfer to ice, and cool for 5 min.
2. Mix 22.5 μL of the denatured glycoproteins with 3 μL 10× GlycoBuffer 3, 3 μL ultrapure water, and 1.5 μL Endo H.
3. For the control reaction, replace the 1.5 μL Endo H with ultrapure water.
4. Incubate for 90 min at 37 °C and stop the reaction by heating to 50 °C for 5 min.
5. Subject samples to SDS–PAGE and perform immunoblotting, for example, with anti-BRI1 antibody (Fig. 3a) to distinguish between ER-retained and secreted BRI1 variants.
6. To confirm the presence of oligomannosidic N-glycans, glycoproteins can be purified and subjected to mass spectrometry.

   The attachment of the HDEL or KDEL ER-retrieval peptide sequence to the C-terminal end leads to the steady-state accumulation of proteins in the ER. An example for such a protein is a chimeric protein consisting of an N-terminal signal peptide, the Fc domain region from human IgG, GFP, and the HDEL peptide (SP-Fc-GFP-HDEL). This protein can be expressed transiently in *N. benthamiana* leaves as described in Subheading 3.2 and ER accumulation can be monitored by confocal microscopy. Efficient one-step purification of this protein is achieved by affinity chromatography using binding of the Fc domain region to bacterial Protein A (*see* **Note 10**).
7. Harvest 1 g of leaves 2 dpi and extract proteins in RIPA buffer as described in Subheading 3.2.
8. Add 50 μL protein A Sepharose that has been washed several times with 1× PBS.

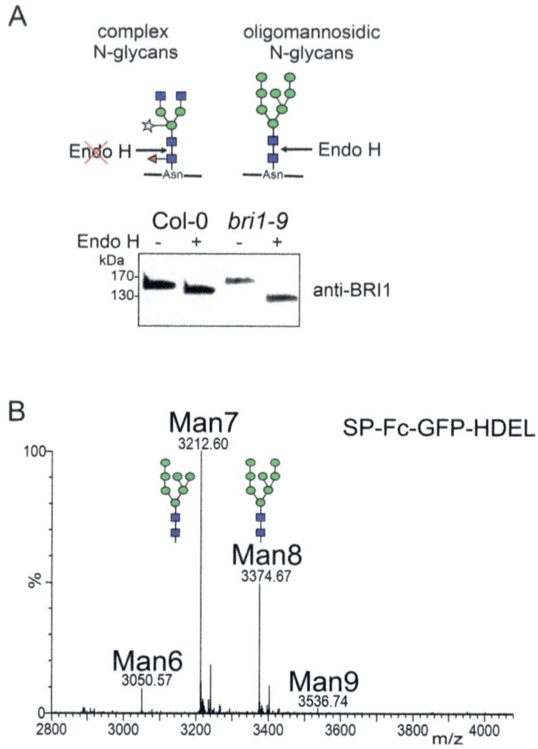


**Fig. 3** (**a**) Endo H digestion and detection of BRI1 glycosylation in protein extracts from Col-0 wild-type and the *bri1-9* mutant. BRI1 in Col-0 is only partially deglycosylated (few oligomannosidic N-glycans are present) as visible by the minor shift in mobility (band >130 kDa). Endo H treatment of BRI1 in the mutant results in complete deglycosylation (band <130 kDa). The cartoon illustrates the cleavage specificity of Endo H. For description of sugar symbols, *see* Fig. 1. (**b**) Transiently expressed SP-Fc-GFP-HDEL was purified and the glycopeptide (peptide sequence: TKPREEQYNSTYR – N-glycosylation site is underlined) from the Fc domain was subjected to LC–ESI–MS analysis. Peak labels were made according to the ProGlycAn system (www.proglycan.com)

9. Incubate for 1.5 h at 4 °C using an orbital shaker, spin down briefly, and discard the supernatant.
10. Wash the Sepharose three times with 250 μL 1× PBS using micro spin columns.
11. Elute the Fc domain-containing protein from the column by adding 50 μL 1× Laemmli buffer.
12. Elute as described for the RFP-Trap®_A purification and separate by SDS–PAGE.
13. Stain the gel with Coomassie Brilliant Blue and excise the glycoprotein band from the gel. Perform S-alkylation and digestion with trypsin (*see* **Note 11**) and analyze the resulting glycopeptides by liquid chromatography–electrospray ionization–mass spectrometry (LC–ESI–MS) (Fig. 3b). The

detailed procedure for this MS-based glycopeptide analysis has been described recently [31].

### 3.4 Monitoring of Glycan-Dependent ERAD

Glycan-dependent ERAD is a major pathway for disposal of misfolded glycoproteins from the ER. To monitor whether a protein is subjected to glycan-dependent ERAD, the class I α-mannosidase inhibitor kifunensine can be used. Kifunensine blocks the trimming of terminal α-linked mannose residues which is mediated by class I α-mannosidases [32]. The exposure of an α1,6-linked mannose residue (Fig. 1b) to the ERAD machinery (including proteins like OS9, SEL1L, and HRD1) is a highly conserved hallmark of glycan-dependent ERAD in different species [24, 33–35]. Kifunensine blocks the formation of the exposed α1,6-linked mannose residue and thus blocks the degradation of ERAD substrates (Fig. 4a) (*see* **Note 12**).

1. Detach leaves from 4-week-old *A. thaliana* plants, cut them into small pieces (alternatively whole seedlings can be used in the same way), and submerge in 0.5× MS medium supplemented with 20 μM kifunensine.
2. Incubate at 22 °C with gentle shaking under long day conditions (16 h light/8 h dark) for 24 h.
3. Harvest seedlings, remove excess liquid, and extract protein as described under Subheading 3.1. To monitor the accumulation of the ERAD substrate, perform immunoblotting with appropriate antibodies (Fig. 4b).
4. An increased signal in the presence of the inhibitor indicates the involvement of a glycan-dependent degradation pathway.
5. To substantiate the finding, the fate of the misfolded glycoprotein can be analyzed in different mutants with distinct defects in glycan-dependent ERAD [24, 26].

## 4 Notes

1. Apart from Asn-X-Ser/Thr, there is the rare possibility for the use of noncanonical glycosylation sites like Asn-X-Cys [36, 37]. The sequon is necessary but not sufficient for N-glycosylation. N-Glycosylation efficiency is dependent on many factors involving the amino acid sequence close to the N-glycosylation site, the secondary structure, the positioning of the site within the protein [13], and organism-specific difference in the composition and function of the OST subunits [12].
2. A shift in mobility can also be caused by other posttranslational modifications including proteolytic processing. Digestion of protein extracts with endoglycosidases that remove oligomannosidic or complex N-glycans is a way to support the

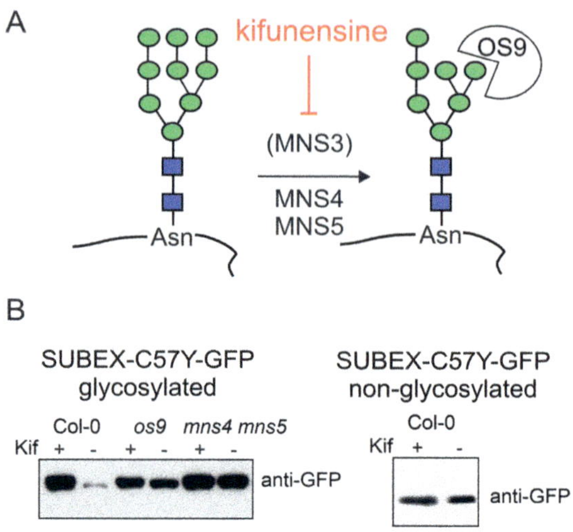


**Fig. 4** (**a**) Illustration of the processing by MNS4/MNS5 which generates the N-glycan that serves as degradation signal for misfolded glycoproteins. The class I α-mannosidase inhibitor kifunensine (Kif) blocks the mannose trimming and leads to accumulation of misfolded glycoproteins (ERAD substrates) in the ER. Likewise, OS9 or MNS4/MNS5 deficiency blocks the degradation. MNS3 may act on the $Man_9GlcNAc_2$ N-glycan but is not necessary for glycan-mediated ERAD [24]. (**b**) The ERAD substrate SUBEX-C57Y-GFP consisting of a misfolded variant of the glycosylated extracellular domain from STRUBBELIG fused to GFP is degraded in Col-0 wild-type and accumulates in the *os9* or *mns4 mns5* mutants as well as in the presence of Kif [42]. A non-glycosylated variant of the misfolded protein is not degraded by the same pathway highlighting the glycan dependency of the protein disposal pathway

observation of underglycosylation. However, complex N-glycans carrying core α1,3-fucose cannot be efficiently removed from whole proteins using endoglycosidases which can complicate the interpretation of the results. Additional blots with antibodies against complex/paucimannosidic N-glycans and lectins that bind to mannose residues can be used to further support a protein underglycosylation defect (see also the procedure for detection of glycoproteins that are recognized by ConA). Apart from the described immunoblot-based protocols, purification of the protein with subsequent MS analysis and quantification of glycosylated versus non-glycosylated peptide should be used to confirm the underglycosylation defect [31].

3. In some cases, a clear protein band cannot be detected anymore in a protein underglycosylation mutant (e.g., KORRIGAN1 in *stt3a-2*) [7]. Most likely, the absence of distinct N-glycans interferes with proper folding leading to degradation or aggregation (exposure of hydrophobic sequence stretches that are otherwise shielded by N-glycans) of these proteins.

4. These differences depend on the number of N-glycans and other protein intrinsic features that are involved in N-glycosylation. For example, some of the non-catalytic OST subunits have accessory function and can slow down protein folding to keep the polypeptide in a glycosylation-competent state [13].
5. Additional approaches can be used that suggest a possible reduction in glycosylation efficiency. These approaches involve treatment of seedlings with tunicamycin. Tunicamycin is an inhibitor that blocks N-glycosylation by interfering with one of the early steps of the lipid-linked oligosaccharide biosynthesis pathway. Plants with defects in N-glycosylation are typically more sensitive to tunicamycin [38].
6. An altered signal with ConA could also result from changes in complex N-glycan formation (e.g., increased signal in the *cgl1* mutant that lacks complex N-glycans). It is therefore recommended to combine the ConA lectin blot with immunoblots using antibodies against complex N-glycans [39].
7. In an alternative approach, protein underglycosylation can be detected in *A. thaliana* by restoration of complex N-glycan formation in the *cgl1* mutant. This approach has been described in detail recently [39].
8. While co-IP can be helpful to identify the components of a protein complex, it does not provide spatiotemporal information on direct interactions of the investigated proteins. Therefore, alternative approaches like FRET-FLIM should be applied for monitoring of real-time protein–protein interactions in cells [40].
9. Oligomannosidic N-glycans can also be found on secreted proteins that are not trafficking through the Golgi (e.g., due to direct ER-to-vacuole protein trafficking). In addition, some N-glycans are not accessible for Golgi-mediated N-glycan processing (e.g., steric hindrance of N-glycan processing) resulting in the presence of oligomannosidic N-glycans on some secreted or plasma membrane-anchored proteins (see also Endo H digestion of BRI1 in Col-0 in Fig. 3a). It is therefore necessary to perform additional experiments (e.g., confocal microscopy of fluorescent fusion proteins) to confirm ER localization.
10. Purification with an immobilized antibody against GFP could also be used instead of binding to Protein A. While such protocols work equally well with our fusion proteins, there are more suppliers for Protein A-related reagents than for reagents that can be used for purification of GFP-tagged proteins. Consequently, Protein A-based purification is typically less expensive than purifications based on GFP-binding proteins. Alternatively, an antibody specific for an (ER-resident) glycoprotein can be used for purification and subsequent N-glycan

analysis. Due to the low expression levels, such protocols typically require extensive optimization.

11. The choice of the protease is dependent on the amino acid sequence. Separate or sequential digestion with alternative enzymes, such as Glu-C, Asp-N, or chymotrypsin, is sometimes necessary to improve the detection of glycopeptides [31].

12. Although the components of the glycan-dependent ERAD pathway are already well known, the steps that lead to the recognition of aberrant glycoproteins as well as the downstream steps that lead to their disposal are less clear. In mammals and yeast, ERAD involves translocation into the cytoplasm, ubiquitination, and proteasomal degradation. In plants, the involvement and role of the proteasome in degradation of glycoprotein ERAD substrates is not so well established [41, 42]. While distinct misfolded glycoproteins are subjected to proteasome-dependent degradation in plants [43], aberrant glycoproteins with powerful ER export signals are subjected to alternative pathways including vacuolar degradation [44, 45].

## Acknowledgments

We would like to thank Friedrich Altmann and Daniel Maresch (both from the Department of Chemistry) for LC–ESI–MS analysis. This work was supported by grants from the Austrian Science Fund (FWF): P31920 and P31921.

### References

1. Xu C, Ng DT (2015) O-mannosylation: the other glycan player of ER quality control. Semin Cell Dev Biol 41:129–134
2. Gill DJ, Clausen H, Bard F (2011) Location, location, location: new insights into O-GalNAc protein glycosylation. Trends Cell Biol 21: 149–158
3. Strasser R, Seifert G, Doblin MS, Johnson KL, Ruprecht C, Pfrengle F, Bacic A, Estevez JM (2021) Cracking the "sugar code": a snapshot of *N*- and *O*-glycosylation pathways and functions in plants cells. Front Plant Sci 12:640919
4. Strasser R (2016) Plant protein glycosylation. Glycobiology 26:926–939
5. Aebi M (2013) N-linked protein glycosylation in the ER. Biochim Biophys Acta 1833:2430–2437
6. Aebi M, Bernasconi R, Clerc S, Molinari M (2010) N-glycan structures: recognition and processing in the ER. Trends Biochem Sci 35: 74–82
7. Kang J, Frank J, Kang C, Kajiura H, Vikram M, Ueda A, Kim S, Bahk J, Triplett B, Fujiyama K, Lee S, von Schaewen A, Koiwa H (2008) Salt tolerance of *Arabidopsis thaliana* requires maturation of N-glycosylated proteins in the Golgi apparatus. Proc Natl Acad Sci U S A 105: 5933–5938
8. Fanata WI, Lee KH, Son BH, Yoo JY, Harmoko R, Ko KS, Ramasamy NK, Kim KH, Oh DB, Jung HS, Kim JY, Lee SY, Lee KO (2013) N-glycan maturation is crucial for cytokinin-mediated development and cellulose synthesis in *Oryza sativa*. Plant J 73:966–979
9. Strasser R (2022) Recent developments in deciphering the biological role of plant complex *N*-Glycans. Front Plant Sci 13:897549
10. Lehle L, Strahl S, Tanner W (2006) Protein glycosylation, conserved from yeast to man: a

model organism helps elucidate congenital human diseases. Angew Chem Int Ed Engl 45:6802–6818

11. Farid A, Pabst M, Schoberer J, Altmann F, Glössl J, Strasser R (2011) *Arabidopsis thaliana* alpha1,2-glucosyltransferase (ALG10) is required for efficient N-glycosylation and leaf growth. Plant J 68:314–325
12. Kelleher D, Gilmore R (2006) An evolving view of the eukaryotic oligosaccharyltransferase. Glycobiology 16:47R–62R
13. Shrimal S, Gilmore R (2019) Oligosaccharyltransferase structures provide novel insight into the mechanism of asparagine-linked glycosylation in prokaryotic and eukaryotic cells. Glycobiology 29:288–297
14. Jeong IS, Lee S, Bonkhofer F, Tolley J, Fukudome A, Nagashima Y, May K, Rips S, Lee SY, Gallois P, Russell WK, Jung HS, von Schaewen A, Koiwa H (2018) Purification and characterization of Arabidopsis thaliana oligosaccharyltransferase complexes from the native host: a protein super-expression system for structural studies. Plant J 94:131–145
15. Koiwa H, Li F, McCully M, Mendoza I, Koizumi N, Manabe Y, Nakagawa Y, Zhu J, Rus A, Pardo J, Bressan R, Hasegawa P (2003) The STT3a subunit isoform of the Arabidopsis oligosaccharyltransferase controls adaptive responses to salt/osmotic stress. Plant Cell 15:2273–2284
16. Nekrasov V, Li J, Batoux M, Roux M, Chu Z, Lacombe S, Rougon A, Bittel P, Kiss-Papp M, Chinchilla D, van Esse H, Jorda L, Schwessinger B, Nicaise V, Thomma B, Molina A, Jones J, Zipfel C (2009) Control of the pattern-recognition receptor EFR by an ER protein complex in plant immunity. EMBO J 28:3428–3438
17. Saijo Y, Tintor N, Lu X, Rauf P, Pajerowska-Mukhtar K, Häweker H, Dong X, Robatzek S, Schulze-Lefert P (2009) Receptor quality control in the endoplasmic reticulum for plant innate immunity. EMBO J 28:3439–3449
18. Lerouxel O, Mouille G, Andème-Onzighi C, Bruyant M, Séveno M, Loutelier-Bourhis C, Driouich A, Höfte H, Lerouge P (2005) Mutants in defective glycosylation, an Arabidopsis homolog of an oligosaccharyltransferase complex subunit, show protein underglycosylation and defects in cell differentiation and growth. Plant J 42:455–468
19. Farid A, Malinovsky FG, Veit C, Schoberer J, Zipfel C, Strasser R (2013) Specialized roles of the conserved subunit OST3/6 of the oligosaccharyltransferase complex in innate immunity and tolerance to abiotic stresses. Plant Physiol 162:24–38
20. Müller LM, Lindner H, Pires ND, Gagliardini V, Grossniklaus U (2016) A subunit of the oligosaccharyltransferase complex is required for interspecific gametophyte recognition in Arabidopsis. Nat Commun 7:10826
21. Caramelo JJ, Parodi AJ (2015) A sweet code for glycoprotein folding. FEBS Lett 589:3379–3387
22. Schoberer J, König J, Veit C, Vavra U, Liebminger E, Botchway SW, Altmann F, Kriechbaumer V, Hawes C, Strasser R (2019) A signal motif retains Arabidopsis ER-α-mannosidase I in the cis-Golgi and prevents enhanced glycoprotein ERAD. Nat Commun 10:3701
23. Vembar S, Brodsky J (2008) One step at a time: endoplasmic reticulum-associated degradation. Nat Rev Mol Cell Biol 9:944–957
24. Hüttner S, Veit C, Vavra U, Schoberer J, Liebminger E, Maresch D, Grass J, Altmann F, Mach L, Strasser R (2014) Arabidopsis class I α-mannosidases MNS4 and MNS5 are involved in endoplasmic reticulum-associated degradation of misfolded glycoproteins. Plant Cell 26:1712–1728
25. Liu Y, Li J (2014) Endoplasmic reticulum-mediated protein quality control in Arabidopsis. Front Plant Sci 5:162
26. Hüttner S, Veit C, Schoberer J, Grass J, Strasser R (2012) Unraveling the function of *Arabidopsis thaliana* OS9 in the endoplasmic reticulum-associated degradation of glycoproteins. Plant Mol Biol 79:21–33
27. Hamilton CM (1997) A binary-BAC system for plant transformation with high-molecular-weight DNA. Gene 200:107–116
28. Liebminger E, Grass J, Jez J, Neumann L, Altmann F, Strasser R (2012) Myrosinases TGG1 and TGG2 from Arabidopsis thaliana contain exclusively oligomannosidic N-glycans. Phytochemistry 84:24–30
29. Schoberer J, Runions J, Steinkellner H, Strasser R, Hawes C, Osterrieder A (2010) Sequential depletion and acquisition of proteins during Golgi stack disassembly and reformation. Traffic 11:1429–1444
30. Jin H, Yan Z, Nam K, Li J (2007) Allele-specific suppression of a defective brassinosteroid receptor reveals a physiological role of UGGT in ER quality control. Mol Cell 26:821–830
31. Gruber C, Altmann F (2015) Site-specific glycosylation profiling using liquid chromatography-tandem mass spectrometry (LC-MS). Methods Mol Biol 1321:407–415
32. Elbein AD, Tropea JE, Mitchell M, Kaushal GP (1990) Kifunensine, a potent inhibitor of the

glycoprotein processing mannosidase I. J Biol Chem 265:15599–15605
33. Clerc S, Hirsch C, Oggier D, Deprez P, Jakob C, Sommer T, Aebi M (2009) Htm1 protein generates the N-glycan signal for glycoprotein degradation in the endoplasmic reticulum. J Cell Biol 184:159–172
34. Ninagawa S, Okada T, Sumitomo Y, Kamiya Y, Kato K, Horimoto S, Ishikawa T, Takeda S, Sakuma T, Yamamoto T, Mori K (2014) EDEM2 initiates mammalian glycoprotein ERAD by catalyzing the first mannose trimming step. J Cell Biol 206:347–356
35. Su W, Liu Y, Xia Y, Hong Z, Li J (2012) The Arabidopsis homolog of the mammalian OS-9 protein plays a key role in the endoplasmic reticulum-associated degradation of misfolded receptor-like kinases. Mol Plant 5:929–940
36. Zielinska DF, Gnad F, Schropp K, Wiśniewski JR, Mann M (2012) Mapping N-glycosylation sites across seven evolutionarily distant species reveals a divergent substrate proteome despite a common core machinery. Mol Cell 46:542–548
37. de Oliveira MVV, Xu G, Li B, de Souza VL, Meng X, Chen X, Yu X, de Souza SA, Intorne AC, de A. Manhães AME, Musinsky AL, Koiwa H, de Souza Filho GA, Shan L, He P (2016) Specific control of Arabidopsis BAK1/SERK4-regulated cell death by protein glycosylation. Nat Plants 2:15218
38. Hori H, Elbein AD (1981) Tunicamycin inhibits protein glycosylation in suspension cultured soybean cells. Plant Physiol 67:882–886
39. Veit C, Vavra U, Strasser R (2015) N-glycosylation and plant cell growth. Methods Mol Biol 1242:183–194
40. Schoberer J, Liebminger E, Vavra U, Veit C, Castilho A, Dicker M, Maresch D, Altmann F, Hawes C, Botchway SW, Strasser R (2014) The transmembrane domain of N -acetylglucosaminyltransferase I is the key determinant for its Golgi subcompartmentation. Plant J 80:809–822
41. Hong Z, Jin H, Tzfira T, Li J (2008) Multiple mechanism-mediated retention of a defective brassinosteroid receptor in the endoplasmic reticulum of Arabidopsis. Plant Cell 20:3418–3429
42. Hüttner S, Veit C, Vavra U, Schoberer J, Dicker M, Maresch D, Altmann F, Strasser R (2014) A context-independent N-glycan signal targets the misfolded extracellular domain of Arabidopsis STRUBBELIG to endoplasmic-reticulum-associated degradation. Biochem J 464:401–411
43. Niemann MCE, Weber H, Hluska T, Leonte G, Anderson SM, Novák O, Senes A, Werner T (2018) The cytokinin oxidase/dehydrogenase CKX1 is a membrane-bound protein requiring homooligomerization in the endoplasmic reticulum for its cellular activity. Plant Physiol 176:2024–2039
44. Shin YJ, Vavra U, Veit C, Strasser R (2018) The glycan-dependent ERAD machinery degrades topologically diverse misfolded proteins. Plant J 94:246–259
45. Shin YJ, Vavra U, Strasser R (2021) Proper protein folding in the endoplasmic reticulum is required for attachment of a glycosylphosphatidylinositol anchor in plants. Plant Physiol 186:1878–1892

# Chapter 17

# Unfolded Protein Response in *Arabidopsis*

## Cristina Ruberti and Federica Brandizzi

## Abstract

The unfolded protein response (UPR) is a highly regulated signaling pathway that is largely conserved across eukaryotes. It is essential for cell homeostasis under environmental and physiological conditions that perturb the protein folding in the endoplasmic reticulum (ER). *Arabidopsis* is one of the outstanding multicellular model systems in which to investigate the UPR. Here, we described a protocol to induce the UPR in plants, specifically *Arabidopsis*, and to estimate their ability to cope with ER stress through the quantification of physiological parameters.

**Key words** Unfolded protein response, Endoplasmic reticulum stress, Temporary ER stress, Prolonged ER stress, Tunicamycin, *Arabidopsis*

## 1 Introduction

Adverse environmental changes and physiological conditions of growth can interfere with crucial biosynthetic processes and disturb proper protein folding in the endoplasmic reticulum (ER), leading to a potentially lethal condition generally termed "ER stress" [1, 2]. To overcome ER stress, the unfolded protein response (UPR), a highly regulated signaling pathway largely conserved across eukaryotes, is actuated [3]. The UPR activates a network of pathways, which generally promote cell adaptation and restore ER homeostasis; however, under unmitigated stress conditions, they also can ignite programmed cell death (PCD) both in animals [1, 4] and plants [5, 6]. Despite the relevance of UPR in cell growth and stress adaptation and abundant information on the function of the canonical stress sensors and transducers in the UPR [5, 7, 8], we lack a complete understanding of the molecular players involved in the temporally distinct UPR-related events (i.e., UPR initiation, recovery from temporary ER stress, survival from mild and prolonged ER stress, and ignition of PCD under unmitigated ER stress), especially in plants [9].

Verena Kriechbaumer (ed.), *The Plant Endoplasmic Reticulum: Methods and Protocols*, Methods in Molecular Biology, vol. 2772, https://doi.org/10.1007/978-1-0716-3710-4_17,

Depending on the duration of ER stress, three main experimental approaches have been implemented to investigate the contribution of the UPR players in ER stress resolution. The downstream transcriptional initiation of the adaptive UPR signaling response can be monitored at a molecular level by assaying the expression levels of the UPR marker genes in time-course treatment with ER stress-inducing agents (for detailed methods, *see* refs. [3, 10]). The plant's ability to recover after temporary ER stress or to cope with prolonged and mild ER stress is evaluated at phenotypic level on seedlings transferred or germinated and grown on ER stress inducer-containing media [5, 6]. Finally, ER stress-induced cell death can be evoked by growing seedlings under prolonged and severe ER stress conditions and quantified by standard methods to test for PCD (ion leakage and TUNEL staining) [6, 11].

Common chemical agents used to experimentally induce ER stress and activate the UPR in plants are tunicamycin (Tm) or dithiothreitol (DTT) [5, 12, 13]. Tm inhibits the N-acetylglucosamine phosphotransferases; this prevents N-linked glycosylation of nascent polypeptides in the ER lumen. DTT is a potent reducing agent that prevents disulfide bond formation during folding of polypeptides in the ER (*see* **Note 1**). Here, we describe phenotypic analyses most commonly adopted to evaluate ER stress sensitivity of *Arabidopsis* seedlings after a short ER stress-inducing treatment or during prolonged ER stress.

## 2 Materials

1. Basic equipment for plant sterile tissue culture.
2. Plant growth chamber conditions: continuous white light at 21 °C, 100 mEinstein/$m^2$ s, 65% humidity.
3. Plant growth medium: half-strength (1/2) Linsmaier and Skoog (LS) medium, 1% sucrose, pH 5.7.
4. Tunicamycin (Tm) from *Streptomyces* sp. (*see* **Note 2**): prepare 10 mg/mL of Tm stock solution by dissolving 0.005 g Tm in 0.5 mL of DMSO yielding a clear solution (*see* **Notes 3** and **4**).
5. Dimethyl sulfoxide (DMSO).
6. 70% ethanol.
7. 5% commercial bleach freshly prepared.
8. Acetone: 80% solution in water.
9. Petri dishes, square with grids, 100 mm × 100 mm, sterile, polystyrene.
10. Micropore gas-permeable surgical tape.
11. Sterilized toothpicks.

12. Razor blades.
13. Analytical balance.
14. Microplate, 96-well, flat bottom, clear.
15. Microplate reader equipped with 663 and 645 nm wavelengths.

## 3 Methods

### 3.1 Recovery After Temporary ER Stress Treatment

Carry out all procedures at room temperature and if aseptic conditions are required, work in a laminar flow hood.

1. Surface-sterilize the *Arabidopsis* seeds (*see* **Note 5**): in a microcentrifuge tube, soak the seeds in 1 mL 70% ethanol for 1 min, suspend them by turning the tube or vortex, pour out the liquid, rinse two times in distilled sterile water, suspend the seeds in 1 mL 5% commercial bleach for 1 min, and then rinse three times in sterile distilled water (*see* **Note 6**). Store the seeds at 4 °C for 2 days to synchronize them.
2. Autoclave for 25 min the plant growth medium with 12 g/L agar (*see* **Note 7**) and pour 40 mL into each square Petri dish.
3. Use a sterile toothpick to sow the *Arabidopsis* seeds on the solid plant growth medium, placing each seed at an equally spaced distance (Fig. 1a).
4. Seal the plates with one layer of micropore gas-permeable surgical tape to allow gaseous exchange.
5. Germinate and grow the seeds in vertical position in a plant growth chamber for 5 days.
6. Autoclave for 25 min the plant growth medium and store it at room temperature.
7. Prepare Tm stock solution (*see* **Notes 3** and **4**).
8. Freshly prepare 0.075, 0.15, and 0.30 mg/mL Tm stock solutions by 133, 66, and 33× dilution, respectively, of 10 mg/mL Tm sock solution using 1/2 LS liquid medium.
9. Autoclave for 25 min the plant growth medium with 12 g/L agar (*see* **Note 5**). Cool down the medium until 50 °C (*see* **Note 8**) and keep it homogeneously mixed with a magnetic stir bar.
10. Add 0.075, 0.15, and 0.30 mg/mL Tm stock solutions, respectively, to the cooled down 1/2 LS medium by 1000 dilution to make plant growth medium containing 0.075, 0.15, and 0.30 μg/mL Tm (*see* **Notes 9** and **10**).

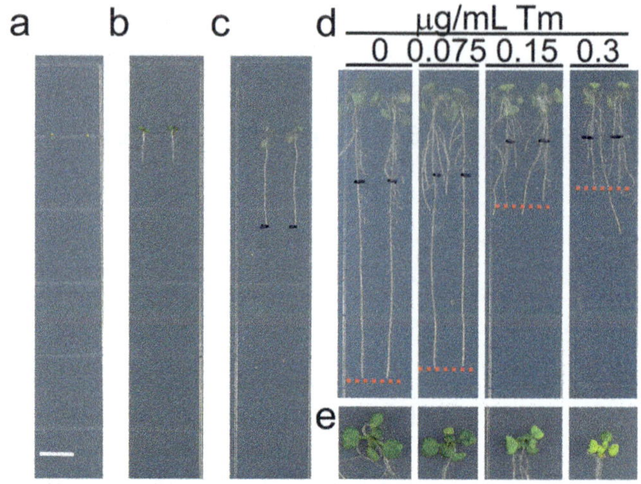


**Fig. 1** *Arabidopsis* Col-0 growth phenotype after recovery from short-term Tm treatment. (**a**) *Arabidopsis* seeds sown at an equally spaced distance on 1/2 LS solid medium for vertical growth. Scale bar: 0.65 mm. (**b**) Five-day-old seedlings were transferred to 1/2 LS solid medium plates containing Tm or DMSO for short-term ER stress treatment. (**c**) Eight-day-old seedlings were transferred from Tm plates to 1/2 LS solid plates and the root tips were marked on the back of the plate using a black extra-fine point marker. (**d**) Root growth inhibition of Tm-treated seedlings compared to untreated seedlings after 3 days of recovery from short-term Tm treatment. Red lines show the primary root tip. (**e**) Close-up view of the shoot of seedlings shown in (**d**)

11. As mock control, the same preparation procedure is carried out replacing the Tm in the 1/2 LS medium with 0.0005% DMSO.
12. Transfer the 5-day-old seedlings using a sterile toothpick to plant growth medium plates containing DMSO, 0.075, 0.15, or 0.30 µg/mL Tm (Fig. 1b). Seal the plates with one layer of micropore gas-permeable surgical tapes to permit gaseous exchange and place them on vertical position into the plant growth chamber.
13. After 3 days, use a sterile toothpick to transfer the seedlings to plant growth medium plates without Tm (or DMSO). Mark the root tip on the back of the plate using an extra-fine point permanent marker and place the plates on vertical position into the plant growth chamber (Fig. 1c).
14. After 3 days, observe the growth phenotype (Fig. 1d, e) and quantify the plant growth parameters as described below.
15. If using Col-0 ecotype wild-type *Arabidopsis*, the plants show growth defects already at 0.15 µg/mL Tm (Fig. 1d, e).

### 3.2 Prolonged ER Stress Treatment

Carry out all procedures at room temperature and if aseptic conditions are required, work in a laminar flow hood.

1. Autoclave for 25 min the plant growth medium and store it at room temperature.
2. Autoclave for 25 min the plant growth medium with 12 g/L agar (*see* **Note 5**).
3. Freshly prepare 0.04, 0.06, and 0.08 mg/mL Tm stock solutions by 250,166, and 125× dilution of 10 mg/mL Tm sock solution, respectively, using 1/2 LS liquid medium.
4. Add 0.04, 0.06, and 0.08 mg/mL Tm stock solutions, respectively, to the cooled down 1/2 LS medium by 1000 dilution to make plant growth medium containing 40, 60, and 80 ng/mL Tm (*see* **Notes 9** and **11**). As mock control, the same preparation procedure is carried out replacing the Tm in the 1/2 LS medium with 0.0005% DMSO.
5. Surface-sterilize the *Arabidopsis* seeds (*see* **Note 3**) and store them at 4 °C for 2 days.
6. Use a sterile toothpick to sow the surface-sterilized seeds on the 1/2 LS medium containing DMSO, 40, 60, and 80 ng/mL Tm. Place each seed at an equally spaced distance (Fig. 1a).
7. Seal the plates with one layer of micropore gas-permeable surgical tape to allow gaseous exchange.
8. Germinate and grow the seeds in vertical position in a plant growth chamber.
9. After 5 days, mark the root tip on the back of the plate using an extra-fine point marker (Fig. 2a) and place the plates back on vertical position into the plant growth chamber.
10. Observe the growth phenotype 7–14 days after germination (Fig. 2b, c), but check the plant growth each day during the assay to precisely monitor the ER stress sensitivity of the seedlings. Quantify the plant growth parameters as described below.
11. If using Col-0 ecotype wild-type *Arabidopsis*, the plants show growth defects starting from 40 ng/mL Tm (Fig. 2b, c).

### 3.3 Plant Growth Parameters

1. The effect of the prolonged ER stress on *Arabidopsis* seeds can be quantified through the quantification of emergence of the true leaves (Fig. 2c). Count the number of seedlings having true leaves on Tm-containing plates and normalize this value with respect to the total number of all sown seeds. Similarly, count the number of seedlings having true leaves on DMSO-containing plates and normalize this value on the total number of all sown seeds [12].

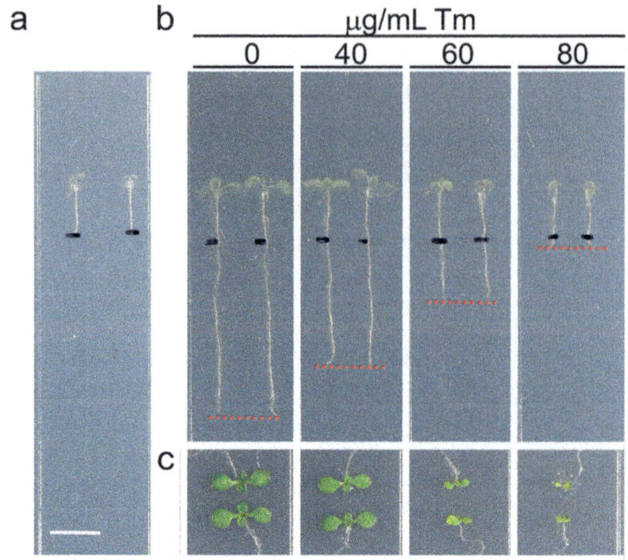


**Fig. 2** *Arabidopsis* Col-0 growth phenotype after prolonged Tm treatment. (**a**) Root tips of 5-day-old seedlings were marked on the back of the plate using a black extra-fine point marker. Scale bar: 0.65 mm. (**b**) Root growth inhibition of Tm-treated seedlings compared to untreated seedlings grown for 10 days vertically. Red lines show the primary root tip. (**c**) Close-up view of the shoot of seedlings shown in (**b**)

2. Fresh shoot weight is an easy readout of plant sensitivity of both short-term ER stress and prolonged ER stress. Using a razor blade, cut the plant shoots at a consistent cut-position on the hypocotyl. Gently dry the shoots on a paper towel and measure the weight with an analytical balance. To quantify the relative growth values, the shoot weight of the seedlings grown on Tm is divided by the shoot weight of the seedlings grown on DMSO plates. For each measurement, use a pool of six to ten shoots for each of the at least six technical replicates for each of the three biological replicates.
3. Chlorophyll (Chl) content of the leaves can be used as an indicator of overall plant health after short-term or prolonged Tm treatment. After measuring fresh shoot weight, place the pool of leaves into a microcentrifuge tube and extract chlorophyll using 1 mL cold 80% acetone at 4 °C for 24 h (*see* **Note 12**) in the dark (*see* **Note 13**). Quickly transfer 0.3 mL of the supernatant to a flat-bottom 96-well microplate and use 0.3 mL 80% acetone blank as reference. Measure the absorbance (A) of the chlorophyll content using a microplate reader at 663 nm and 645 nm nm wavelengths. The chlorophyll concentration is calculated as follows: Chl

(mg/g) = [8.02 × $A_{663}$ + 20.20x$A_{645}$] × $V$/1000 × $W$, where $V$ = volume of the extract (mL) and $W$ = weight of fresh leaves (g) [13, 14].

4. The primary root elongation declines as the concentration of the ER stress inducer increases into the growth media in a dose-dependent manner (Figs. 1d and 2b) [15]. For this reason, ER stress sensitivity can be quantified measuring the primary root length of seedlings grown on Tm plates. Take pictures of the plants treated as described in Subheading 3.1, **steps 1–14** or Subheading 3.2, **steps 1–10** (*see* **Note 14**). Measure the root length from the root tip to the mark made in Subheading 3.1, **step 13** or Subheading 3.2, **step 9**. To quantify the relative root elongation value, the root length of the seedlings grown on Tm is divided by the root length of the seedlings grown on DMSO plates.

## 4 Notes

1. Even if DDT is commonly used to cause protein misfolding in the ER, it is not a specific ER stress inducer, since it blocks disulfide bond formation both in the ER and in the cytosol, and causes oxidative stress and the expression of defense genes [12, 15].
2. The hazard category for Tm is acute toxicity, oral (category 2) H300 (fatal if swallowed). Tm is light sensitive.
3. To avoid formation of aerosols, directly inject the DMSO solvent using a 1 mL syringe with a needle through the lid of the tube containing the solution and mix.
4. To avoid freezing and thawing, aliquot the 10 mg/mL stock solution into relatively small amounts (0.01 mL aliquots) and store them at −20 °C freezer in the dark.
5. ER stress sensitivity is highly affected by the seeds' quality. Use seeds with high germination quality and freshly collected from healthy plants.
6. Do not leave seeds in ethanol or bleach for too long, because the treatment may be toxic to the seeds.
7. Mix homogeneously the solution with a magnetic stir bar before and after to autoclave, to avoid settling of unmelted agar at the bottom of the bottle.
8. Tm is sensitive to high temperature.
9. Prepare the Tm-containing medium plates right before using them.
10. The optimal range of concentration of Tm has to be experimentally determined. As initial screen, use 0.075 and 0.15 μg/

mL Tm and then decrease or increase the Tm concentration depending on the degree of Tm sensitivity of the analyzed *Arabidopsis* ecotype.

11. The optimal range of concentration of Tm has to be experimentally determined. As initial screen, use 20 and 40 ng/mL Tm and then decrease or increase the Tm concentration depending on the degree of Tm sensitivity of the analyzed *Arabidopsis* ecotype and on the purpose of your assay. To evaluate Tm sensitivity under mild ER stress conditions, use low Tm concentration that can be increased to get unmitigated ER stress conditions and consequent cell death.
12. Complete chlorophyll extraction is fulfilled when the shoots are completely white. If the leaves are still partially green, keep the tissues at 4 °C in the darkness for longer time.
13. During chlorophyll extraction and analyses, keep the samples in the dark, because chlorophylls are light-sensitive pigments.
14. To better distinguish the roots, take pictures of the back of the plates.

## Acknowledgments

This study was primarily supported by National Institutes of Health (GM101038 and R35GM136637) with contributing support from the Chemical Sciences, Geosciences, and Biosciences Division, Office of Basic Energy Sciences, Office of Science, US Department of Energy (award number DE-FG02-91ER20021) and AgBioResearch (MICL02598).

### References

1. Chen Y, Brandizzi F (2013) IRE1: ER stress sensor and cell fate executor. Trends Cell Biol 23:547–555
2. Liu J-X, Howell SH (2016) Managing the protein folding demands in the endoplasmic reticulum of plants. New Phytol. https://doi.org/10.1111/nph.13915
3. Ruberti C, Brandizzi F (2014) Conserved and plant-unique strategies for overcoming endoplasmic reticulum stress. Front Plant Sci 5:69
4. Hetz C (2012) The unfolded protein response: controlling cell fate decisions under ER stress and beyond. Nat Rev Mol Cell Biol 13:89–102
5. Chen Y, Brandizzi F (2012) AtIRE1A/AtIRE1B and AGB1 independently control two essential unfolded protein response pathways in Arabidopsis. Plant J Cell Mol Biol 69:266–277
6. Mishiba K et al (2013) Defects in IRE1 enhance cell death and fail to degrade mRNAs encoding secretory pathway proteins in the Arabidopsis unfolded protein response. Proc Natl Acad Sci U S A 110:5713–5718
7. Liu J-X, Srivastava R, Che P, Howell SH (2007) An endoplasmic reticulum stress response in Arabidopsis is mediated by proteolytic processing and nuclear relocation of a membrane-associated transcription factor, bZIP28. Plant Cell 19:4111–4119
8. Iwata Y, Fedoroff NV, Koizumi N (2008) Arabidopsis bZIP60 is a proteolysis-activated transcription factor involved in the endoplasmic reticulum stress response. Plant Cell 20:3107–3121
9. Ruberti C, Kim S-J, Stefano G, Brandizzi F (2015) Unfolded protein response in plants:

one master, many questions. Curr Opin Plant Biol 27:59–66

10. Chen Y, Brandizzi F (2013) Analysis of unfolded protein response in Arabidopsis. Methods Mol Biol Clifton NJ 1043:73–80
11. Yang Z-T et al (2014) The Membrane-Associated Transcription Factor NAC089 controls ER-stress-induced programmed cell death in plants. PLoS Genet 10(3):e1004243
12. Sun L et al (2013) The lumen-facing domain is important for the biological function and organelle-to-organelle movement of bZIP28 during ER stress in Arabidopsis. Mol Plant 6: 1605–1615
13. Ritchie RJ (2006) Consistent sets of spectrophotometric chlorophyll equations for acetone, methanol and ethanol solvents. Photosynth Res 89:27–41
14. Ni Z, Kim E-D, Chen ZJ (2009) Chlorophyll and starch assays. Available at: https://doi.org/10.1038/nprot.2009.12
15. Deng Y, Srivastava R, Howell SH (2013) Protein kinase and ribonuclease domains of IRE1 confer stress tolerance, vegetative growth, and reproductive development in Arabidopsis. Proc Natl Acad Sci 110:19633–19638

# Chapter 18

# Imaging the ER and Endomembrane System in Cereal Endosperm

**Verena Ibl, Jenny Peters, Eva Stoger, and Elsa Arcalís**

## Abstract

The cereal endosperm is a complex structure comprising distinct cell types, characterized by specialized organelles for the accumulation of storage proteins. Protein trafficking in these cells is complicated by the presence of several different storage organelles including protein bodies (PBs) derived from the endoplasmic reticulum (ER) and dynamic protein storage vacuoles (PSVs). In addition, trafficking may follow a number of different routes depending on developmental stage, showing that the endomembrane system is capable of massive reorganization. Thus, developmental sequences involve progressive changes of the endomembrane system of endosperm tissue and are characterized by a high structural plasticity and endosomal activity.

Given the technical dexterity required to access endosperm tissue and study subcellular structures and SSP trafficking in cereal seeds, static images are the state of the art providing a bulk of information concerning the cellular composition of seed tissue. In view of the highly dynamic endomembrane system in cereal endosperm cells, it is reasonable to expect that live cell imaging will help to characterize the spatial and temporal changes of the endomembrane system. The high resolution achieved with electron microscopy perfectly complements the live cell imaging.

We therefore established an imaging platform for TEM as well as for live cell imaging. Here, we describe the preparation of different cereal seed tissues for live cell imaging concomitant with immunolocalization studies and ultrastructure.

**Key words** Cereal endosperm, Endomembrane system, Electron microscopy, Live cell imaging, Fluorescent organelle markers

## 1 Introduction

The endosperm of cereal seeds is a highly differentiated tissue containing specialized cells that are responsible for the synthesis and storage of proteins. The high specialization of the tissue is reflected in the endomembrane system. Storage protein synthesis requires a well-developed endoplasmic reticulum (ER) and involves different storage organelles that will form de novo during seed development [1, 2]. ER-derived protein bodies (PBs) contain

Verena Kriechbaumer (ed.), *The Plant Endoplasmic Reticulum: Methods and Protocols*, Methods in Molecular Biology, vol. 2772,
https://doi.org/10.1007/978-1-0716-3710-4_18,

prolamin aggregates, while the protein storage vacuole (PSV), far from being a plane post-Golgi compartment, accumulates different storage protein classes and often incorporates the content of ER-derived storage bodies in an autophagy-like process [3–6]. At the peak of storage protein production (as soon as 16 dap in maize seeds), endosperm cells undergo programmed cell death (PCD) and generate a starchy endosperm core [7]. This sequence of intracellular changes in developing seeds challenges the endomembrane system that is a plastic and dynamic system capable of rapid morphological and functional adaptation (Fig. 1a). The massive reorganization of the endomembrane system during seed development involves significant morphological changes to the storage organelles and also affects the trafficking and distribution of storage proteins [5, 8–12].

The study of such a dynamic system can be approached by different, complemental imaging methods. Live cell imaging combined with fluorescent membrane markers can be used to follow the morphological changes in situ. This was reported in the context of barley endosperm development, comparing PSVs in the aleurone, subaleurone, and central starchy endosperm layers [10]. The use of TIP3-GFP marker lines, together with the use of fluorescent organelle dyes such as ER-Tracker®, allowed the visualization of morphological changes in the PSVs along development and a spatiotemporal pattern could be stablished [10]. ER-Tracker® and neutral red, as well as other organelle dyes such as FM4-64, BCECF, etc., are very convenient tools to identify and track organelles in cereal endosperm. It should be noted, however, that the staining patterns may be unexpected, due to the specialization of the endomembrane system, and results should be interpreted with some caution and considering appropriate controls (Fig. 1e–m).

Protein trafficking and organelle reshaping can also be investigated by looking at the behavior of recombinant proteins, including storage proteins fused to fluorescent tags [5, 13–15]. Endogenous glycoproteins are rarely found in the endosperm [16] and are difficult to follow due to the lack of appropriate detection antibodies, so recombinant glycoproteins produced in the context of molecular farming are particularly useful because they can be traced by immunolocalization and by characterizing their glycan modifications. Accordingly, a significant reduction in the number of PSVs was observed during maize seed development, and the fate of the recombinant glycoprotein phytase and the endogenous vacuolar maize proteins corn legumin-1 and α-globulin was found to change during endosperm development. All three proteins were found to locate to the PSVs in early development, while closer to maturity, they were found at the periphery of the ER-derived protein bodies [17]. This example reflects the flexibility and dynamics of a system, where interactions between

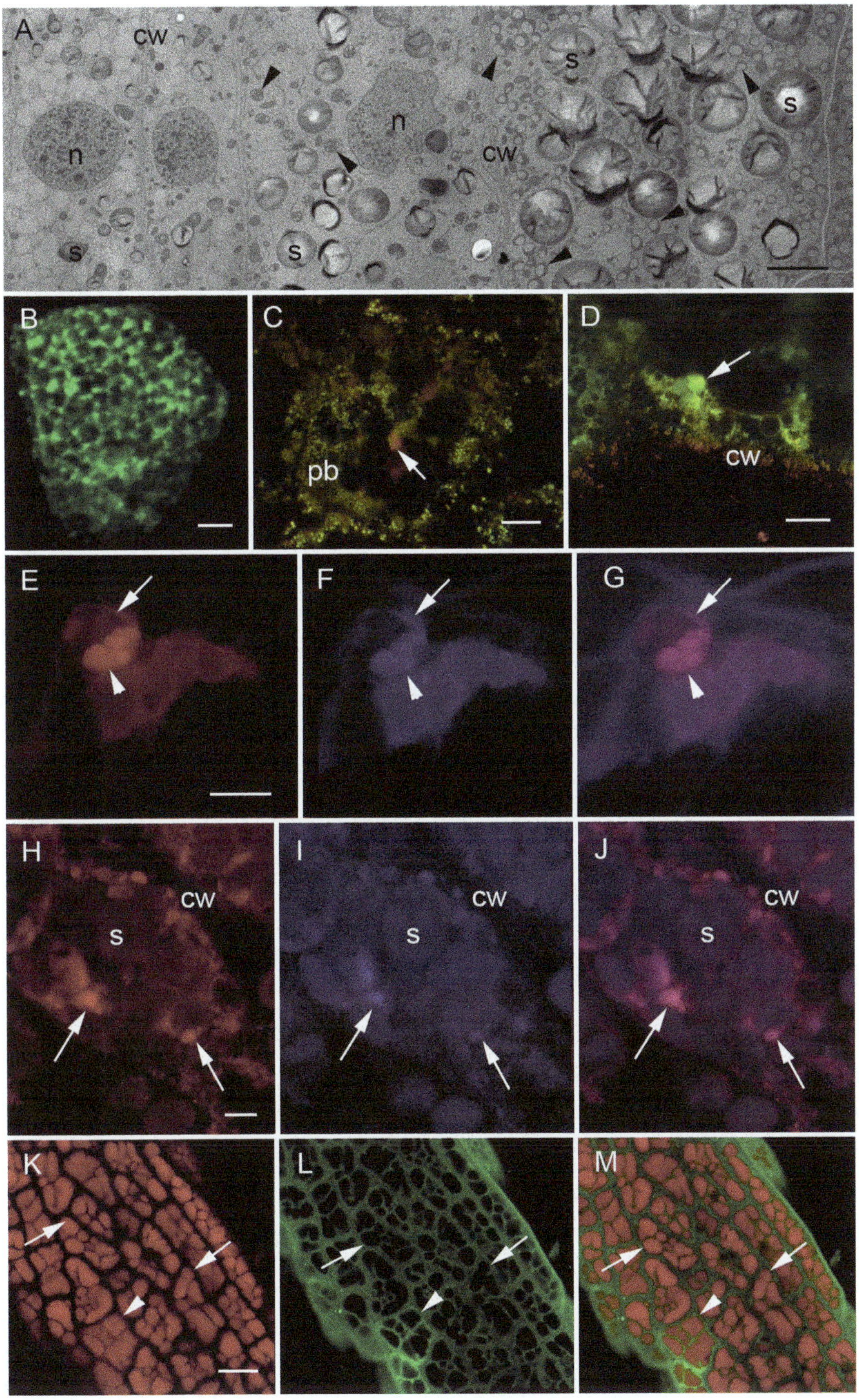


**Fig. 1** Imaging the endomembrane system of the cereal endosperm. (**a**) Electron microscopy. Outer layers of maize endosperm (mid-developmental stage). See the changes in the ultrastructure of maize endosperm, younger cells (left) contain a nucleus (n) and a few starch grains (s), while the older cells (right) are packed with starch grains (s) and zein bodies (arrowheads). Cw, cell wall. B-M. CLSM. (**b**) Maize endosperm. Transient expression of OsTIP3::PDIL1;1-GFP. See the fully developed ER network. (**c**) Maize endosperm. Immunolocalization of corn legumin-1 in a young endosperm cell (arrow). As counterstain, ER-Tracker® green stains the

organelles and changes in protein trafficking fulfill distinct developmental functions [18].

Cereal endosperm can be accessed by various imaging techniques, which differ from each other and offer specific advantages and disadvantages. With live cell imaging, for example, it has become possible to monitor dynamic endomembrane processes in real time, however at limited resolution. Electron microscopy on the other hand provides high resolution, but requires sample fixation and the information must be derived from static images. A combination of approaches, as here described, is therefore the best choice for obtaining maximum information.

## 2 Material

### 2.1 Plant Material

Maize plants (*Zea mays* L.), cv. Hi II (A188XB73), were grown in a climate-controlled growth chamber at 28 °C/24 °C day/night temperature with a 14-h photoperiod. Barley (*Hordeum vulgare* L., cv. Golden Promise) was grown at 16 °C and 70% relative humidity under a 16-h photoperiod. After tillering, cultivation was continued at 22 °C. Seeds were collected at different developmental stages, prior to desiccation, typically from 8 days after pollination (daps) on.

### 2.2 Cereal Transformation for Endosperm ER and Endomembrane System Visualization

#### 2.2.1 Reporter Gene Constructs

1. OsTIP3::PDIL1;1-GFP (ER lumen marker) [19].
2. OsTIP3::CherrySec61 (ER membrane marker) [10, 19].
3. OsTIP3::TiP3-GFP (tonoplast marker) [19].

#### 2.2.2 Endosperm Transient Transformation

1. Particle accelerator gun and accessories.
2. Temperature-controlled light incubator.
3. Vortexer.
4. Microfuge.

**Fig. 1** (continued) zein bodies (pb). (**d**) Barley endosperm. ER-Tracker® green and red perfectly co-localize and label a protein body (arrow). (**e**–**g**). Barley endosperm. (**e**) Neutral red, (**f**) Autofluorescence, (**g**) Merged. See that neutral red stains the protein body (arrowhead) within the storage vacuole (arrow). (**h**–**j**) Barley endosperm. (**h**) ER-Tracker® red, (**i**). Autofluorescence, (**j**) Merged. See that ER-Tracker® red stains the protein bodies (arrows). *Cw* cell wall, *s* starch. (**k**–**m**) Controls. Arabidopsis embryo. (**k**) Neutral red. (**l**) ER-Tracker® green, (**m**) Merged. Note that in *Arabidopsis* embryos, neutral red stains the lumen of the storage vacuoles (arrows), while ER-Tracker green® reveals the presence of a fine ER network (arrowhead). Bars 5 μm (**a**), 10 μm (**b**–**m**)

5. Stock solution of plasmid DNA carrying the transgene of interest for study of the cereal endosperm ER and endomembrane system.
6. Gold particles (0.7–0.9 μm; Bio-Rad).
7. 100 μM spermidine.
8. 2.5 μm $CaCl_2$.
9. Ice-cold 100% (v/v) ethanol.
10. Fine forceps and scalpel.
11. Petri dishes.
12. Culture pots and tubes.
13. Parafilm.
14. Sterile, double-distilled water.
15. 5% (w/v) sodium hypochlorite solution.
16. Cereal endosperm bombardment medium pH 5.8 (KOH): MS (Murashige and Skoog) salts 4.3 g/L (*see* **Note 1**), sucrose 30 g/L, Gelrite 3 g/L.

### 2.3 Fluorescent Dyes for Endosperm ER and Endomembrane System Visualization

1. Confocal laser scanning microscope (CLSM).
2. Free software ImageJ.
3. Razor blades.
4. Forceps.
5. Reagent tubes (*see* **Note 2**).
6. Cover glasses (high precision: 170 ± 5 μm).
7. Petroleum jelly.
8. DMSO.
9. 4 μM neutral red (*see* **Note 3**).
10. ER-Tracker® stock solution (*see* **Note 4**).

### 2.4 Immunolocalization of Endosperm ER and Endomembrane System Markers for CLSM

1. Vibratome.
2. Razor blades.
3. Embryo dishes.
4. Fine forceps.
5. Cyanoacrylate glue (*see* **Note 5**).
6. 0.1 M phosphate buffer (pH 7.4): 10.9 g/l $Na_2HPO_4$ (anhydrous), 3.2 g/l $Na_2HPO_4$.
7. Fixative: 4% paraformaldehyde in phosphate buffer 0,1 M pH 7.4 (*see* **Notes 6** and 7).
8. PBT: phosphate buffer plus 0.25% Tween 20.
9. Microscopy glass slides.

10. Cover glasses (high precision: 170 ± 5 μm).
11. Blocking solution: 5% BSA (bovine serum albumin) in phosphate buffer, pH 7.4.

***2.5 Electron Microscopy***

1. Transmission electron microscope.
2. Ultramicrotome.
3. Fine forceps (*see* **Note 8**) and razor blades.
4. Diamond knife.
5. EM grids (*see* **Notes 9** and **10**).
6. 0.1 M cacodylate buffer, pH 7.4.
7. Fixative: 2% paraformaldehyde plus 2.5% glutaraldehyde in cacodylate buffer, pH 7.4.
8. Secondary fixative (ultrastructure): 1% OsO4 plus 0.8% KCNFe in water (*see* **Note 11**).
9. 0.5% aqueous uranyl acetate.
10. Epoxy resin (*see* **Note 12**).
11. Beem capsules (size 00).

## 3 Methods

***3.1 Cereal Transformation***

*3.1.1 Transient Transformation (Fig. 1b)*

– Coating of gold particles for particle bombardment:

1. Mix 5 mg of sterile gold particles (mixture of 0.9 and 0.7 μm diameter) with 20 μg circular plasmid DNA.
2. Add sterile water to a final volume of 50 μL.
3. Add 50 μL of 0.1 M spermidine (*see* **Note 13**) and further vortex for 2 min.
4. Slowly add 50 μL of 2.5 M $CaCl_2$ and further vortex for 10 min.
5. Centrifuge.
6. Wash the gold pellet twice with ice-cold 100% (v/v) ethanol.
7. Resuspend in 500 μL ice-cold ethanol.
8. Store at −20 °C until further use.

*3.1.2 Particle Bombardment of Young Endosperm Tissue*

1. Collect immature ears (*see* **Note 14**) from growth chamber-grown cereals.
2. Remove the kernels from the ears and surface-sterilize in 5% (v/v) NaClO for 20 min at room temperature (*see* **Note 15**).
3. Wash several times the kernels in sterile water.

4. Cut the kernels in thin sections (*see* **Note 16**) with a sharp razor blade and place the sections in the center of plates containing cereal endosperm bombarding medium.
5. Spot 50 μL of the prepared DNA–gold particle suspension in the core of a microcarrier.
6. Let the ethanol evaporate and assemble the gene gun according to the manufacturer's guidelines.
7. Use the appropriate rupture disk for bombardment (*see* **Notes 17** and **18**).
8. Reporter gene expression can be already detected 2 days after bombardment, depending on the experimental conditions.

#### *3.2 Fluorescent Dyes for Endosperm ER and Endomembrane Visualization (Fig. 1d–j)*

1. Collect immature ears from growth chamber-grown cereals (*see* **Note 19**).
2. Cut the kernels in thin sections (0.5–1 mm) with a razor blade.
3. Wash the sections with tap water (electrical conductivity: 280 μS/cm; pH 7.9) (*see* **Note 20**).
4. Transfer the sections to a new tube and soak them in the fluorescent dyes in the dark at room temperature.
5. Mount the sections in tap water (*see* **Notes 21** and **22**).
6. Immediately analyze the section by CLSM (*see* **Notes 23–25**).

#### *3.3 Immuno-localization of Endosperm ER and Endomembrane System Markers for CLSM (Fig. 1c)*

1. Harvest kernels at the appropriate developmental stage and cut in small pieces with a sharp razor blade, around 2 $mm^3$ (*see* **Notes 26** and **27**).
2. Fix the samples in 4% paraformaldehyde in phosphate buffer for 2 h at RT (*see* **Notes 28** and **29**).
3. Wash at least three times (10 min each) with phosphate buffer 0.1 M, pH 7.4.
4. Glue the sample with cyanoacrylate glue on the vibratome plate (*see* **Notes 30** and **31**).
5. Obtain vibratome sections and transfer them to a glass embryo dish (*see* **Notes 32** and **33**).
6. Dehydrate and rehydrate the sections through ethanol series (30%, 50%, 70%, 100%), 5′ steps.
7. Apply 2% cellulase in phosphate buffer 0.1 M, pH 7.4, for 1 h at RT to digest the cell wall.
8. Remove the cellulase solution and add 0.5% Triton X-100 in phosphate buffer 0.1 M, pH 7.4 (*see* **Notes 34** and **35**).
9. Block the nonspecific binding sites with a solution of 5% BSA in phosphate buffer 0.1 M, pH 7.4, for 10 min.

10. Incubate sections with the appropriate antibody diluted in phosphate buffer 0.1 M, pH 7.4, for 2 h at RT (*see* **Notes 36** and **37**).
11. Wash three times (10 min each) in PBT 0.1 M, pH 7.4.
12. Incubation sections with the secondary antibody diluted in phosphate buffer 0.1 M, pH 7.4, for 1 h at RT in the dark (*see* **Notes 38** and **39**).
13. Wash three times (10 min each) in PB 0.1 M, pH 7.4.
14. Mount and observe (*see* **Notes 40–42**).

### 3.4 Electron Microscopy

Fixation and embedding: samples for fixation should not be bigger than 1 $mm^3$.

#### 3.4.1 Ultrastructural Studies of Endosperm ER and Endomembrane System

1. Immerse the tissue in a drop of fixative and cut small pieces (*see* **Notes 26** and **27**).
2. Fix the samples for 2 h at RT (*see* **Note 43**).
3. Wash the samples at least four times (10 min each) with cacodylate buffer 0.1 M, pH 7.4, at 4 °C.
4. Post-fix the samples with 1% aqueous OsO4 and 0.8% KFeCN for 3 h at 4 °C.
5. Wash the samples several times (minimum five changes, 15 min each) with $ddH_2O$.
6. Incubate the samples with 0.5% aqueous uranyl acetate, 4 °C overnight.
7. Wash the samples at least four times (10 min each) with $ddH_2O$.
8. Dehydrate the samples in acetone series, keep the samples at 4 °C (10 min in 50% acetone, 10 min in 70% acetone (three times), 10 min in 90% acetone (three times), 10 min in 96% acetone (three times), in 100% acetone (three times)).
9. Infiltrate the samples with epoxy resin. Incubate samples (overnight in 25% resin in 100% acetone at 4 °C, 3 h in 50% resin in 100% acetone, 3 h in 75% resin in 100% acetone).
10. Incubate overnight in pure resin at 4 °C.
11. Change to fresh resin and incubate for another 3 h.
12. Place the infiltrated tissue samples in beem capsules filled with resin. Polymerization of the resin takes 48 h at 65 °C (*see* **Notes 44** and **45**).
13. Sectioning: sections showing silver interferences (~90 nm) are collected on 200 mesh hexagonal grids. Sections can be directly observed under the electron microscope (*see* **Note 46**).

## 4 Notes

1. Can be purchased as ready-mixed powders from many suppliers (e.g., Duchefa, Sigma).
2. We prefer 2 mL Eppendorf tubes.
3. A 4 mM stock solution in distilled water can be aliquoted and kept for longer periods of time at −20 °C.
4. ER-Tracker® is available in red, green, and blue. Follow manufacturer instructions to prepare the working solutions. Stock solutions can be stored for longer periods of time at −20 °C. We are using ER-Tracker red and green depending on the experimental conditions, and we have not observed differences in the staining patterns of both (Fig. 1d).
5. Can be found in hardware stores and should be stored at 4 °C.
6. Cacodylate buffer 0.1 M, pH 7.4 and 1xPBS, pH 7.4, are also suitable buffers.
7. A 10% stock solution can be aliquoted and kept at −20 °C.
8. Number 3–5 nonmagnetic forceps are recommended.
9. Gold or copper grids are suitable. To improve the stability of the sections under the electron beam, we recommend to use grids coated with a formvar membrane.
10. Grids are available in different meshes and shapes. Grids with a high mesh number offer better support but less open area. We prefer 200 hexagonal mesh because they offer a good compromise between support and open area.
11. A stock solution of 4% OsO4 plus 3.2% KFeCN in water can be kept at 4 °C in a sealed glass vial for longer periods of time. Discard blackened solutions.
12. Epoxy resins facilitate the sectioning and are more stable under the electron beam. Spurr's resin has been widely used for embedding; however, since one of its components has been retired from the market, it is not any more available. Low-viscosity resin kits can be purchase from many suppliers.
13. Drop the solution into the mixture under constant vortexing.
14. 10–14 daps for maize and 8–12 daps for barley.
15. Move to a sterile bench for the following steps.
16. 1 mm in thickness sections are optimal for bombardment.
17. Depending on the species and the developmental stage of the tissue, different rupture disks will be needed. Usual ranges are 600 to 1100 psi. An optimization of the protocol for specific needs is recommended.
18. Bombard each plate twice.
19. The optimal developmental stage depends on the experimental purposes. Seeds close to maturation have a higher content in

starch grains that can be difficult in handling, labeling, and imaging. For consistency between experiments, cut sections on the same area, and we prefer cross sections on the medial axis of the seed.

20. Transfer the sections to a new specimen holder before washing them.
21. Sections can be also mounted on fresh fluorescent dye.
22. Use some petroleum jelly along the borders to avoid floating and to stabilize the cover glasses.
23. Neutral red and ER-Tracker need short incubation times on cereal endosperm (10–15 min).
24. As the section is mounted in tap water for live cell imaging, the appropriate objective is a water objective.
25. It is important to use the laser settings as low as possible to avoid any damages of the sample and causing bleed through if more fluorescent dyes will be used in parallel. High-sensitivity detectors are recommended.
26. Use a piece of dental wax, put a drop of fixative on the wax sheet, and place the seed in it prior to cutting. The wax will prevent squeezing and the cut surface will be immediately in contact with the fixative improving its preservation.
27. Cut 2-mm-thick slices from the endosperm and then cut them in pieces so that each piece of tissue includes the aleurone and pericarp. For consistency between experiments, we cut cross sections on the medial axis of the seed.
28. Work under a fume hood when using paraformaldehyde.
29. Alternatively overnight at 4 °C.
30. Avoid placing the tissue samples on the center of the plate; preferably place them close to the border of the plate. More than one sample can be glued together, with 120° separation in between.
31. For optimal results, glue the samples so that the pericarp faces the blade.
32. Optimally 30 μm, can be up to 70 μm for cereals.
33. Endosperm sections are very delicate due to the thin cell walls of endosperm cells. Using embryo dishes will increase the volume of reagents consumed but facilitates the handling of the sections.
34. The digestion of the cell walls and the disorganization of the plasma membrane induced by Triton X-100 improve the penetration of the antibodies used for localization. However, these treatments can induce many artifacts and destroy the tissue. Depending on the experimental purposes, both steps can be shortened or even skipped.

35. At this point, sections can be difficult to see. Laying a black cardboard under the embryo dish and working with a binocular can ease the process of changing reagents.
36. Alternatively overnight at 4 °C.
37. The optimal dilution depends on the antibody used. Commonly, dilutions range between 1:100 and 1:500.
38. Cover the embryo dish with aluminum foil.
39. While any fluorophore-labeled antibody is suitable for confocal microscopy, we prefer Alexa Fluor® -labeled antibodies because they are very bright and do not face easily.
40. Mount in phosphate buffer or in phosphate buffer/glycerol 1: 1 (v/v). Using a bit of petroleum jelly in each corner of the coverslip will help to fix the preparation and prevents the sample from squeezing.
41. The laser as well as the detector settings depends on the concentration of the epitope to detect and the fluorophore used.
42. The cereal endosperm matures centripetally, which means that the cells under the aleurone are much younger than the ones in the deeper layers of the starchy endosperm. For consistency between experiments, we like to work always with the same cell layers.
43. Alternatively, overnight at 4 °C.
44. A thin paper stripe label can be embedded together with the sample for easy identification of the resulting blocks.
45. Put one drop of resin in the capsules before adding the tissue piece to avoid the formation of air bubbles at the block tip. It is not necessary to embed more than one piece of sample per capsule.
46. *See* **Note 42**.

## Acknowledgments

The authors would like to thank the BOKU-VIBT Imaging Centre for access and expertise.

### References

1. Muntz K (1998) Deposition of storage proteins. Plant Mol Biol 38:77–99
2. Shewry PR, Halford NG (2002) Cereal seed storage proteins: structures, properties and role in grain utilization. J Exp Bot 53:947–958
3. Shewry PR, Napier JA, Tatham AS (1995) Seed storage proteins: structures and biosynthesis. Plant Cell 7:945–956
4. Galili G (2004) ER-derived compartments are formed by highly regulated processes and have special functions in plants. Plant Physiol 136: 3411–3413
5. Tosi P et al (2009) Trafficking of storage proteins in developing grain of wheat. J Exp Bot 60:979–991

6. Levanony H et al (1992) Evidence for a novel route of wheat storage proteins to vacuoles. J Cell Biol 119:1117–1128
7. Sabelli PA, Larkins BA (2009) The development of endosperm in grasses. Plant Physiol 149:14–26
8. Hoh B et al (1995) Protein storage vacuoles form de novo during pea cotyledon development. J Cell Sci 108:299–310
9. Wang Y et al (2010) OsRab5a regulates endomembrane organization and storage protein trafficking in rice endosperm cells. Plant J 64: 812–824
10. Ibl V et al (2014) Fusion, rupture, and degeneration: the fate of in vivo-labelled PSVs in developing barley endosperm. J Exp Bot 65: 3249–3261
11. Shy G et al (2001) Expression patterns of genes encoding endomembrane proteins support a reduced function of the Golgi in wheat endosperm during the onset of storage protein deposition. J Exp Bot 52:2387–2388
12. Vitale A, Hinz G (2005) Sorting of proteins to storage vacuoles: how many mechanisms? Trends Plant Sci 10:316–323
13. Saito Y et al (2012) Formation mechanism of the internal structure of type I protein bodies in rice endosperm: relationship between the localization of prolamin species and the expression of individual genes. Plant J 70:1043–1045
14. Shigemitsu T et al (2013) Accumulation of rice prolamin-GFP fusion proteins induces ER-derived protein bodies in transgenic rice calli. Plant Cell Rep 32:389–399
15. Wakasa Y, Takaiwa F (2013) The use of rice seeds to produce human pharmaceuticals for oral therapy. Biotechnol J 8:1133–1143
16. Woo YM et al (2001) Genomics analysis of genes expressed in maize endosperm identifies novel seed proteins and clarifies patterns of zein gene expression. Plant Cell 13:2297–2317
17. Arcalis E et al (2010) The changing fate of a secretory glycoprotein in developing maize endosperm. Plant Physiol 153:693–702
18. Arcalis E et al (2014) The dynamic behavior of storage organelles in developing cereal seeds and its impact on the production of recombinant proteins. Front Plant Sci 5:439
19. Onda Y, Kumamaru T, Kawagoe Y (2009) ER membrane-localized oxidoreductase Ero1 is required for disulfide bond formation in the rice endosperm. Proc Natl Acad Sci U S A 106:14156–14161

# Chapter 19

# Multi-omics Resources for Understanding Gene Regulation in Response to ER Stress in Plants

**Dae Kwan Ko and Federica Brandizzi**


## Abstract

Proteotoxic stress of the endoplasmic reticulum (ER) is a potentially lethal condition that ensues when the biosynthetic capacity of the ER is overwhelmed. A sophisticated and largely conserved signaling, known as the unfolded protein response (UPR), is designed to monitor and alleviate ER stress. In plants, the emerging picture of gene regulation by the UPR now appears to be more complex than ever before, requiring multi-omics-enabled network-level approaches to be untangled. In the past decade, with an increasing access and decreasing costs of next-generation sequencing (NGS) and high-throughput protein–DNA interaction (PDI) screening technologies, multitudes of global molecular measurements, known as omics, have been generated and analyzed by the research community to investigate the complex gene regulation of plant UPR. In this chapter, we present a comprehensive catalog of omics resources at different molecular levels (transcriptomes, protein–DNA interactomes, and networks) along with the introduction of key concepts in experimental and computational tools in data generation and analyses. This chapter will serve as a starting point for both experimentalists and bioinformaticians to explore diverse omics datasets for their biological questions in the plant UPR, with likely applications also in other species for conserved mechanisms.


**Key words** Gene regulation, The unfolded protein response, Transcription, Multi-omics, Systems biology, Networks

## 1 Introduction

Because plants are sessile organisms, they are exposed to a wide array of environmental stresses throughout their life cycles. As such, plants have evolved unique and shared mechanisms to ensure cell resilience against such situations and growth. For instance, environmental and physiological conditions of growth can lead to the accumulation of misfolded proteins in the endoplasmic reticulum (ER), a condition known as ER stress. The unfolded protein response (UPR) is designed to alleviate ER stress in an adaptive period, and if ER stress is not resolved, the UPR proceeds into a pro-cell death phase [1]. While the UPR is a coordinated action of

Verena Kriechbaumer (ed.), *The Plant Endoplasmic Reticulum: Methods and Protocols*, Methods in Molecular Biology, vol. 2772, https://doi.org/10.1007/978-1-0716-3710-4_19,

multiple signaling pathways at different molecular levels, transcriptional regulation is considered a primary response. The UPR in plants has been mainly studied in the multicellular model plant species *Arabidopsis thaliana* (hereafter Arabidopsis) with highly effective and specific ER stress inducers, such as tunicamycin (Tm). In Arabidopsis, gene expression changes in the actuation and termination of the UPR are largely affected by two basic leucine zipper transcription factors (TFs): an ER membrane-associated bZIP28 [ACTIVATING TRANSCRIPTION FACTOR 6 (ATF6) in mammals] and the protein kinase and ribonuclease, inositol-requiring enzyme 1 (IRE1), and its splicing target *bZIP60* [*X-box binding protein 1* (*XBP1*) in metazoans; *HAC1* in yeast] [2–5]. Upon ER stress, bZIP28 is translocated to the Golgi where it is processed by S1P and S2P proteases to free the N-terminus component of bZIP28 into the cytosol, which subsequently relocates to the nucleus. In parallel, IRE1 unconventionally splices *bZIP60* transcripts to create the active form of bZIP60 (sbZIP60) which subsequently moves to the nucleus. The nucleus-localized bZIP28 and sbZIP60 (hereafter, UPR-bZIP TFs) coordinately regulate the expression of genes necessary for proper ER protein folding or degradation [2, 6–8]. Recent studies have shown that transcriptional activities of the UPR bZIP TFs are modulated by other TFs or transcriptional regulators [9–14], indicating the existence of complex regulatory mechanisms coordinated by the UPR to balance gene expression induction and suppression. The identification of ER-to-nucleus signaling pathways and availability of the highly effective ER stress inducers and genetic materials collectively make the plant UPR a fantastic model system not only for investigating the UPR but also for analyzing stress-related gene regulation. However, how the plant UPR orchestrates dynamic gene expression changes and is integrated with regulators of other stress responses is largely unknown and has been an area of active investigation in the research community in recent years.

With the advent of genomics technological advances, the plant UPR research community has made significant efforts in generating and analyzing substantial amounts of multi-omics datasets at different molecular levels over the last decades. Those datasets have served as a foundational resource for the research and will get more attention due to the development of new computational data mining tools. The purpose of this chapter is to [1] outline the currently available plant UPR-related omics datasets suitable for applications of various computational analyses at systems levels and [2] briefly highlight studies in which the omics datasets were applied in the context of plant UPR. Having these goals in mind, we made our list of UPR-related multi-omics datasets public online (https://brandizzilab.natsci.msu.edu/resources/upromics.aspx) intending to update it in the future (*see* **Note 1**). Therefore, this chapter serves as a guide and a pivotal resource for the plant UPR research community.

## 2 Transcriptomes in the Plant UPR

Gene expression is a complex cellular process in which the information encoded in the DNA is transcribed into RNA to determine cellular properties and functions. A collection of RNAs transcribed in a certain tissue, time point, and developmental or physiological condition is defined as the transcriptome and reflects a current cellular state. Thus, it can unravel molecular mechanisms underlying cellular and organismal phenotypes. Profiling transcriptome changes also allows identifying DNA sequences on the gene promoters that may function as regulatory components, known as *cis*-regulatory elements (CREs), and thus infer DNA-interacting TFs through the TF-binding motif database [15]. Here, we summarize transcriptome analysis studies in plant UPR addressing diverse biological questions ranging from how transcriptomes are changed upon ER stress and how transcriptomes are affected during persistent ER stress to how the UPR bZIP TFs regulate transcriptomic changes in ER stress resolution. We also highlight the contribution of the transcriptomic data to advancing our understanding of gene regulation in the plant UPR.

### 2.1 The Early Stage of ER Stress Transcriptomics: DNA Microarray

Like many other fields, initial transcriptomic studies in the field of plant UPR relied on hybridization-based DNA microarray technologies that utilize the ability to deposit up to tens of thousands of different DNA sequences on a small glass slide to which cDNA labeled with fluorochrome dyes are hybridized. The importance of understanding ER stress-responsive transcriptome changes was well-recognized and characterized in yeast [8] and mammals [16], but was not in plants until two transcriptome analysis studies independently reported global gene expression changes under ER stress using microarrays on Affymetrix GeneChips containing probe sets representing approximately one-third of *Arabidopsis* genes (~8000) [6, 17]. Soon later, a similar transcriptome study was reported using a fluid microarray-to-DNA microarray method by which a larger number of ER stress-responsive genes compared to the earlier studies were obtained and subsequently verified on DNA microarrays, generating a high-confidence set of ER stress-responsive genes [7]. While the exact numbers of ER stress-responsive genes varied, these earlier studies consistently revealed global alteration of transcripts encoding ER-resident molecular chaperones, foldases, ER-associated degradation (ERAD) components, and secretory pathway proteins. The existence of ER stress-responsive CREs in the promoters of stress-responsive genes was also identified. These earlier studies made a strong contribution to understanding gene responses to ER stress, yet there were still fundamental knowledge gaps. For example, while different biological processes were found to be associated with either upregulated or downregulated genes, the function of those genes in cellular responses to ER stress could not be fully explored because the studies could offer only a single snapshot during the progress of ER stress. Five years later, a time-series transcriptome analysis in ER stress was reported using Agilent Arabidopsis 2 Oligo Microarrays [18], covering a larger number of genes (21,500) than the ones used in the previous studies. The time-series datasets (2, 5, and 10 h of Tm treatment) allowed for hierarchical clustering analyses of 259 Tm-responsive genes, a number that is substantially higher than in the previous studies (~70). Based on the expression patterns, the study identified six gene clusters that displayed unique expression patterns. The ER stress hallmark genes encoding three BINDING PROTEINs (BiPs) and three ER-localized J domain-containing proteins (ERdj3s) belonged to an early upregulated cluster, while genes associated with cellular damage caused by ER stress, such as autophagy-associated ATG8, were contained in a later upregulated cluster. The analyses also revealed the upregulation of many TF genes, indicating the actuation of transcriptional signaling cascades and a possibility that TFs other than the UPR-bZIP TFs could be involved in the UPR gene regulatory dynamics. All these transcriptome studies commonly discovered the downregulation of genes encoding secretory pathway proteins (i.e., proteins entering the ER and being secreted, or proteins targeted to the vacuole through the secretory pathway). Later, Mishiba et al. [19] found that, among the 125 genes downregulated in wild type (Col-0) in the previous studies, the transcript abundance of 120 genes (96%) was not altered in a functional knockdown transfer DNA (T-DNA) mutant of IRE1, *ire1a/b*. In combination with an enhanced cell death of *ire1a/b*, but not of a knockout *bZIP60* mutant, *bzip60-2*, these data suggested alternative functions of IRE1 other than activating bZIP60 in the UPR including regulated IRE1-dependent decay (RIDD). Thus, in this study, global RIDD targets were investigated in Col-0 and *ire1a/b* treated with a transcription inhibitor drug, actinomycin D, before Tm treatment, to prevent de novo transcription, followed by transcriptome analyses using Affymetrix Arabidopsis ATH1 Genome Array. In the degraded mRNA pool, the analyses identified a significant proportion of transcripts of genes encoding secretory pathway proteins, demonstrating that the downregulation of those genes, which was also initially found in the previous studies [6, 7, 17], was due to IRE's RIDD activity. Therefore, collectively, microarray technologies enabled the investigation of ER stress-induced transcriptomics, which was not feasible before. Nevertheless, microarray methods had inevitable technical issues such as a high level of signal-to-noise rate and indistinguishable homolog genes, which were later resolved by NGS technologies, as described below.

### 2.2 Advent of the NGS Technologies: RNA-seq

In contrast to microarray methods, sequencing-by-synthesis approaches (e.g., Sanger sequencing, tag-based methods, and RNA-seq) directly determine the cDNA sequences derived from mRNA. Among them, RNA-seq has the most advantages (e.g., high throughput, highly accurate, quantitative, and less expensive for sequencing depth per nucleotide) over other earlier methods and has revolutionized the way transcriptomes can be profiled in any organism. The increasing access and decreasing costs of RNA-seq, development of multiplex technologies, and availability of high-quality plant genomes have collectively allowed for the widespread application of RNA-seq approaches in studying the plant UPR. In 2018, two RNA-seq analysis studies profiled ER stress-inducible transcriptome changes in Arabidopsis and maize, respectively. Kim et al. [20] performed transcriptome profiling in ER stress-treated Arabidopsis Col-0 and three double mutants of *bZIP17*, which is another ER membrane-associated TF, *bZIP28*, and *bZIP60* (i.e., *bzip17/60*, *bzip28/60*, and *bzip17/28*), highlighting the regulatory interactions among the three bZIP TFs in the UPR. In a following study, a genetic mutant that suppresses the root growth defects in *bzip17/28* was identified, and the corresponding T-DNA triple mutant was subjected to RNA-seq analyses along with Col-0 and its parent line, *bzip17/28*, under ER stress [13]. The expression induction of hundreds of genes seen in *bzip17/28* was substantially attenuated in the triple mutant, indicating that the phenotypic suppression in the triple mutant was associated with the ER stress-inducible transcriptomic changes. Srivastava et al. [21] investigated transcriptomic changes during persistent ER stress in maize by performing a time-series RNA-seq analysis in maize. The high-resolution set of temporal time points (0, 3, 6, 12, 24, 36, and 60 h), which covered pro-survival to pro-death stages of the UPR, allowed for dynamic clustering of genes differentially regulated, revealing multi-phased patterns of transcriptomic changes as ER stress progressed. The two transcriptome studies described above were limited to study in either UPR mutants (functional genomics) or time-series samples (temporal dynamics). Later, the two approaches were combined in a recent study where ER stress-responsive gene reprogramming was investigated in Col-0 and single knockout mutants of *bZIP28* and *bZIP60* (*bzip28-2* and *bzip60-2*, respectively) during a time course of recovery from ER stress to synergistically investigate biological questions on the complex gene regulation in the plant UPR [22]. Consistent with the phenotypic difference of the two single mutants displayed in ER stress recovery (less recovery rate of *bzip28-2* than *bzip60-2*), hundreds of genes were regulated in either bZIP28- or bZIP60-dependent manner during the analyzed recovery time points, indicating that bZIP28 and bZIP60 are individually necessary for growth establishment in recovery from ER stress. In Arabidopsis, PHOSPHATASE TYPE 2CA-INTERACTING RING FINGER PROTEIN 1 (PIR1), an E3 ligase effector, operates downstream of IRE1a/b. Notably, the knockout mutant of *PIR1* exhibited a heightened genome-wide transcriptional induction compared to Col-0, suggesting a potential role for PIR1 in modulating the stability of transcriptional activators [9]. This transcriptomic observation ultimately led to the identification of ABI5 as a pro-survival factor in the UPR. Additionally, RNA-seq analysis was employed to profile transcriptomic changes in potato (*Solanum tuberosum* L.) under adaptive ER stress, thereby broadening our comprehension of UPR gene regulation to include the non-model plant system [23].

## 3 TF–DNA Interactomes in the Plant UPR

While gene regulation is a remarkably intricate biochemical process tightly coordinated by multiple regulatory factors at different molecular levels, the regulatory dynamics is mainly controlled by DNA sequences near the transcriptional start site of a gene (i.e., CREs), and trans-regulatory elements that bind to CREs such as TFs, suggesting that interplays between CREs and TFs at gene promoters encode the blueprints of dynamic gene expression. Thus, identifying global interactions between DNA (e.g., gene promoters) and proteins (e.g., TFs) is a fundamental step to understand the molecular mechanisms by which UPR genes are timely modulated. There have been multiple studies to obtain global PDIs in efforts to understand gene regulatory mechanisms and comprehensive TF–DNA interaction landscapes, which have substantially advanced our understanding of gene regulation in plant UPR. Here, we introduce PDI datasets associated with plant UPR. In general, there are two types of PDI analyses: gene-centered and TF-centered approaches as described below.

### 3.1 Gene-Centered Approaches

Gene-centered approaches aim to discover transcriptional regulators upstream of target genes of interest. Yeast one-hybrid (Y1H) is the most widely used technique [24]. Briefly, a DNA fragment (e.g., promoter region) is cloned upstream of two different reporters, and these reporter constructs are integrated into the genome of a yeast strain. Next, plasmids expressing TFs as hybrid proteins (hence the name of the assay) fused with the strong transcriptional activation domain (AD) of the yeast Gal4 transcription factor are introduced into a yeast strain. When a TF interacts with the DNA fragment of interest in yeast, the AD moiety activates reporter expression in yeast regardless of whether the TF is an activator or repressor in vivo. Sanger sequencing the plasmid in each of these colonies can reveal the identity of the TFs that bind the DNA fragment. An updated version, enhanced Y1H (eY1H) [25] is now available with two important technical advances that make it a feasible approach to identify potential global TF–DNA interactions underlying plant UPR: high-throughput technologies allowing for identifying thousands of potential interactions and availability of gold-standard TF collections in Arabidopsis (~2000 TFs) and maize (~500 TFs). Ko and Brandizzi [26] analyzed the maize time-series transcriptome data published earlier [21] and employed six biomarker genes for an eY1H screen. The selected genes could function in the multiphasic process from pro-survival to pro-death activities in response to persistent ER stress. The maize eY1H screen revealed 831 PDIs, consisting of 262 TFs, in a highly interconnected network underlying the UPR. The results indicated that the UPR could be controlled by a broad array of multifunctional TFs that could functionally link the UPR with numerous biological processes with which the TFs are associated. Subsequent network analyses further led to the identification of candidate regulators for functional analyses, demonstrating the effectiveness of eY1H application in investigating gene regulation in plant UPR. In a separate study, eY1H was applied to identify upstream regulators of the UPR biomarker gene (*BiP3*) and the UPR-bZIP TF genes (*bZIP28* and *bZIP60*) in Arabidopsis, discovering 603 PDIs consisting of 285 TFs [10]. This result was further used to construct a TF network map that provided critical data for further gene characterization, which will be described in Subheading 4.2. Despite its successful application in understanding gene regulation in plant UPR, eY1H has unavoidable technical downsides including expected false positives and the necessity of validation of the eY1H-resulting PDIs by other techniques in vitro, in vivo, or in planta.

### 3.2 TF-Centered Approaches

While gene-centered approaches are designed to identify upstream TFs of target genes, TF-centered approaches flip the coin, focusing on downstream targets of TFs of interest. Chromatin immunoprecipitation followed by sequencing (ChIP-seq) is the main tool in TF-centered approaches [27]. In ChIP-seq, chromatin regions interacting with proteins are cross-linked by fixatives (e.g., formaldehyde) and fragmented by sonication. The DNA fragments bound by TFs are immunoprecipitated, purified, and used for the NGS sequencing library construction. Although ChIP-seq has multiple requirements for its success (e.g., good quality ChIP-seq libraries and antibodies for a TF or protein tag which is fused with a TF of interest in transgenic lines), it has been the most powerful tool to determine global maps of TF binding in many biological systems.

The first ChIP-seq study of bZIP28 and bZIP60 under ER stress was reported using transgenic lines expressing a green fluorescent protein (GFP)-tagged fusion to either bZIP28 or bZIP60 in *bzip60-2* during recovery from ER stress (0, 12, and 24 h) [22]. In line with the notion that bZIP28 and bZIP60 underwent ER stress-specific protease cleavage [4] and unconventional splicing by IRE1 [27], respectively, for activation as TFs, GFP-bZIP28 and GFP-bZIP60 proteins bound the promoters of UPR biomarker genes specifically in Tm treatment but not in mock treatment (DMSO). The time-series ChIP-seq analyses not only provided global binding profiles of bZIP28 and bZIP60 but also revealed the temporal binding patterns of these TFs and new potential CREs enriched in the binding peaks other than the canonical ER stress-responsive ones. Furthermore, the pairwise comparison of bZIP28- and bZIP60-bound genes identified a significant overlap of genes associated with ER stress-associated GO terms, as expected, along with GO terms specific to bZIP28-bound genes and bZIP60-bound genes, implying distinctive roles of the UPR-bZIP TFs when ER stress is resolved in line with the finding in the RNA-seq analyses performed in the same study described in Subheading 2.2 [22]. Using this approach, the abscisic acid-associated TF, G-class bZIP TF (GBF2), was identified and shown to play a significant role as a repressor of DNA-binding activities of the UPR-bZIP TFs [10]. In this study, the genomic regions bound by GBF2 were mapped after 6 h of ER stress by performing ChIP-seq analysis using a *GBF2* native promoter-driven yellow fluorescent protein for energy transfer (YPet)-tagged GBF2 line (*pGBF2:GBF2-YPet*), revealing a significant overlap of its bound genes with ones bound by bZIP28 and/or bZIP60. In addition to PDI profiling under ER stress conditions, which is the main focus of this section, Srivastava et al. [21] overexpressed GFP-tagged ZmbZIP17 (maize ortholog of bZIP28) and ZmbZIP60 (maize ortholog of bZIP60) in maize mesophyll protoplasts (plant cells without cell wall) and performed ChIP-seq under non-stress conditions, revealing their global binding profiles. Zhang et al. [3] investigated global DNA-binding interactions of bZIP28 in a constitutive promoter-driven MYC-tagged bZIP28 expression transgenic lines under heat stress conditions where bZIP28 was activated. Another high-throughput TF-centered approach that has been increasingly used is DNA affinity purification sequencing (DAP-seq) which maps genomic DNA regions interacting with TFs that are expressed and purified in vitro [15]. The Arabidopsis DAP-seq study revealed global in vitro DNA-binding profiles of 529 TFs including bZIP28 but not bZIP60. These examples highlight that TF-centered approaches are powerful for the discovery of critical elements contributing to ER stress management which could not be otherwise easily discovered using gene-centered approaches.

## 4 Network-Level Resources in the Plant UPR

While the high-throughput experimental datasets described above are critical to generate and test hypotheses or provide additional ways to confirm results obtained in an independent study, it is still challenging to extract significant biological information from a matrix of expression levels of thousands of genes or coordination of hundreds of binding regions of TFs for a deeper understanding of the underlying molecular mechanisms [29]. It is increasingly recognized that to better understand the function of a cell, tissue, or organism, molecular components (e.g., genes and proteins) and their activities can be conceptualized as systems of interacting elements, creating biological networks. A transcriptional network is generally defined as a collection of genes and TFs, which interact with each other, and visualized with nodes (genes or TFs) and edges (interactions) in a map. In recent years, with the advent of omics datasets, transcriptional networks have gained popularity to better understand the complex gene regulatory dynamics of plant UPR, leading to the discovery of novel regulators in the UPR and the underlying mechanism [10, 22, 26]. In the following section, we introduce transcriptional network resources for investigating plant UPR in two major forms: coexpression networks and gene regulatory networks (GRNs).

### 4.1 Coexpression Networks in the Plant UPR

A coexpression network is a type of transcriptional network constructed solely based on similarity in gene expression across samples (e.g., time points and genotypes). The main principle for coexpression network analyses is "guilt-by-association" based on which genes belonging to the same gene module are regarded as coregulated by common regulators or shared by similar functions. Distinct from hierarchical clustering analyses, the power of coexpression network analyses is to identify the correlation patterns among hundreds to thousands of genes across samples, cluster those with similar expression patterns into modules, and pinpoint the most interacting genes known as regulatory hub genes within the single computational pipeline. In the time-course RNA-seq analyses during recovery from ER stress in Col-0, *bzip28-2*, and *bzip60-2* described earlier, Ko and Brandizzi performed weighted correlation network analysis (WGCNA), a popular modeling algorithm for creating coexpression networks [30], and constructed 14 coexpression modules, which showed distinct expression pattern and biological processes significantly enriched [22]. While, among them, only one module was selected for functional characterization in the study, the rest of the coexpression modules have remained unexplored and could serve as a starting point for investigation of how the plant UPR mediate transcriptomes associated with significant biological processes such as photosynthesis and secondary metabolism when ER stress is resolved. A step-by-step instruction of WGCNA analyses using the UPR RNA-seq dataset has recently been reported (*see* **Note 2**) [31].

### 4.2 GRNs in Plant UPR

GRNs are constructed based on interactions between genes and TFs experimentally identified by high-throughput methods such as ChIP-seq and eY1H, describing the hierarchical architecture of gene regulation. Importantly, the information on gene–TF interactions can be further integrated with transcriptome data (e.g., coexpression networks), creating multilayered networks which can be more informative to interrogate complex gene regulatory mechanisms. For example, the ChIP-seq profiles of the UPR-bZIP TFs during recovery from ER stress were introduced into one of the coexpression modules into a GRN, uncovering that the GRN consists of two subnetworks, each of which is associated more closely with bZIP28 or bZIP60 [22]. In other examples, the TF–DNA interactomes of Arabidopsis and maize UPR biomarker genes obtained by eY1H screens [10, 26] were integrated with the corresponding time-course transcriptome data during ER stress recovery in Arabidopsis [10] and persistent ER stress in maize [21], respectively. Both studies revealed large-scale GRNs where the promoters of bait genes were bound by hundreds of multifunctional TFs. While the TF interactions need to be further validated in planta for further analyses, these GRNs could serve as a playground for the plant UPR community to unravel the complex gene regulation and address their biological questions.

## 5 Conclusions

This chapter summarizes recent updates and the status of multi-omics resources in the field of the plant UPR. The application of multi-omics approaches has been increasingly appreciated by the community in recent years and has led to the critical discovery of novel regulators and mechanisms in the past decades. While generating new datasets is important and often necessary to address hypotheses, it is also critical to effectively process and mine the existing multi-omics datasets into informative forms easily recognizable for discovering new biological information or enabling a deeper understanding of molecular mechanisms underlying significant biological traits. Both approaches (generation of new datasets and data mining) represent a promising route toward a comprehensive understanding of the complex molecular mechanisms behind gene regulation in the plant UPR.

## 6 Notes

1. We have created a catalog of multi-omics datasets up to date (https://brandizzilab.natsci.msu.edu/resources/upromics.aspx), which will be updated on a regular basis for the research community.
2. The step-by-step instruction of WGCNA using the time-course transcriptome data in recovery from ER stress is published and scripts are available [31].

## Acknowledgments

This study was supported primarily by the National Institutes of Health (R35GM136637) with contributing support from the Great Lakes Bioenergy Research Center, US Department of Energy, Office of Science, Office of Biological and Environmental Research (DE-SC0018409), Chemical Sciences, Geoscience and Biosciences Division, Office of Basic Energy Sciences, Office of Science, US Department of Energy (DE-FG02-91ER20021), and MSU AgBioResearch (MICL02598).

### References

1. Pastor-Cantizano N, Ko DK, Angelos E, Pu Y, Brandizzi F (2020) Functional diversification of ER stress responses in Arabidopsis. Trends Biochem Sci 45(2):123–136. https://doi.org/10.1016/j.tibs.2019.10.008
2. Song ZT, Sun L, Lu SJ, Tian Y, Ding Y, Liu JX (2015) Transcription factor interaction with COMPASS-like complex regulates histone H3K4 trimethylation for specific gene expression in plants. Proc Natl Acad Sci U S A 112(9):2900–2905. https://doi.org/10.1073/pnas.1419703112
3. Zhang SS, Yang H, Ding L, Song ZT, Ma H, Chang F et al (2017) Tissue-specific Transcriptomics reveals an important role of the unfolded protein response in maintaining fertility upon heat stress in Arabidopsis. Plant Cell 29(5):1007–1023. https://doi.org/10.1105/tpc.16.00916
4. Liu JX, Srivastava R, Che P, Howell SH (2007) An endoplasmic reticulum stress response in Arabidopsis is mediated by proteolytic processing and nuclear relocation of a membrane-associated transcription factor, bZIP28. Plant Cell 19(12):4111–4119. https://doi.org/10.1105/tpc.106.050021
5. Iwata Y, Fedoroff NV, Koizumi N (2008) Arabidopsis bZIP60 is a proteolysis-activated transcription factor involved in the endoplasmic reticulum stress response. Plant Cell 20(11):3107–3121. https://doi.org/10.1105/tpc.108.061002
6. Martínez IM, Chrispeels MJ (2003) Genomic analysis of the unfolded protein response in Arabidopsis shows its connection to important cellular processes. Plant Cell 15(2):561–576. https://doi.org/10.1105/tpc.007609
7. Kamauchi S, Nakatani H, Nakano C, Urade R (2005) Gene expression in response to endoplasmic reticulum stress in Arabidopsis thaliana. FEBS J 272(13):3461–3476. https://doi.org/10.1111/j.1742-4658.2005.04770.x
8. Travers KJ, Patil CK, Wodicka L, Lockhart DJ, Weissman JS, Walter P (2000) Functional and genomic analyses reveal an essential coordination between the unfolded protein response and ER-associated degradation. Cell 101(3):249–258. https://doi.org/10.1016/s0092-8674(00)80835-1
9. Ko DK, Kim JY, Thibault EA, Brandizzi F (2023) An IRE1-proteasome system signalling cohort controls cell fate determination in unresolved proteotoxic stress of the plant endoplasmic reticulum. Nat Plants 9(8):1333–1346. https://doi.org/10.1038/s41477-023-01480-3

10. Ko DK, Brandizzi F (2022) Transcriptional competition shapes proteotoxic ER stress resolution. Nat Plants 8(5):481–490
11. Lai YS, Renna L, Yarema J, Ruberti C, He SY, Brandizzi F (2018) Salicylic acid-independent role of NPR1 is required for protection from proteotoxic stress in the plant endoplasmic reticulum. Proc Natl Acad Sci U S A 115(22): E5203–E5E12. https://doi.org/10.1073/pnas.1802254115
12. Nawkar GM, Kang CH, Maibam P, Park JH, Jung YJ, Chae HB et al (2017) HY5, a positive regulator of light signaling, negatively controls the unfolded protein response in. Proc Natl Acad Sci U S A 114(8):2084–2089. https://doi.org/10.1073/pnas.1609844114
13. Kim JS, Sakamoto Y, Takahashi F, Shibata M, Urano K, Matsunaga S et al (2022) TBP-ASSOCIATED FACTOR 12 ortholog NOBIRO6 controls root elongation with unfolded protein response cofactor activity. Proc Natl Acad Sci U S A 119(6): e2120219119. https://doi.org/10.1073/pnas.2120219119
14. Meng Z, Ruberti C, Gong Z, Brandizzi F (2017) CPR5 modulates salicylic acid and the unfolded protein response to manage tradeoffs between plant growth and stress responses. Plant J 89(3):486–501. https://doi.org/10.1111/tpj.13397
15. O'Malley RC, Huang SC, Song L, Lewsey MG, Bartlett A, Nery JR et al (2016) Cistrome and epicistrome features shape the regulatory DNA landscape. Cell 165(5):1280–1292. https://doi.org/10.1016/j.cell.2016.04.038
16. Okada T, Yoshida H, Akazawa R, Negishi M, Mori K (2002) Distinct roles of activating transcription factor 6 (ATF6) and double-stranded RNA-activated protein kinase-like endoplasmic reticulum kinase (PERK) in transcription during the mammalian unfolded protein response. Biochem J 366(2):588–594. https://doi.org/10.1042/bj20020391
17. Noh S-J, Kwon CS, Oh D-H, Moon JS, Chung W-I (2003) Expression of an evolutionarily distinct novel BiP gene during the unfolded protein response in Arabidopsis thaliana. Gene 311:81–91
18. Iwata Y, Sakiyama M, Lee M-H, Koizumi N (2010) Transcriptomic response of Arabidopsis thaliana to tunicamycin-induced endoplasmic reticulum stress. Plant Biotechnol 27(2): 161–171
19. Mishiba K, Nagashima Y, Suzuki E, Hayashi N, Ogata Y, Shimada Y et al (2013) Defects in IRE1 enhance cell death and fail to degrade mRNAs encoding secretory pathway proteins in the Arabidopsis unfolded protein response. Proc Natl Acad Sci U S A 110(14):5713–5718. https://doi.org/10.1073/pnas.1219047110
20. Kim JS, Yamaguchi-Shinozaki K, Shinozaki K (2018) ER-anchored transcription factors bZIP17 and bZIP28 regulate root elongation. Plant Physiol 176(3):2221–2230. https://doi.org/10.1104/pp.17.01414
21. Srivastava R, Li Z, Russo G, Tang J, Bi R, Muppirala U et al (2018) Response to persistent ER stress in plants: a multiphasic process that transitions cells from prosurvival activities to cell death. Plant Cell 30(6):1220–1242. https://doi.org/10.1105/tpc.18.00153
22. Ko DK, Brandizzi F (2022) Advanced genomics identifies growth effectors for proteotoxic ER stress recovery in Arabidopsis thaliana. Commun Biol 5(16):16
23. Herath V, Verchot J (2022) Comprehensive transcriptome analysis reveals genome-wide changes associated with endoplasmic reticulum (ER) stress in potato (*Solanum tuberosum* L.). Int J Mol Sci 23(22):13795. https://doi.org/10.3390/ijms232213795
24. Li JJ, Herskowitz I (1993) Isolation of ORC6, a component of the yeast origin recognition complex by a one-hybrid system. Science 262(5141):1870–1874
25. Gaudinier A, Zhang L, Reece-Hoyes JS, Taylor-Teeples M, Pu L, Liu Z et al (2011) Enhanced Y1H assays for Arabidopsis. Nat Methods 8(12):1053–1055. https://doi.org/10.1038/nmeth.1750
26. Ko DK, Brandizzi F (2020) A temporal hierarchy underpins the transcription factor-DNA interactome of the maize UPR. Plant J 105(1):254. https://doi.org/10.1111/tpj.15044
27. Park PJ (2009) ChIP-seq: advantages and challenges of a maturing technology. Nat Rev Genet 10(10):669–680. https://doi.org/10.1038/nrg2641
28. Deng Y, Humbert S, Liu JX, Srivastava R, Rothstein SJ, Howell SH (2011) Heat induces the splicing by IRE1 of a mRNA encoding a transcription factor involved in the unfolded protein response in Arabidopsis. Proc Natl Acad Sci U S A 108(17):7247–7252. https://doi.org/10.1073/pnas.1102117108
29. Ko DK, Brandizzi F (2020) Network-based approaches for understanding gene regulation and function in plants. Plant J 104:302
30. Langfelder P, Horvath S (2008) WGCNA: an R package for weighted correlation network analysis. BMC Bioinform 9:559. https://doi.org/10.1186/1471-2105-9-559
31. Ko DK, Brandizzi F (2023) Coexpression network construction and visualization from transcriptomes underlying ER stress responses. Methods Mol Biol 2581:385–401. https://doi.org/10.1007/978-1-0716-2784-6_27

# Chapter 20

# Variable-Angle Epifluorescence Microscopy for Single-Particle Tracking in the Plant ER

**Charlotte Pain, Christopher Tynan, Stanley W. Botchway, and Verena Kriechbaumer**

## Abstract

Single-particle tracking (SPT) of biomolecules in the plant endoplasmic reticulum has the potential to inform on the formation of protein–protein complexes, metabolons, and the transport of molecules through both the ER membrane and lumen. Plant cells are particularly challenging for observing and tracking single molecules due to their unique structure, size, and considerable autofluorescence. However, by using variable-angle or highly inclined epifluorescence microscopy (VAEM) and transient expression in tobacco, it is possible to observe single-particle dynamics in the ER. Selecting the appropriate fluorophore, and ensuring the correct fluorophore density in the ER, is essential for successful SPT. By using tuneable fluorophores, which can be photoconverted and photoactivated, it is possible to vary the density of visible fluorophores in the ER dynamically. Here we describe methods to prepare plant samples for VAEM and two methods for determining and analyzing single-particle tracks from VAEM time series.

**Key words** Single-particle tracking, VAEM, Endoplasmic reticulum, TrackMate, Automated single-particle tracking, paGFP, mEOS

## 1 Introduction

Single-particle tracking (SPT) uses live cell imaging to characterize the dynamics of single molecules or small protein aggregates as they move through different cellular structures. SPT requires microscopical techniques with high sensitivity, capable of detecting the fluorescence of either individual or small groups of fluorophores. Two techniques commonly combined to perform SPT are sptPALM (single-particle tracking photoactivated localization microscopy) [1, 2] and VAEM (variable-angle epifluorescence microscopy) [3], which can be achieved using commercially available TIRF microscope systems. sptPALM is an adaptation of PALM, which achieves super-resolution by continuously photoactivating specific subset of fluorophores in the cell. This separates

Verena Kriechbaumer (ed.), *The Plant Endoplasmic Reticulum: Methods and Protocols*, Methods in Molecular Biology, vol. 2772, https://doi.org/10.1007/978-1-0716-3710-4_20,

particles temporally, allowing for the detection of individual particles that otherwise cannot be distinguished from each other spatially. PALM uses the locations of photoactivated static fluorophores which are used to build a super-resolved map of the protein distribution, whereas sptPALM tracks the position of moving photoactivated single molecules. The second technique, VAEM, employs an epifluorescence microscope equipped with ultrasensitive cameras that is typically combined a total internal reflection fluorescence system (TIRF system) to vary the angle of light incident on the sample. TIRF microscopy excites the sample with an evanescent field that penetrates only a few hundred nanometers beyond an interface with a refractive index mismatch. This enables the detection of single molecules by providing a large reduction in background fluorescence but requires the excitation laser to be incident on a suitable interface beyond the critical angle for total internal reflection. VAEM works in a similar manner, but uses a subcritical angled laser to produce a slanted beam of light that penetrates to variable depths within the cell [4, 5]. This beam of light only activates fluorophores in a thin (typically <1 μm) layer of the sample while ensuring that fluorescent molecules outside this area remain dark. In plant cells, these techniques are commonly used alongside fluorescently tagged proteins, either fused directly to a protein of interest or introduced by SNAP tagging [6]. While in mammalian cells, small peptide tags covalently labeled with fluorescent dyes [7] and quantum dots [8] have been also used to assess particle dynamics in different cellular structures. The selection of fluorescent probes is particularly important for SPT, as detection of single particles requires control over the density of active probes and the fluorophore must be sufficiently "bright" to obtain a detectable signal above the sample fluorescence background. Also desirable for SPT is high photostability and the absence of observable blinking kinetics so that individual molecules can be tracked continuously for as long as possible. To achieve sufficient control of fluorophore density when using fluorescent fusion proteins, "tuneable" versions of these proteins that can be photoactivated or converted as needed are particularly useful. One example is photoactivatable GFP (paGFP) of which only 1% of molecules are spontaneously activated under 488 nm excitation [8]. With the addition of intense activating illumination at 413 nm, its fluorescence is increased a 100-fold [9]. Another tuneable fluorophore excited at 488 nm is mEOS. This is a photoconvertible fluorophore that emits green fluorescence in its native state (excitation 488, emission 516 nm) and is converted to a red (581 nm) fluorophore after exposure to near-UV light (approximately 390 nm) [10, 11]. An alternative way to reduce fluorophore density is to photobleach excess fluorophores using prolonged exposure to the excitation laser under normal imaging conditions or increasing the laser intensity up to threefold. However, this can have negative physiological effects through the increased risk of phototoxicity and sample heating.

Plant cells are uniquely challenging for SPT due to their structure and composition. Plant cells are particularly bulky when compared to their mammalian cell counterparts and are composed of structures with variable refractive indices [12]. The presence of a cell wall and waxy cuticle at the cell surface can result in light scattering, obscuring the fluorescence of single molecules. In addition, images produced of plant cells often have a low signal-to-background ratio [13] due to autofluorescence, generated by common cellular components such as chlorophyll and lignin [14].

It is easiest to perform SPT near the cell surface, either at the plasma membranes or at structures just below the cell surface including the endoplasmic reticulum. This limits the distance over which light scattering may occur and minimizes its negative impact on the signal-to-background ratio. In plant cells, SPT has been used to investigate the dynamics of single molecules across the plasma membrane [3], quantify the effect of the cell wall on plasma membrane nanodomains [15, 16], and analyze the effect of the cytoskeleton on plasma membrane-localized proteins [17]. The plasma membrane represents a relatively simple organelle for this kind of analysis, as it is part of a large membrane expanse, which is relatively flat and can be found nearby the coverslip, thus removing the need for deep laser penetration.

SPT on the ER is more challenging due to both its depth within a cell and its 3D structure. Compared to the plasma membrane, which is composed of a lipid bilayer approximately 2.5–5 nm wide [18], the ER is composed of two lipid bilayers, enclosing the ER lumen, and has a width of approximately 40 nm [19] . SPT in mammalian ER is challenging due to the three-dimensional nature of the ER network, and although plant ER is functionally two-dimensional, it is not necessarily a completely flat surface and often not at the optimal orientation for VAEM. As a result, single molecules are often out of focus during microscopy, increasing the background intensity and reducing the contrast between in-focus molecules and the background. However, SPT in the ER of mammalian cells and plant cells has been achieved despite these challenges. This includes measurements of ER lumen dynamics [20] and the localization and formation of ER–plasma membrane contact site complexes [8].

The potential for SPT at the ER has only just begun to be explored but it could soon be applied to measurements of metabolon formation on the ER surface, signaling receptor coordination at the ER surface, and cellular responses to pathogen attack and to establish different functional roles of the different ER structural subdomains.

The methods required for SPT on the plant ER described here are:

(a) Selecting fluorophores for VAEM SPT.
(b) Mounting samples and collecting data.
(c) Analyzing particle trajectories with TrackMate.
(d) Automatic particle detection and analysis.

## 2 Materials

### 2.1 Single-Particle Imaging by VAEM Microscopy

1. 5–6-week-old *Nicotiana tabacum* or *Nicotiana benthamiana* plants, grown at 21 °C, 14 h light, 10 h dark, transiently expressing the protein of interest (*see* Chap. 1 for *Agrobacterium*-mediated transient transformation) fused to an appropriate fluorophore (*see* Subheading 3.1 for details).
2. Deionized water for mounting.
3. Glass microscopy slides.
4. Glass coverslips, ideally #1.5 high precision (0.17 ± 0.05 mm) (*see* **Note 1**).
5. Micropore tape to secure coverslips to glass slides.
6. Razor blade.
7. Tweezers.

### 2.2 Single-Particle Tracking

1. System requirements: Windows 7–11, mac OS X 10.8 or later, or Linux on amd64 or x86 architectures.
2. Fiji ImageJ 2.1.0/1.53q.
3. Java 1.8.0.
4. TrackMate 7.6.1.
5. S3 software, available in the supplementary data of Wilson et al. (2016).
6. MATLAB 2017b (may work with other versions).

## 3 Methods

### 3.1 Selecting Fluorophores for VAEM SPT

One of the most challenging aspects of single-particle tracking in the ER is controlling fluorophore density. Insufficient fluorophore density makes identification of single fluorophore particles against the background fluorescence of the cell challenging, while a too high density of fluorophores results in single particles becoming indistinguishable from the others (Fig. 1). Several techniques exist for controlling the density of fluorophores in the ER:

- Lowering the optical density of the Agrobacteria infiltration medium.
- The use of "tuneable" fluorophores.

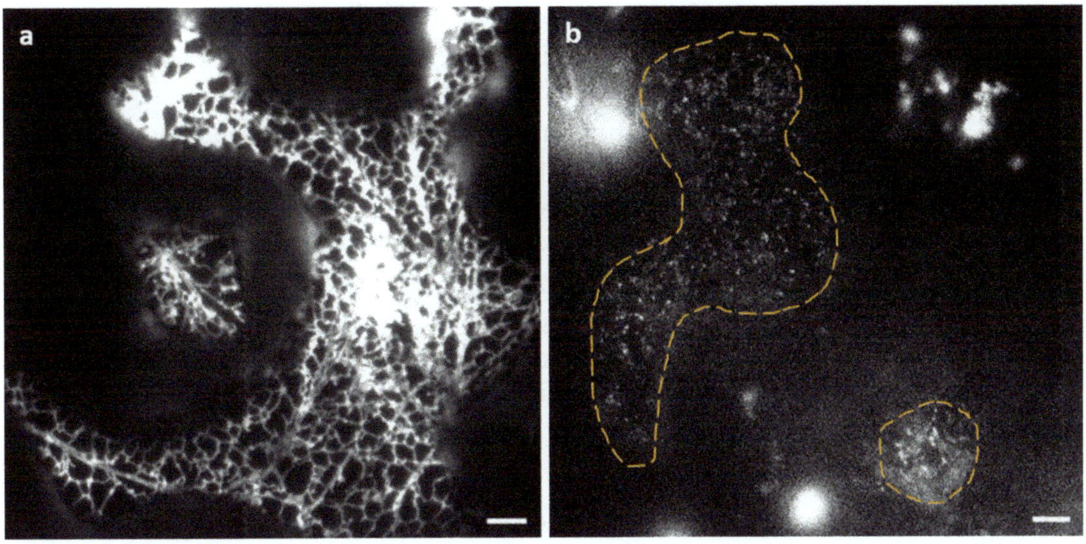


**Fig. 1** Optimizing fluorophore density for SPT. Transient overexpression of paGFP-CXN after infiltration of Agrobacteria cultures at an optical density of (**a**) $OD_{600} = 0.1$ and (**b**) $OD_{600} = 0.001$. Cell outline shown where required using an orange dashed line overlay. Scale bars = 5 μm

In order to establish which fluorophores and optical densities may work best for a given protein of interest, we recommend the following procedure:

1. Select the desired fluorophore (*see* **Note 2**).
2. Fuse to a protein of interest under an overexpression promoter, for example, the 35S promoter derived from the cauliflower mosaic virus [21] or the ubiquitin-10 promoter [22].
3. Transform this construct into the *Agrobacterium* strain GV3101.
4. Using the Agrobacteria-mediated transient expression technique (described in Chap. 1), infiltrate 5–6-week-old *Nicotiana tabacum* or *Nicotiana benthamiana* leaves with the Agrobacteria in infiltration medium at a range of optical densities (*see* **Note 3**).

### 3.2 Mounting Samples and Collecting Data

1. Pipette 10 μL deionized water onto a slide.
2. Using a razor blade, cut an approximately 25 mm$^2$ piece of tobacco leaf and transfer it to the slide using tweezers, taking care to damage the leaf tissue as little as possible.
3. Mount the leaf piece abaxial (underside) facing upward.
4. Pipette a small droplet of water onto the surface of the leaf.

5. Apply the coverslip over the leaf.
6. Tape the coverslip in place, ensuring that the leaf piece is gently pressed against the coverslip (*see* **Note 4**).
7. Transfer the sample to the microscope and mount on the microscope as appropriate.
8. Focus on a cell and then angle the beam to a subcritical angle suitable for single molecule imaging (*see* **Note 5**).
9. Set gain values to maximum.
10. Acquire frames with the maximum frame rates possible for at least 60 s. We recommend 0.05 s per frame (or 20 frames per second) when using an emCCD.

### 3.3 Analyzing Particle Trajectories with TrackMate

TrackMate [23] is a Fiji plugin [24] that can be used to track discrete objects (*see* **Note 6**). It is most often applied to organelles such as mitochondria [25], lysosomes [26], and fusiform bodies through the ER [27]. However, it has previously been applied to single-particle tracking of single molecules, including the movement of formin molecules [28] and receptors at the plasma membranes in plants [29]. The method included here is specific for single-particle tracking using VAEM in the plant ER (*see* **Note 7**).

1. Download Fiji and Java.
2. Open Fiji and import images by dragging and dropping onto the Fiji program.
3. Ensure that autoscale is selected when importing images and click okay to import them.
4. If there are particularly bright objects (saturated pixels) in the image frame that are not to be included in the analysis, create an ROI around the edge of the cell of interest (using either polygon selection or rectangle selection).
5. Launch TrackMate by selecting Plugins > Tracking > TrackMate.
6. TrackMate should automatically import pixel width, height, and frame intervals from the image metadata. However, if these fields are not automatically populated, input this information manually.
7. Ensure LoG detection is selected (*see* **Note 8**) and then click "Next" to move to the next stage.
8. Set the estimated diameter of the particle and the threshold level. These settings will depend on how images are acquired.

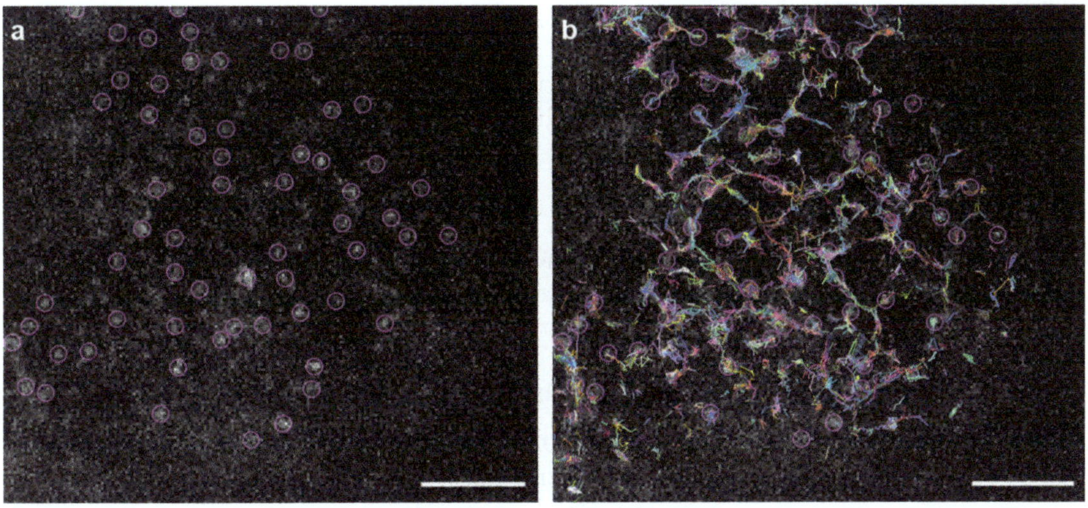


**Fig. 2** Example SPT output from TrackMate. (**a**) Single particles of paGFP-CXN in the ER detected by TrackMate (outlines in purple) and (**b**) the tracks linking the detected single particles, color coded in order of detection. Scale bars = 5 μm

9. Use the Preview function to show where particles have been detected (Fig. 2a).
10. Adjust threshold and diameter parameters until all particles have been identified accurately. Click "Next" to move to the next stage.
11. Remove poor quality particles, with a high signal-to-noise ratio by selecting auto. Click next to move to the next stage.
12. Navigate through the time series to ensure that single-particle detection has worked well, i.e., all or almost all of the particles in an image have been detected with minimal background interference. If not, repeat the previous stages and adjust the input parameters to minimize the background. It is also possible at this point to recolor the identified particles at this stage. Click "Next" to move to the next stage.
13. Ensure simple LAP tracker is selected to allow connection of particle tracks (*see* **Note 9**). Click "Next" to move to the next stage.
14. Set the appropriate parameters for linking detected spots together; again this will depend on sample specifics but can be easily modified later. It is recommended that the linking and gap-closing max distances are set to no more than double the diameter of a particle, and the gap-closing max frame gap is set to 2. Click "Next" to move to the next stage.

15. Display details of the produced tracks and assess for issues in linking particles to form tracks (Fig. 2b). Tracks with large jumps and particularly jagged trajectories can be an indication of poor particle linkage.
16. Select Tracks to open the particle tracking data and save the output results.

### 3.4 Automatic Particle Detection and Analysis

An alternative to TrackMate is an automated particle tracking and detection software produced by Wilson et al. (2016). Its automated capabilities make it more appropriate for larger datasets, allowing for rapid analysis of a large volume of data. Documentation is provided alongside the software, however in brief:

1. Download the software, included in the supplementary data of the original paper [30].
2. Launch the software using the MATLAB command line, typing "ParticleTracking();".
3. Navigate to the folder containing the single-particle time series.
4. Ensure correct image import by setting the first and last frame, alongside the filename prefix.
5. Estimate the detection parameters for the given time series, including the number of particles in an image, the mean particle size, the minimum particle size, the signal-to-noise ratio and the particle density.
6. Check the detected single particles and adjust the original parameters if needed until particle detection is accurate. If needed, the advanced parameter settings can be modified to threshold the background or smooth the image. Fine-tuning these detection parameters depends heavily on the particular image acquisition pipeline chosen, but our workflow and analysis showed the software performed well with minimal modifications.
7. Perform the full-particle detection routine. This generally take several minutes.
8. Set up the linking parameters, specifically the maximum displacement between frames for a given particle and the minimum length of track. It is recommended to observe the experimental data to estimate values for good analysis, typically no more than double-particle diameter for maximum displacement and a minimum length of track of 3 work well.
9. Save and export the data, including the image of particle tracks, density, speed, and direction of particle movement, which are displayed automatically (Fig. 3).

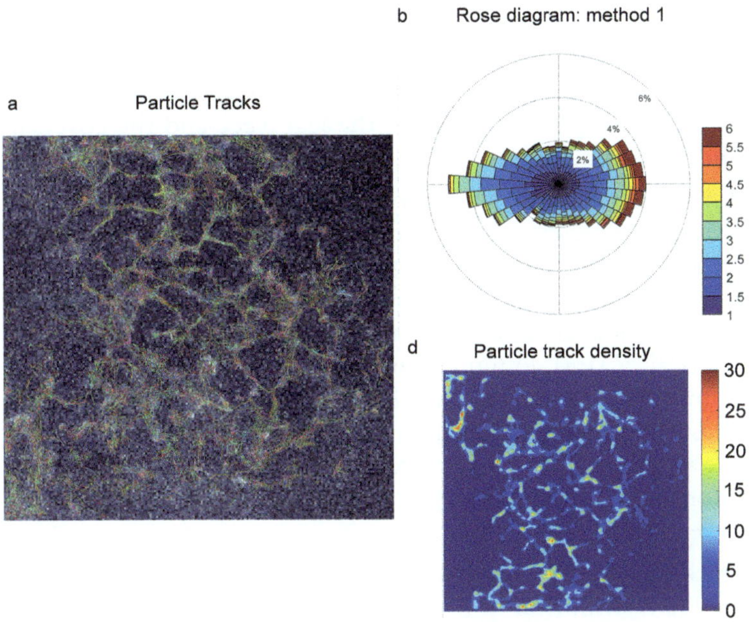


**Fig. 3** Example output of the S3 software. An example output produced by the S3 software including (**a**) detected particle tracks, (**b**, **c**) two rose diagrams of the direction of particle movement, (**d**) a heatmap of the density of particle tracks, and (**e**) speed of particle tracks

## 4 Notes

1. #1.5H glass coverslips are available, e.g., from Thorlabs and Fisher Scientific.
2. Up-to-date information on the excitation/emission spectra, brightness, and in some cases, photostability of fluorophores is available on FPbase (www.fpbase.org). This online software tool allows selection of the most appropriate fluorophore for a given application. As a starting point in plant cells, we recommend paGFP and mEOS.
3. Infiltrations with too high optical densities, and therefore too densely packed single particles, result in cells where individual molecules cannot be distinguished from each other (Fig. 1a). Instead, the goal is individually visible single particles that can be distinguished from the image background (Fig. 1b).

   By reducing the density of the Agrobacteria infiltrated into the leaf, each cell will likely have fewer transformation events, resulting in reduced expression of a given construct. This reduces the density of the fluorescent molecules in each cell, making it easier to distinguish individual fluorescent particles. As a starting point, we recommend testing infiltration $OD_{600}$ = 0.1, 0.01, 0.005, and 0.001. Depending on the individual expression of each construct, typically infiltrations with an $OD_{600}$ of 0.01 to 0.001 work best.

4. Ensuring the leaf section is closely appressed to the coverslip is essential. Air pockets between the leaf piece and the coverslip will inhibit imaging.
5. We recommend an angle of approximately 60° or wherever gives the greatest contrast in the sample against the background. It may take some time to establish the best angle for a given sample, and the angle may need to be changed as the sample drifts or as samples are changed. If possible, ring or elliptical VAEM illumination is generally better than point illumination as it will avoid shadowing of the illuminated area. Often, two identical emission filters have to be used or one with a higher optical density in order to sufficiently reject chloroplast autofluorescence and observe single fluorescent molecules.
6. TrackMate is a free, easy-to-use plugin that is available as part of the Fiji distribution of ImageJ.
7. More information on the full capabilities of TrackMate is available at https://imagej.net/plugins/trackmate/getting-started.
8. Several particle detection methods are available, depending on image structure and type, and it may be more appropriate to use a different detector.
9. Other particle joiners are available but LAP tracker works well for single-particle tracking.

## Acknowledgments

This work was supported by a program access grant to the Central Laser Facility at Harwell (ID 2202).

### References

1. Bayle V, Fiche J-B, Burny C, Platre MP, Nollmann M, Martinière A et al (2021) Single-particle tracking photoactivated localization microscopy of membrane proteins in living plant tissues. Nat Protoc 16:1600–1628
2. Manley S, Gillette JM, Lippincott-Schwartz J (2010) Single-particle tracking Photoactivated localization microscopy for mapping single-molecule dynamics. Methods Enzymol 475: 109–120
3. Wan Y, Ash WM III, Fan L, Hao H, Kim MK, Lin J (2011) Variable-angle total internal reflection fluorescence microscopy of intact cells of Arabidopsis thaliana. Plant Methods 7: 1–7
4. Chen T, Ji D, Tian S (2018) Variable-angle epifluorescence microscopy characterizes protein dynamics in the vicinity of plasma membrane in plant cells. BMC Plant Biol 18(43): 1–17
5. Konopka CA, Bednarek SY (2008) Variable-angle epifluorescence microscopy: a new way to look at protein dynamics in the plant cell cortex. Plant J 53(1):186–196
6. Bosch PJ, Correa IR Jr, Sonntag MH, Ibach J, Brunsveld L, Kanger JS et al (2014) Evaluation

of fluorophores to label SNAP-tag fused proteins for multicolor single-molecule tracking microscopy in live cells. Biophys J 107(4): 803–814
7. Clarke DT, Martin-Fernandez ML (2019) A brief history of single-particle tracking of the epidermal growth factor receptor. Methods Protoc 2(1):12
8. Li H, Dou S-X, Liu Y-R, Li W, Xie P, Wang WC et al (2015) Mapping intracellular diffusion distribution using single quantum dot tracking: compartmentalized diffusion defined by endoplasmic reticulum. J Am Chem Soc 137: 436–444
9. Patterson GH, Lippincott-Schwartz J (2002) A photoactivatable GFP for selective photolabeling of proteins and cells. Science 297(5588): 1873–1877
10. McKinney SA, Murphy CS, Hazelwood KL, Davidson MW, Looger LL (2009) A bright and photostable photoconvertible fluorescent protein. Nat Methods 6(2):131–133
11. Wiedenmann J, Ivanchenko S, Oswald F, Schmitt F, Röcker C, Salih A et al (2004) EosFP, a fluorescent marker protein with UV-inducible green-to-red fluorescence conversion. Proc Natl Acad Sci U S A 101(45): 15905–15910
12. Komis G, Mistrik M, Šamajová O, Ovecka M, Bartek J, Šamaj J (2015) Superresolution live imaging of plant cells using structured illumination microscopy. Nat Protoc 10(8): 1248–1263
13. Guo A-Y, Zhang Y-M, Wang L, Bai D, Xu Y-P, Wu W-Q (2021) Single-molecule imaging in living plant cells: a methodological review. Int J Mol Sci 22(10):5071–5083
14. Donaldson L (2020) Autofluorescence in plants. Molecules 25(10):2393–2413
15. McKenna JF, Rolfe DJ, Webb SED, Tolmie AF, Botchway SW, Martin-Fernandez ML et al (2019) The cell wall regulates dynamics and size of plasma-membrane nanodomains in Arabidopsis. PNAS 116(26):12857–12862
16. Martinière A, Lavagi I, Nageswaran G, Rolfe DJ, Maneta-Peyret L, Luu D-T et al (2012) Cell wall constrains lateral diffusion of plant plasma-membrane proteins. PNAS 109(31): 12805–12810
17. Hutten SJ, Hamers DS, den Toorn MA, van Esse W, Nolles A, Bücherl CA et al (2017) Visualization of BRI1 and SERK3/BAK1 nanoclusters in Arabidopsis roots. PLoS One 12(1):1–19
18. Yamamoto T (1963) On the thickness of the unit membrane. J Cell Biol 17:413–421
19. Pain C, Kriechbaumer V, Kittelmann M, Hawes C, Fricker M (2019) Quantitative analysis of plant ER architecture and dynamics. Nat Commun 10(1):984
20. Holcman D, Parutto P, Chambers JE, Fantham M, Young LJ, Marciniak SJ et al (2018) Single particle trajectories reveal active endoplasmic reticulum luminal flow. Nat Cell Biol 20(10):1118–1125
21. Amack SC, Antunes MS (2020) CaMV35S promoter – a plant biology and biotechnology workhorse in the era of synthetic biology. Curr Plant Biol 24:100179
22. Grefen C, Donald N, Hashimoto K, Kudla J, Schumacher K, Blatt MR (2010) A ubiquitin-10 promoter-based vector set for fluorescent protein tagging facilitates temporal stability and native protein distribution in transient and stable expression studies. Plant J 64(2): 355–365
23. Tineveza J-Y, Perry N, Schindelin J, Hoopes GM, Reynolds GD, Laplantine E et al (2017) TrackMate: an open and extensible platform for single-particle tracking. Methods 115:80–90
24. Schindelin J, Arganda-Carreras I, Frise E, Kaynig V, Longair M, Pietzsch T et al (2012) Fiji: an open-source platform for biological-image analysis. Nat Methods 9: 676–682
25. Chustecki JM, Etherington RD, Gibbs DJ, Johnston IG (2022) Altered collective mitochondrial dynamics in the Arabidopsis msh1 mutant compromising organelle DNA maintenance. J Exp Bot 73(16):5428–5439
26. de Castro IF, Tenorio R, Ortega-González P, Knowlton JJ, Zamora PF, Lee CH et al (2020) A modified lysosomal organelle mediates nonlytic egress of reovirus. J Cell Biol 219(7):1–19
27. Pain C, Kriechbaumer V (2020) Defining the dance: quantification and classification of endoplasmic reticulum dynamics. J Exp Bot 71(6):1757–1762
28. Wu S-Z, Bezanilla M (2018) Actin and microtubule cross talk mediates persistent polarized growth. J Cell Biol 217(10):3531–3544
29. Bücherl CA, Jarsch IK, Schudoma C, Segonzac C, Mbengue M, Robatzek S et al (2017) Plant immune and growth receptors share common signalling components but localise to distinct plasma membrane nanodomains. elife 6:1–28
30. Wilson RS, Yang L, Dun A, Smyth AM, Duncan RR, Rickman C et al (2016) Automated single particle detection and tracking for large microscopy datasets. R Soc Open Sci 3(5):1–13

# Chapter 21

# Use of Super-Resolution Live Cell Imaging to Distinguish Endoplasmic Reticulum: Nuclear Envelope Subcellular Localization

**Nadine Field and Katja Graumann**


## Abstract

A distinguishing feature of eukaryotes is the presence of a nuclear envelope (NE) and endomembrane system. The NE is a double-membrane system that surrounds chromatin and is continuous with the endoplasmic reticulum (ER). This interface is crucial in various processes such as calcium signaling and ER-associated degradation. The outer nuclear membrane and ER share a multitude of proteins although some are only functional in one domain, whereas the inner nuclear membrane has its own unique proteome. Until recently, it was not possible to distinguish between the inner and outer nuclear membranes as well as perinuclear ER using light microscopy – only electron microscopy was suitable for this. Now, however, using super-resolution live cell imaging, this can be achieved while still observing protein and membrane dynamics in real time. The protocols described here will allow researchers to determine subcellular localization of potential NE/ER proteins in live plant cells, helping to gain new insights into protein functionality.


**Key words** Nuclear envelope, Endoplasmic reticulum, Confocal microscopy, Proximity distance, Fluorescence intensity ratio

## 1 Introduction

Conserved in all eukaryotic organisms, the nuclear envelope (NE) is a double-membrane structure consisting of an inner and an outer phospholipid bilayer, referred to as the inner nuclear membrane (INM) and outer nuclear membrane (ONM), respectively, with the two separated by a lumenal space called the perinuclear space [1]. Nuclear pore complexes (NPCs) are embedded in the pore membrane, which connects INM and ONM, and mediate bidirectional nuclear transport in and out of the nucleus [2].

While continuous with each other, the INM is physically distinct from the ONM, with a subset of unique membrane proteins functioning in regulating chromatin organization and other

Verena Kriechbaumer (ed.), *The Plant Endoplasmic Reticulum: Methods and Protocols*, Methods in Molecular Biology, vol. 2772, https://doi.org/10.1007/978-1-0716-3710-4_21,

nuclear processes [3]. In addition, a specific set of INM and ONM proteins can interact to form complexes across the perinuclear space. C-terminal Sad1/Unc84 (SUN)-domain proteins (C-ter SUNs) are located at the INM and interact with Klarsicht/Anc1/Syne homology (KASH) proteins on the ONM forming linker of nucleoskeleton and cytoskeleton (LINC) complexes [4]. These have a multitude of nuclear and cellular roles such as functioning in homologous recombination, nuclei morphology, and nuclei movement [4].

The ONM is continuous with the membrane of the endoplasmic reticulum (ER) – a large and dynamic organelle with a variety of functions, including the synthesis, transport, and folding of proteins [5, 6]. This NE–ER interface is essential for various cellular processes across all eukaryotic organisms.

ONM–ER contact sites play a role in calcium signaling and ER-associated degradation (ERAD) of proteins as part of quality control and functional regulation [1].

Due to the continuous nature of the NE and the ER, some proteins are present in both structures, although only enriched in the domain in which they are functional; in live plant cell imaging, C-ter SUNs and KASH proteins will be visible both in the NE and the ER, but are only enriched in the NE. In comparison, calnexin will also be visible in both structures but is an ER-enriched protein [4].

Consequently, establishing the exact subcellular localization of such proteins is important to gain insights into their functions and the molecular mechanisms behind them. Before recent developments in imaging, it was not possible to distinguish between INM and ONM using live cell imaging, and electron microscopy had to be used exclusively with fixed and stained samples. However, using super-resolution confocal microscopy, this is now possible, so protein and membrane dynamics can be explored in real time [7].

In this chapter, we detail two image analysis methodologies to establish subcellular localization using data from live cell confocal imaging in plants. The first determines ER/NE fluorescence intensity (FI) ratios as an indicator of which membrane domain the protein is enriched in. The second is based on proximity distance across the NE, which can be used to determine whether proteins are INM, ONM, or PNER localized.

## 2 Materials

1. A confocal microscope with the capacity for super-resolution imaging is required, i.e., ZEISS LSM 880 with AiryScan detector.
2. Pipette and appropriate tip.

3. Surgical or electrical tape.
4. Microscope glass slides.
5. No. 1.5 coverslips.
6. Fiji-equipped ImageJ (v1.53n) with the Radial Profile Extended plugin (*see* **Note 1**).
7. Microsoft Excel.
8. Plant tissue sample expressing transient ER or NE markers (*see* **Note 2**).

## 3 Methods

### 3.1 Mounting of Samples for Imaging

1. Mount plant tissue sample (e.g., leaf or seedling sample), used for all imaging, by placing an 80 μL bead of water on a microscope slide using a pipette.
2. Place the plant sample on top of this using forceps or a razor blade.
3. Using the pipette, place a bead of water in the middle of a no. 1.5 coverslip and lower this onto the sample.
4. Use 3M micropore surgical tape wrapped around each side of the coverslip to ensure the coverslip stays in place and movement is reduced.

### 3.2 ER/NE Fluorescence Intensity Analysis

1. Find cells in which to collect images of NE and ER fluorescence in the same optical plane where the NE is in the optimal midsection (*see* **Note 3**). Save images.
2. In Fiji, open the images of interest and using the "Freehand line" tool, draw a continuous line across the NE and ER ensuring that the line crosses the NE and ER membranes multiple times (Fig. 1).
3. Select "Analyze" and "Plot Profile," to plot the fluorescence of the NE and ER (Fig. 1).
4. In the resulting plot window, select "data" and save the data. Open this file in Excel.
5. In the data table, identify the peak values for the NE and ER sections.
6. In Excel, calculate the average NE and ER peak fluorescence.
7. Using the average NE and ER peak fluorescence, calculate the NE/ER ratio. A ratio closer to 1 indicates NE enrichment, whereas a value closer to 0 indicates ER enrichment (*see* **Note 4**).

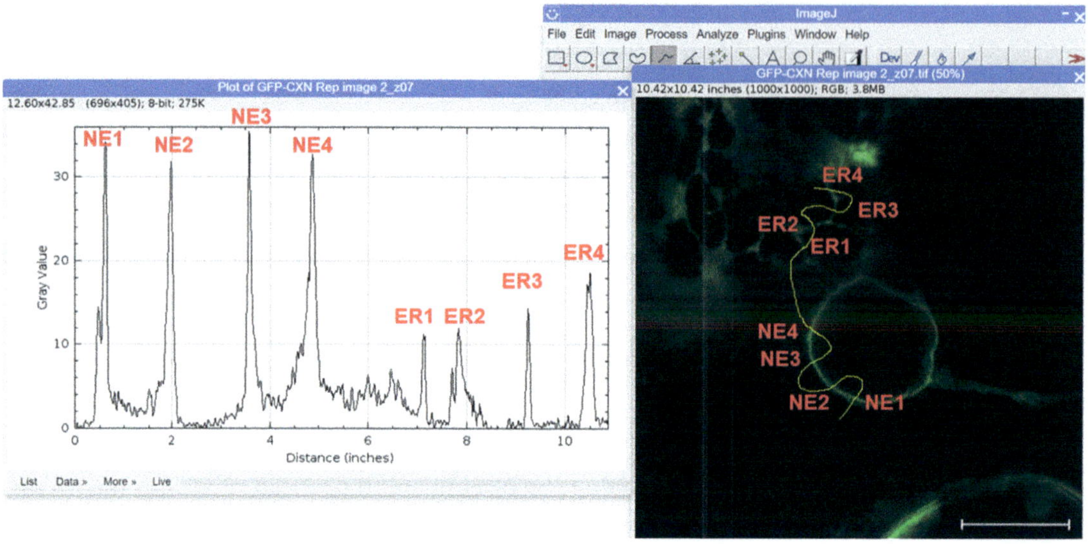


**Fig. 1** Determining NE/ER ratio to examine protein enrichment in the NE and ER membranes. Average NE and ER fluorescent values from the same optical plane can be used to calculate NE/ER ratios

### 3.3 Radial Line Profiles to Determine Proximity Distance

1. Find nuclei that express both a fluorescently tagged ONM/ER marker protein and a fluorescently tagged protein of interest.
2. Collect Z-stacks of nuclei with 1 μm distance between each slice with a ×100 oil immersion lens at a zoom factor of 4.5.
3. In Fiji, open the desired Z-stack and navigate to the medial slice of the nucleus.
4. Select "Plugins," select "Radial Profile Angle," and set the "Integration Angle" to 30°, which will set a triangular ROI (Fig. 2).
5. Place the point of the triangle in the estimated center of the nucleus, use the mouse wheel to adjust the size of the ROI, and adjust the "Starting Angle" to move the ROI around its point (Fig. 2).
6. Tick "Use Spatial Calibration" and click "Calculate Radial Profile" to generate a plot, and click "Data" and "Save Data" to save as an Excel spreadsheet (Fig. 2).
7. In each Excel spreadsheet, use the formula =INDEX(A2:AX, MATCH(MAX(B2:BX),B2:BX,0)) (where X = the max number of cells in the column) to return the radius in microns at the maximum FI.
8. Calculate an average radius in microns for each ONM/ER marker protein and each protein of interest, and subtract the protein of interest radius from the ONM/ER marker radius.

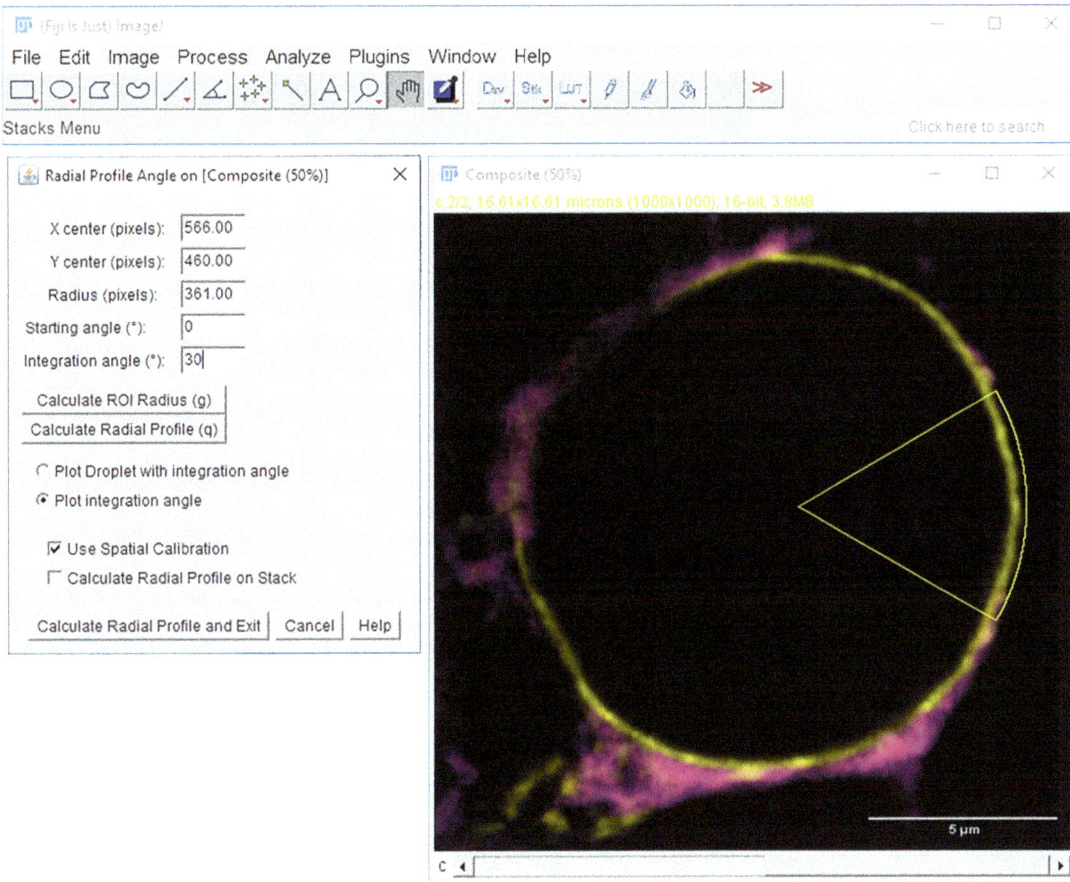


**Fig. 2** Determining proximity distance of the INM and ONM membranes. Maximum fluorescence intensity values from the same optical plane can be used to calculate proximity distance

9. A negative distance in microns indicates the protein is ONM/ER enriched, whereas a positive distance indicates the protein is INM enriched.

## 4 Notes

1. This plugin allows the choice of starting and integration angle over which the integration on the ROI is done. This is available for download from the ImageJ website (https://imagej.nih.gov/ij/plugins/radial-profile-ext.html).
2. In our work, we usually use 1 $cm^2$ leaf sections of *N. benthamiana* or *N. tabacum* 6-week-old plants grown on Levingtons F2+sand in 16:8 light–dark cycle and transiently expressing fluorescent markers as described in Sparkes et al [8].

3. Don't use the optical section where the nuclei poles are in focus as they are sheetlike in appearance
4. Appropriate ER and NE markers should be used for NE/ER ratio control values. LBR-GFP or SUN1-FP are useful NE markers; calnexin-FP and RTN-FP are good ER markers.

## References

1. Mehrtash AB, Hochstrasser M (2019) Ubiquitin-dependent protein degradation at the endoplasmic reticulum and nuclear envelope. Semin Cell Dev Biol 93:111–124
2. Webster BM, Lusk CP (2016) Border safety: quality control at the nuclear envelope. Trends Cell Biol 26:29–39
3. Groves NR, McKenna JF, Evans DE, Graumann K, Meier I (2019) A nuclear localization signal targets tail-anchored membrane proteins to the inner nuclear envelope in plants. J Cell Sci 132:jcs226134
4. Graumann K (2014) Evidence for LINC1-SUN associations at the plant nuclear periphery. PLoS One 9:e93406
5. Jan CH, Williams CC, Weissman JS (2014) Principles of ER cotranslational translocation revealed by proximity-specific ribosome profiling. Science 346:1257521
6. Schwarz DS, Blower MD (2016) The endoplasmic reticulum: structure, function and response to cellular signaling. Cell Mol Life Sci 73:79–94
7. Dumur T, Duncan S, Graumann K, Desset S, Randall RS, Scheid OM, Prodanov D, Tatout C, Baroux C (2019) Probing the 3D architecture of the plant nucleus with microscopy approaches: challenges and solutions. Nucleus 10:181–212
8. Sparkes IA, Runions J, Kearns A, Hawes C. (2006) Rapid, transient expression of fluorescent fusion proteins in tobacco plants and generation of stably transformed plants. Nature Protocols 1(4):2019–25. https://doi.org/10.1038/nprot.2006.286

# Chapter 22

# Using ER-Targeted Photoconvertible Fluorescent Proteins in Living Plant Cells

**Jaideep Mathur and Puja Puspa Ghosh**


## Abstract

Photoconvertible fluorescent proteins (pcFPs) enable differential coloring of a single organelle. Several pcFP-based probes have been targeted to the endoplasmic reticulum (ER) and can serve as useful tools to study ER dynamics and interactions with other organelles. Here, we describe the procedure to conduct live-cell imaging experiments using ER-targeted pcFP-based probes. Potential problems that might occur during the experiments, their solutions, and several ways to improve the experiments are discussed.


**Key words** Photoconvertible fluorescent proteins, mEos, Dendra, mMaple, Kaede, Organellar dynamics, Live-cell imaging, Endoplasmic reticulum

## 1 Introduction

Fluorescent proteins, spanning the entire color spectrum, are widely used tools for investigating subcellular and sub-organellar events in living cells [1]. The use of single-color fluorescent proteins (FP) has led to major insights on understanding the dynamic behavior of organelles in the plant cell [2] investigating membrane-trafficking pathways and protein–protein interactions [3] and [4–6]. The FP toolbox has been further enriched by the discovery of optical highlighter proteins whose spectral characteristics change upon irradiation by specific wavelengths. Optical highlighters include photo-inducible proteins that are initially nonfluorescent and emit fluorescence only after irradiation, photoconvertible proteins (pcFPs) that fluoresce in a particular color in their non-photoconverted form but whose fluorescence emission changes irreversibly upon irradiation, and photo-switchable proteins that can be alterably switched between different fluorescent states [1, 7, 8].

The most commonly used pcFPs, such as Dendra [9], Eos [10], Kaede [11], and Maple [12] (Table 1), change irreversibly from

Verena Kriechbaumer (ed.), *The Plant Endoplasmic Reticulum: Methods and Protocols*, Methods in Molecular Biology, vol. 2772, https://doi.org/10.1007/978-1-0716-3710-4_22,

**Table 1**
**A list of photoconvertible proteins commonly used in plants**

| Protein | Source organism | Color before photo-conversion $\lambda_{max}$ ex/em $_{nm}$ | Color after photo-conversion $\lambda_{max}$ ex/em nm | State | References |
|---|---|---|---|---|---|
| Kaede | *Trachyphyllia geoffroyi* | Green 508/518 | Red 572/580 | Tetramer | [11] |
| EosFP wild type (WT) | *Lobophyllia hemprichii* | Green 506/516 | Red 571/581 | Tetramer | [10] |
| mEos | Modified from WT EosFP | Green 505/516 | Red 569/581 | Monomer | [10] |
| PS-CFP | Synthetic construct | Cyan 400/468 | Green 490/511 | Monomer | [13] |
| KikGR | *Favia favus* | Green 507/517 | Red 583/593 | Tetramer | [14] |
| tdEos | Modified from WT EosFP | Green 505/516 | Red 569/581 | Tandem dimer | [15] |
| Dendra2 | *Dendronephthya* sp. | Green 490/507 | Red 553/573 | Monomer | [9] |
| mKikGR | Modified from tetrameric KikGR | Green 505/515 | Red 580/591 | Monomer | [16] |
| IrisFP | Modified from WT EosFP | Green 488/516 | Red 551/580 | Tetramer | [17] |
| mEosFP2 | Modified from WT EosFP | Green 506/519 | Red 573/584 | Monomer | [18] |
| mClavGR2 | Modified from *Clauveria* sp. derived CyanFP | Green 488/504 | Red 566/583 | Monomer | [19] |
| mIrisFP | Modified from | Green 486/516 | Red 546/578 | Monomer | [20] |
| mMaple | Modified from mClavGR2 | Green 489/505 | Red 566/583 | Monomer | [12] |

green to red following a brief exposure to 390–405 nm wavelengths [10, 21]. Several optical highlighter-based probes have been developed for use in plants [22, 23] and shown to work without interfering with subcellular dynamics and normal plant development [24–26]. Moreover, both the green and red forms of pcFPs can be used together with other single-colored fluorescent proteins. pcFP-based probes have been targeted to different organelles and structures in plants, such as plasma membrane [27], assorted vesicles and vacuoles [23], F-actin [23, 28] microtubules

[23], mitochondria [25, 29], peroxisomes [30], plastids [31], ER [22, 23, 25], and nucleus [32–34].

Several pcFP-based probes have been specifically targeted to the ER and show high potential for assessing ER membrane dynamics and ER interactions with other organelles. The ER is also recognized as a major player in forming membrane contact sites (MCSs) with other organelles [35, 36]. The use of ER-targeted pcFPs to investigate MCS formation in living plant cells responding to environmental cues is a promising area for exploration.

Here, we list the available ER-targeted monomeric (m)EosFP [10]- and mMapleFP [25]-based probes and simple protocols for their photo-conversion and subsequent visualization. While our lab has optimized protocols for use with mEosFP-based probes, the methods are equally applicable to tandem dimeric and tetrameric versions of EosFP [10] as well as to Kaede [11], Dendra [9], and mMaple [12] FPs expressed in plants.

Different FPs such as $CX_{TMD}$ -mEosFP have been targeted to the ER membrane as fusions to the calnexin (CX) transmembrane domain (TMD) [22, 23, 37, 38]. Targeting mEosFP to the ER lumen has been achieved by adding a signal sequence at the N-terminus and an ER retention motif HDEL [39, 40] at the C-terminus [23]. Both ER membrane and lumen-targeted probes can be used in transient assays involving DNA-coated gold/tungsten particle bombardment, PEG-mediated protoplast transformation, or *Agrobacterium tumefaciens* infiltration. Alternatively, stable *Arabidopsis* and tobacco transgenic plants can be generated following standard protocols.

ER-targeted mEosFP expression in a plant cell labels a dynamic, fluorescent green mesh. Photo-conversion from green to red fluorescence is achieved by excitation with blue–violet light spanning the 390–410 nm range or by a dedicated 405 nm laser line (Fig. 1). Straightforward applications of pcFPs utilizing their color-changing property have included tracking of organelles and their interactions [24, 27, 30, 31]. Another property with high but still underutilized potential is their irreversibility following photo-conversion. Similar to FRAP [41], this property has resulted in the color recovery after photo-conversion (CRAP) method [23, 25]. In CRAP, the reversion to a green form of the pcFP suggests either displacement of the photoconverted red form to another region (s) of the cell or the formation of fresh protein to add to the red form and change the spectral property again into the green part of the spectrum [23, 42]. CRAP creates a milder photo-conversion step compared to the harsh photobleaching step required in FRAP. It also allows for simultaneous observations on both non-photoconverted and photoconverted forms of the protein [23].

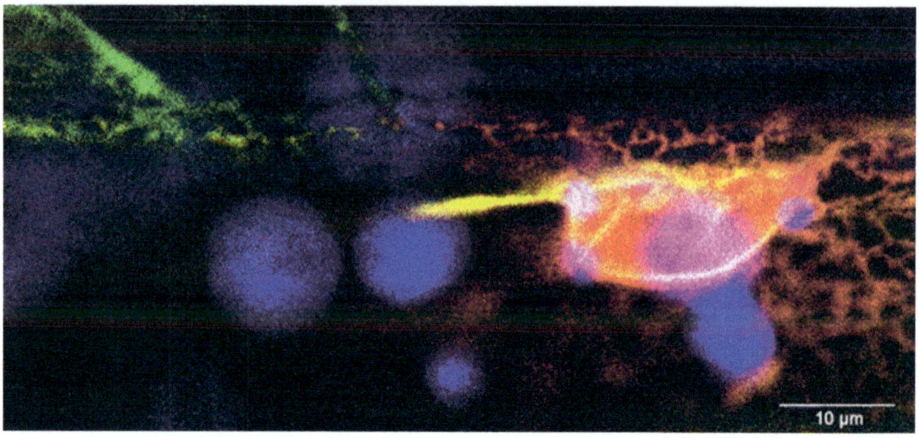


**Fig. 1** A representative image showing of cells expressing $CX_{TMD}$-mEosFP showing areas (top left) with non-photoconverted, green fusion protein and orange–red (center-right) area after photo-conversion. Chlorophyll fluorescence false colored blue. The probe dissipates quickly in a healthy living plant cell with active cytoplasmic streaming to progressively show up as yellow fluorescence and reverts completely to green after some time

**Table 2**
**Optical highlighters targeted to the ER in plants**

| Fluorescent probe | Target/targeting sequence | References |
|---|---|---|
| $CX_{TMD}$-PAGFP (photoactivable GFP) | ER membrane/transmembrane domain (TMD) of *Arabidopsis thaliana* calnexin; At5g61790 | [22] |
| $CX_{TMD}$-mEos-FP | ER membrane/TMD of *A. thaliana* calnexin | [23] |
| $CX_{TMD}$-mMaple | ER membrane/TMD of *A. thaliana* calnexin | [25] |
| mEosFP-HDEL | ER lumen/ER retention peptide HDEL at C-terminus | Mathur et al. unpublished |

The pcFP probes developed for the ER (Table 2) can now be used to obtain further insights on ER dynamics and protein dispersal within the ER lumen and between ER derivative organelles such as ER bodies, as well as ER relationship with other organelles at assorted membrane contact sites.

## 2 Materials

### 2.1 Epifluorescence Microscopy Setup with a Digital Camera

1. A high-pressure mercury plasma arc-discharge lamp (e.g., Mercury Short ARC #HBO 103 W/2 (OSRAM GmbH, Steinerne Furt 62, 86167 Augsburg, Germany). Alternative lamp source 380–660 nm: Sola light engine https://lumencor.com/products/sola-light-engine).

2. Glass filter cubes: DAPI/Hoechst/AMCA filter (excitation, 340–380 nm; 400 DCLP; emission, 435–485 nm; chroma filter # 31000v2) or a "D" filter (excitation filter, 355–425 nm; dichroic mirror 455 nm; suppression filter LP 470 nm) (*see* **Note 1**), an Endow GFP band-pass emission filter (HQ 470/40X; Q 495/LP; HQ 525/50 m; chroma filter #41017), a TRITC (green) filter (HQ535/30x; Q570LP; HQ620/60 m; e.g., chroma filter # 41002c) (*see* **Note 2**).
3. Lens: Standard lens of 10x, 20x, 40x, and 63x for epifluorescence microscopy with maximum numerical aperture are advocated. For imaging living cells using an upright microscope setup, 40X and 63X water immersion lens work very well.

### 2.2 Confocal Laser Scanning Microscope Setup

1. Lasers 405 nm, 488 nm Ar/Kr, 543 nm He/Ne.
2. Proprietary software from all CLSM vendors provides precise control over laser functioning and image acquisition and should be appropriately adjusted for visualization of green and red fluorescence with minimal bleed-through. If a 405 nm laser is unavailable, a hybrid approach using an epifluorescence microscope and confocal setup is possible (*see* **Note 3**), but lacks the precision afforded by software-controlled lasers.

## 3 Method

1. Place a plant sample mounted on a slide (*see* **Notes 1** and **2**) on the microscope stage and use brightfield to identify and focus on a region of interest (ROI).
2. If required, use the FITC glass filter for a short-duration observation of green fluorescence (*see* **Notes 2** and **3**).
3. Acquire the first pre-photoconversion images using 488 and 543 nm excitation. At this point, only observe the green fluorescence of the non-photoconverted pcFP. There should be no bleed-through in the red channel of CLSM or under the TRITC filter of epifluorescence microscope (*see* **Note 3**).
4. Depending upon the strength of the 405 nm laser or the light intensity coming through a glass D filter (or similar), irradiate the region of interest for 3–10 s. Photo-conversion will occur (*see* **Notes 4** and **5**).
5. Change to the TRITC filter to view a photoconverted ROI, or on a CLSM, switch to the 488 and 543 nm laser-scanning modes to simultaneously capture images in the green and red parts of the spectrum. Photoconverted form of protein will fluoresce red (*see* **Notes 5** and **6**).

6. To observe photoconverted and non-photoconverted areas, overlay the images captured in the green and red channels (Fig. 1; *see* **Note 7**).
7. Depending on the experiment, acquire image series in xyz, xyt, or xyzt dimensions.
8. For data collection and post-acquisition image processing, use for examples free software, such as ImageJ (http://rsb.info.nih.gov/ij), Fiji (http://mac.softpedia.com/get/Graphics/Fiji.shtml), and Adobe Photoshop (http://www.adobe.com).

## 4 Notes

1. Use intact tissue or seedlings when possible, taking care to not observe cells in areas that were used to pick up the sample; avoid touching plant tissue/areas to be observed when mounting samples on a slide. Even minimal disturbances disturb and change ER dynamics!
2. Many artifacts are created during sample preparation. For many plants, including tobacco and *Arabidopsis*, wounded/mutilated cells and semidry tissue start fluorescing green–yellow. In some plants, roots auto-fluoresce green and it is important to check this phenomenon beforehand. Borderline green fluorescence also becomes common in cells with photobleached plastids (*see* **Note 4**).
3. Depression slides maintain moisture around the plant sample during live-cell imaging experiments and are recommended. For the same reason, the use of larger 24 × 60 mm premium glass coverslips is advised for observing live plant cells.
4. Minimize the use of UV–blue light for initial focusing. Photobleaching creates a strong artifact that can interfere with your overall interpretation on pcFPs. Therefore, for initial focusing and observations, it is important to consider using a glass filter that also allows chlorophyll autofluorescence (seen as red upon blue light excitation) to be observed simultaneously. Many filters dedicated to observing GFP are designed to cut off the chlorophyll fluorescence completely and thus do not allow a researcher to estimate the health of their sample. The emission spectrum from wounded and photobleached plant tissue shifts progressively into yellow and orange regions of the spectrum. Blue light excitation-induced artifacts can be easily misinterpreted as the results of photo-conversion.
5. Although an epifluorescence microscope equipped with a digital camera can easily visualize and acquire images using pcFP-based probes, a CLSM setup with software-controlled 405, 488, 514, and 543 nm lasers ensures better quality and

precision due to automation of the photo-conversion step, followed by quick image acquisition, and data processing. This is especially important for investigations on the ER as the photoconverted protein diffuses very rapidly in the dynamic, continuously reorganizing ER. However, if equipping the CLSM system with a diode 50 mW 405 nm violet laser is not economically feasible, a hybrid approach can be adopted. In this approach, a DAPI or D glass filter can be used for photo-conversion manually on the epifluorescence microscope. Generally, a standard microscope diaphragm contains pinhole sizes ranging from 7 mm to 500 μm. However, most microscope companies can create supplementary pinholes of up to 50 μm diameter at a relatively low cost compared to installing a 405 nm laser. The broader bandwidth through the D glass filter makes it an efficient tool for the green to red photo-conversion compared to the single 405 nm laser line. However, a major limitation of the glass filter-based photo-conversion alternative is the lack of precision and automation compared to the software-controlled operation of the 405 nm laser system.

6. Depending on the protein expression levels, all pcFPs are susceptible to aggregation-induced artifacts that become especially prominent upon confinement by membranes. High expression of ER membrane-targeted $CX_{TMD}$ -mEosFP sometimes results in artifacts such as enlarged cisternae and circular bodies. Protein aggregates have been observed photoconverting under prolonged exposure to white fluorescent lamps that may have a small component of the violet–blue region of the spectrum.

7. Imaging after photo-conversion follows routine procedures involved in fluorescence microscopy where different emission spectra are recorded. The use of pcFPs creates the possibility of incomplete photo-conversion. Under this situation, the emissions can be recorded in both the green and red channels. To avoid this situation, ensure complete photo-conversion of the pcFP within the ROI and ensure that there is only minimal green fluorescence in that region. However, partial photo-conversion can also be applied usefully.

## References


1. Shaner NC, Patterson GH, Davidson MW (2007) Advances in fluorescent protein technology. J Cell Sci 120:4247–4260. https://doi.org/10.1242/jcs.005801
2. Held MA, Boulaflous A, Brandizzi F (2008) Advances in fluorescent protein-based imaging for the analysis of plant endomembranes. Plant Physiol 147:1469–1481
3. Benvenuto G, Formiggini F, Laflamme P et al (2002) The photomorphogenesis regulator DET1 binds the aminoterminal tail of histone H2B in a nucleosome context. Curr Biol 12: 1529–1534

4. Batoko H, Zheng HQ, Hawes C et al (2000) A rab1 GTPase is required for transport between the endoplasmic reticulum and Golgi apparatus and for normal Golgi movement in plants. Plant Cell 12:2201–2218
5. Lee MH, Min MK, Lee YJ et al (2002) ADP-ribosylation factor 1 of Arabidopsis plays a critical role in intracellular trafficking and maintenance of endoplasmic reticulum morphology in Arabidopsis. Plant Physiol 129: 1507–1520
6. Saint-Jore CM, Evins J, Batoko H et al (2002) Redistribution of membrane proteins between the Golgi apparatus and endoplasmic reticulum in plants is reversible and not dependent on cytoskeletal networks. Plant J 29:661–678
7. Ai HW, Henderson JN, Remington SJ, Campbell RE (2006) Directed evolution of a monomeric, bright and photostable version of Clavularia cyan fluorescent protein: structural characterization and applications in fluorescence imaging. Biochem J 400:531–540
8. Wiedenmann J, Oswald F, Nienhaus GU (2009) Fluorescent proteins for live cell imaging: opportunities, limitations, and challenges. IUBMB Life 61:1029–1042
9. Gurskaya NG, Verkhusha VV, Shcheglov AS et al (2006) Engineering of a monomeric green-to-red photoactivatable fluorescent protein induced by blue light. Nat Biotechnol 24: 461–465
10. Wiedenmann J, Ivanchenko S, Oswald F et al (2004) EosFP, a fluorescent marker protein with UV-inducible green-to-red fluorescence conversion. Proc Natl Acad Sci U S A 101: 15905–15910
11. Ando R, Hama H, Yamamoto-Hino M et al (2002) An optical marker based on the UV-induced green-to-red photoconversion of a fluorescent protein. Proc Natl Acad Sci U S A 99:12651–12656
12. McEvoy AL, Hoi H, Bates M, Platonova E, Cranfill PJ et al (2012) mMaple: a Photoconvertible fluorescent protein for use in multiple imaging modalities. PLoS One 7(12):e51314
13. Chudakov DM, Lukyanov S, Lukyanov KA (2007) Tracking intracellular protein movements using photoswitchable fluorescent proteins PS-CFP2 and Dendra2. Nat Protoc 2: 2024–2032
14. Tsutsui H, Karasawa S, Shimizu H et al (2005) Semi-rational engineering of a coral fluorescent protein into an efficient highlighter. EMBO Rep 6:233–238
15. Nienhaus GU, Nienhaus K, Holzle A et al (2006) Photoconvertible fluorescent protein EosFP: biophysical properties and cell biology applications. Photochem Photobiol 82:351–358
16. Habuchi S, Tsutsui H, Kochaniak AB et al (2008) mKikGR, a monomeric photoswitchable fluorescent protein. PLoS One 3(12): e3394
17. Adam V, Lelimousin M, Boehme S et al (2008) Structural characterization of IrisFP, an optical highlighter undergoing multiple photo-induced transformations. Proc Natl Acad Sci U S A 105(47):18343–18348
18. McKinney SA, Murphy CS, Hazelwood KL et al (2009) A bright and photostable photoconvertible florescent protein for fusion tags. Nat Methods 6(2):131–133
19. Hoi H, Shaner NC, Davidson MW et al (2010) A monomeric photoconvertible fluorescent protein for imaging of dynamic protein localization. J Mol Biol 410:776–791
20. Fuchs J, Böhme S, Oswald F et al (2010) A photoactivatable marker protein for pulse-chase imaging with superresolution. Nat Methods 7(8):627–630
21. Nienhaus K, Nienhaus GU, Wiedenmann J et al (2005) Structural basis for photo-induced protein cleavage and green-to-red conversion of fluorescent protein EosFP. Proc Natl Acad Sci U S A 102:9156–9159
22. Runions J, Brach T, Kuhner S et al (2006) Photoactivation of GFP reveals protein dynamics within the endoplasmic reticulum membrane. J Exp Bot 57:43–50
23. Mathur J, Radhamony R, Sinclair AM et al (2010) mEosFP-based green-to-red photoconvertible subcellular probes for plants. Plant Physiol 154(4):1573–1587
24. Jaipargas EA, Barton KA, Mathur N et al (2015) Mitochondrial pleomorphy in plant cells is driven by contiguous ER dynamics. Front Plant Sci 6:783
25. Griffiths N, Jaipargas EA, Wozny MR et al (2016) Photo-convertible fluorescent proteins as tools for fresh insights on subcellular interactions in plants. J Microsc 263(2):148–157
26. Delfosse K, Wozny MR, Barton KA et al (2018) Plastid envelope-localized proteins exhibit a stochastic spatiotemporal relationship to stromules. Front Plant Sci 9:754
27. Dhonukshe P, Aniento F, Hwang I et al (2007) Clathrin-mediated constitutive endocytosis of PIN auxin efflux carriers in Arabidopsis. Curr Biol 17:520–527
28. Schenkel M, Sinclair AM, Johnstone D et al (2008) Visualizing the actin cytoskeleton in living plant cells using a photo-convertible-mEos::FABD-mTn fluorescent fusion protein. Plant Methods 4:21

29. Arimura S, Yamamoto J, Aida GP et al (2004) Frequent fusion and fission of plant mitochondria with unequal nucleoid distribution. Proc Natl Acad Sci USA 101:7805–7808
30. Sinclair AM, Trobacher CP, Mathur N et al (2009) Peroxule extension over ER-defined paths constitutes a rapid subcellular response to hydroxyl stress. Plant J 59:231–242
31. Schattat MH, Griffiths S, Mathur N et al (2012) Differential coloring reveals that plastids do not form networks for exchanging macromolecules. Plant Cell 24:1465–1477
32. Martin K, Kopperud K, Chakrabarty R, Banerjee R, Brooks R, Goodin MM (2009) Transient expression in Nicotiana benthamiana fluorescent marker lines provides enhanced definition of protein localization, movement and interactions in planta. Plant J 59:150–162
33. Rausin G, Tillemans V, Stankovic N et al (2010) Dynamic nucleocytoplasmic shuttling of an arabidopsis SR splicing factor: role of the RNA-binding domains. Plant Physiol 153: 273–284
34. Wu S, Koizumi K, MacRae-Crerar A et al (2011) Assessing the utility of photoswitchable fluorescent proteins for tracking intercellular protein movement in the arabidopsis root. PLoS One 6:e27536
35. Prinz WA (2014) Bridging the gap: membrane contact sites in signaling, metabolism, and organelle dynamics. J Cell Biol 205(6): 759–769
36. Mathur J, Kroeker OF, Lobbezoo M et al (2022) The ER is a common mediator for the behavior and interactions of other organelles. Front Plant Sci 13:846970
37. Huang L, Franklin AE, Hoffman NE (1993) Primary structure and characterization of an Arabidopsis thaliana calnexin-like protein. J Biol Chem 268:6560–6566
38. Irons S, Evans D, Brandizzi F (2003) The first 238 amino acids of the human lamin B receptor are targeted to the nuclear envelop in plants. J Exp Bot 54:943–950
39. Boevink P, Santa Cruz S, Hawes C et al (1996) Virus-mediated delivery of the green fluorescent protein to the endoplasmic reticulum of plant cells. Plant J 10:935–941
40. Haseloff J, Siemering K, Prasher D et al (1997) Removal of a cryptic intron and subcellular localization of green fluorescent protein are required to mark transgenic Arabidopsis plants brightly. Proc Natl Acad Sci USA 94:2122–2127
41. Axelrod D, Koppel DE, Schlessinger J et al (1976) Mobility measurement by analysis of fluorescence photobleaching recovery kinetics. Biophys J 16:1055–1069
42. Wozny M, Schattat MH, Mathur N et al (2012) Color recovery after photoconversion of H2B::mEosFP allows detection of increased nuclear DNA content in developing plant cells. Plant Physiol 158:95–106

# Chapter 23

# In Vivo Analysis of ER-Associated Protein Degradation and Ubiquitination in *Arabidopsis thaliana*

**Jiaqi Sun and Huanquan Zheng**


## Abstract

The endoplasmic reticulum (ER) is the cellular site for the biosynthesis of proteins and lipids. The ER is highly dynamic, whose homeostasis is maintained by proper ER shaping, unfolded protein response (UPR), ER-associated degradation (ERAD), and selective autophagy of the ER (ER-phagy). In ERAD and ER-phagy, unfolded/misfolded proteins are degraded in the 26S proteasome and the vacuole, respectively. Both processes are vital for normal plant development and plant responses to environmental stresses. While it is known that ubiquitination of a protein initiates EARD, recent research indicated that ubiquitination of a protein also promotes the turnover of the protein through ER-phagy. In this chapter, we describe in detail two in vivo methods for investigating (1) the degradation efficiency and (2) ubiquitination level of an ER-associated protein in *Arabidopsis thaliana*.


**Key words** Endoplasmic reticulum, ER-associated protein degradation, ER-phagy, Ubiquitination, In vivo analysis

## 1 Introduction

The endoplasmic reticulum (ER) is an interconnected membranous network composed of cisternae/sheets and tubules in eukaryotic cells including plant cells [1]. The ER extensively stretches throughout whole cells and is the largest organelle in eukaryotic cells [2, 3]. The ER plays essential roles in protein synthesis, folding and modification, lipid synthesis, and cellular calcium homeostasis [4]. There are ever-changing demands for protein biosynthesis during cell development and in responses to different environmental stresses. To meet the complicated biosynthetic demands, the ER constantly changes its shape and size [5], increases its capacity for protein biosynthesis through unfolded protein response (UPR) [6], and undergoes ER-associated protein degradation (ERAD) [7] and autophagy of the ER (ER-phagy) [8, 9].

Verena Kriechbaumer (ed.), *The Plant Endoplasmic Reticulum: Methods and Protocols*, Methods in Molecular Biology, vol. 2772, https://doi.org/10.1007/978-1-0716-3710-4_23,

For ER shaping, recent advances revealed that ER membrane fusogens – atlastins in animals, ROOT HAIR DEFECTIVE3 (RHD3) in plants and Sey1 in yeasts, and membrane-bending proteins reticulons (RTNs) make profound contributions to the ER morphology [10–14]. Atlastins/RHD3/Sey1 are dynamin-like large GTPases. They promote the efficient ER membrane fusion through a dimerization formation and the action of the C-terminal amphipathic helix [15, 16]. RTNs have two transmembrane domains that can form a wedge-like hairpin structure in the ER membrane to generate high local membrane curvature [17, 18]. To maintain homeostasis of the ER, the actions of these ER-shaping proteins are precisely modulated in cells. Overexpression of RHD3 leads to excessive membrane fusion, generating sheet-like ER which is composed of intensive three-way junctions [19], while overexpression of RTNs constricts the lumen of the ER leading to an extensive remodeling of ER tubules [14, 20, 21].

ERAD is a major ER-related protein quality control system that facilitates translocation of unfolded/misfolded proteins in the ER into the cytosol for their degradation [6, 7]. ERAD involves ubiquitination and retro-translocation of the substrate and degradation of the substrate in the cytosolic 26S proteasome [7]. When the accumulation of unfolded/misfolded proteins in the ER is beyond the capacity of ERAD, ER-phagy system will be activated [22]. ER-phagy selectively transports and degrades ER domains including proteins, membranes, and luminal contents in the vacuole [6, 22]. The specificity of ER-phagy is determined by various ER-phagy receptors, which could be activated under different stresses [6]. Both ERAD and ER-phagy might be activated by the ubiquitination of proteins [9].

Ubiquitination is a posttranslational protein modification in which single or multiple ubiquitin molecules are covalently added to the substrate protein [23]. Ubiquitination requires three concerted steps catalyzed by three enzymes: ubiquitin-activating enzyme (E1), ubiquitin-conjugating enzyme (E2), and ubiquitin protein ligase (E3) [24]. The ubiquitination of a protein begins with formation of a thioester bond between ubiquitin and an E1 enzyme [25]. Then the ubiquitin molecule is transferred to a cysteine residue on an E2 enzyme, and the E2 interacts with an E3 to transfer ubiquitin to the E3-bound substrate [25]. E3 ligases have a specific molecular recognition system to determine which proteins are ubiquitinated and degraded [26]. Recent work also showed that E2 may also directly interact with the substrate [27]. Several ER localized E2s and E3s have been recently identified that have essential roles in plant development, environmental stress responses, and hormone signaling [9]. For example, an E2 protein UBC32 directly modulates the degradation of AtOS9, an ER luminal lectin that participates in ERAD [27]. Rma1H1, an ER-localized hot pepper (*Capsicum annuum*) E3 ligase, ubiquitinates aquaporin PIP2;1 to enhance the plant drought tolerance [28]. Recently, Lunapark (LNP) proteins have been shown to function as a novel E3 ligase without a classic E3 ligase domain, ubiquitinating RHD3 to promote the degradation of the protein [19]. Interestingly, the degradation of RHD3 can be inhibited by MG132, a potent 26S proteasome inhibitor, as well as by concanamycin A (ConcA), a vacuolar V-ATPase inhibitor that neutralizes the vacuolar pH and blocks the vacuolar degradation, suggesting that both the 26 s proteasome and ER-phagy are involved in LNP-mediated degradation of RHD3 [19].

The degradation of ER-associated proteins has been investigated in vitro, semi-in vivo, and in vivo in E2/E3 mutants [29, 30] and/or in tobacco with co-overexpression of an E2 or E3 with the substrate [19, 31]. Ubiquitination of an ER protein by either E2 or E3 has also been investigated in vivo in E2/E3 mutants and/or in the tobacco-based overexpression system [19, 31]. Furthermore, to demonstrate the specificity of ubiquitination by an E2 or E3, *Arabidopsis*-derived E1, E2, and E3 and the substrate can be purified and mixed for in vitro ubiquitination test [24]. Recently, all the components of the ubiquitination reaction (including ubiquitin, E1, E2, E3, and ubiquitination substrates) are also reconstituted in an *E. coli* system, which bypassed the complex of protein purification to investigate the specificity of the ubiquitination of a protein by an E2 or E3 [32].

In this chapter, we describe in detail two methods regarding (1) how to quantitatively investigate the protein degradation efficiency in vivo and (2) how to detect the ubiquitination level of a protein by an E2 or E3 in vivo. These two methods are performed both in *Arabidopsis* seedlings in the background of wild-type and an E2/E3 mutant. The two methods have been successfully used to investigate the degradation efficiency and ubiquitination level of RHD3 by LNPs [19]. An antibody against either the native protein or a tag (e.g., GFP, if the substrate is fused with GFP) can be used. In the following sections, we will only describe the method for detecting GFP-tagged substrate.

## 2 Materials

### 2.1 In Vivo Protein Degradation Assay

1. *Arabidopsis thaliana* seedlings: All plant lines should be derived from the same ecotype, e.g., Columiba-0 (Col-0). The mutants can be T-DNA insertional mutants ordered from the Arabidopsis Biological Resource Center. In this case, PCR and RT-PCR should be carried out to identify the homozygous lines. Transgenic *Arabidopsis* lines expressing a GFP-fused substrate protein driven by its native promoter in the corresponding mutant (the complementary line, considered as wild type) and in the E2 or E3 mutant background can be

made by the flower dip method [33]. All seeds are sterilized and sown on the 1/2 Murashige and Skoog (MS) medium. After the cold treatment at 4 °C for 3 days, the plates are moved to a growth chamber at 22 °C under continuous light with 60% humidity.

2. 1/2 MS medium: 2.165 g/L MS basal salt sand 10 g/L sucrose, pH 5.7 (use KOH/HCl to adjust pH), 10 g/L agar (*see* **Note 1**).
3. 10 mg/ml cycloheximide (CHX, a protein translation inhibitor) stock solution: prepared in DMSO and stored in −20 °C.
4. 10 mM MG132 stock solution: prepared in DMSO and stored in −20 °C.
5. 10 mM concanamycin A (ConcA) stock solution: prepared in DMSO and stored in −20 °C.
6. 2X SDS loading buffer: 100 mM Tris–Cl (pH 6.8), 4% (w/v) SDS (sodium dodecyl sulfate), 0.2% (w/v) bromophenol blue, 20% (v/v) glycerol, 200 mM DTT (dithiothreitol) (*see* **Note 2**).
7. SDS–PAGE gel: acrylamide (30%), 1.5 M Tris–HCl (pH 8.8), 10% (w/v) SDS, 10% (w/v) ammonium persulfate (AP), and TEMED (*see* **Note 3**).
8. SDS running buffer: 25 mM Tris–HCl, 200 mM glycine, 0.1% (w/v) SDS.
9. Polyvinylidene difluoride (PVDF) membranes.
10. Extra thick blot filter paper (Bio-Rad).
11. Semi-transfer buffer: 10% methanol, 24 mM Tris–HCl, 194 mM glycine.
12. Antibodies: rabbit anti-GFP (Abcam; 1:5000 dilution), rabbit anti-actin (Agrisera; 1:4000 dilution) or mouse anti-tubulin (Sigma-Aldrich; 1:4000), goat anti-rabbit IgG–peroxidase (Sigma-Aldrich, 1:10,000 dilution), or goat anti-mouse IgG–peroxidase (Sigma-Aldrich; 1:10,000 dilution) for enhanced chemiluminescence (ECL).
13. Liquid nitrogen.
14. The ChemiDoc Imaging System (Bio-Rad).

### 2.2 In Vivo Protein Ubiquitination Assay

1. *Arabidopsis thaliana* seedlings: as described in Subheading 2.1, **item 1**.
2. Protein extraction buffer: 50 mM Tris–HCl pH 7.5, 150 mM NaCl, and 10% (v/v) glycerol, and 0.5% IGEPAL® CA-630 (NP40). Store at room temperature. Add 1/100 volume of protease inhibitor cocktail, 50 mM MG132, and 10 mM N-ethylmaleimide (NEM) immediately prior to use (*see* **Note 4**).

3. GFP-Trap®_MA beads (ChromoTek).
4. Wash buffer: 10 mM Tris–HCl pH 7.5; 150 mM NaCl, 0.5 mM EDTA.
5. Antibodies: rabbit anti-GFP (Abcam; 1:5000 dilution), rabbit anti-ubiquitin (UB) (Agrisera; 1:1000 dilution), anti-rabbit IgG–peroxidase (Sigma-Aldrich, 1:10,000 dilution).
6. Magnetic bead separation rack.

## 3 Methods

### 3.1 In Vivo Protein Degradation Assay

In the assay described below, no protein purification is required. Total proteins are extracted at different time points after the CHX treatment that blocks protein biosynthesis; thus, the turnover of a protein is investigated in living conditions.

1. Collect 7-day-old wild-type *Arabidopsis thaliana* seedlings with or without the expression of GFP-tag fused protein for the drug treatment in **step 2** (*see* **Note 5**).
2. Immerse the seedlings in liquid 1/2 MS medium with 200 μM cycloheximide (CHX), 50 μM MG132, and/or 0.5 μM concanamycin A (ConcA). CHX is a translation inhibitor that suppresses new protein synthesis so that the degradation efficiency of synthesized proteins can be observed in vivo [30]. MG132 is a potent 26S proteasome inhibitor [34]. ConcA is a vacuolar V-ATPase inhibitor that neutralizes the vacuolar pH and blocks the vacuolar degradation (*see* **Note 6**) [35].
3. Collect the seedlings (about 0.5 g) after drug treatment at 0, 6, 16, and 24 h (*see* **Note 7**).
4. Grind the seedlings into fine powders in liquid nitrogen.
5. Extract total proteins with 2X SDS loading buffer (1 g tissue/1 ml SDS loading buffer), boil the samples at 95 °C for 10 mins, and subsequently perform SDS–PAGE gel electrophoresis.
6. After electrophoresis, transfer the SDS–PAGE gel to a PVDF membrane followed by immunoblotting with antibodies specific to the tag or the native protein. After the incubation of peroxidase-conjugated secondary antibody, ECL is then applied. The blot is visualized with the ChemiDoc Imaging System (Bio-Rad). The expose time may range from 3 to 50 s until the desired band intensity appears. Either actin or tubulin can be used as the loading control.
7. Quantify the intensity of blot bands relative to the loading control with ImageJ. The protein degradation rate can be visualized with the quantification of bands at different time points relative to the band at 0 h (set as 1).

### 3.2 In Vivo Protein Ubiquitination Assay

The ubiquitination of a substrate protein by an E2/E3 ligase can be verified by detecting the ubiquitination level of the substrate protein in wild-type and the E2 or E3 ligase mutant background (*see* **Note 8**). In the method described below, the substrate needs to be purified.

1. Grow *Arabidopsis* seedlings on the 1/2 MS medium for 7 days.
2. Grind seedlings (about 0.5 g) into a fine powder in liquid nitrogen and transfer to a pre-weighed tube.
3. Add the protein extraction buffer (1.5 mL extraction buffer/ 0.5 g seedlings) and vortex.
4. Then pour the extracts (approximately 1 mL) to a new mortar and further grind until no clumps remained (*see* **Note 9**).
5. Transfer the extracts to the tube by pipetting and incubate on the ice for 30 min.
6. Centrifuge the extracts at 17,000 g for 10 min at 4 °C.
7. Transfer the supernatant to a new 1.5 mL tube without touching the pellet (*see* **Note 10**).
8. Take 50 μL supernatant for Western blot analysis (input sample).
9. Wash 25 μL of GFP-Trap®_MA beads with 500 μL ice-cold wash buffer and magnetically separate beads until supernatant is clear. Discard the supernatant and repeat wash step twice.
10. Mix the protein lysate (about 1 mL) with beads and incubate on the rotator for 2 h at 4 °C.
11. Magnetically separate beads until supernatant is clear. Save 50 μL supernatant for Western blot analysis (unbound sample).
12. Resuspend beads in 500 μL wash buffer. Magnetically separate beads until supernatant is clear. Discard the supernatant and repeat wash twice.
13. Resuspend beads in 100 μL 2X SDS loading buffer (IP sample). Boil at 95 °C for 10 min.
14. Centrifuge all samples for 1 min at 17,000 g and load 20 μL supernatant of each sample for SDS–PAGE.
15. Western blot using anti-GFP for quantification of the purified protein and anti-UB for the ubiquitination of the protein. The fraction of input, unbound, and IP samples can be loaded separately to check the efficiency of protein purification. The purified substrate in wild-type and the E2/E3 mutant background should be adjusted to the same amount for the ubiquitination detection (*see* **Notes 11** and **12**). Usually, different lengths of polyubiquitin chains are added to the substrate; therefore, a smeared band above the predicted size of the substrate is often observed in the wild-type background, while a less smeared band is observed in the E2/E3 mutant background [19].

## 4 Notes

1. For the preparation of liquid 1/2 MS medium, agar should not be added.
2. Store 2X SDS loading buffer without DTT at room temperature and add DTT just before using.
3. Choose the proper concentration of acrylamide in SDS–PAGE gel according to protein size. Here is the suggested range: 7% acrylamide for 50 kDa–500 kDa, 10% acrylamide for 20 kDa–300 kDa, 12% acrylamide for 10 kDa–200 kDa, and 15% acrylamide for 3 kDa–100 kDa.
4. NEM is a potent deubiquitylation inhibitor, which can preserve the ubiquitination during the purification [36].
5. Mutants with and without the tagged protein should be included in the drug treatment and the in vivo degradation assay.
6. The concentration of CHX, MG132, and ConcA can be adjusted according to the degradation efficiency of a given protein.
7. The lifetime of different ER-associated proteins varies, and many of them are relatively stable so that longer CHX treatment may be required to detect the protein degradation.
8. The ubiquitination of a protein can also be verified by the transient expression of both the substrate protein and the E3 ligase in *Nicotiana benthamiana* leaves.
9. The extracts should be ground to a mixture without clogs after pipetting.
10. If the supernatant is not clear, the centrifuge step can be repeated a few times more.
11. For the ubiquitination test, samples should be freshly prepared, because ubiquitinated proteins are more sensitive to degradation.
12. To test the ubiquitination level under various stresses, this assay can be performed after stress treatment of plants.

## Acknowledgments

The research in the Sun lab is supported by a grant from the National Natural Science Foundation of China (no. 32200287), a Qilu Scholarship from Shandong University (Qingdao, China), and a financial aid from the Key Laboratory of Plant Development and Environmental Adaptation Biology, Ministry of Education, School of Life Sciences, Shandong University. The research in the Zheng lab is supported by a discovery grant (no. 315863) from the *Natural Sciences and Engineering Research Council* of Canada (*NSERC*), Canada.

## References


1. Chen S, Novick P, Ferro-Novick S (2013) ER structure and function. Curr Opin Cell Biol 25(4):428–433
2. Voeltz GK, Prinz WA (2007) Sheets, ribbons and tubules – How organelles get their shape. Nat Rev Mol Cell Biol 8(3):258–264
3. Lin S, Sun S, Hu J (2012) Molecular basis for sculpting the endoplasmic reticulum membrane. Int J Biochem Cell Biol 44(9): 1436–1443
4. Angelos E, Ruberti C, Kim SJ, Brandizzi F (2017) Maintaining the factory: the roles of the unfolded protein response in cellular homeostasis in plants. Plant J 90(4):671–682
5. Sparkes I, Runions J, Hawes C, Griffing L (2009) Movement and remodeling of the endoplasmic reticulum in nondividing cells of tobacco leaves. Plant Cell 21(12):3937–3949
6. Brandizzi F (2021) Maintaining the structural and functional homeostasis of the plant endoplasmic reticulum. Dev Cell 56(7):919–932
7. Strasser R (2018) Protein quality control in the endoplasmic reticulum of plants. Annu Rev Plant Biol 69:147–172
8. Sun J, Wang W, Zheng H (2022) ROOT HAIR DEFECTIVE3 is a receptor for selective autophagy of the endoplasmic reticulum in Arabidopsis. Frontiers in Plant Sci 13:817251
9. Liu R, Xia R, Xie Q, Wu Y (2021) Endoplasmic reticulum-related E3 ubiquitin ligases: key regulators of plant growth and stress responses. Plant Commun 2(3):100186
10. Zhang M, Wu F, Shi J, Zhu Y, Zhu Z, Gong Q, Hu J (2013) RHD3 family of dynamin-like GTPases mediates homotypic endoplasmic reticulum fusion and is essential for Arabidopsis development. Plant Physiol 163(2):713–720
11. Lee H, Sparkes I, Gattolin S, Dzimitrowicz N, Roberts LM, Hawes C, Frigerio L (2013) An Arabidopsis reticulon and the atlastin homologue RHD3-like2 act together in shaping the tubular endoplasmic reticulum. New Phytol 197(2):481–489
12. Sun J, Zheng H (2018) Efficient ER fusion requires a dimerization and C-terminal tail mediated membrane anchoring of RHD3. Plant Physiol 176:406
13. Kriechbaumer V, Botchway SW, Slade SE, Knox K, Frigerio L, Oparka K, Hawes C (2015) Reticulomics: protein-protein interaction studies with two Plasmodesmata-localized Reticulon family proteins identify binding partners enriched at Plasmodesmata, endoplasmic reticulum, and the plasma membrane. Plant Physiol 169(3):1933–1945
14. Sparkes I, Tolley N, Aller I, Svozil J, Osterrieder A, Botchway S, Mueller C, Frigerio L, Hawes C (2010) Five Arabidopsis reticulon isoforms share endoplasmic reticulum location, topology, and membrane-shaping properties. Plant Cell 22(4): 1333–1343
15. Sun J, Zheng H (2018) Efficient ER fusion requires a dimerization and a C-terminal tail mediated membrane anchoring of RHD3. Plant Physiol 176(1):406–417
16. Hu J, Shibata Y, Zhu PP, Voss C, Rismanchi N, Prinz WA, Rapoport TA, Blackstone C (2009) A class of dynamin-like GTPases involved in the generation of the tubular ER network. Cell 138(3):549–561
17. Hu J, Shibata Y, Voss C, Shemesh T, Li Z, Coughlin M, Kozlov MM, Rapoport TA, Prinz WA (2008) Membrane proteins of the endoplasmic reticulum induce high-curvature tubules. Science 319(5867):1247–1250
18. Hu J, Prinz WA, Rapoport TA (2011) Weaving the web of ER tubules. Cell 147(6): 1226–1231
19. Sun J, Movahed N, Zheng H (2020) LUNAPARK is an E3 ligase that mediates degradation of ROOT HAIR DEFECTIVE3 to maintain a tubular ER network in Arabidopsis. Plant Cell 32(9):2964–2978
20. Tolley N, Sparkes I, Craddock CP, Eastmond PJ, Runions J, Hawes C, Frigerio L (2010) Transmembrane domain length is responsible for the ability of a plant reticulon to shape endoplasmic reticulum tubules in vivo. Plant J 64(3):411–418
21. Stefano G, Renna L, Lai Y, Slabaugh E, Mannino N, Buono RA, Otegui MS, Brandizzi F (2015) ER network homeostasis is critical for plant endosome streaming and endocytosis. Cell Discov 1:15033
22. Gubas A, Dikic I (2022) ER remodeling via ER-phagy. Mol Cell 82(8):1492–1500
23. Carvalho P, Goder V, Rapoport TA (2006) Distinct ubiquitin-ligase complexes define

convergent pathways for the degradation of ER proteins. Cell 126(2):361–373

24. Zhao Q, Tian M, Li Q, Cui F, Liu L, Yin B, Xie Q (2013) A plant-specific in vitro ubiquitination analysis system. Plant J 74(3):524–533
25. Hellmann H, Estelle M (2002) Plant development: regulation by protein degradation. Science 297(5582):793–797
26. Cowan AD, Ciulli A (2022) Driving E3 ligase substrate specificity for targeted protein degradation: lessons from nature and the laboratory. Annu Rev Biochem 91:295–319
27. Chen Q, Liu R, Wang Q, Xie Q (2017) ERAD tuning of the HRD1 complex component AtOS9 is modulated by an ER-bound E2, UBC32. Mol Plant 10(6):891–894
28. Lee HK, Cho SK, Son O, Xu Z, Hwang I, Kim WT (2009) Drought stress-induced Rma1H1, a RING membrane-anchor E3 ubiquitin ligase homolog, regulates aquaporin levels via ubiquitination in transgenic Arabidopsis plants. Plant Cell 21(2):622–641
29. Chen Q, Zhong Y, Wu Y, Liu L, Wang P, Liu R, Cui F, Li Q, Yang X, Fang S, Xie Q (2016) HRD1-mediated ERAD tuning of ER-bound E2 is conserved between plants and mammals. Nat Plants 2:16094
30. Liu YD, Zhang CC, Wang DH, Su W, Liu LC, Wang MY, Li JM (2015) EBS7 is a plant-specific component of a highly conserved endoplasmic reticulum-associated degradation system in Arabidopsis. Proc Natl Acad Sci USA 112(39):12205–12210
31. Liu L, Zhang Y, Tang S, Zhao Q, Zhang Z, Zhang H, Dong L, Guo H, Xie Q (2010) An efficient system to detect protein ubiquitination by agroinfiltration in Nicotiana benthamiana. Plant J 61(5):893–903
32. Han Y, Sun J, Yang J, Tan Z, Luo J, Lu D (2017) Reconstitution of the plant ubiquitination cascade in bacteria using a synthetic biology approach. Plant J 91(4):766–776
33. Clough SJ, Bent AF (1998) Floral dip: a simplified method for agrobacterium-mediated transformation of Arabidopsis thaliana. Plant J 16(6):735–743
34. Wang F, Zhu D, Huang X, Li S, Gong Y, Yao Q, Fu X, Fan LM, Deng XW (2009) Biochemical insights on degradation of Arabidopsis DELLA proteins gained from a cell-free assay system. Plant Cell 21(8):2378–2390
35. Yoshimoto K, Hanaoka H, Sato S, Kato T, Tabata S, Noda T, Ohsumi Y (2004) Processing of ATG8s, ubiquitin-like proteins, and their deconjugation by ATG4s are essential for plant autophagy. Plant Cell 16(11):2967–2983
36. Emmerich CH, Cohen P (2015) Optimising methods for the preservation, capture and identification of ubiquitin chains and ubiquitylated proteins by immunoblotting. Biochem Biophys Res Commun 466(1):1–14

# Chapter 24

# Overexpression of Endoplasmic Reticulum Proteins from *Arabidopsis thaliana* in Baculovirus

Victor M. Bolanos-Garcia

## Abstract

The overproduction of proteins of the endoplasmic reticulum (ER) of plant cells in prokaryotic heterologous gene expression system remains a technical challenge. Recent advances in genetically modified insect cell technology and virus engineering methods have paved the way to produce recombinant ER plant proteins, including those harboring posttranslational modifications, and therefore, to yield ER plant proteins that are natively folded and fully functional. The present contribution focuses on the baculovirus-expression system flashBAC, which overcomes certain technical hurdles found in other insect cell-based expression systems such as the generation of a bacmid and the negative selection of recombinant clones.

**Key words** Endoplasmic reticulum, Heterologous gene expression, *Arabidopsis thaliana*, Protein overproduction, flashBAC, Baculovirus, Posttranslational modifications

## 1 Introduction

Insect cells are a widely used system for the overproduction of recombinant proteins of different origin, ranging from green plants to vertebrate and invertebrate animals. A key advantage over other expression systems such as bacterial cells is the possibility to produce fully functional proteins that undergo posttranslational modifications including glycosylation, phosphorylation, and acetylation, to name a few. Indeed, early studies showed that insect cells are a suitable system for the overexpression of endoplasmic reticulum (ER) plant proteins [1, 2] and more recently, of proteins associated with the ER, paving the way to enhance our understanding of the molecular mechanisms underlying ER function [3].

Baculovirus are insect viruses that primarily infect insect larvae of the order Lepidoptera (e.g., butterflies and moths) [4] and constitute the most frequently used vehicle to produce recombinant proteins in insect cells. To this aim, a recombinant baculovirus

Verena Kriechbaumer (ed.), *The Plant Endoplasmic Reticulum: Methods and Protocols*, Methods in Molecular Biology, vol. 2772, https://doi.org/10.1007/978-1-0716-3710-4_24,

is genetically modified to harbor a foreign gene of interest, which is subsequently expressed in a host insect cell line under control of a baculovirus gene promoter. The most extensively used baculovirus is based on the genetically modified genome *Autographa californica* Multicapsid Nucleopolyhedrovirus (AcMNPV). This is a double-stranded, supercoiled DNA genome of approximately 130 kb in length that is packaged into a rod-shaped nucleocapsid [5, 6]. Although this nucleocapsid can incorporate large gene inserts due to the property of expanding lengthways, the AcMNPV genome is considered to be too large for direct foreign gene insertion. For this reason, the gene of interest is usually cloned into a transfer plasmid that contains sequences flanking the *polh* gene in the virus genome. Following co-transfection (e.g., simultaneous introduction) of the virus genome and the transfer plasmid into host insect cells, exchange of DNA with the insertion of the gene of interest into the viral genome at the *polh* locus takes place through homologous recombination. Subsequently, a recombinant baculovirus is generated and the virus genome replicated. The recombinant virus can be easily harvested from the culture medium by centrifugation. The most frequently utilized lepidopteran insect cell lines for baculovirus expression system are derived from *Spodoptera frugiperda* (*Sf9* and *Sf21*) and *Trichoplusia ni (T. ni). These cell lines can grow in suspension or adherent cell culture in supplied or serum-free medium containing phosphate buffer to adjust the pH. These cell lines grow optimally at 28 °C without the requirement of* $CO_2$, *rendering protein production scale-up feasible and economic* [7].

The flashBAC™ baculovirus expression system is an advanced technology to produce recombinant baculoviruses. flashBAC™ builds on BacPAK6 technology in which a *lacZ* gene has been inserted at the AcMNPV *polh* gene locus and a Bsu361 restriction site introduced in each side of *lacZ* [8]. One important feature of the flashBAC™ system is the addition of a bacterial artificial chromosome (BAC) at the *polh* gene locus and the partial deletion of the essential *ORF1629* replication gene. The *polh* gene facilitates the maintenance of the virus genome in bacterial cells as a bacmid, while *ORF1629* prevents viral replication in insect cells. Production of recombinant virus in insect cells is attained through co-transfection of the AcMNPV genome (flashBAC™ DNA) with an appropriate transfer plasmid harboring the foreign gene of interest. Homologous recombination between ORF603 and ORF1629 in flashBAC™ DNA and the transfer plasmid replaces the BAC replicon with the gene of interest under the polh promoter, while simultaneously restoring ORF1629, thus enabling replication of recombinant virus containing the gene of interest (Fig. 1). An important advantage of the flashBAC system is that separation from the parental virus is not required because of the incapability of nonrecombinant virus to replicate. Consequently, the

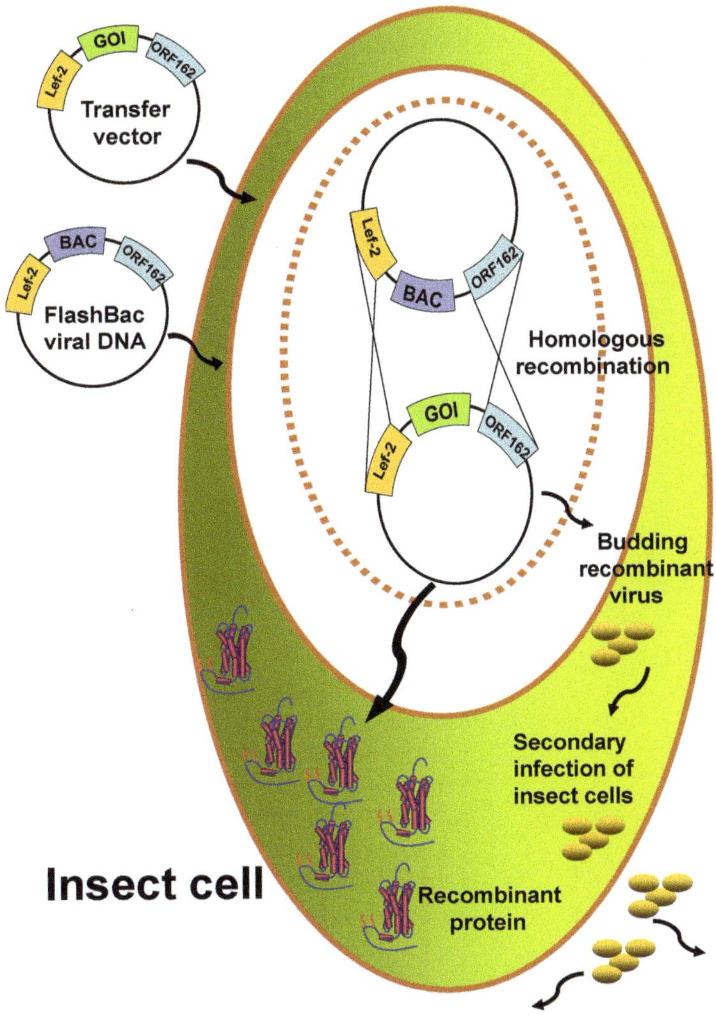


**Fig. 1** Schematic representation of the flashBAC™ system. The modified AcMNPV genome contains a BAC at *polh* gene locus with part of *ORF1629* gene deleted. Insect cells are co-transfected with *flash*BAC DNA and the transfer vector containing the gene of interest. Restoration of gene function and insertion of the foreign gene into the virus DNA under *polh* promoter takes place through homologous recombination. Baculovirus is produced as long as the recombinant virus replicates. Baculovirus is harvested and used to infect a new batch of cells to produce the recombinant protein of interest

production of recombinant virus is reduced to a single-step procedure, turning it amenable to high-throughput gene expression platforms [9].

The baculovirus pOET5 vector (Oxford Expression Technologies, OET) is a dual promoter, compatible transfer plasmid for the flashBAC expression system (Fig. 2). We used recently this vector and Sf9 and *T. ni* cells to overexpress proteins that control cell

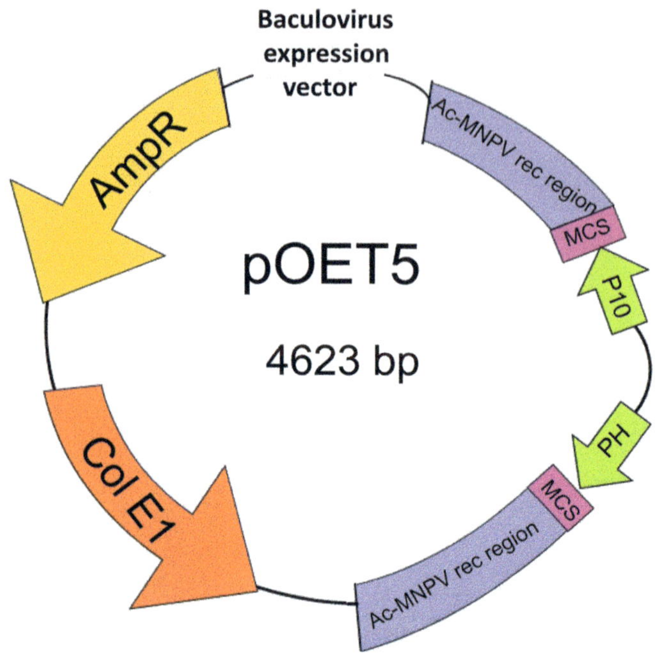


**Fig. 2** pOET5 vector map. The pOET5 transfer vector (4590 bp) has a Col E1 bacterial origin of replication and an ampicillin-resistant gene for selection in *E. coli*. Two multiple cloning sites (MCS) comprising unique restriction sites for foreign gene insertion in the correct orientation have replaced a segment the polyhedrin (*polh*) and p10 sequences. The *Autographa californica* Multicapsid Nucleopolyhedrovirus (AcMNPV) gene sequences flanking the gene in the transfer vector MCS enable homologous recombination with the viral DNA in insect cells facilitating expression cassette insertion into the locus

division in green plants [10]. The dual promoter feature of this transfer plasmid enables the simultaneous expression of two foreign genes under the strong AcMNPV polh promoter and the very late promoter p10. pOET5 contains a bacterial origin of replication and an ampicillin resistance gene, allowing plasmid propagation and selection in *E. coli* cells. The two multiple cloning sites (MCS) that have replaced the *polh* sequences harbor unique restriction sites that enable the insertion of a foreign gene in the correct orientation for its expression.

## 2 Materials

All solutions are prepared at room temperature (unless indicated otherwise using ultrapure, deionized water (e.g., purified water with a conductivity of 18 MΩ-cm at 25 °C) and analytical grade reagents.

#### 2.1 SDS (Sodium Dodecyl Sulphate)–Polyacrylamide Gel Electrophoresis (SDS–PAGE)

1. 4–12% NuPAGE pre-cast gels 1.5 mm, 15 wells.
2. SDS–PAGE running buffer: MOPS 50 mM,Tris base 50 mM, SDS 3.46 mM, EDTA 1 mM (*see* **Note 1**).
3. NuPAGE gel tank.
4. Plus Protein All Blue molecular weight ladder or equivalent.
5. Laemmli buffer 5× (Tris–HCl 125 mM pH 6.8, glycerol 10% (v/v), SDS 4% (w/v), β-mercaptoethanol 2% (v/v), bromophenol blue 0.02% (w/v).

#### 2.2 Immunoblotting

1. Nitrocellulose membranes.
2. Western blot transfer buffer: Tris–HCl 25 mM, glycine 192 mM, SDS 0.01%, methanol 20% (v/v).
3. PBS buffer: NaCl 150 mM, Na phosphates 50 mM, pH 7.4.
4. PBS containing 0.05% Tween 20 (PBST).
5. Blocking solution: 5% milk in PBST. Store at 4 °C.
6. Thick blot filter paper for Western blot, 9.5 × 15.2 cm.
7. Alkaline phosphatase (AP) or horseradish peroxidase (HRP) anti-6x His tag antibodies. Secondary antibody that is HRP- or AP-conjugated.
8. Nitro blue tetrazolium (NBT)/5-bromo-4-chloro-3-indolyl phosphate (BCIP) 1-Step™ ready-made mix or Clarity western ECL ready-made mix.

#### 2.3 Insect Cell Work

1. 35 mm (6-well) plates for tissue culture.
2. Serum-free TC100 medium.
3. SF 921 protein-free cell culture medium.
4. pOET5 vector.
5. baculoFECTIN II transfection reagent
6. flashBAC™ kit.
7. 5-Bromo-4-chloro-3-indolyl-β-D galactopyranoside (X-gal).
8. N,N-Dimethylformamide (DMF).
9. Fetal bovine serum (FBS).
10. Penicillin/streptomycin mix (10,000 IU and10,000 μg/ml, respectively).
11. PBS buffer: NaCl 150 mM, Na phosphates 50 mM, pH 7.4.
12. Neutral red dye solution diluted (1:20) in sterile phosphate-buffered saline (PBS).

## 3 Methods

### 3.1 Cloning of Heterologous Plant ER Genes

1. Add a 5′ sequence encoding for a hexahistidine tandem after the start codon of the heterologue gene sequence to facilitate the purification of the gene expression product by immobilized metal affinity chromatography (IMAC). References [11, 12] present a revision of the chemical principles and applications underpinning IMAC.
2. Following gene amplification by PCR, the amplicon of interest is purified and cloned into the appropriate restriction sites of the pOET5 vector (a combination we use when possibly is BamHI and SalI) using standard molecular cloning protocols.
3. Confirm the appropriate cloning of the heterologue gene (e.g., in-frame, mutation-free) in pOET5 by DNA Sanger sequencing using appropriate pOET sequencing primers.

### 3.2 Insect Cell Transfection

Insect cells co-ttransfection to produce seed stocks (P0) of recombinant baculovirus. The entirety of the following technique is carried Out Under aseptic conditions.

1. Grow *Spodoptera frugiperda 9* (*Sf9*) cells in serum-free TC100 medium as suspension cultures in shake flasks (28 °C, 120 rpm).
2. At least 1 h prior to transfection, use *Sf9* cells in the log phase of growth and with at least 95% viability to seed 35 mm cell culture dishes with 2 ml of Sf9 cells in serum-free TC100 growth medium at a cell density of $0.5 \times 10^6$ cells/ml.
3. Incubate at 28 °C for 1 h.
4. Meanwhile, prepare the co-transfection mix of DNA and transfection reagent. For each co-transfection, use sterile tubes and add 100 μL of serum-free TC100 medium, followed by 100 ng virus DNA (as supplied in the *flash*BAC™ kit: 5 μL at 20 ng/μL).
5. Add 500 ng of transfer vector containing gene of interest or 500 ng of control plasmid (lacZ-positive control as supplied in the *flash*BAC™ kit: 5 μL at 100 ng/μL).
6. Add 1.2 μL of baculoFECTIN and gently mix to prevent shearing of the DNA.
7. Prepare a control (mock) transfection by omitting the addition of DNA to the mix.
8. Incubate the transfection mix for 15 min at room temperature to allow the nanoparticle–DNA complex to form.
9. Gently remove 1 mL of growth medium from each well of the 6-well plates and discard, leaving only 1 mL of serum-free TC100 medium in each well.

10. Add dropwise the co-transfection mix into the corresponding well, ensuring an even distribution of the mixture.
11. Incubate the plates overnight at 28 °C.
12. Following overnight incubation, add 1 mL of ESF 921 protein-free cell culture medium, so each well had a total volume of 2 mL.
13. Incubate the 6-well plates at 28 °C for 4 more days.
14. Harvest the seed stock (P0) from each well into sterile tubes and store in the dark at 4 °C.
15. For the lacZ-positive control to confirm transfection efficiency, stain the infected cells using X-gal. To this aim, add 1 mL of ESF 921 protein-free cell culture medium containing 15 μL X-gal (2% w/v in DMF) to the well and incubate at 28 °C for 5 h.
16. Formation of a blue color confirms lacZ expression in the recombinant virus.

### 3.3 Baculovirus Amplification

1. The entirety of the following procedure is carried out under aseptic conditions.
2. Generate a Passage 1 (P1) virus stock by taking 0.5 mL of the P0 stock virus to infect 50 mL of Sf9 cells at a density of $2 \times 10^6$ cells/ml.
3. Incubate the cells for 5 days, 28 °C, 120 rpm.
4. Harvest by centrifugation (20 min, 4 °C, 3000 rpm). Recover the supernatant and store in the dark at 4 °C. For long-term storage at −80 °C, add 5% v/v of fetal bovine serum (FBS).
5. Determine the titer of the virus P1 stock by plaque assay.

### 3.4 Plaque Assay

1. The entirety of the following procedure is carried out under aseptic conditions.
2. Seed 12-well or 6-well plates with either Sf9 in ESF 921 protein-free cell culture medium or *Spodoptera frugiperda* 21 (Sf21) insect cells in TC100 medium supplemented with 10% FBS. The recommended cell density is shown in Table 1.

**Table 1**
**Recommended cell densities for insect cell lines Sf9 and Sf21 in 6- or 12-well plates**

| Insect cell line | Plate type | Cell density per well (cells/mL) | Volume per well (mL) |
|---|---|---|---|
| *Sf9* | 6-well | $0.35 \times 10^6$ | 2 |
| *Sf9* | 12-well | $0.4 \times 10^6$ | 1 |
| *Sf21* | 6-well | $0.75 \times 10^6$ | 2 |
| *Sf21* | 12-well | $0.4 \times 10^6$ | 1 |

3. Incubate plates overnight at 28 °C to obtain anticipated confluency ≥70%.
4. Perform serial dilutions (1:10) of virus stock by tenfold steps ($10^{-1}$–$10^{-7}$) in TC100 medium+10% FBS using 50 μL of virus and 450 μL of growth medium as diluent at each step, ensuring to mix thoroughly.
5. Aspire medium from wells and gently add 100 μL of virus dilutions dropwise to the center of each well.
6. Incubate plates for 45 min at room temperature with agitation to allow the virus to adsorb. During the 45-min incubation period, the overlay is prepared as follows:
7. Warm low melting temperature agarose (Sigma-Aldrich 2% w/v solution in deionized water) to hand hot ~50–55 °C and dilute 1:2 with pre-warmed (28 °C) TC100 medium+10% FBS or ESF 921 protein-free cell culture medium (if using *Sf21* or *Sf9* cells, respectively). Keep the mix in an oven set at 55 °C to prevent setting.
8. Following the incubation period, aspire the virus suspension and add 1 mL or 2 mL of overlay to 12-well or 6-well plates, respectively, allowing the overlay to flow down the side of the well and spread slowly over the cell monolayer.
9. Once the agarose has set at room temperature, add 1 mL of antibiotic mix (10,000 IU penicillin/10,000 μg/mL streptomycin) in TC100 medium+10% FBS or ESF 921 protein-free cell culture medium (if using *Sf21* or *Sf9* cells, respectively) to the cells and incubate the plate(s) at 28 °C for 3/4 days for *Sf21*/*Sf9* cells, respectively.
10. In order to visualize the virus plaques, stain the cells with a previously sterilized neutral red dye solution diluted (1:20) in sterile PBS.
11. Incubate plates overnight at 28 °C, then tip the dye off and count the plaques.
12. Determine the virus titer from the following equation:

$$\text{Average plaque count} * \text{Dilution factor} * 10 = \text{Virus titre (pfu/mL)}$$

13. Where dilution factor is the inverse of the dilution and the multiplication by 10 is because 0.1 ml of virus was added to each well.

### 3.5 Time-Course Protein Production

Once a virus titer of high infectivity (typically $10^7$ to $10^8$ pfu/mL) is obtained, time-course expression tests are conducted in insect cell lines such as Sf9 and *T. ni* at three different multiplicity of infection (MOI) levels and for up to 96 h post-infection.

1. For small-scale production trials of heterologue plant endothelium reticulum proteins, infect cell lines Sf9 and *Trichoplusia ni* (*T. ni*) at a cell density of 1.5 × $10^6$ cells/mL and 1.0 × $10^6$ cells/mL, respectively, with the P1 virus stock at 0.1, 1.0, and 5.0 multiplicity of infection (MOI) titers.
2. Collect samples of 1 mL from the different MOIs after 24 h, 48 h, 72 h, and 96 h and harvest by centrifugation at 12,000 rpm for 10 min. The pellet and supernatant can be stored at −20 °C for further analyses. To breakdown the insect cells, use a lysis buffer solution consisting of Tris–HCl 50 mM pH 8.0, β-mercaptoethanol5 mM, KCl 100 mM, Nonidet P-40 1% (v/v), cOmplete™, EDTA-free protease inhibitor cocktail tablets (Roche) (*see* **Note 2**).
3. Remove cell membrane fragments and other debris by centrifugation at 12,000 rpm, at 2–8 °C for 15 min.
4. Keep an aliquot of the supernatant (e.g., the soluble fraction) and the pellet (the insoluble fraction) for SDS–PAGE and Western blotting analysis.
5. For large-scale expression, infect 1 to 4 L of Sf9 or *T. ni* insect cells in culture at 1.5 × $10^6$ cells/mL and incubate at 28 °C according to the optimized conditions found in the small-scale time-course expression tests.
6. Once the maximal expression hours is determined (usually, after 2–3 days of incubation post-infection), harvest the cells by centrifugation at 4000 rpm at 2–8 °C for 20 min (Beckman, J2–21) and the pellets stored at −20 °C.
7. Cell lysis is carried out as described above. Remove intact cells, cell membrane fragments, and other debris by centrifugation at 12,000 rpm at 2–8 °C for 45 min (Beckman, J2–21).
8. The soluble fractions are recovered and used for SDS–PAGE and Western blotting analyses.

### 3.6 SDS–Page

1. Mix aliquots of the collected soluble fractions from the time-course expression tests with SDS–PAGE loading dye Laemmli buffer 5× and heat at 95 °C for 5 min.
2. Assemble the protein gel system with 15-well precast NuPAGE 4–12% Bis-Tris–HCl gels (1.0 mm thick). Fill the gel tank with the running buffer (e.g., MOPS 50 mM,Tris base 50 mM, SDS 3.46 mM, EDTA 1 mM).
3. Run gels at 120 V until the dye front reaches near the bottom of the gel (~40 min). Ensure the gel is running during this time (bubbles should be visible coming off the bottom electrode) (*see* **Note 3**).

### 3.7 Western Blotting

Once the SDS–PAGE run is completed, rinse the gel with deionized water and transfer to a small plastic box containing enough transfer buffer solution to cover the gel.

1. Cut a piece of polyvinylidene fluoride (PVDF) membrane (Bio-Rad®) of a size similar to that of the protein gel (*see* **Note 4**).
2. Soak the PVDF membrane in pure methanol for 1 min (*see* **Note 5**).
3. Transfer the membrane to a small plastic box containing enough transfer buffer solution to cover the PVDF membrane and incubate for 5 min (*see* **Note 6**).
4. During the equilibration of the PVDF membrane in the transfer buffer, assemble the blotting sandwich as follows: Whatman filter–membrane–gel–Whatman filter. Place the gel on the PVDF membrane in such a way it covers the membrane completely. Use a 5 or 10 mL pipette to roll out air bubbles from the gel–membrane sandwich prior to placing in transfer cassette. This will prevent the trapping of air bubbles in between the gel and the PVDF membrane.
5. Use a semidry transfer blot to transfer the proteins from the gel to the PVDF membrane. Typically, this step is performed at 25 V for 5 min per protein gel.
6. Following transfer, the membrane is blocked in PBS, 5% (w/v) skim milk powder, 0.1% Tween 20 overnight at 4 °C with rotation.
7. Dilute the primary antibody according to supplier's recommendations in blocking solution.
8. Incubate the membrane with the primary antibody diluted in PBST for 1–2 h at 4 °C with rotation. The use of alkaline phosphatase (AP) or horseradish peroxidase (HRP) anti-6x His tag antibody is recommended to speed up the process.
9. Remove the primary antibody by six 5-min washes with PBST followed by a 1-h incubation (4 °C with rotation) with secondary antibody that is HRP- or AP-conjugated (the dilution is according to supplier's recommendations) in PBST.
10. Following six 5-min washes as in **step 9** above, for HRP-conjugated antibody, apply the colorimetric 1-Step NBT/NCIP ready-made mix to the membrane for the detection of proteins on the blots. For alkaline phosphatase-conjugated antibody, use the 1-Step™ NBT/BCIP ready-made mix (*see* **Note 7**).
11. Capture the blot images are captured on a ChemiDoc (Bio-Rad) instrument or alike.

## 4 Notes

1. Unpolymerized polyacrylamide is a neurotoxin, so you need to wear gloves during this procedure.
2. Cell lysis solution can be irritant to the skin, mucous membranes, and eyes. In case of accidental contact, rinse cautiously with water for several minutes.
3. The electrophoresis and transfer apparatus uses high voltages.
4. Avoid touching the PVDF membrane with bare fingers as the proteins left on it (mainly keratin) may cross-react with the antibodies.
5. Methanol is a flammable and harmful solvent that may cause blindness or death if swallowed. Store in a cool, well-ventilated area. Use PPE-based containers to handle it. Keep away from strong oxidizing agents. Avoid heat/ignition sources.
6. It is very important that during the Western blotting procedure everything is kept damp. Pouring a small volume of transfer buffer onto the stack to keep it moist during assembling it should suffice. It is also essential that there are no air bubbles between any of the layers; otherwise, transfer will not occur.
7. The NBT/BCIP staining solution is a flammable liquid and vapor. Harmful in contact with the skin. Causes serious eye irritation.
8. The Tris–glycine–methanol blotting buffer solution must be disposed of as hazardous waste.
9. Western blotting bands fade away when stored for long time periods (e.g., a few days) even when the blot is covered with water. Therefore, it is best to image the blots immediately after performing the chemiluminescence reaction.

### References

1. Macdonald H, Henderson J, Napier RM, Venis MA, Hawes C, Lazarus CM (1994) Authentic processing and targeting of active maize auxin-binding protein in the baculovirus expression system. Plant Physiol 105:1049–1057
2. Henderson J, Macdonald H, Lazarus CM, Napier RM, Hawes CR (1996) Protein retention in the endoplasmic reticulum of insect cells is not compromised by baculovirus infection. Cell Biol Int 20:413–422
3. Bréhélin C, Yurchenko O, de Vries J, Valerius O, Braus GH, Ischebeck T, Chapman KD, Dyer JM, Mullen RT (2021) LDIP cooperates with SEIPIN and LDAP to facilitate lipid droplet biogenesis in Arabidopsis. Plant Cell 33:3076–3103
4. Van Regenmortal MHV, Fauquet CM, Bishop DHL, Cartsens EB, Estes MK, Lemon SM, Maniloff J, Mayo MA, McGeoch DJ, Pringle CR, Wickner RB (2000) Virus taxonomy: Classification and nomenclature of viruses. Seventh report of the international committee on taxonomy of viruses. Academic, San Diego. ISBN-13 978-0123702005
5. Smith GE, Summers MD, Fraser MJ (1983) Production of human beta interferon in insect cells infected with a baculovirus expression vector. Mol Cell Biol 3:2156–2165

6. Pennock GD, Shoemaker C, Miller LK (1984) Strong and regulated expression of Escherichia coli beta-galactosidase in insect cells with a baculovirus vector. Mol Cell Biol 4:399–406
7. Liu F, Wu X, Li L, Liu Z, Wang Z (2013) Use of baculovirus expression system for generation of virus-like particles: successes and challenges. Protein Expr Purif 90:104–116
8. Kitts PA, Possee R (1993) A method for producing recombinant baculovirus expression vectors at high frequency. BioTechniques 14: 810–817
9. Chambers A, Aksular M, Graves L, Irons S, Possee R, King L (2018) Overview of the baculovirus expression system. Curr Protoc Protein Sci 91:5.4.1–5.4.6
10. Cosma MA, Curtis NL, Pain C, Kriechbaumer V, Bolanos-Garcia VM (2022) Biochemical, biophysical, and functional characterisation of the E3 ubiquitin ligase APC/C regulator CDC20 from Arabidopsis thaliana. Front Physiol 13:938688
11. Bolanos-Garcia VM, Davies OR (2006) Structural analysis and classification of native proteins from E. coli commonly co-purified by immobilised metal affinity chromatography. Biochim Biophys Acta 1760:1304–1313
12. Cheung RC, Wong JH, Ng TB (2012) Immobilized metal ion affinity chromatography: a review on its applications. Appl Microbiol Biotechnol 96:1411–1420

# Chapter 25

# Observing ER Dynamics over Long Timescales Using Light Sheet Fluorescence Microscopy

**Charlotte Pain, Verena Kriechbaumer, and Alessia Candeo**


## Abstract

The recent significant progress in developmental bio-imaging of live multicellular organisms has been greatly facilitated by the development of light sheet fluorescence microscopy (LSFM). Both commercial and custom LSFM systems offer the best means for long-term rapid data collection over a wide field of view at single-cell resolution. This is thanks to the low light exposure required for imaging and consequent limited photodamage to the biological sample, and the development of custom holders and mounting techniques that allow for specimens to be imaged in near-normal physiological conditions. This method has been successfully applied to plant cell biology and is currently seen as one of the most efficient techniques for 3D time-lapse imaging for quantitative studies. LSFM allows one to capture and quantify dynamic processes across various levels, from plant subcellular compartments to whole cells, tissues, and entire plant organs. Here we present a method to carry out LSFM on Arabidopsis leaves expressing fluorescent markers targeted to the ER. We will focus on a protocol to mount the sample, test the phototoxicity of the LSFM system, set up a LSFM experiment, and monitor the dynamics of the ER during heat shock.


**Key words** Light sheet fluorescence microscopy (LSFM), Arabidopsis, GFP-HDEL, Phototoxicity, Photobleaching, FIJI

## 1 Introduction

Over the past decades, light sheet fluorescence microscopy (LSFM) has demonstrated its unique ability to image large biological specimens with single-cell resolution and minimum damage, for long time periods, shedding new light on biological functions. LSFM relies on a unique illumination technique to generate a thin sheet of light through a sample (Fig. 1). The technique was initially developed in 1902 [2], but was redeveloped as a technique suitable for cell biology in 2004 as SPIM (single plane illumination microscopy) [3]. Due to the many advantages of LSFM for long-term imaging of biological samples, this technique extensively adopted across a range of biological disciplines.

Verena Kriechbaumer (ed.), *The Plant Endoplasmic Reticulum: Methods and Protocols*, Methods in Molecular Biology, vol. 2772,
https://doi.org/10.1007/978-1-0716-3710-4_25,

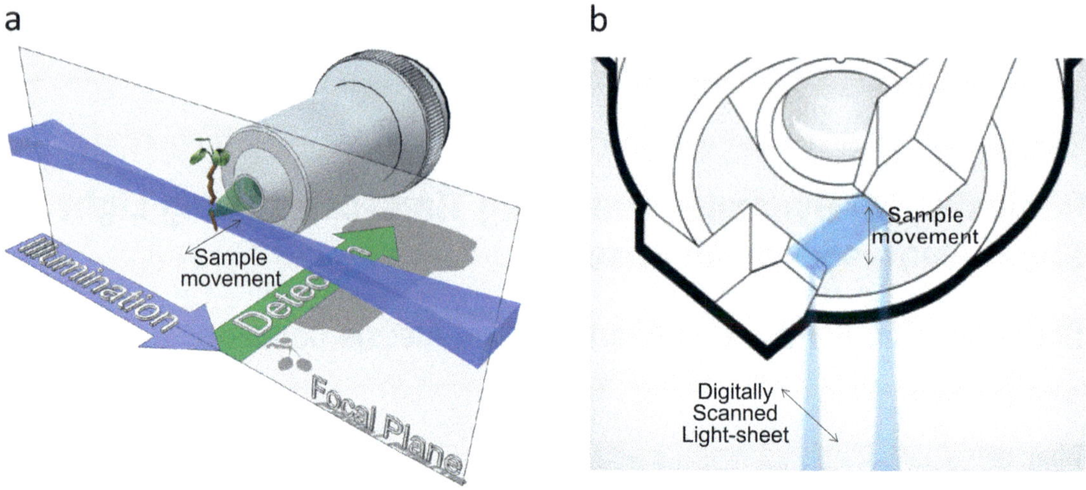


**Fig. 1** LSFM working principle to obtain optical sectioning. A thin sheet of laser light is used to excite the fluorophores present in a whole plane of a sample, which is collected in one shot by a detection objective held orthogonally to the excitation axis. The fluorescence is then suitably filtered and recorded by a high-speed camera. To record a 3D volume, the sample can be scanned through the light sheet or, vice versa, the light sheet can be moved through the sample. Variations of the method have been devised, but the two main families of methods can be distinguished based on the light sheet generation method. Panel (**a**) shows a whole plane of light created with, for example, a cylindrical lens, like in the openSPIM approach [1], while (**b**) displays the so-called digitally scanned light sheet or DSLM, in which a focused beam is scanned faster than the camera exposure time to virtually create a light sheet, like in the Leica SP8 with DLS that we are going to use below

The key advantage of LSFM is that, by exciting only a small portion of the sample using a thin plane of light, the sample is exposed to significantly less energy when compared to conventional fluorescence microscopy techniques. For comparison, when considering imaging of a zebra fish embryo, it is estimated that LSFM exposes the sample to 300× less energy for a similar quality image created using a conventional microscope and up to 5000× less energy than a confocal microscope [4]. This has numerous advantages for long-term imaging. Firstly, the sample is exposed to physiologically suitable level of light, similar to the level of light experienced by an organism when exposed to normal sunlight [5]. This reduces the risk of physiological changes in the sample due to overexposure to bright light. Secondly, practically, lower light energy results in less photobleaching of fluorophores, whereby the brightness of fluorophores decreases over time, allowing for longer periods of imaging. Finally, a light sheet by its nature provides good optical sectioning of the sample [5], limiting blurring artifacts present due to out-of-focus light affecting the sample.

LSFM in plants faces much the same challenges as any fluorescence microscopy technique. Firstly, plants cells are relatively bulky compared to mammalian cells that are surrounded by a thick cell

that can cause the scattering of light. In older LSFMs, the thickness of specimens can be particularly challenging for light scattering; however, modern LSFMs are now designed to compensate for these issues [6, 7]. In addition plant cells often contain fluorescent compounds and structures, such as chlorophyll and lignin [8], which can result in background fluorescence that obscures the structure of interest. A challenge in all LSFM, particularly when imaging over a long period of time, is the requirement that the sample remains stable for the entire duration of imaging. This can require the design and development of specific specimen holders that are able to support a given tissue type in near physiological conditions for an extended period of time [9].

Long-term imaging comes with an increased risk of phototoxic effects on specimens. Phototoxicity is described as cellular damage that occurs in cells due to the toxic effects of bright light. This is primarily caused by the impact of the formation of reactive oxygen species (ROS) [10], which can cause oxidation of proteins, lipids, and DNA as well as affect cellular signaling [11] and the cell cycle [12]. In addition to ROS production, damage can occur by sample heating in the path of the laser and, if UV or near-UV light is used, this in itself can cause DNA damage and modified cell physiology [13]. How exactly plant cells are affected by phototoxicity, and how this damage is manifested and can be observed, is not well documented. This is because the specific impacts of phototoxicity are dependent on the cell type exposed to the light, the intensity of the light, and the imaging modality used [14]. However, in *Saccharomyces cerevisiae*, it has been observed that phototoxicity slows the rate of the cell cycle [15], and in mammalian cells, phototoxicity is correlated with a reduced rate of microtubule growth [16] and reduced cell motility [17].

LSFM has been used to measure plant development and processes across both cellular and subcellular levels. The long and gentle imaging of LSFM allows for image capture of developing plant organs, including Arabidopsis flowers and germline differentiation [18], and primary roots and lateral root primordia, over several hours [19, 20]. At the subcellular level, the role of the Arabidopsis secretory machinery in cell plate formation [21] and calcium dynamics during root growth [22] has been characterized.

Here we describe a setup for LSFM on Arabidopsis cotyledons and cut leaf pieces from tobacco using a commercial light sheet system, the Leica SP8 with light sheet module (Fig. 1b), and include methods for:

(a) Mounting a sample for stability
(b) Testing for phototoxicity and photobleaching
(c) Setting up a light sheet experiment
(d) Monitoring ER during heat treatment

## 2 Materials

**2.1 Mounting Samples for Stability**

1. Stably transformed Arabidopsis or transiently transformed tobacco leaf piece.
2. Razor.
3. Tweezer.
4. 1/2 MS media (2.4 g/L Murashige and Skoog salt mixture, 0.8% phytoagar in deionized water pH 5.6–5.8, autoclaved).
5. Glass bottom dish, 35 mm diameter.
6. Pipette P1000 and tips.
7. Microwave.

## 3 Methods

**3.1 Mounting a Sample for Stability**

There are a variety of techniques and holders available for the mounting of plant samples for LSFM [22, 23]. They depend on the organ of interest, the expected amount of time the sample will be imaged over, and where it might be appropriate to apply a small amount of pressure on the sample. Presented here is an example of a simple mounting method, appropriate for the imaging of Arabidopsis cotyledons and cut tobacco leaf pieces with a horizontal light sheet beam, like the one of the Leica SP8 with DLS depicted in Fig. 1b.

1. Heat 1/2 MS media in the microwave until fully melted (*see* **Note 1**).
2. Cool the media well, so that it is still liquid but as cool as possible (*see* **Note 2**).
3. Using a pipette, make a line of 1/2 MS along the base of the dish, ensuring that it is well supported at the edges of the dish to prevent it becoming loose (Fig. 2).
4. Cute a leaf piece with a razor blade or, using tweezers, transfer a whole seedling to the 1/2 MS before it has completely solidified and lightly embed the sample in the media (*see* **Note 3**).
5. Once the sample is fully cooled and secure, transfer the dish to the microscope for imaging.

**3.2 Testing for Phototoxicity and Photobleaching**

Fluorescence microscopy can have negative impacts on sample physiology and fluorescent marker brightness. It is important to perform some initial phototoxicity and photobleaching tests to ensure that the sample is not degrading during the experiment and to ensure that experimental design is optimal to prevent sample damage.

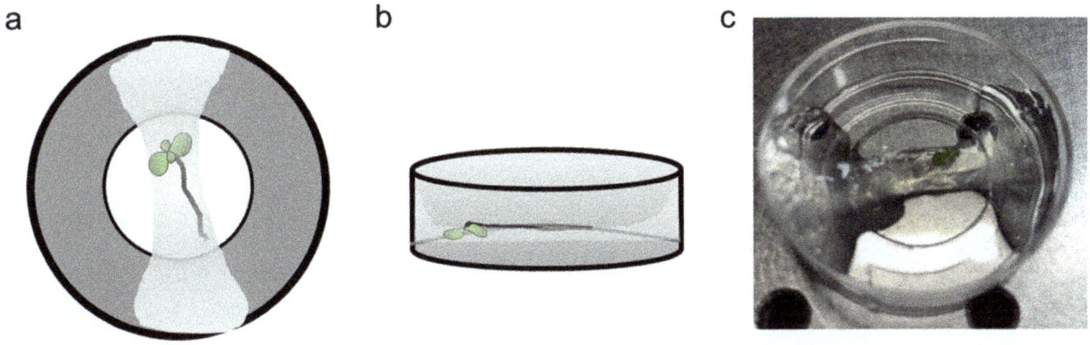


**Fig. 2** Example mounting method for Arabidopsis seedlings and cut leaf pieces. Schematic representation of an Arabidopsis whole seedling mounted in 1/2 MS in a dish viewed form (**a**) the top and (**b**) the side. (**c**) A photographic example of a mounted, 7-day-old Arabidopsis seedling

1. Prepare the sample as usual.
2. Calculate approximately how long a sample will need to remain in the microscope for a standard experiment, and perform the tests over at least that length of time.
3. Collect images over the expected period of time at the most appropriate frame rate and laser power.
4. To monitor phototoxicity, ensure that cellular morphology is as expected, if examining cellular dynamics, ensure they continue at an approximately consistent speed throughout the experiment, and if monitoring cell division, ensure that continues as appropriate (*see* **Note 4**).
5. Measure photobleaching by opening the timeseries with the open-source software Fiji [24] (*see* **Note 5**) and ensure that the FILM macro set (Facility for Imaging by Light Microscopy, Imperial College, developed by Stephen Rothery) is downloaded and installed from www.imperial.ac.uk/medicine/facility-for-imaging-by-light-microscopy/software/fiji/ (Fig. 3a).
6. Identify an area of interest that is visible and in focus for the duration of the timeseries (Fig. 3b, c).
7. Using the "macro toolset switcher" (Fig. 3a), open the "Intensity" macro.
8. From here, select the rectangular ROI button, then draw an ROI across a region in the area of interest (Fig. 3d), and add to the ROI manager by pressing "ctrl + T" (*see* **Note 6**).
9. Select the ROI button (Fig. 3a) to measure image intensity across the timeseries. Results are produced as a graph of intensity across time (Fig. 3e) and as a table of values (Fig. 3f). Compare the intensity of the sample at the beginning and end of the timeseries to look for large drops in intensity that are indicative of photobleaching (*see* **Note 7**).

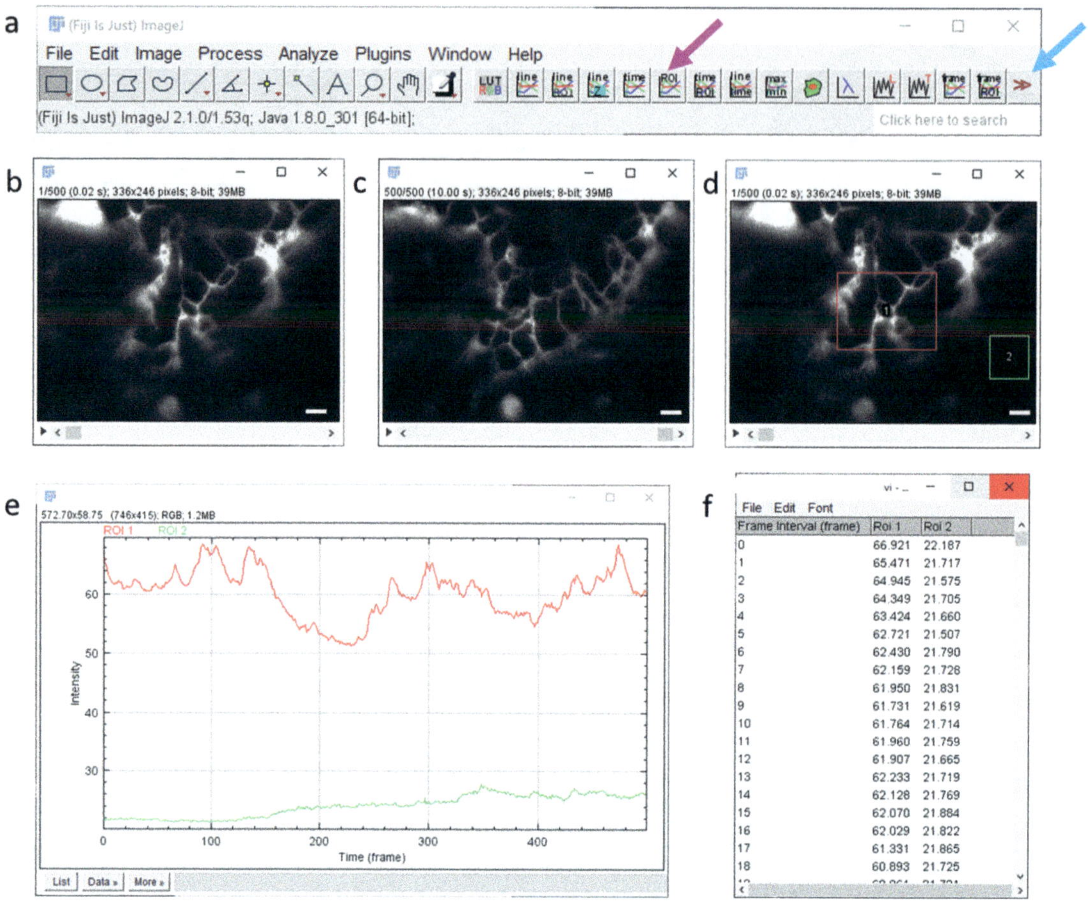


**Fig. 3** Example photobleaching analysis, intensity of an ER marker over 10 s. (**a**) The Fiji toolbar showing the macro toolset switcher (blue arrow) and the ROI intensity profile FILM macro (magenta arrow). The first (**b**) and last (**c**) frame of the example timeseries, with (**d**) the sample region of interest (red box, 1) and background region of interest (green box, 2). Intensity of the sample and background region for the duration of the timeseries output as a graph (**e**) and as a data table (**f**). Scale bar 10 μm

### *3.3 Setting Up a Light Sheet Experiment*

Even though a variety of experimental light sheet fluorescence microscopy setups exists, some of which even tailored around the specific biological question, we will focus on the use of a commercial microscope, the Leica TCS SP8 DLS (digital light sheet), for the observation of ER dynamics in Arabidopsis and tobacco leaves.

1. Consider the field of view (FOV) dimension and the required resolution for ER observation and chose a suitable detection objective. These parameters are linked to the objective magnification and numerical aperture.
2. Chose the optical configuration of the illumination arm that generates a light sheet with a thickness (two times the beam waist; *see* **Note 8**) and a Rayleigh length (i.e., the distance from the beam waist where the area of the laser beam doubles and

indeed over which the beam can be considered homogeneous) compatible with the detection objective depth of field and FOV, respectively (*see* **Note 9**).

3. Place the sample on the stage and lower the microscope head and the detection objective, carefully checking that the mirrors end up on the sides of the sample itself (*see* **Note 10**). Then using a bright-field lamp, illuminate the sample and adjust its position in the center of the FOV.
4. Adjust the position of the detection objective in order to have the sample in focus (*see* **Note 11**).
5. Turn the laser on, keeping it to the minimum power which enables the observation of the fluorescent markers, as derived from the phototoxicity assessment. Consider performing the next three steps on a sacrificial area of the sample in order to avoid overexposing the sample.
6. Insert in the detection path the correct emission filter (*see* **Note 12**).
7. Match the light sheet with the focal plane of the detection objective. The two are matched when the image is the sharpest. However, the image will likely be blurred as the beam waist of the light sheet will likely be out of the FOV.
8. Match the beam waist of the light sheet with the FOV. This is achieved when the image is the sharpest (*see* **Note 13** for other systems).
9. Iterate points 6 to 9 until the image is as sharp and in focus as possible.
10. Find a region of interest to image (*see* **Note 14**).
11. Consider the speed of the dynamic process that should be imaged, as it will help in deciding the following parameters.
12. For volumetric time-lapse imaging, evaluate the volume dimension to be imaged and fix the first and the last plane to acquire (*see* **Note 15**).
13. Decide the step between planes to be acquired, keeping in mind that the smaller the step, the higher is the axial resolution (*see* **Note 16**).
14. Consider again the expected speed of the temporal dynamic to be observed and decide for how long to image the sample and how often. The lower limit is the acquisition time of a single volume, but one should also consider the amount of generated data. Keep in mind it is easy to generate several GB of data (*see* **Note 17**).
15. Test the acquisition parameter for a short amount of time.

16. Check that there is enough memory in the PC.
17. Run the experiment.

### 3.4 Monitoring ER During Heat Treatment

In this section, we report on a method to monitor changes in the structure and dynamics of the in ER in *Arabidopsis thaliana* leaves stably expressing the ER marker, GFP-HDEL, during a heat treatment (i.e., variation of the environmental temperature from room temperature to 45 °C).

1. Place the microscope in an enclosure with a temperature control and ensure that the temperature can be monitored for the (duration of the experiment *see* **Note 18**).
2. To observe the ER in detail, use a HC FLUOTAR 25×/0.95 water immersion objective or similar. With a setup that mounts a 2× tube lens in front of the camera, an observed FOV of 295 μm × 295 μm with a lateral resolution at a wavelength of 350 nm of 340 nm is possible.
3. For sample illumination, use a PL FLUOTAR 2.5×/0.07 objective or similar, which will allow for the generation of a 3.7-μm-thick light sheet homogeneously propagating for 239 μm, covering most of the FOV.
4. Set up the TwinFlect mirrors 5 mm apart, allowing enough space to adjust the sample.
5. Mount the Arabidopsis leaf as described previously and place it in between the mirrors, focusing on it with the detection objective and after matching the light sheet with the detection objective focal plane.
6. Select an imaging area, making sure that the leaf is slightly angled with respect to the light sheet.
7. Collect images as rapidly as possible to capture the dynamics of the ER, over as long a timespan as possible. Depending on the system used, imaging one volume every 10 s for 419 times, with volumetric scan limits to acquire 290 μm overall at a step of 1.5 μm, should be possible (*see* **Note 19**).
8. Commence imaging while the enclosure is at room temperature (22 °C). After 15 min, change the set temperature in the enclosure to 45 °C. Record images until the temperature reaches this temperature, and for some time afterward (Fig. 4).
9. Analyze the changes in ER dynamics using an appropriate software, for example, AnalyzER [25] (*see* **Note 20**).

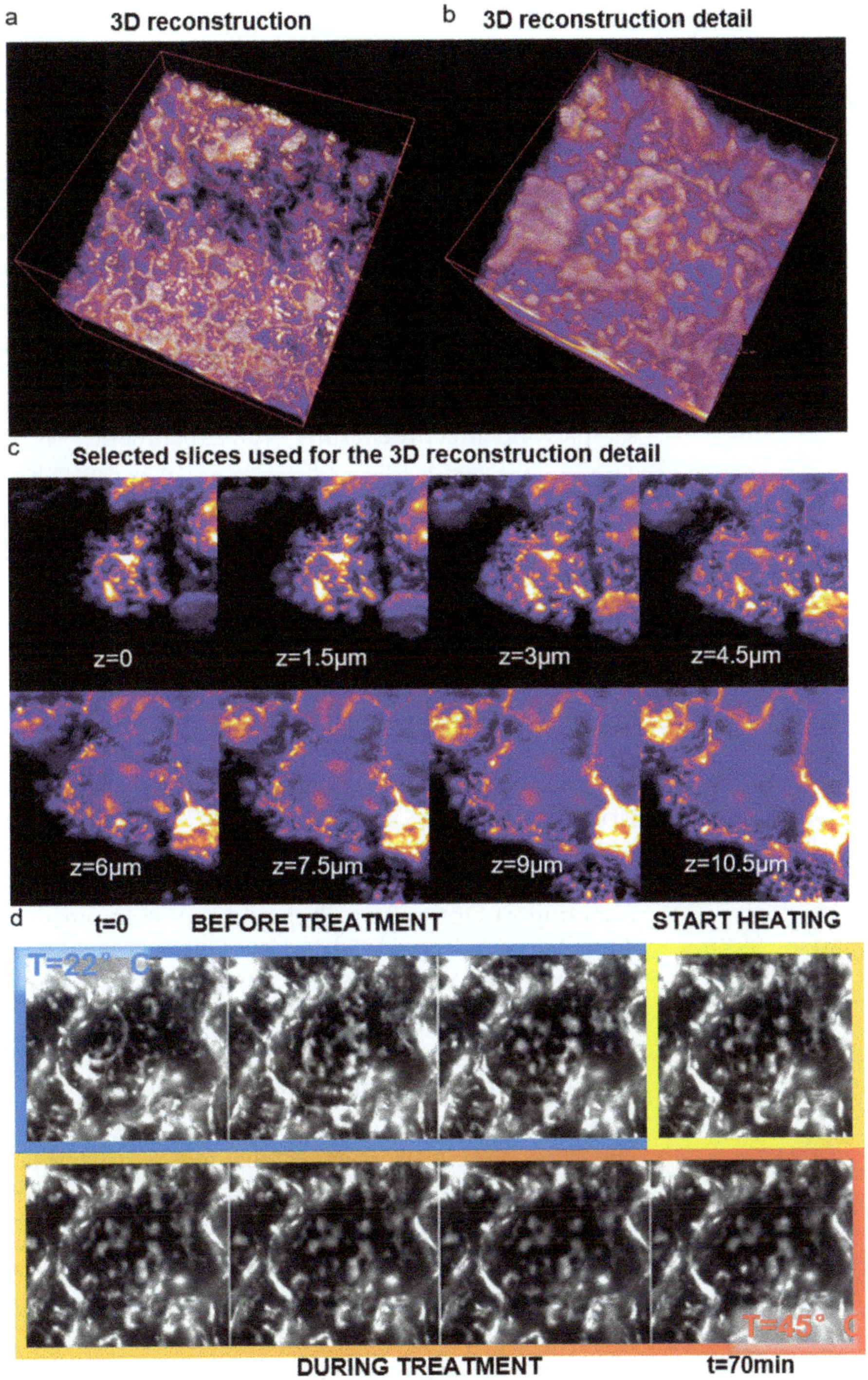


**Fig. 4** 3D imaging of HDEL *Arabidopsis thaliana* leaf during heat treatment. (**a**) 3D reconstruction of one single volume of 295 μm × 295 μm × 290 μm of the leaf showing mostly one cell layer. (**b**) 3D reconstruction of a volume of 75 μm × 75 μm × 70 μm cropped from the main one highlighting the details that could be obtained. (**c**) Selected planes used for the reconstruction in (**b**), each one spaced of 1.5 μm. (**d**) Registered MIPs (maximum intensity projections, i.e., an image whose pixel intensity consists of the highest intensity value found in the volume at that pixel position) of a zoomed-in area at different timepoints, each spaced of 10 min. The first three MIPs were acquired before the heat treatment and show an evolving ER. The fourth one was acquired 5 min after the heating treatment was started, and still shows some evolution, while after 15 min from the treatment, the ER appears more and more static, with large cisternae agglomerates

## 4 Notes

1. The use of solid medium for mounting is essential for LSFM as mounting in liquid on a standard slide will usually result in a lot of evaporation over the course of the experiment. This in turn may cause the sample to move, resulting in substantial drift during image collection.
2. Ensuring that the 1/2 MS media is well cooled before applying it to the sample ensures that the sample is not exposed to excess heat before imaging. This can affect the physiology of the samples and affect the results of an experiment.
3. Depending on the imaging requirements, PBS or another appropriate liquid medium can be added to the dish. This can be used to transfer drugs to the sample or to heat or cool the sample as required.
4. None of these measures will guarantee that there is no impact of phototoxicity on the samples; however, it will give an indication as to the health of the sample during imaging.
5. LSFM data can be particularly large, and as such may be challenging to import into Fiji directly. To deal with these issues, use the virtual stack option. This can be done by selecting virtual stack on the import screen or by importing timeseries by using "File" > "Import" > "Image Sequence" and ensuring that virtual stack is selected. Alternatively using the "Bio-Formats Import Option" which should appear automatically when dragging and dropping data into Fiji, it is possible to select "Specify range for each series" and/or "Crop on import" to reduce the size of the timeseries being imported.
6. We recommend also adding an ROI in the background to measure overall image intensity changes during image capture.
7. Some fluctuations in sample intensity are expected, especially if observing subcellular structures that may move during the time course of experiment, and if there is any evidence of sample drift.
8. The light sheet thickness determines the sectioning capability.
9. If the light sheet is too thick (i.e., the numerical aperture of the illumination optics is too small), then the sectioning will be worse and the acquired images will show out-of-focus light, while if the light sheet is too thin (i.e., the numerical aperture of the illumination optics is too large), then the Rayleigh length could be too short and the recorded image could present out-of-focus areas at the edges.

10. In the case of the Leica SP8 light sheet, the combination of detection on illumination objective determines also the distance between the two mirrors (called TwinFlect mirrors) used to create the light sheet and that are positioned on the sides of the sample. Indeed, the sample mounting needs to be compatible with the mirrors' distance.
11. For a different light sheet setup, the detection objective might be fixed; hence, the sample must be moved to reach the focal plane of the detection objective.
12. The filter should block the excitation laser and the undesirable fluorescence, such as autofluorescence.
13. In LSFM setups where a cylindrical lens is use for the light sheet creation, like in the openSPIM [1], to match the position of the beam waist with the FOV, the cylindrical lens is rotated in order to be able to observe the light sheet from its side, so that it is more obvious where the beam waist is positioned. Then, the lens is rotated at 90° to create the flat light sheet. In LSFM setups where the digitally scanned light sheet can be kept still by halting the galvanometric scanners, the light sheet waist is made evident by stopping the scanning of the beam. In both cases, to accommodate the waist position, the last lens or a combination of intermediate lenses must be adjusted.
14. Leaves highly absorb and scatter visible light; hence, the laser light penetration is limited to almost one or very few cell layers, depending on the leaf stage. This means that in order to observe the ER from many cells in the FOV, the leaf needs to be angled of few degrees with respect to the light sheet, so that the light is incident to the surface at a very shallow angle. In this way, the ER from many cells can be imaged by moving the sample through the light sheet.
15. Remember that in general the larger the volume, the slower the acquisition will be.
16. Ultimately, the axial resolution is limited by the depth of field of the objective and by the light sheet thickness. As a rule of thumb, a step smaller than the light sheet beam waist provides oversampling. In addition, it is important to consider that the larger the number of planes to be acquired, the slower the acquisition will be, meaning that for dynamic experiments, the temporal resolution will be worse.
17. Commonly, in LSFM, the cameras that are used are 16-bit CMOS camera, with about 2048×2048 pixels, like in the case of the frequently used Hamamatsu ORCA-Flash 4.0. This means that as a rule of thumb, one image corresponds to 8 MB. Indeed, if we acquire a volume of 125 planes, we end up recording 1 GB. Indeed, if we acquire 1000 timepoints, one can easily generate 1 TB of data, which requires not only a lot

of computer space but also an enormous amount of computational power to be opened and analyzed.

18. As the sample is mounted in water and observed with a water dipping objective, to have a correct feedback on the temperature perceived by the specimen itself, it is good practice to use an additional temperature controller equipped with a thermocouple that can be inserted directly in the petri dish holding the sample.
19. The setting detailed here mean we recorded 128 slices for 419 times, which adds up to about 450 GB of data. If this dataset would be too large, it is important to adjust these settings.
20. The increase in temperature can cause a drift in the sample while being imaged, which required registration of the timelapse after the computation of all the MIPs of the timepoints. All the analysis was performed with Fiji.

## Acknowledgements

This work was supported by a direct access grant to the Central Laser Facility at Harwell (19130002) to VK and AC.

### References

1. Pitrone PG, Schindelin J, Stuyvenberg L, Preibisch S, Weber M, Eliceiri KW et al (2013) OpenSPIM: an open-access light-sheet microscopy platform. Nat Methods 10:598–599
2. Siedentopf H, Zsigmondy R (1902) Uber Sichtbarmachung und Größenbestimmung ultramikoskopischer Teilchen, mit besonderer Anwendung auf Goldrubingläser. Ann Phys 1: 1–39
3. Huisken J, Swoger J, del Bene F, Witterbrodt J, Stelzer EHK (2004) Optical sectioning deep inside live embryos by selective plane illumination microscopy. Science 305(5686): 1007–1009
4. Keller PJ, Schmidt AD, Wittbrodt J, Stelzer HK (2008) Reconstruction of zebrafish early embryonic development by scanned light sheet microscopy. Science 322(5904):1065–1069
5. Stelzer EHK (2015) Light-sheet fluorescence microscopy for quantitative biology. Nat Methods 12(1):23–26
6. de Medeiros G, Norlin N, Gunther S, Albert M, Panavaite L, Fiuza U-M et al (2015) Confocal multiview light-sheet microscopy. Nature Communications 6:1–8
7. Krzic U, Gunther S, Saunders TE, Streichan SJ, Hufnagel L (2012) Multiview light-sheet microscope for rapid in toto imaging. Nat Methods 9:730–733
8. Donaldson L (2020) Autofluorescence in plants. Molecules 25(10):2393–2413
9. Laroche T, Burri O, Dubey LK, Seitz A (2019) Development of sample-adaptable holders for lightsheet microscopy. Front Neuroanat 13(26):1–6
10. Tosheva KL, Yuan Y, Pereira PM, Culley S, Henriques R (2020) Between life and death: strategies to reduce phototoxicity in super-resolution microscopy. J Phys D Appl Phys 53(163001):1–14
11. Laissue PP, Alghamdi RA, Tomancak P, Reynaud EG, Shroff H (2017) Assessing phototoxicity in live fluorescence imaging. Nat Methods 14:657–661
12. Burhans WC, Heintz NH (2009) The cell cycle is a redox cycle: linking phase-specific targets to cell fate. Free Radic Biol Med 47(9): 1282–1293
13. Hollósy F (2002) Effects of ultraviolet radiation in plant cells. Micron 33(2):179–197

14. Icha J, Weber M, Waters JC, Norden C (2017) Phototoxicity in live fluorescence microscopy, and how to avoid it. BioEssays 39(8):1–15
15. Carlton PM, Boulanger J, Kervrann C, Sibarita J-B, Salamero J, Gordon-Messer S et al (2010) Fast live simultaneous multiwavelength four-dimensional optical microscopy. PNAS 107(37):16016–16022
16. Wäldchen S, Lehmann J, Klein T, van de Linde S, Sauer M (2015) Light-induced cell damage in live-cell super-resolution microscopy. Sci Rep 5:1–12
17. Alghamdi RA, Exposito-Rodriguez M, Mullineaux PM, Brooke GN, Laissue PP (2021) Assessing phototoxicity in a mammalian cell line: how low levels of blue light affect motility in PC3 cells. Front Cell Dev Biol 9: 1–11
18. Valuchova S, Mikulkova P, Pecinkova J, Klimova J, Krumnikl M, Bainar P et al (2020) Imaging plant germline differentiation within Arabidopsis flowers by light sheet microscopy. elife 9:e52546
19. Maizel A, von Wangenheim D, Federici F, Haseloff J, Stelzer EHK (2011) High-resolution live imaging of plant growth in near physiological bright conditions using light sheet fluorescence microscopy. Plant J 68(2):377–385
20. de Luis Balaguer MA, Ramos-Pezzotti M, Rahhal MB, Melvin CE, Johannes E, Horn TJ et al (2016) Multi-sample Arabidopsis Growth and Imaging Chamber (MAGIC) for long term imaging in the ZEISS Lightsheet Z.1. Dev Biol 419(1):19–25
21. Berson T, von Wangenheim D, Takáč T, Šamajová O, Rosero A, Ovečka M et al (2014) Trans-Golgi network localized small GTPase RabA1d is involved in cell plate formation and oscillatory root hair growth. BMC Plant Biol 14(252):1–16
22. Candeo A, Doccula FG, Valentini G, Bassi A, Costa A (2017) Light sheet fluorescence microscopy quantifies calcium oscillations in root hairs of Arabidopsis thaliana. Plant Cell Physiol 58(7):1161–1172
23. Berthet B, Maizel A (2016) Light sheet microscopy and live imaging of plants. J Microsc 263(2):158–164
24. Schindelin J, Arganda-Carreras I, Frise E, Kaynig V, Longair M, Pietzsch T et al (2012) Fiji: an open-source platform for biological-image analysis. Nat Methods 9: 676–682
25. Pain C, Kriechbaumer V, Kittelmann M, Hawes C, Fricker M (2019) Quantitative analysis of plant ER architecture and dynamics. Nature. Communications 10:10

# Chapter 26

# Gene Stacking and Stoichiometric Expression of ER-Targeted Constructs Using "2A" Self-Cleaving Peptides

**Tatiana Spatola Rossi, Mark Fricker, and Verena Kriechbaumer**

## Abstract

Simultaneous stoichiometric expression of multiple genes plays a major part in modern research and biotechnology. Traditional methods for incorporating multiple transgenes (or "gene stacking") have drawbacks such as long time frames, uneven gene expression, gene silencing, and segregation derived from the use of multiple promoters. 2A self-cleaving peptides have emerged over the last two decades as a functional gene stacking method and have been used in plants for the co-expression of multiple genes under a single promoter. Here we describe design features of multicistronic polyproteins using 2A peptides for co-expression in plant cells and targeting to the endoplasmic reticulum (ER). We designed up to quad-cistronic vectors that could target proteins in tandem to the ER. We also exemplify the incorporation of self-excising intein domains within 2A polypeptides, to remove residue additions. These features could aid in the design of stoichiometric protein co-expression strategies in plants in combination with targeting to different subcellular compartments.

**Key words** Self-cleaving peptides, Gene stacking, Subcellular targeting, Recombinant protein, Endoplasmic reticulum

## 1 Methods for Gene Stacking

Multigene engineering is essential in modern research for the production of novel traits, the modification of metabolic pathways, and the expression of protein complexes, among others [1, 2]. Historically, gene stacking in plants was achieved by crossing transgenic lines harboring individual transgenes. However, this method is very time consuming and labor intensive and can lead to the segregation of transgenes in subsequent generations due to integration at different genomic loci. Co-transformation with separate plasmids is an alternative strategy that can reduce the time requirement, although this method presents the drawbacks of requiring multiple selection markers, and a lower probability of co-transformation as the gene

Verena Kriechbaumer (ed.), *The Plant Endoplasmic Reticulum: Methods and Protocols*, Methods in Molecular Biology, vol. 2772, https://doi.org/10.1007/978-1-0716-3710-4_26,

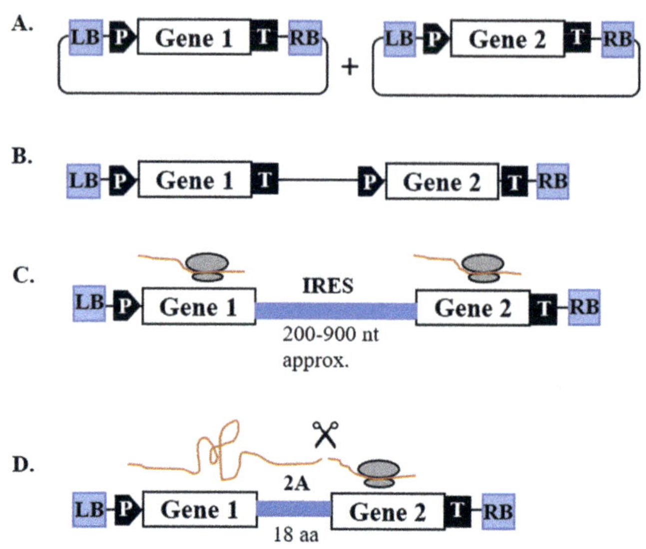


**Fig. 1** Methods for multigene engineering in plants through Agrobacterium-mediated transformation
Schematic diagram showing different co-expression techniques: (**a**) Co-expression from separate plasmids, (**b**) co-expression within a same T-DNA region, (**c**) polypeptide mediated by IRES with two different translation sites, (**d**) polypeptide mediated by self-cleaving peptide with co-translational cleavage. *LB* left border, *RB* right border, *P* promoter, *T* terminator

number increases (Fig. 1a). Alternatively, vectors containing multiple expression cassettes have been developed (Fig. 1b), such as Modular Cloning (MoClo) [3], MultiSite Gateway Technology [4], or “2in1” vectors [5], among others, which allow the incorporation of two or more genes in a single transformation event. These methods have been used successfully for genetic engineering. However, they sometimes result in uncoordinated gene expression due to the use of separate promoters, or gene silencing. Therefore, methods based on polyproteins have been developed, which allow co-expression under a single promoter and co-translational cleavage to give separate proteins. These include the use of internal ribosome entry sites (IRES) and 2A self-cleaving peptides (Fig. 1c, d).

IRES sequences are present in the 5′UTR region of some genes (mainly in viruses) and mediate a cap-independent initiation of translation [6]. IRES vary in length and neither their amino acid sequence nor their secondary structure is conserved between organisms, so their activity has to be determined empirically [7]. Synthetic polypeptides containing IRES sequences have been used and shown to mediate the translation of two genes [6, 8]. However, these sequences have the drawbacks of being fairly long (ranging from 130 bp to 1 kb, but typically around 500 bp or more) and of having a relatively low translational efficiency. For example, the downstream gene has been reported to translate at levels of around 10% of the upstream gene [1, 9].

2A self-cleaving peptides are short (18–22-amino acid) viral sequences that produce translation of two or more proteins with co-translational cleavage via a "ribosomal skipping" mechanism [10]. A conserved sequence in their C-terminus is thought to induce hydrolysis of a peptidyl(2A)-tRNAGly ester bond, releasing the nascent polypeptide while continuing with downstream translation [10]. The most commonly used 2A peptide derives from the foot-and-mouth disease virus (FMDV), termed F2A. Self-cleaving peptides from other viruses have also been identified, such as E2A (from equine rhinitis virus), P2A (from porcine teschovirus 1), and T2A (from *Thosea asigna* virus) [11]. All 2A peptides contain the conserved C-terminal sequence –GDVExNPGP, with cleavage occurring between the terminal glycine and proline residues [11]. This process leaves the final proline residue attached to the N-terminus of the downstream protein, and the remainder of the self-cleaving peptide fused to the C-terminus of the upstream protein.

2A peptides have a high cleavage efficiency and have been used successfully for polyprotein construction in a variety of systems including mammalian, insect, plant, fungal, and yeast cells, for biomedical and biotechnological purposes [12]. It has been widely reported that this system is more effective and efficient for co-expression of genes than the use of IRES, due to their shorter length, higher translational levels of downstream proteins, functionality across different cell types, and ability to concatenate more than two transgenes [1, 9, 12–14].

## 2 Use of 2A Self-Cleaving Peptides in Plants

Gene stacking in plants mediated by self-cleaving peptides was tested more than two decades ago, by co-expressing the chloramphenicol acetyltransferase (CAT) and β-glucuronidase (GUS) reporters in wheat [15]. Since then, others have made use of the system in plants to express biosynthetic enzymes, to produce protein complexes, and to study protein trafficking through target-reporter systems.

As an example of introducing novel biosynthetic pathways, Ha and co-workers expressed a bicistronic vector containing the carotenoid biosynthetic genes *Psy* and *CrtI* used in Golden Rice [16, 17] linked by F2A in rice endosperm [13]. They obtained β-carotene production as 44% of total carotenoid content, showing that the 2A system is viable for the co-expression of biosynthetic enzymes [13]. Other studies used 2A sequences to co-express a larger number of genes. For example, Liao et al. used 2As for metabolic engineering of the production of health-promoting natural products, by co-expressing six biosynthetic enzymes in tobacco and *Arabidopsis* [18]. Another project aimed at biofortification in maize expressed 11 transgenes in the anthocyanin biosynthesis pathway using 2As, to produce "Purple Embryo Maize" [19]. A recent study used 2A to further the development of gene editing in plants. They created Cas9-P2A-GFP to monitor the levels of Cas9 in *Arabidopsis* and create Cas9-free CRISPR plants [20].

2A peptides are attractive for the production of protein complexes, where a 1:1 stoichiometry between protein subunits is often needed. In this regard, F2A has been used to produce recombinant Ebola monoclonal antibodies in tobacco [21]. Similar expression levels of the heavy and light chains were reported, and a greater yield of assembled antibodies compared to expressing the heavy and light chains from separate transcription units. The authors hypothesized that the coordinated expression could be the cause of the higher yields.

It would be expected for proteins in a polypeptide to be produced at nearly 1:1 stoichiometry when present in the same transcriptional unit. However, this is not always the case, as protein quantities can be affected by other processes such as proteasomal degradation, ribosome drop-off along the length of the transcript, the gene order, the order and type of self-cleaving peptides, and the nature of protein additions adjacent to self-cleaving peptides, among others [11, 22]. Nevertheless, even if the expression ratio is not precisely 1:1, the stoichiometry tends to be fixed and stable, conferring a much higher coordination of gene expression compared to other co-expression strategies.

Due to this characteristic, 2As have been used in the development of ratiometric reporter systems. For example, Samalova and co-workers studied membrane trafficking in tobacco by linking a trafficked marker to a reporter marker (i.e., YFP-2A-secGFP), in the presence or absence of dominant mutants of effector proteins [23]. The 2A system allowed the expression of the reference fluorophore (YFP) at a constant ratio to the trafficked marker (secGFP). This was useful to normalize fluorescence of the trafficked marker to the reference and in this way, account for cell-to-cell variability in expression levels. The 2A system has been used to further study membrane trafficking in plants by linking reporter proteins to small GTPase proteins [24]. Similarly, it has been used for ratiometric analysis to investigate auxin dynamics in *Arabidopsis* protoplasts [25] and to study the degradation of karrikins in *Nicotiana benthamiana* [26]. Khosla and co-workers have developed a system of Gateway-compatible vectors termed pRATIO, which include a Gateway cassette for cloning a gene of interest upstream of a target fluorophore and a reference fluorophore separated by F2A [26]. These may facilitate the generation of constructs for ratiometric studies using the 2A system.

### 2.1 2A Self-Cleaving Peptides and Subcellular Targeting

During the design of polypeptides, the compatibility between self-cleaving peptides and different subcellular targeting signals must be considered when optimizing a gene stacking strategy. El Amrani and co-workers expressed an ER-targeted protein with a cytosolic protein in tobacco (erGFP-F2A-ble) and found that the dual localization was compatible with the 2A system [27]. They also tested the co-expression of a cytosolic protein, with a protein anchored to the ER membrane in the downstream position (GUS-F2A-F5H). With this, they showed that the 2A system allowed targeting to the ER when the ER-targeted protein was found either in the anterior or posterior position of the polypeptide [27]. Additionally, they tested the localization of a chloroplast-targeted protein placed both in the anterior and posterior positions in combination with a cytosolic protein and found correct subcellular localization of both proteins [27].

Samalova and co-workers used F2A to study membrane trafficking and suggested a model in which when the upstream protein localizes to the cytosol and the downstream protein has a signal peptide directing it to the ER membrane, the signal peptide is sufficient to target and cleave, without the requirement of including a self-cleaving peptide [23]. However, if both proteins translocate through the ER membrane, they found that the self-cleaving peptide was necessary in order to avoid degradation. They suggested that the self-cleaving peptide orients the second signal peptide correctly to allow translocation through the ER to continue [23]. In addition, they found that ER-targeted RFP was missorted to the vacuole in the presence of 2A (spRFP-F2A). This should be considered during polypeptide design. It is not known whether other 2A sequences (see Table 1) produce missorting when fused C-terminally to an RFP moiety. A later study, which also used 2As to study membrane trafficking, found that bicistronic 2A polypeptides could target to the Golgi and to the ER [24].

**Table 1**
**Commonly used self-cleaving peptide sequences for biotechnology**

| Name | Derived from | Sequence |
|---|---|---|
| F2A | Foot-and-mouth disease virus | (GSG) VKQTLNFDLLKLAGDVESNPG*P |
| P2A | Porcine teschovirus-1 | (GSG) ATNFSLLKQAGDVEENPG*P |
| T2A | Thosea asigna virus | (GSG) EGRGSLLTCGDVEENPG*P |
| E2A | Equine rhinitis A virus | (GSG) QCTNYALLKLAGDVESNPG*P |

Different self-cleaving peptide sequences [28] showing the conserved C-terminal region (marked in red) and the cleavage site (marked with an asterisk). The GSG spacer [29] may be included to increase cleavage efficiency

## 3 Design Considerations of 2A Polypeptides

### 3.1 Combining Multiple 2A Sequences

Several 2A sequences of different viral origin have been identified (Table 1) [28]. The effectiveness of self-cleaving peptides for protein co-expression lies in their ability to efficiently produce cleavage between their C-terminal glycine and proline residues and to continue translation. In most cases, the cleavage efficiency is below 100%, yielding some proportion of the uncleaved product, which may be subject to degradation or comprise a dysfunctional fusion protein. Moreover, after cleavage, ribosome drop-off may occur, resulting in lower translation of the downstream protein. In spite of these, 2As are regarded as a highly efficient system for co-expression, with high translation rates of the downstream protein, compared with other systems such as IRES [9].

The first identified and most commonly used peptide for synthetic biology purposes is F2A [30]. T2A and P2A have later emerged as sequences with higher cleavage and translation efficiencies in bicistronic constructs [14, 31]. They have also frequently been used in combination to generate tri-cistronic vectors [32]. Liu and co-workers performed a systematic comparison varying the gene position and the type of 2A peptide in bi-, tri-, and quad-cistronic constructs using mammalian cell lines [11]. They found that in bi-cistronic constructs, the highest efficiency of expression was obtained using T2A, but in all cases, the protein in the second position was found to have substantial lower expression. Therefore, the design of polypeptides could take into account the generally stronger expression of the first position. In the case of tri-cistronic constructs, they also found the strongest expression of the first position and the best results obtained with the use of T2A followed by P2A [11]. Finally, for quad-cistronic constructs, they obtained the best expression of all proteins using T2A, followed by P2A and finally E2A (the same combination shown in Fig. 2), and they found a negative effect on repeating a same 2A sequence within a polypeptide [11]. Therefore, the repetition of the same self-cleaving peptide multiple times is not advised as it may hinder expression levels and result challenging during synthesis or cloning.

Cleavage and expression efficiency of self-cleaving peptides has also been reported to be potentially affected by the identity of the adjacent sequences and therefore may result case-specific [9, 11]. In particular, the addition of a glycine–serine–glycine (GSG) spacer following the upstream gene (Table 1) has been shown to increase the cleavage efficiency of 2A sequences [29]. In view of the context-dependent results of 2As, testing the polypeptide with the genes of interest and fluorescent proteins (as shown in Fig. 2) may give an idea of the efficacy of the polypeptide strategy for the specific genes of interest. Moreover, Western blots can be performed to determine the cleavage efficiency of the polypeptide (by observing or not

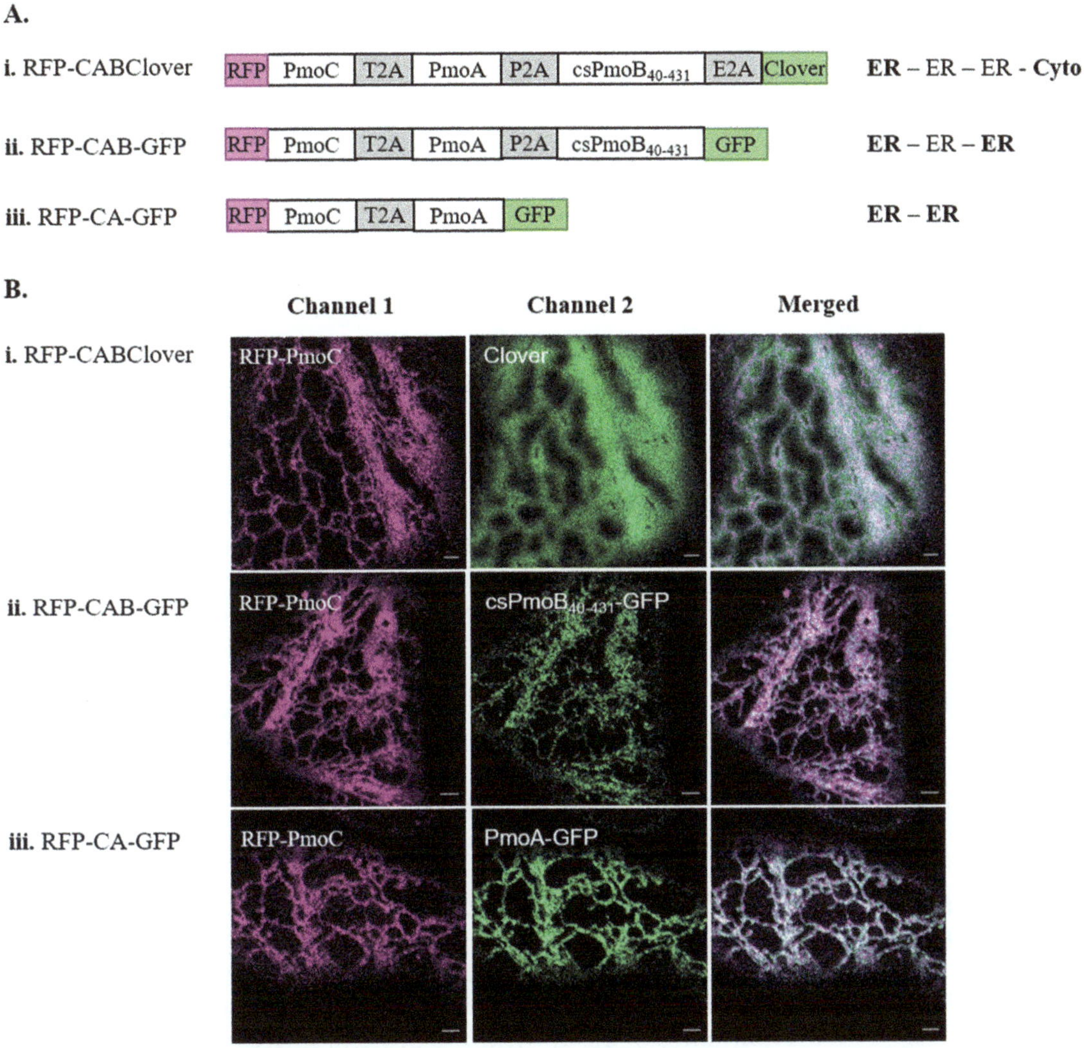


**Fig. 2** Example of subcellular targeting to the ER using self-cleaving peptides
(**a** i) Diagram of polypeptide containing all three subunits of the bacterial pMMO complex and Clover, with RFP fused to the N-terminal protein. The expected localization of each protein is specified to the right, and in bold are the localizations that can be verified by microscopy. (**a** ii) Polypeptide containing all three subunits with fusion of GFP to the ER-targeted pmoB (csPmoB$_{40\text{-}431}$). (**a** iii) Polypeptide containing the first two subunits with fusion of GFP to PmoA. (**b** i) Confocal images of RFP-CABClover following transient transformation of tobacco, showing localization in the ER of PmoC (magenta), and cytosolic localization of Clover (green), indicating correct cleavage of E2A. (**b** ii) Confocal images of RFP-CAB-GFP showing localization in the ER of both PmoC (magenta) and csPmoB$_{40\text{-}431}$ (green). (**b** iii) Confocal images of RFP-CA-GFP showing localization in the ER of both PmoC (magenta) and PmoA (green). Overall, this shows that the three subunits can be targeted to the ER in tandem, and Clover can act as a cytosolic marker for expression that is not attached to any subunits. Scale bar = 2 μm

a band corresponding to the uncleaved product). Western blots can be carried out using antibodies against the proteins of interest; alternatively, anti-2A antibodies are available which bind to the C-terminal conserved region of 2A sequences.

Finally, the polypeptide design should also take into account a small number of added residues which become incorporated to the upstream protein (expanded in Subheading 3.2).

We explored the use of different 2A sequences to recombinantly co-express three integral membrane proteins from the bacterial particulate methane monooxygenase (pMMO) enzyme in tobacco plants (proteins PmoC, PmoA, and PmoB—in the order they appear in the bacterial operon). PmoC and PmoA are localized to the ER when recombinantly expressed in tobacco, due to their targeting and transmembrane domains. However, PmoB was incorrectly localized; therefore, its bacterial signal peptide was replaced by a plant signal peptide from pumpkin (*Cucurbita* sp.) 2S albumin [33] to target it to the ER, yielding csPmoB$_{40\text{-}431}$. We produced three polypeptides containing the pMMO subunits separated by self-cleaving peptides and found that all three subunits are localized to the ER (Fig. 2). In one of the polypeptides (RFP-CABClover), a C-terminal Clover fluorophore acted as a cytosolic reporter to confirm expression of the complete polypeptide (Fig. 2). This showed that tandem targeting to the secretory pathway is compatible with the 2A system, and the cytosolic expression of Clover showed translation throughout the quad-cistronic polypeptide as well as correct cleavage.

### 3.2 Evaluation and Removal of Protein Additions

The mode of function of 2A self-cleaving sequences is such that a short number of residues (all residues in the sequence except the C-terminal proline) become incorporated to the C-terminus of the upstream protein, while the proline residue becomes added to the N-terminus of the downstream protein. Therefore, the polypeptide design must contemplate whether these residue additions may impact upon protein structure or hinder function in any way, such as by interfering with binding to co-factors, disrupting protein–protein interactions, or altering subcellular targeting.

As an example, crystallography data shows that the pMMO enzyme forms a heterotrimer, which is repeated three times to form a nonamer [34]. The sequence of the RFP-CABClover polypeptide (Fig. 2) implies that each of the three subunits will have 20–22 added residues to its C-terminus, as they are each upstream of a self-cleaving peptide. Therefore, analysis of the quaternary structure of pMMO was carried out to evaluate whether these additions could potentially interfere with the complex assembly and consequently hinder enzymatic function.

Protein modeling was performed to predict the structure of the subunits with the added residues. One approach used SWISS-MODEL. However, this approach only produced a structure for the regions of the sequence which aligned sufficiently with the template (PDB ID 3RFR, [35]), and the protein additions were not modeled. Other bioinformatic tools used tested included I-TASSER and Phyre2, which combine template-based modeling with ab initio folding, for regions without a template. These software packages produced full models of the subunits with the additions; however, the confidence score for the additional regions was extremely low. Unsurprisingly, ab initio modeling is less reliable for protein structure prediction in comparison to template-based models [36]. This is especially true for highly flexible regions which can adopt multiple conformations, such as is the case for the self-cleaving peptides, which are short and have a high content of glycine, serine, and alanine residues. Therefore, modeling approaches were not considered appropriate to provide a confident result for the folding of the subunits with the protein additions. Instead, the quaternary structure of pMMO was examined using PyMOL to evaluate whether the additions would be located at sterically restricted regions which could interfere with the native interactions.

The C-termini of PmoC and PmoB face the outside of the nonamer structure; thus, additions to these protein termini would likely not compromise interactions with the other subunits (Fig. 3a). On the other hand, the C-terminus of PmoA faces the core of the nonamer and is in close proximity to PmoC and PmoB from the adjacent trimer (Fig. 3a).This suggested that protein additions on this position might interfere with the ability of PmoA to interact properly in the complex. To remove the added residues, the polypeptide sequence was modified to include an intein domain upstream of the P2A peptide (Fig. 3b). The intein domain possesses hyper-N-terminal autocleavage capacity and if fused upstream of a 2A peptide constitutes a self-excising fragment which leaves no added residues to the proteins [37].

In the case of the pMMO proteins, the N-terminus of PmoB is also a region where additional residues could be detrimental, as the N-terminal histidine is part of a copper-binding site within the protein structure. The signal peptide from *Cucurbita* sp. (Cs) used to target PmoB to the ER leaves three added residues to its N-terminus (Fig. 3c). Therefore, a different signal peptide was evaluated, derived from tobacco pathogenesis-related protein 1a (PR1a), to avoid residue additions in this critical region. The signal peptide cleavage site was assessed using **SignalP 6.0** [38]. The sequence was codon optimized for tobacco and synthesized before cloning into a plant expression vector.

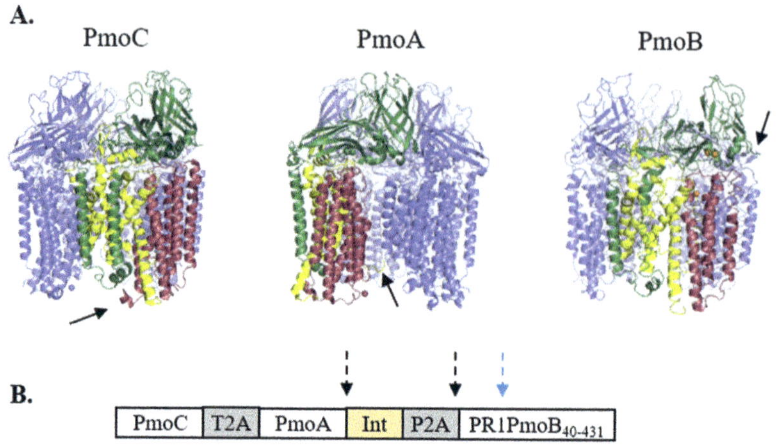


C. Cs MARLTSIIALFAVALLVADAYA*YRT
PR1 MGFVLFSQLPSFLLVSTLLLFLVISHSCRA*

**Fig. 3** Removal of added 2A sequence residues
(**a**) pMMO nonamer structure (PDB ID 3RFR), with one heterotrimer colored with PmoC in dark red, PmoA in yellow, and PmoB in green; the other two trimers are colored in slate. Arrows indicate the position of the C-terminus of each subunit. (**b**) Modified pMMO polypeptide which contains a self-excising intein domain [37]. Black dotted arrows indicate the cleavage sites of the self-excising domain and the blue dotted arrow indicates the cleavage site of the PR1 signal peptide. (**c**) Amino acid sequences of the signal peptide from pumpkin 2S albumin (Cs) and from tobacco pathogenesis-related protein 1a (PR1) for targeting to the secretory pathway, showing the cleavage sites with an asterisk

## 4 Protocol for the Expression of Polypeptide Vectors Containing Fluorescent Proteins and Ratiometric Analysis in Plant Cells

1. Design polypeptide taking into account the combination of self-cleaving peptides (T2A, P2A, and E2A are recommended for quad-cistronic vectors) and peptide additions to the C-terminus of proteins upstream of a 2A sequence. Add targeting signals to the individual proteins if needed (SignalP 6.0 [38] and MULocDeep [39] can be used to predict subcellular localization).
2. Codon-optimize the sequence and synthesize.
3. For cloning using Gateway technology [40], amplify the synthesis product with primers containing Gateway overhangs (forward primer, 5′- GGGGACAAGTTTGTACAAAAAAG CAGGC -TNN-(template-specific sequence)-3′; reverse primer, 5′- GGGGACCACTTTGTACAAGAAAGCTGG - GTN-(template-specific sequence)-3) (*see* **Note 1**).

4. Purify the amplicon by running the PCR product on a 0.7% agarose gel, cutting the band with a clean scalpel and performing gel extraction (i.e., using the NEB Monarch® DNA Gel Extraction Kit) (*see* **Note 2**).
5. Perform a BP recombination reaction to incorporate the amplicon to a pDONR221 vector following instructions on the Gateway technology manual [40] (*see* **Note 3**).
6. Stop BP reaction by adding 0.5 μL of proteinase K. Transform NEB 5-alpha competent *E. coli* cells adding kanamycin (50 mg/L) as selection agent.
7. Check the presence of the insert using colony PCR.
8. Grow a positive colony in LB medium and extract plasmid DNA using a Miniprep Kit. It is recommended to send the sample for sequencing at this step for verification.
9. Perform an LR recombination reaction, similarly to **step 5**, but using 1 μL of a destination vector (pDEST) of choice (i.e., containing an N-terminal fluorescent protein fusion, C-terminal fluorescent protein fusion, or no fluorophore) (*see* **Note 4**), 1 μL of pENTRY vector, 1 μL LR of clonase and 2 μL of $dH_2O$. This reaction yields an expression vector.
10. Stop the LR reaction and transform *E. coli* as in **step 6**, using the appropriate antibiotic as a selection agent (this should be different to kanamycin).
11. Repeat **steps 7** and **8** (without sequencing).
12. Transform competent Agrobacterium cells (strain GV3101) with the expression vector (*see* **Note 5** for generation of competent Agrobacterium cells). For transformation, thaw an aliquot of 50 μL cells on ice and add approximately 250 ng of DNA. Incubate cells on ice for 5 min, subsequently at −80 °C for 3 min, and then incubate at a 37 °C water bath for 4 min. Add 1 mL of LB media without antibiotics and incubate at 28 °C while shaking for 2 h. Spread on plates containing 25 mg/L rifampicin and the appropriate plasmid selection agent. Incubate plates at 28 °C for 3 days.
13. Grow a single Agrobacterium colony in 3 mL LB medium containing 25 mg/L rifampicin and the selection agent overnight at 28 °C with shaking at 180 rpm.
14. Perform transient transformation of tobacco via agroinfiltration as described in [41]. A summarized protocol with basic steps can also be found in [42].
15. Image cells using a confocal laser scanning microscope (CLSM). Cut a small leaf piece (approximately 6 $mm^2$) using a scalpel and place on a glass slide with the abaxial side facing up. Pipette water on top of the specimen and cover with a

coverslip. If two fluorophores are present in the construct, a ratiometric analysis based on fluorescence can be carried out.

16. For two-channel imaging using a Zeiss LSM880, set up a two-track mode to reduce spectral bleed-through with line switching and set excitation and detection wavelengths as specified for each fluorophore (i.e., for GFP excitation at 488 nm and detection at 495–550 nm and for RFP excitation at 561 nm with detection at 570–615 nm) (*see* **Note 6**).
17. Acquire the two-track images adjusting the laser power and gain to capture the full dynamic range (to carry out ratiometric analysis, it is important to avoid saturated pixels in either channel and to maintain the same imaging parameters in different sessions) (*see* **Notes 7–10**).
18. Carry out image analysis using ImageJ. Split the two channel images and perform background subtraction, followed by thresholding of the areas of interest (optional) (*see* **Note 11**).
19. Measure the fluorescence intensity (on an 8-bit scale) for each channel. The ratio of GFP/RFP fluorescence can be calculated and compared between samples.

## 5 Notes

1. Make sure that the additional nucleotides in the forward primer (NN) do not create a stop codon. Use Q5 High-Fidelity DNA Polymerase for this step.
2. For large DNA fragments (>8 kb), it is recommended to add additional water to the dissolving buffer (see NEB Gel Extraction Kit instructions).
3. Volumes can be used at 1/2 of those specified in the manual, i.e., 1 μL pDONR vector (150 ng/μL), 1 μL PCR product (150 ng/μl), 1 μL BP clonase (1 μL), and 2 μL dH2O. Reactions are incubated at 25 °C for 1 h or overnight and yield a pENTRY vector.
4. Karimi and co-workers created a set of commonly used Gateway-compatible destination vectors for Agrobacterium-mediated plant transformation [43]
5. Competent *Agrobacterium tumefaciens* cells (strain GV3101) are made starting with a 5 mL pre-culture of cells in LB media incubated overnight at 28 °C and 180 rpm, with 25 μg $mL^{-1}$ rifampicin. This is used to inoculate a larger volume (200 mL) of LB in a sterile beaker, which is incubated overnight in the same conditions, and grown to an OD600 of 0.5–1. The culture is split into 50 mL falcon tubes, chilled on ice for 5 min, and centrifuged at 3500 rpm for 30 min at 4 °C. Pellets

are resuspended in 15 mL of ice-cold $CaCl_2$ solution (20 mM, 15% glycerol), then spun at 5000 rpm for 10 min at 4 °C, and again resuspended in 15 mL of ice-cold $CaCl_2$ solution. The competent cells are aliquoted, snap-frozen in liquid nitrogen, and stored at −80 °C.

6. Typically, a Zeiss PlanApo 63×/1.46 NA oil immersion objective or a PlanApo 100×/1.46 NA oil immersion objective on the LSM880 is used. Airyscan mode can be used to increase resolution. Typically, 512 × 512 images are collected in 8-bit with 2-line averaging at an (x,y) pixel spacing of 20–80 nm.
7. Imaging at 2 days after infiltration is usually sufficient to observe fluorescence in the ER.
8. Protein expression levels may vary. Large constructs or functional proteins may induce silencing and be expressed at lower levels compared to marker proteins (such as GFP-HDEL).
9. If expression levels are very low, consider co-expressing with a silencing suppressor such as P19 in *Nicotiana benthamiana* [44].
10. Replicas can be taken by infiltrating different plants (biological replicas) and acquiring various images within a same leaf piece (technical replicas) (e.g., 5–20, depending on the observed level of variability within the sample, if there is a lot of cell-to-cell variability, a greater number of technical replicas should be acquired, or ROIs containing several cells can be sampled, to obtain a more accurate representation of the sample).
11. Different thresholding algorithms are available in ImageJ which can be tried before deciding which one segments the areas of interest better.

## Acknowledgments

This work was supported by a BBSRC Doctoral Training Program PhD studentship to TSR as well as a Leverhulme Trust fund (RGP-2015-1927) and a BBSRC Engineering Biology Breakthrough Award (BB/W011166/1) to VK.

### References

1. Halpin C (2005) Gene stacking in transgenic plants – the challenge for 21st century plant biotechnology. Plant Biotechnol J 3:141–155
2. Ceccon CC, Caverzan A, Margis R, Salvadori JR, Grando MF (2020) Gene stacking as a strategy to confer characteristics of agronomic importance in plants by genetic engineering. Ciência Rural 50
3. Weber E, Engler C, Gruetzner R, Werner S, Marillonnet S (2011) A modular cloning system for standardized assembly of multigene constructs. PLoS One 6:e16765
4. Cheo DL, Titus SA, Byrd DRN, Hartley JL, Temple GF, Brasch MA (2004) Concerted assembly and cloning of multiple DNA segments using in vitro site-specific

recombination: functional analysis of multi-segment expression clones. Genome Res 14: 2111
5. Hecker A, Wallmeroth N, Peter S, Blatt MR, Harter K, Grefen C (2015) Binary 2in1 vectors improve in planta (co)localization and dynamic protein interaction studies. Plant Physiol 168: 776–787
6. Urwin P, Yi L, Martin H, Atkinson H, Gilmartin PM (2000) Functional characterization of the EMCV IRES in plants. Plant J 24:583–589
7. Ngoi S, Chien A, Lee C (2004) Exploiting internal ribosome entry sites in gene therapy vector design. Curr Gene Ther 4:15–31
8. Gouiaa S, Khoudi H, Leidi EO, Pardo JM, Masmoudi K (2012) Expression of wheat Na +/H + antiporter TNHXS1 and H +- pyrophosphatase TVP1 genes in tobacco from a bicistronic transcriptional unit improves salt tolerance. Plant Mol Biol 79:137–155
9. Minskaia E, Ryan MD (2013) Protein coexpression using FMDV 2A: effect of "linker" residues. Biomed Res Int:1–12
10. Donnelly MLL, Luke G, Mehrotra A, Li X, Hughes LE, Gani D et al (2001) Analysis of the aphthovirus 2A/2B polyprotein "cleavage" mechanism indicates not a proteolytic reaction, but a novel translational effect: a putative ribosomal "skip". J Gen Virol 82:1013–1025
11. Liu Z, Chen O, Wall JBJ, Zheng M, Zhou Y, Wang L et al (2017) Systematic comparison of 2A peptides for cloning multi-genes in a polycistronic vector. Sci Rep 7:1–9
12. Luke GA, Roulston C, Tilsner J, Ryan MD (2015) Growing uses of 2A in plant biotechnology. Biotechnology InTech
13. Ha SH, Liang YS, Jung H, Ahn MJ, Suh SC, Kweon SJ et al (2010) Application of two bicistronic systems involving 2A and IRES sequences to the biosynthesis of carotenoids in rice endosperm. Plant Biotechnol J 8:928–938
14. Kim JH, Lee SR, Li LH, Park HJ, Park JH, Lee KY et al (2011) High cleavage efficiency of a 2A peptide derived from porcine teschovirus-1 in human cell lines, zebrafish and mice. PLoS One 6
15. Halpin C, Cooke SE, Barakate A, Amrani AE, Ryan MD (1999) Self-processing 2A-polyproteins – a system for co-ordinate expression of multiple proteins in transgenic plants. Plant J 17:453–459
16. Ye X, Al-Babili S, Klöti A, Zhang J, Lucca P, Beyer P et al (2000) Engineering the provitamin A (beta-carotene) biosynthetic pathway into (carotenoid-free) rice endosperm. Science 287:303–305
17. Paine JA, Shipton CA, Chaggar S, Howells RM, Kennedy MJ, Vernon G et al (2005) Improving the nutritional value of Golden Rice through increased pro-vitamin A content. Nat Biotechnol 23:482–487
18. Liao J, Liu T, Xie L, Mo C, Huang X, Cui S et al (2022) Plant metabolic engineering by multigene stacking: synthesis of diverse mogrosides. Int J Mol Sci 23:10422
19. Liu X, Yang W, Mu B, Li S, Li Y, Zhou X et al (2018) Engineering of 'Purple Embryo Maize' with a multigene expression system derived from a bidirectional promoter and self-cleaving 2A peptides. Plant Biotechnol J 6:1107–1109
20. Wang J, Chen H (2020) A novel CRISPR/Cas9 system for efficiently generating Cas9-free multiplex mutants in Arabidopsis. aBIOTECH 1:6–14
21. Lin Y, Hung CY, Bhattacharya C, Nichols S, Rahimuddin H, Kittur FS et al (2018) An effective way of producing fully assembled antibody in transgenic tobacco plants by linking heavy and light chains via a self-cleaving 2A peptide. Front Plant Sci 9:1379
22. Reinhardt A, Kagawa H, Woltjen K (2020) N-terminal amino acids determine KLF4 protein stability in 2A peptide-linked polycistronic reprogramming constructs. Stem Cell Rep 14: 520–527
23. Samalova M, Fricker M, Moore I (2006) Ratiometric fluorescence-imaging assays of plant membrane traffic using polyproteins. Traffic 7: 1701–1723
24. Burén S, Ortega-Villasante C, Ötvös K, Samuelsson G, Bakó L, Villarejo A (2012) Use of the Foot-and-Mouth Disease Virus 2A peptide co-expression system to study intracellular protein trafficking in Arabidopsis. Caplan S, editor. PLoS One 7:e51973
25. Wend S, Dal Bosco C, Kämpf MM, Ren F, Palme K, Weber W et al (2013) A quantitative ratiometric sensor for time-resolved analysis of auxin dynamics. Sci Rep 31(3):1–7
26. Khosla A, Rodriguez-Furlan C, Kapoor S, Van Norman JM, Nelson DC (2020) A series of dual-reporter vectors for ratiometric analysis of protein abundance in plants. Plant Direct 4:e00231
27. El Amrani A, Barakate A, Askari BM, Li X, Roberts AG, Ryan MD et al (2004) Coordinate expression and independent subcellular targeting of multiple proteins from a single transgene. Plant Physiol 135:16–24
28. Szymczak AL, Vignali DAA (2005) Development of 2A peptide-based strategies in the design of multicistronic vectors. Expert Opin Biol Ther 5:627–638

29. Wang Y, Wang F, Wang R, Zhao P, Xia Q (2015) 2A self-cleaving peptide-based multi-gene expression system in the silkworm *Bombyx mori*. Sci Rep 5:16273
30. Ryan MD, King AMQ, Thomas GP (1991) Cleavage of foot-and-mouth disease virus poly-protein is mediated by residues located within a 19 amino acid sequence. J Gen Virol 72:2727–2732
31. Szymczak AL, Workman CJ, Wang Y, Vignali KM, Dilioglou S, Vanin EF et al (2004) Correction of multi-gene deficiency in vivo using a single "self-cleaving" 2A peptide–based retroviral vector. Nat Biotechnol 22:589–594
32. Wang L, Liu Z, Yin C, Asfour H, Chen O, Li Y et al (2015) Stoichiometry of Gata4, Mef2c, and Tbx5 influences the efficiency and quality of induced cardiac myocyte reprogramming. Circ Res 116:237–244
33. Hara-Hishimura I, Takeuchi Y, Inoue K, Nishimura M (1993) Vesicle transport and processing of the precursor to 2S albumin in pumpkin. Plant J 4:793–800
34. Lieberman RL, Rosenzweig AC (2005) Crystal structure of a membrane-bound metalloenzyme that catalyses the biological oxidation of methane. Nature 43:177–182
35. Smith SM, Rawat S, Telser J, Hoffman BM, Stemmler TL, Rosenzweig AC (2011) Crystal structure and characterization of particulate methane monooxygenase from *Methylocystis* species strain M. Biochemistry 50:10231–10240
36. Lee J, Freddolino PL, Zhang Y (2017) Ab initio protein structure prediction. In: From protein structure to function with bioinformatics: second edition, pp 3–35
37. Zhang B, Rapolu M, Kumar S, Gupta M, Liang Z, Han Z et al (2017) Coordinated protein co-expression in plants by harnessing the synergy between an intein and a viral 2A peptide. Plant Biotechnol J 15:718–728
38. Teufel F, Almagro Armenteros JJ, Johansen AR, Gíslason MH, Pihl SI, Tsirigos KD et al (2022) SignalP 6.0 predicts all five types of signal peptides using protein language models. Nat Biotechnol 40:1023–1025
39. Jiang Y, Wang D, Yao Y, Eubel H, Künzler P, Møller IM et al (2021) MULocDeep: a deep-learning framework for protein subcellular and suborganellar localization prediction with residue-level interpretation. Comput Struct Biotechnol J 19:4825–4839
40. Gateway ® Technology (2003) A universal technology to clone DNA sequences for functional analysis and expression in multiple systems, vol 70. Invitrogen
41. Sparkes IA, Runions J, Kearns A, Hawes C (2006) Rapid, transient expression of fluorescent fusion proteins in tobacco plants and generation of stably transformed plants. Nat Protoc 1:2019–2025
42. Spatola Rossi T, Pain C, Botchway SW, Kriechbaumer V (2022) FRET-FLIM to determine protein interactions and membrane topology of enzyme complexes. Curr Protoc 2:e598
43. Karimi M, Inzé D, Depicker A (2002) GATEWAY™ vectors for Agrobacterium-mediated plant transformation. Trends Plant Sci 7:193–195
44. Voinnet O, Rivas S, Mestre P, Baulcombe D (2003) An enhanced transient expression system in plants based on suppression of gene silencing by the p19 protein of tomato bushy stunt virus. Plant J 33:949–956

# Chapter 27

# Automated Segmentation and 3D Reconstruction of Different Membranes from Confocal Z-Stacks

**Maryam Alsadat Zekri and Ingeborg Lang**

## Abstract

Confocal laser scanning microscopy (CLSM) is an advanced microscopy technique based on fluorescence technology which produces sharp images of a specimen in a single focal plane. The optical sectioning by CLSM allows to have z-stacks which can be further processed into 3D reconstructions. These then provide the option of variable perspectives and additional precise data evaluation on structural and anatomical alterations. Here, we used CLSM to image the thylakoids of cyanobacteria and the endoplasmic reticulum (ER) in moss protonemata as an example. Then, out of the confocal z-stacks, we create 3D constructions of the membranes and their alterations to present a holistic, structural view from different angles.

**Key words** Cyanobacteria, Cyanophages, CLSM, DNA, Thylakoids, Amira, 3D reconstruction, Moss protonemata

## 1 Introduction

Over the past few decades, new microscopy technologies have been introduced to take high-quality images of small objects in a very short time. Modern fluorescence microscopy, electron microscopy, and micro-computed tomography not only produce high-resolution images in xy direction but also in z, like focusing through the sample. Reconstructions of such image stacks allow for more in-depth studies of biological and non-biological objects because the samples can be optically rotated and viewed from different angles.

Today, based on fluorescence technology, one of the standard methods is confocal laser scanning microscopy (CLSM) which is an optical imaging technique for increased resolution and contrast by introducing a pinhole to block out-of-focus light. Thus, confocal microscopy employs a smaller, very focused beam of light at a narrow depth level at a time. In CLSM, the laser moves in x- and y-axis to create an optical image of the sample. The scanning

Verena Kriechbaumer (ed.), *The Plant Endoplasmic Reticulum: Methods and Protocols*, Methods in Molecular Biology, vol. 2772, https://doi.org/10.1007/978-1-0716-3710-4_27,

process continues by moving the laser to another focal plane and scanning another optical section which eventually provides a z-stack of the sample. Ideally, the resolution in z direction is the same as in x- and y-axis. Such optical sections can be further used to create a 3D reconstruction of the sample. Powerful graphic cards then allow for rotations of the 3D structures, thereby providing a view of the object from all angles. That way, we gain a better understanding of an object's volume, shape, and geometry [1–4]. Very fast scans even enable us to follow developmental stages [5] or to visualize changing structures like membranes. Additionally, 3D data provide the basis for precise and accurate measurements of an object. For example, from 2D images provided by micro-computed tomography, Theroux-Rancourt et al. [6] applied various methods to calculate the mesophyll cell surface ($S_m$) in poplar leaves from 2D images. They then processed the 2D images to create a 3D structure. The $S_m$ then was calculated from the 3D structure too. The results showed that estimated $S_m$ calculated from 2D images was deviated from 3D values by 10–30%. It showed that the 3D structures provide more accurate measurements than 2D images.

All biological membranes form discrete boundary layers for individual compartments. Some membranes provide a certain internal environment with specific content or pH; others play an essential role by isolating the protoplasm, by hosting signaling proteins for cell communication, or by facilitating ion trafficking. Particularly for imaging dynamic structures and processes like cell membranes or ion flux, CLSM is superior to high-resolution electron microscopy because living samples can be analyzed [7–9]. Moreover, the innovation of labeling techniques with fluorescent dyes facilitates the visualization of a wide range of biomolecules such as proteins and enzymes. That way, it is possible to label specific molecules in a cell and track their responses and behavior [10–12], and we constantly gain a more viable understanding of physiological processes in cells. However, a cell and all the internal structures have a certain volume. To get the holistic picture, we therefore need to image and visualize microscopic structures in 3D so that we can appreciate them from all angles.

Here, we visualize the alteration of thylakoid membranes in a cyanobacterium (*Limnospira fusiformis*) and the DNA in response to cyanophage induction (shift from lysogenic to lytic phage cycle). 3D reconstruction provides a great view of minute changes in thylakoids through the viral induction [13]. Moreover, we show the endoplasmic reticulum (ER) of the moss *Physcomitrium patens* as an example for organization of rapidly changing ER membranes, e.g., osmotic stress [14]. In both cases, the 3D structure of thylakoids and ER was reconstructed and illustrated using the Amira® software.

## 2 Materials

### 2.1 Stock Solutions (All Stock Solutions Should Be Autoclaved)

1. Stock solution I: 25 g/L $K_2HPO_4$. Autoclave.
2. Stock solution II: 12.5 g/L $NaNO_3$ (stock solution II). Autoclave.
3. Stock solution III: 50 g $K_2SO_4$ and 50 g NaCl in 1 L MilliQ water (stock solution III).
4. Stock solution IV: 10 g $MgSO_4.7H_2O$ and 2 g $CaCl.2H_2O$ in 1 L MilliQ water. Autoclave.
5. Stock solution V: 0.5 g $FeSO_4.7H_2O$ and 4g EDTA in 1 L MilliQ water. Autoclave.
6. Micronutrient stock solution I: 2.86 g $H_3BO_3$, 1.81 g $MnCl_2.4H_2O$, 0.222 g $ZnSO_4.4H_2O$, and 0.0177 g $Na_2MoO_4$ in 1 L MilliQ water. Autoclave.
7. Micronutrient stock solution II: Dissolve 79 mg $CuSO_4.5H_2O$ in 1 L MilliQ water (micronutrient stock solution II). Autoclave.
8. 80 g/L $NaHCO_3$ in MilliQ water. Use sterile syringe filters 0.22 um to filter $NaHCO_3$ (*see* **Note 1**).
9. 0.5 μg/μL mitomycin C (*see* **Note 2**).
10. 0.2% v/v glutaraldehyde: 0.8 mL glutaraldehyde 25% is diluted in 100 mL $dH_2O$ (*see* **Note 3**).
11. 10% v/v DAPI in $dH_2O$.

### 2.2 Buffers

1. Zarrouk medium (preparation must be done in laminar flow cabinet; *see* **Note 4**): Mix 20 ml of stock solutions I, II, III, IV, and V with 500 ml MilliQ water. Add 1ml of micronutrient stock solution I and 1ml of micronutrient stock solution II and mix it well. Add 250 ml of $NaHCO_3$ and mix until $NaHCO_3$ dissolves. Adjust the pH to 8–9. Fill the medium up to 1 L with autoclaved MilliQ water (248 mL). Ultrafiltrate the final solution to remove remaining particles.

### 2.3 Equipment

1. Plankton net (mesh size 30 μm).
2. 250 mL Erlenmeyer flask for cyanobacterium cultures.
3. Laboratory glass bottles with screwcap (1 L for Zarrouk medium constituents and 250 mL for the micronutrients).
4. 1 mL Eppendorf tubes.
5. 50 mL falcon tubes.
6. Pipette (2–20 μL and 1 mL).
7. Pipette tips (2 μL and 1 mL).
8. Nail polish.

9. Vaseline
10. 0.22 μm syringe filters, microscope slides, and coverslips.
11. Lens cleaning tissue.
12. Microscope slides and coverslips.
13. pH meter.
14. Analytical balance.
15. Autoclave.
16. Culture cabinet.
17. Laminar flow cabinet.
18. Vacuum filtration set.
19. Confocal laser scanning microscope (e.g., Leica TCS SP5 DM-6000 CS) with the connected LAS AF Software v4.
20. Amira software 6.2.0.

## 3 Methods

All the procedures can be conducted at room temperature.

### 3.1 Cyanophage Induction in L. fusiformis Culture

1. Concentrate the cells of *L. fusiformis* with a plankton net (mesh size 30 μm) from unialgal cultures. Keep the samples in Zarrouk medium at 25 °C, in a 16:8–h light/dark cycle and at 15 μmol photons $m^{-2}$ $h^{-1}$.
2. Take 4 mL of *L. fusiformis* culture and dilute it in 50 mL Zarrouk medium and leave it for 3 days at 32 °C in 100 mL Erlenmeyer flasks under constant light (40 μE $m^{-2}$ $s^{-1}$) and 100 rpm shaking.
3. Divide the well-grown culture into two falcon tubes to have 27 mL culture in each tube.
4. For a final MMC concentration of 1.5 $\mu gml^{-1}$, add 81 μL MMC to one of the falcon tubes and label it with +MMC. The other falcon tube should be labeled as −MMC.
5. Take 980 μL of each sample (+MMC and −MMC), add it to an Eppendorf tube, and fix it with 12 μL of 0.2% glutaraldehyde and label it with time point 0.
6. Repeat this step every 3 h to follow the ultrastructure alteration (*see* **Note 5**).

### 3.2 Sample Preparation of L. fusiformis for CLSM

1. Take one drop of the *L. fusiformis* culture and put it on a microscope slide (*see* **Note 6**) (Fig. 1).
2. Add 2 μL DAPI to it and mix it gently with a pipette tip.
3. Leave it for 5 min at room temperature.

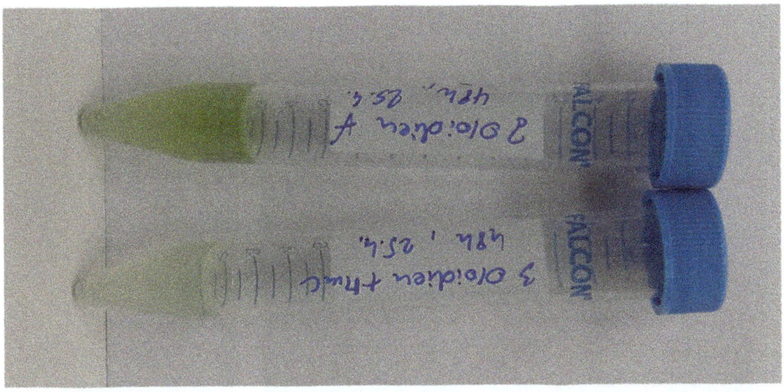


**Fig. 1** The culture treated with MMC (right) shows pale green color due to lower concentration of cyanobacterium. Falcon tube in the left is the negative control

4. Place a rectangular (24 mm × 50 mm) coverslip on the sample and press it down very gently with the thumb (*see* **Note 7**).
5. Remove the extra sample spilling out with a piece of microscope lens cleaning paper (*see* **Note 8**).
6. Seal the edges of the coverslip with a nail polish to fix the coverslip and avoid air bubbles between the coverslip and slide (*see* **Note 9**).

### 3.3 Sample Preparation for P. patens for CLSM

1. Cultivate *P. patens* in sterile cultures in a $PpNH_4$-moss medium [15].
2. For propagation, inoculate small pieces of protonema in new agar plates every 5 weeks.
3. Grow the plantlets in culture cabinet at 20 °C with 50% relative humidity and a 14:10-h light/dark cycle.
4. Take the protonemata of *P. patens* (Funariaceae) expressing mGFP-HDEL [15].
5. Place pieces of protonema tissue, 2–5 mm in size, in a drop of water on a glass slide between two Vaseline stripes (*see* **Note 10**) and cover it with a coverslip.
6. Using the two Vaseline® stripes, create some space between slide and coverslip and prevent squeezing of the cells.

### 3.4 Confocal Laser Scanning Microscopy of L. fusiformis

1. For DAPI imaging, use the UV diode (405 nm, 5% laser power) and a 420–480 nm long-pass filter.
2. Set the scan speed at 400 $ms^{-1}$ per pixel, the pinhole size at 102.86 airy units for the optimal image acquisition and data processing.
3. Adjust the resolution at 1024 × 1024 and the z step size between 0.25 and 0.46 according to the number of individual images.

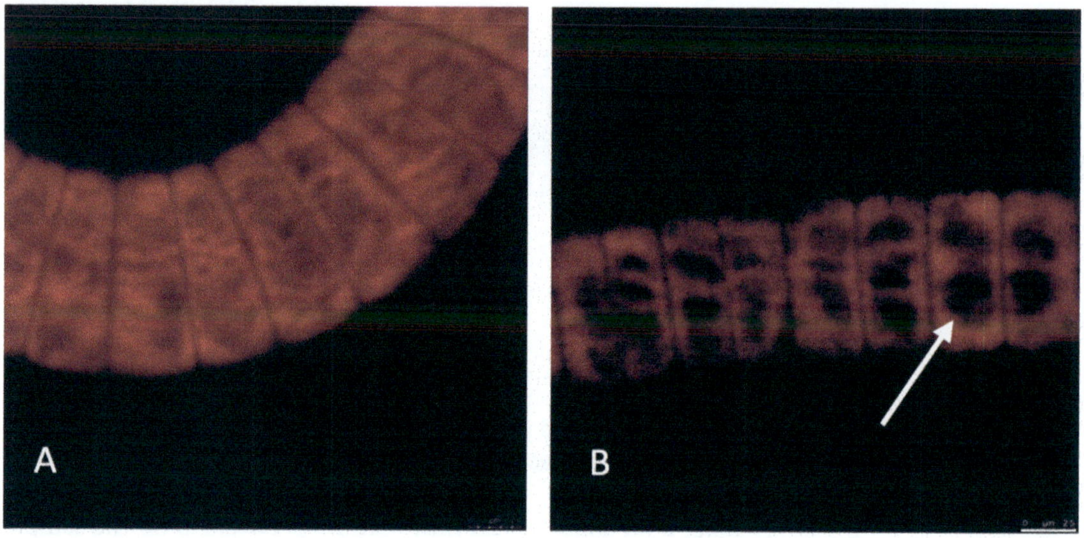


**Fig. 2** Maximum projection of z-stack imaged by CLSM showing the thylakoids of cyanobacterium *L. fusiformis*. (**a**) thylakoids before cyanophage induction. The network of the thylakoids looks uniform. (**b**) Thylakoids after cyanophage induction. White arrow shows the hole-like (?) alterations in the otherwise continuous network of thylakoids. Scale bar: 25 μm

4. Set the number of images to 20–30 to reduce bleaching.
5. Use the argon laser line of 496 nm at the intensity of 10% in combination with 650–700 nm long-pass filter for chlorophyll examination of thylakoids.
6. Set the pinhole size to 102.86 for the optical image acquisition and the scan speed at 400 ms per pixel.
7. Take 20–30 optical serial sections for further image analysis.
8. Adjust the gain to obtain the optimal chlorophyll and DAPI fluorescence image (*see* **Notes 11–15**) (Fig. 2).

### *3.5 Confocal Laser Scanning Microscopy of* P. patens

1. Use a multi-argon laser with an excitation wavelength of 488 nm and emission wavelengths for GFP-ER (495–550 nm) and chlorophyll (670–770 nm).
2. Set the focal depth to 0.5 μm and adjust the pinhole to 1 airy unit.
3. Select 200 or 400 Hz as scanning speed to acquire images with high quality (*see* **Notes 11** and **12**).

### *3.6 3D Reconstruction with Amira® (Threshold Segmentation)*

1. Open the image stack in Start window: Open data. Data will appear in a Project View window. The stacks of thylakoids and DNA (Object) will appear in separated green boxes. Each green box represents one channel corresponding to one of the CLSM channels. Here, DNA appears in channel one, thylakoids appear in channel two, and a bright-field image is shown in channel three (Fig. 3).

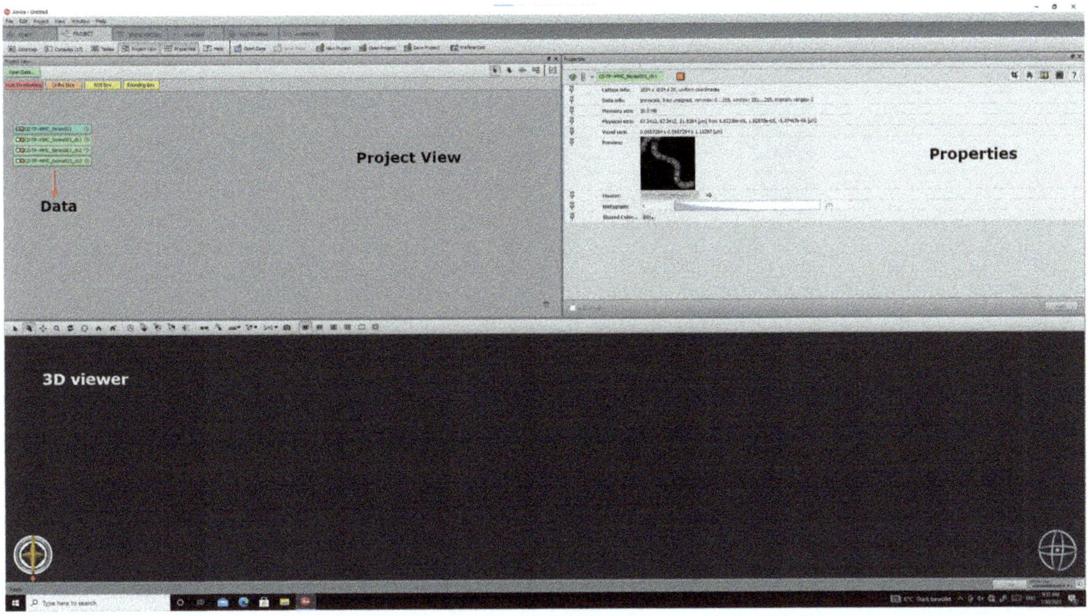


**Fig. 3** Overall view of the Project workroom in Amira. Project View containing the data represented as "objects" (red arrow) and commands which will be attached to the objects. In Properties window, you will find the properties of the data and the settings of the modules. 3D Viewer window shows the objects after reconstruction and is still empty at this stage

2. Select Segmentation workroom of the top bar (red arrow, Fig. 4). To segment the first object, select the channel corresponding to that object in the Image drop-down list (Fig. 4). Here, we start with thylakoids, i.e., the second channel; therefore, we select the second channel in the drop-down menu in Image.
3. Add a Material by clicking on Add (①). Select the color of the segmented object by double-clicking on the color square next to Material (②) (Fig. 5).
4. Click on Threshold bar (.ıl). A slider dialog will open. By sliding the threshold slider to right or left, the object will be saturated with blue color. The saturated area will be reconstructed later. Therefore, it is important to move through the scans to make sure that the object is thoroughly labeled (Fig. 6).
5. Click on Select Mask Voxels and then click on ⊕ to add the mask to the object. By clicking on ⊖, you may remove the mask (Fig. 7).
6. Go back to the Project View in the main menu. You will find a new green box added to other green icons. The name of the new data is the same name of the channel you have masked with the .labels* suffix. Right-click on the new data and select the

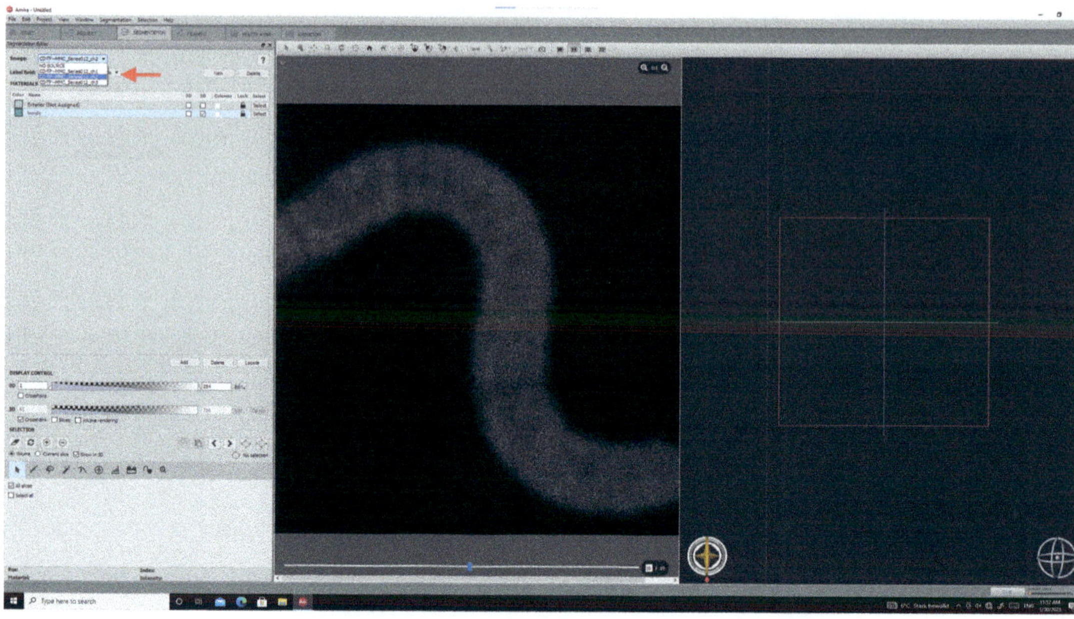

**Fig. 4** Segmentation workroom where you can extract the object. Select the channel (or object) you are willing to extract from Image icon (red arrow)

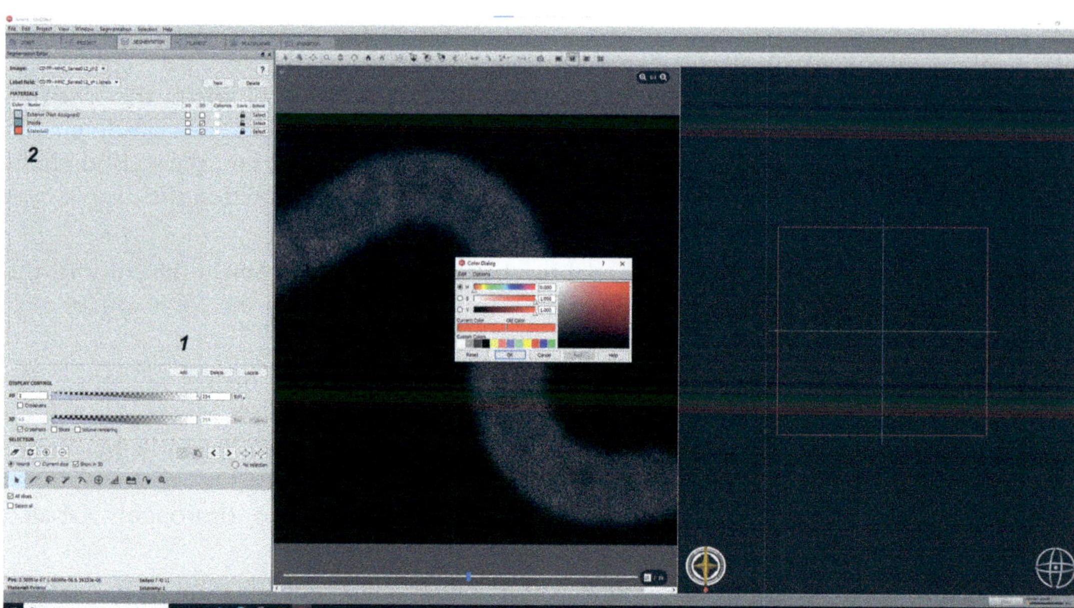


**Fig. 5** Masking the object. Add a Material to the selected object (①) and choose the preferred color (②). Here, we chose red for the thylakoid membranes

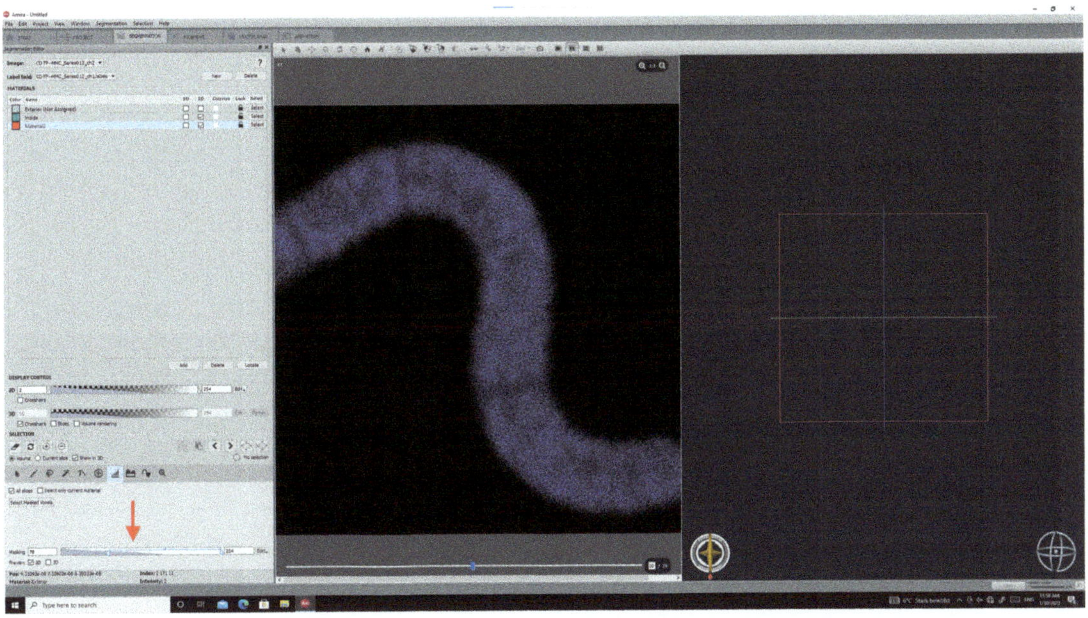

**Fig. 6** Threshold Segmentation. Move the slider (red arrow) to left and right to saturate the object

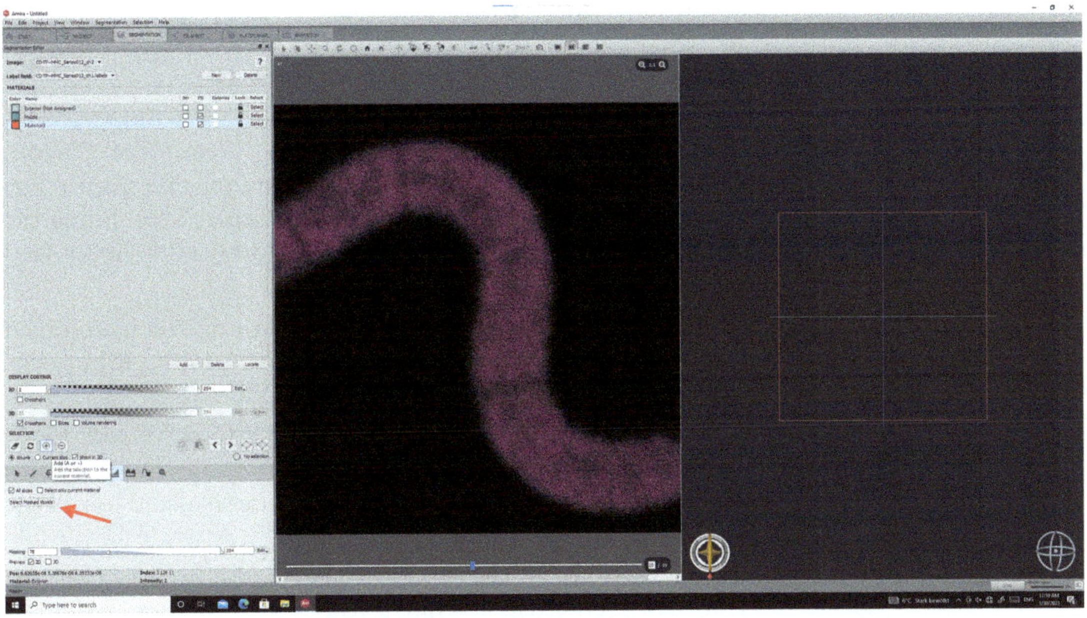

**Fig. 7** Masking the object. Click on Select Mask Voxel to mask the saturated area (red arrow) and add the mask to the object by clicking on ⊕

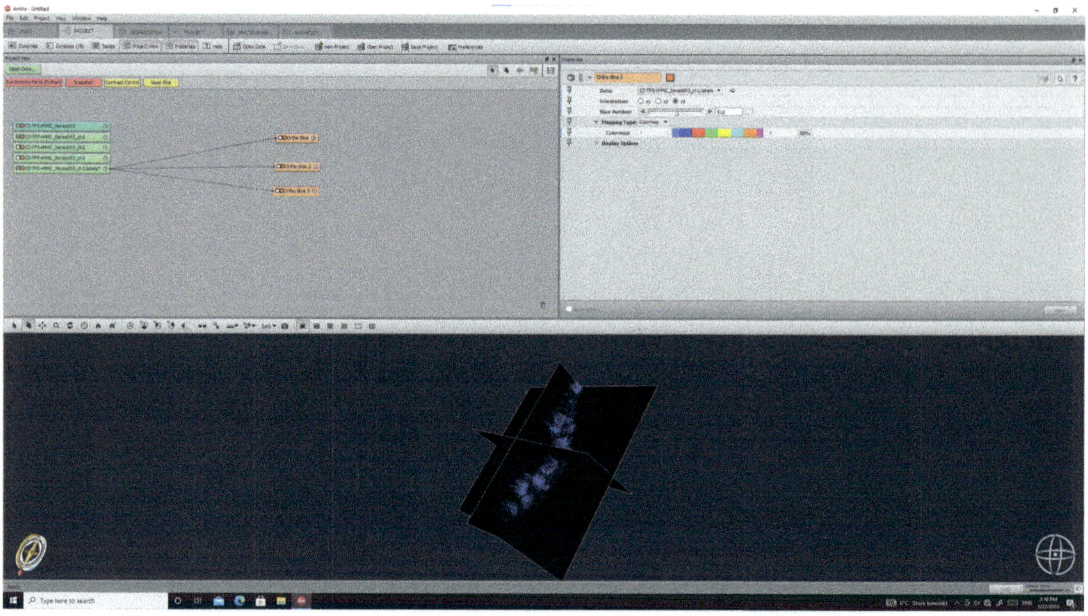

**Fig. 8** The new data is added to the Project View with the .labels* suffix containing the masked object. In our example, three Ortho Slices are added to the labeled object in xy, xz, and yz orientation

Ortho Slice from Display category (Fig. 8). You may attach more than one Ortho Slice to the mask data so you can slice the object in either xy, xz, or yz axes (*see* **step 10**).

7. Right-click on the .labels* data and from the Compute category, select Generate Surface (Fig. 9a). Select the Unconstrained Smoothing in Properties window. To adjust the roughness of the object, set the Smoothing at 5 by sliding the slider (*see* **Note 16**). In Properties window, click Apply (red arrow) (Figs. 9b and 13a–f).
8. In the Project View workroom, another green data icon (green box) is automatically added with the .surf* suffix. Right-click on the .surf* data and select Surface View (Fig. 10a) from Display category. The 3D image of the object will pop up in 3D Viewer window (Fig. 10b).
9. Turn off the Ortho Slices by clicking on the small red box in the Ortho Slice module (Fig. 10a).
10. Take a snapshot of the object by clicking on camera icon (📷) (Fig. 10b, red arrow). White arrows show the invagination in thylakoids, 3 h after cyanophage induction.
11. Add a second object (here, the DNA image stack) to the previous object (i.e., thylakoids), by first turning off the red box in Surface View module attached to the .surf* data.
12. Open the Segmentation workroom. Delete the previous Material.

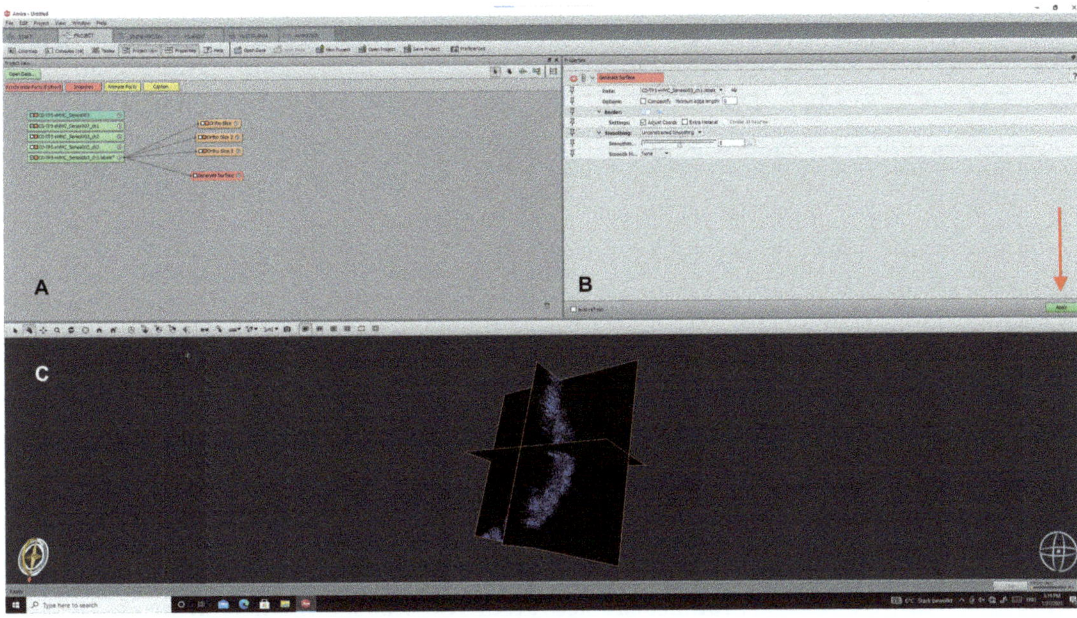


**Fig. 9** Generate Surface module attached to the labeled object (**a**). Click Apply to execute the module (red arrow, **b**). Visualize the object by activating Ortho Slice (**c**)

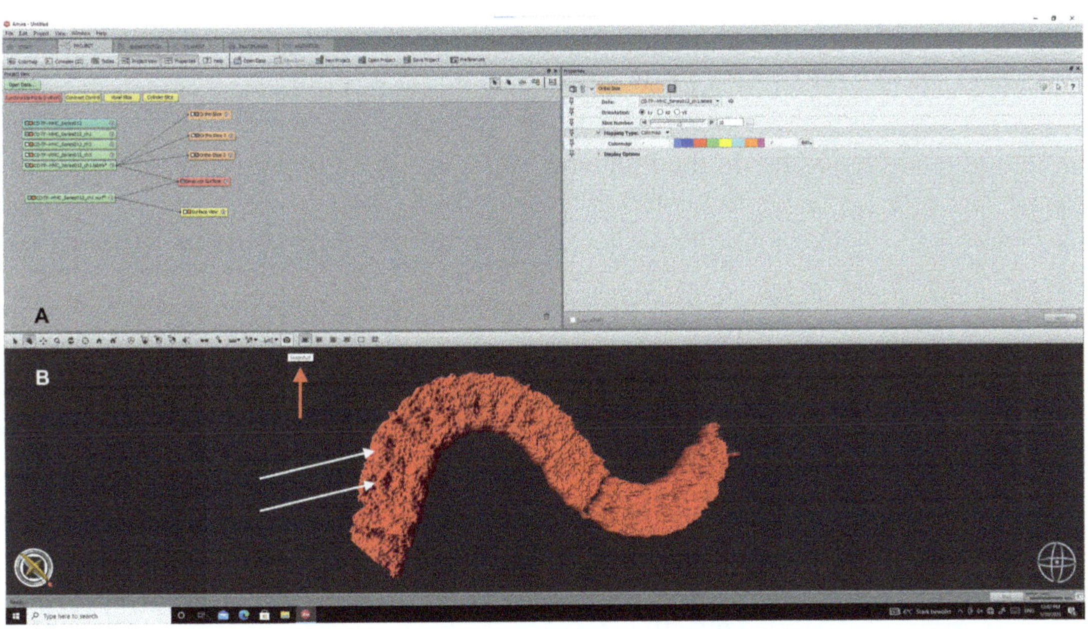


**Fig. 10** Construct a 3D image of the object. (**a**) Attach Surface View to the .surf* data. (**b**) The 3D image of the object will appear in the 3D Viewer window. Red arrow shows the camera icon to take a snapshot of the 3D object. White arrows show the thylakoid invagination

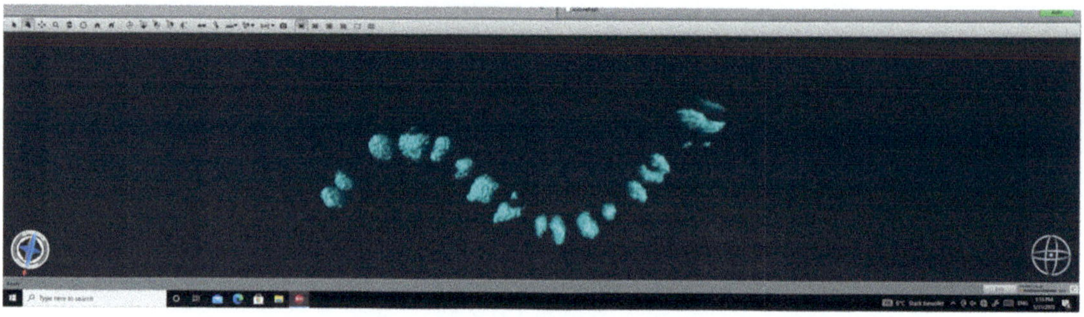

**Fig. 11** DNA within the cyanobacterium after 3D reconstruction, shown in 3D Viewer window

13. Select the second channel from the drop-down menu of Image.
14. Add another material and choose the color as described in **step 3**.
15. Repeat the **steps 4–8**. The 3D image of DNA will appear in 3D Viewer window as shown in Fig. 11. If both Surface View modules are activated, you will see the overlay image of both objects (*see* **Notes 17** and **18**).
16. Attach all the Ortho Slices to the. surf* data to exhibit the inner part of the membrane and internal structures by slicing the object in either longitudinal, lateral, or transversal sections.
17. Make sure that both Surface View modules are activated.
18. Click on the small red boxes in Ortho Slice modules to hide them (Fig. 12a; *see* **Note 19**).
19. In the Properties window of the Ortho Slice with xy orientation, click on the small cube called Clip on the upper right corner (Fig. 12b, red arrow; *see* **Note 20**). Here, you can see through the object in xy orientation (Fig. 12c).
20. Change the slice number by scrolling the Slice Number in Properties window of the Ortho Slice to right and left to move through the object in xy axis. To have a cross-sectional view, repeat this step for the Ortho Slice in xz orientation (Figs. 12d and 13).
21. To provide a movie of the object, slide the curser on 3D viewer in your desired pace and orientation so that the image in 3D Viewer window starts rotating (Fig. 14d).
22. Select Animation from the main menu and then Animation Director.
23. In Animation Director, press on the movie icon (🎞) to open the Movie Creation (Fig. 14c).
24. On the left window of Animation Director, move the slider to left to set the duration of the movie (Fig. 14c, red arrow). The

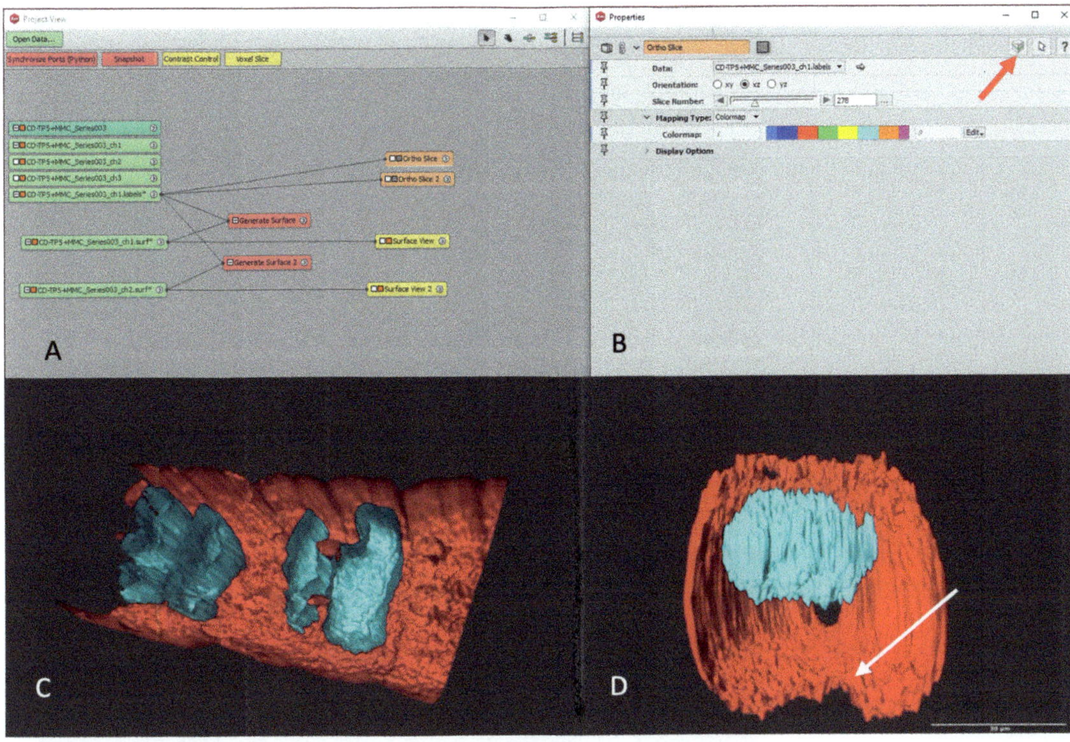


**Fig. 12** Slicing the object to visualize the internal structures. (**a**) The Project View workspace. (**b**) Clip tool to slice the image (red arrow). (**c**) Hiding the top layers to have a longitudinal view into the internal objects. (**d**) A cross section produced by clipping in xz orientation. The invagination of the thylakoid (white arrow) and accumulation of DNA to the other side of the thylakoid are visible in the cross-sectional view

number of frames will be adjusted automatically according to the time; however, you still can change the number of the frames after adjusting the time (Fig. 14c).

25. Choose the directory to save the movie. Set the Quality, Size, and File Format. Click Create Movie (Fig. 14c).
26. To create a movie of the object like "growing" through the scans, repeat **steps 12–14**.
27. On Project View, select an Ortho Slice module with your desired orientation. If you want to move through the object in xy orientation, select the Ortho slice in xy orientation (Fig. 15a).
28. In the Properties window of the selected Orth slice, move the Slice Number to the point that you want the movie to start (Fig. 15b).
29. Click on the ⏰ next to the Slice Number (Fig. 15b, red arrow).
30. One stop point will be added to the beginning time of the movie.

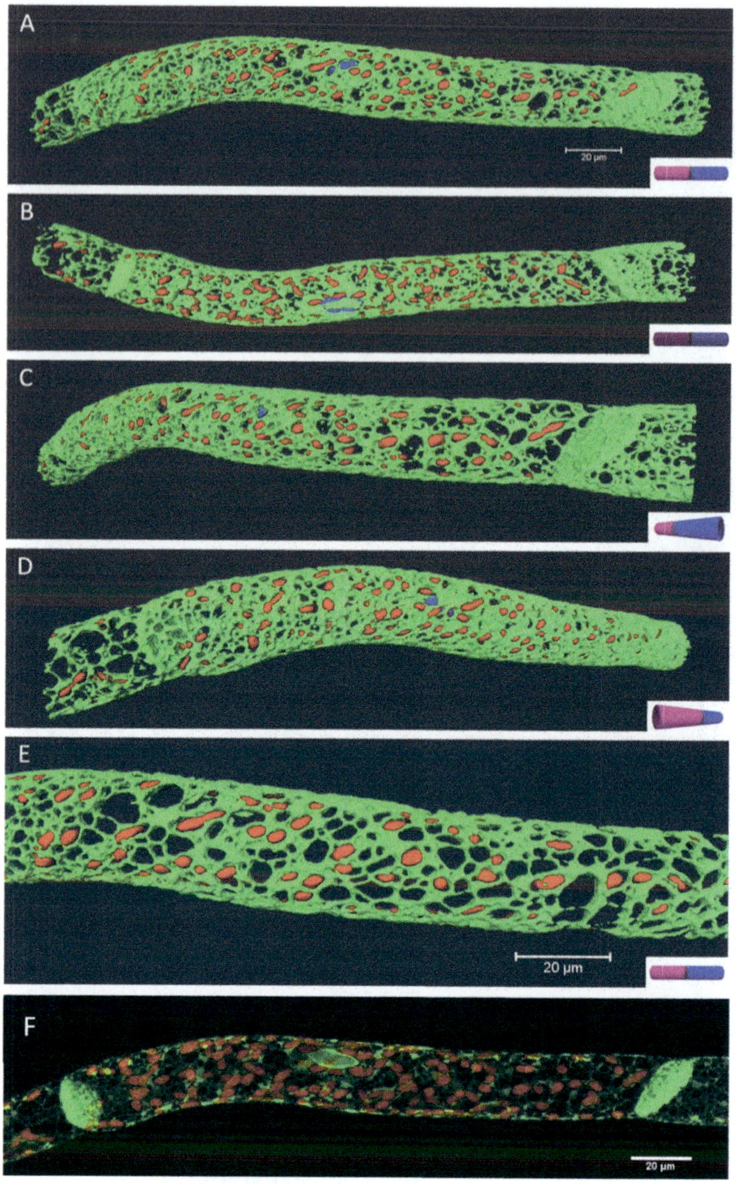


**Fig. 13** Internal structures in a moss protonema, analog to cyanobacteria shown above. (**a–e**) 3D reconstruction of z-stacks from the confocal channels ER (green), chloroplasts (red), and nucleus (blue) shown from different perspectives. In the bottom right corner, false-colored protoplasts in purple and blue correspond to the respective viewing angle. (**f**) Maximum projection of a z-stack imaged by CLSM illustrating the fluorescence of ER (green) and chloroplasts (red)

31. In Animation Directory, move the orange slider to the next time (Fig. 15c, red arrow).
32. Go back to the Ortho Slice properties window, and move the Slice Number to the next slice (or any slice you wish to illustrate. In the movie, the object will grow from the first scan to the second scan you select here)

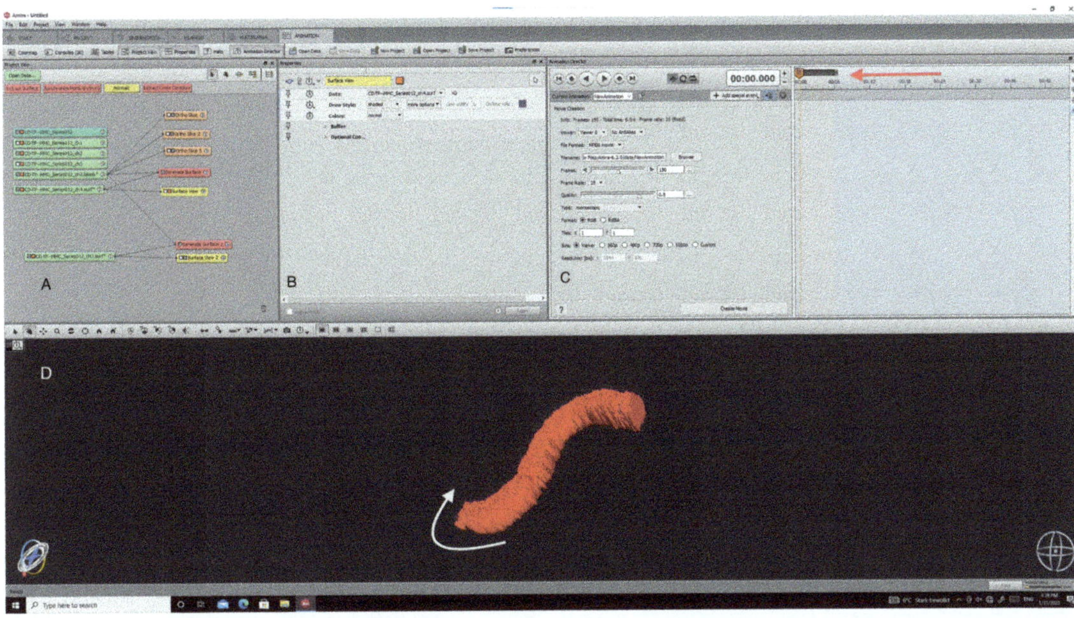


**Fig. 14** Instructions to create a movie. (**a**) Turn off the Ortho Slices to hide them. (**b**) Properties window. (**c**) Animation Directory: red arrow shows the slider to adjust the time. (**d**) 3D viewer: rotate the object by sliding the curser in desired pace and orientation. White arrow shows the rotation of the object

33. Click on the ⏰. Another stop point will be added to the movie where you had moved the orange slider to it.
34. Continue to set time points until you have assigned time points to the all the scans you wish to go through (Fig. 15e). Choose the directory to save the movie. Set the Quality, Size, and File Format. Click Create Movie (Fig. 15c).

## 4 Notes

1. Do not autoclave $NaHCO_3$ since heating denatures the sodium bicarbonate and produces sodium carbonate, carbon dioxide, and water.
2. MMC is a carcinogen and should be handled with Silver Shield or Norfoil gloves.
3. Fixation of the cyanobacteria was necessary to allow penetration of DAPI for DNA staining.
4. Wear gloves and sterile the laminar with 70% ethanol thoroughly before you start.
5. Keep all the cyanobacterial cultures on the constant shaking. Three to six hours after induction with MMC, the difference between two treatments should be visible with the culture induced with MMC exhibiting a pale green color due to the lower concentration of cyanobacteria (Fig. 1).

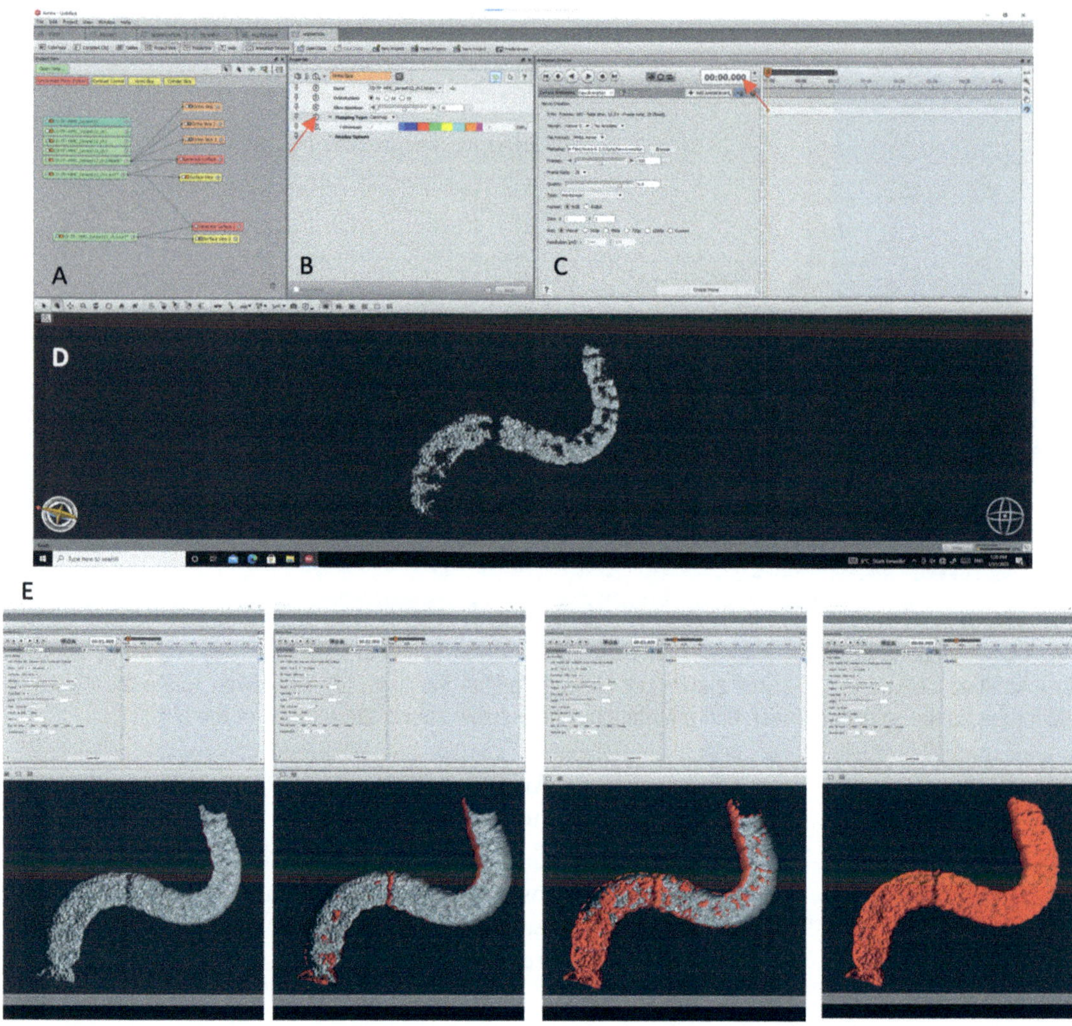


**Fig. 15** Create a movie of the object that is "growing" scan by scan. (**a**) Project View: turn on the Surface View modules and turn off all the Ortho Slices. Click on the Ortho Slice with xy orientation. (**b**) Properties window of the Ortho Slice with xy orientation. Red arrow shows the ⏰ by which you can assign time to each scan. After clicking on the ⏰, a time point will be added to the movie. (**c**) Animation Directory. Red arrow shows the time slider. After selecting the Slide Number in Properties window, move this time slider forward and then click on ⏰. (**d**) 3D Viewer window where you can see the 3D image of the object. (**e**) Assigning time points to the scans. In the movie, the sample will grow from the lower scan to the upper one

6. To prepare the slide, do not put a big drop of culture. The filaments tend to move to the edges, and if the drop of culture is too big, all the filaments will move to the edges or will spill out below the coverslip.
7. Do not press the coverslip too much on the sample since it may cause disruption of the filaments. On the other hand, if the pressure is not enough, the filaments will move around under the microscope and the image acquisition is no longer possible.

It is better to take a small drop and use a rectangular coverslip. Moreover, press the coverslip gently for almost 5–7 s.

8. Only use microscope lens cleaning paper when you prepare the slides. Normal tissue papers or filter papers are strongly absorbent and will absorb all the sample.
9. Hold the coverslip at the edges with two fingers when you are removing the extra liquid or sealing with nail polish. Do not put pressure here to avoid removal of the filaments.
10. Use a brush to leave a strip of Vaseline.
11. All images were taken with a 63× water immersion objective.
12. Because of the autofluorescence characteristic of chlorophyll, there is no need to label the thylakoid.
13. Optical sections should be taken in less than 3 min (best acquisition time is 2 min) to avoid bleaching.
14. Bleaching occurs first on thylakoids and increases the signal from DNA which causes artifacts if the time of scanning takes more than 3 min.
15. Adjust the gain by removing the blue dots to remove the artifact signals.
16. Note that by over-smoothing the object, you may lose some of the details.
17. Notice that there will always be one .labels* data. It means that when you create another material, and label a second object, the previous .labels* data will overwrite with a new labeled object. You will not have specified .labels* data designated to each channel.
18. In our example, the second object is DNA which is enveloped with thylakoids. Therefore, in the overlay image of DNA and thylakoids, we do not see the DNA. To observe the DNA (or any other internal structure covered by membranes), we need to slice the image and remove the layers on top.
19. By hiding the Ortho Slice, it is not deactivated and can be used to visualize the interior part of the object.
20. By clicking on this cube, either the top or bottom of the box will disappear. After removing the upper half of the Clip box, the layers on top of the Ortho Slice will hide.

## Acknowledgments

The study was partly funded by a faculty of life sciences network initiative "Cyanophage Hunters Network" and the Vienna Science and Technology Fund (WWTF) [10.47379/LS19013]. We thank Leo Pokorny for maintaining the cyanobacterial cultures. We are grateful to Dr. Michael Schagerl and the phycology team at the Department of Functional and Evolutionary Ecology, University of Vienna. Many thanks to Mag. Dominik Harant for providing some of the unpublished images of his diploma thesis on moss protonemata. We thank the team of the Shared Service Facility "Cell Imaging and Ultrastructure Research" at the Faculty of Life Sciences, University of Vienna, for their support and the use of the confocal microscopes.

## References

1. Bao M, Xie J, Katoele N, Hu X, Wang B, Piruska A, Huck WTS (2019) Cellular volume and matrix stiffness direct stem cell behavior in a 3D microniche. ACS Appl Mater Interfaces 11(2):1754–1759. https://doi.org/10.1021/acsami.8b19396
2. Calhoun PS, Kuszyk BS, Heath DG, Carley JC, Fishman EK (1999) Three-dimensional volume rendering of spiral CT data: theory and method. RadioGraphics 19(3):745–764. https://doi.org/10.1148/radiographics.19.3.g99ma14745
3. Kumar G, Tison CK, Chatterjee K, Pine PS, McDaniel JH, Salit ML, Young MF, Simon CG (2011) The determination of stem cell fate by 3D scaffold structures through the control of cell shape. Biomaterials 32(35):9188–9196. https://doi.org/10.1016/j.biomaterials.2011.08.054
4. Théroux-Rancourt G, Voggeneder K, Tholen D (2020) Shape matters: the pitfalls of analyzing mesophyll anatomy. New Phytol 225(6):2239–2242. https://doi.org/10.1111/nph.16360
5. Handschuh S, Schwaha T, Neszi NZ, Walzl M, Woess E (2007) Advantages of 3D reconstruction in Bryozoan development research: tissue formation in germinating statoblasts of Plumatella fungosa (Pallas, 1768). Plumatellidae, Phylactolaemata
6. Theroux-Rancourt G, Earles J, Gilbert M, Zwieniecki M, Boyce C, McElrone A, Brodersen C (2017) The bias of a two-dimensional view: comparing two-dimensional and three-dimensional mesophyll surface area estimates using noninvasive imaging. New Phytol 215:1609–1622. https://doi.org/10.1111/nph.14687
7. Reichert U, Linden T, Belfort G, Kula M-R, Thömmes J (2002) Visualising protein adsorption to ion-exchange membranes by confocal microscopy. J Membr Sci 199(1):161–166. https://doi.org/10.1016/S0376-7388(01)00693-7
8. Reingruber H, Zankel A, Mayrhofer C, Poelt P (2011) Quantitative characterization of microfiltration membranes by 3D reconstruction. J Membr Sci 372(1):66–74. https://doi.org/10.1016/j.memsci.2011.01.037
9. Satoh H, Delbridge LM, Blatter LA, Bers DM (1996) Surface:volume relationship in cardiac myocytes studied with confocal microscopy and membrane capacitance measurements: species-dependence and developmental effects. Biophys J 70(3):1494–1504. https://doi.org/10.1016/S0006-3495(96)79711-4
10. Jing C, Cornish VW (2011) Chemical tags for labeling proteins inside living cells. Acc Chem Res 44(9):784–792. https://doi.org/10.1021/ar200099f
11. Kricka LJ, Fortina P (2009) Analytical ancestry: "firsts" in fluorescent labeling of nucleosides, nucleotides, and nucleic acids. Clin Chem 55(4):670–683. https://doi.org/10.1373/clinchem.2008.116152
12. Suzuki T, Matsuzaki T, Hagiwara H, Aoki T, Takata K (2007) Recent advances in fluorescent labeling techniques for fluorescence microscopy. Acta Histochem Cytochem 40(5):131–137. https://doi.org/10.1267/ahc.07023
13. Zekri MA, Schagerl M, Schweichhart J, Lang I (2021) Confocal microscopy reveals alterations of thylakoids in Limnospira fusiformis during prophage induction. Protoplasma 258(6):1251–1259. https://doi.org/10.1007/s00709-021-01656-8
14. Harant D, Lang I (2021) 3D dissection of structural membrane-wall contacts in filamentous moss protonemata. Int J Mol Sci 22(1):158
15. Bezanilla MM. Bezanilla-Lab-homepage. Available online: https://sites.dartmouth.edu/bezanillalab. Accessed 23 Nov 2020

# Chapter 28

# Visualizing Orientation and Topology of ER Membrane Proteins In Planta


## Michelle Schlößer, José M. Ugalde, and Andreas J. Meyer



## Abstract

The orientation of membrane proteins within the lipid bilayer is key to understanding their molecular function. Similarly, the proper topology of multispanning membrane proteins is crucial for their function. Although bioinformatics tools can predict these parameters assessing the presence of hydrophobic protein domains sufficiently long to span the membrane and other structural features, the predictions from different algorithms are often inconsistent. Therefore, experimental analysis becomes mandatory. Redox-based topology analysis exploits the steep gradient in the glutathione redox potential ($E_{GSH}$) across the ER membrane of about 80 mV to visualize the orientation of ER membrane proteins by fusing the $E_{GSH}$ biosensor roGFP2 to either the N- or the C-termini of the investigated protein sequence. Transient expression of these fusion proteins in tobacco leaves allows direct visualization of orientation and topology of ER membrane proteins in planta. The protocol outlined here is based on either a simple merge of the two excitation channels of roGFP2 or a colocalization analysis of the two channels and thus avoids ratiometric analysis of roGFP2 fluorescence.



**Key words** Endoplasmic reticulum, ER membrane protein, Glutathione redox potential, Protein topology, Redox-based topology analysis, roGFP2


## 1 Introduction

Eukaryotic cells are highly compartmentalized to separate biochemical reactions and pathways that evolved for specific purposes, often with incompatible features. Yet, continuous interaction between these compartments is essential, requiring the transport of ions and metabolites across the respective membranes. To enable transport, communication, and physical interaction of compartments, the membranes contain a large number of transmembrane proteins to fulfill the respective functions.

The protein composition of membranes is a primary determinant of organelle identity. In this sense, however, the endoplasmic reticulum (ER) membrane is special because virtually all membrane proteins in the tonoplast and the plasma membrane need to pass the


Verena Kriechbaumer (ed.), *The Plant Endoplasmic Reticulum: Methods and Protocols*, Methods in Molecular Biology, vol. 2772, https://doi.org/10.1007/978-1-0716-3710-4_28,

ER where they are first inserted into the lipid bilayer. While 20–30% of the whole genome in eukaryotes codes for membrane proteins [1], the number of membrane proteins located on the ER surface or passing through the ER membrane is thus particularly high. Membrane proteins consist of one or more hydrophobic transmembrane domains (TMDs). Insertion of these proteins into the membrane typically occurs co-translationally through the SEC61 translocon complex [2, 3] or for tail-anchored proteins post-translationally via the GET pathway [4] Appropriate functioning of membrane proteins can only be ensured if the correct orientation and topology within the membrane are adopted. For this, most multispanning proteins contain positively charged residues and domains that are retained on the cytosolic side of a membrane, which is a characteristic feature of the so-called "positive-inside" rule [5].

Because most membrane proteins work in a polar fashion, functional characterization of unknown membrane proteins requires first exact knowledge of their orientation within the membrane. Despite major advances in recognizing hydrophobic TMDs and using the positive-inside rule for predictions, bioinformatics algorithms frequently fail to provide sufficiently clear and unique information. Hence, experimental analysis of proteins is mandatory.

While biochemical methods such as protease protection assays can be used to analyze the orientation and topology of membrane proteins, more convenient imaging-based approaches have been developed for the visualization of protein orientation and topology in vivo. These include in situ proteinase protection assays, which involve the selective permeabilization of the plasma membrane with digitonin and protein digestion with proteinases [6]. Other methods, such as bimolecular fluorescence complementation approaches [7], require the transformation and expression of two separate constructs or are based on pH quenching of YFP on the apoplastic face of the plasma membrane [8], which depends on a steep pH gradient across the membrane and cannot be applied for proteins in the early secretory pathway. A simpler approach is based on the self-indicating redox-dependent properties of redox-sensitive GFP2 (roGFP2). This protein contains two engineered cysteines in close proximity to the chromophore capable of forming an intramolecular disulfide bridge [9]. The formation of the disulfide bridge causes slight conformational changes that alter the protonation of the chromophore and thus the protonation-dependent excitation properties [10]. When reduced, roGPF2 is predominantly excited at 488 nm with only weak excitation at 405 nm. In the oxidized state, excitation at 405 nm increases, while excitability at 488 nm is diminished. The formation and release of the disulfide bridge is catalyzed by glutaredoxins (GRXs), which mediate redox equilibration between the local glutathione redox potential ($E_{GSH}$) and roGFP2 [11].

The steep $E_{GSH}$ gradient across the ER membrane with values of ~−320 mV in the cytosol and ~−240 mV in the ER lumen allows the utilization of this gradient for visualizing the orientation of proteins within the ER membrane [11–13]. With a midpoint potential of −280 mV and a nearly linear dynamic range of 70 mV, roGFP2 is the ideal candidate for providing a binary read-out of its localization on either the cytosolic or the luminal face of the ER membrane in the redox-based topology assay (ReTA). For visualizing the orientation of membrane proteins, roGFP2 is translationally fused to either the N- or the C-termini of the respective proteins, and the fusion constructs are transiently expressed in tobacco (*Nicotiana tabacum*) leaves (*see* Fig. 1a). Since a binary readout is expected, there is no need for quantitative measurement or elaborate sensor calibrations. The protocol described here thus avoids ratiometric analysis of roGFP2 fluorescence and instead relies on either simple superimposition of images from the two fluorescence channels and qualitative assessment of the merge

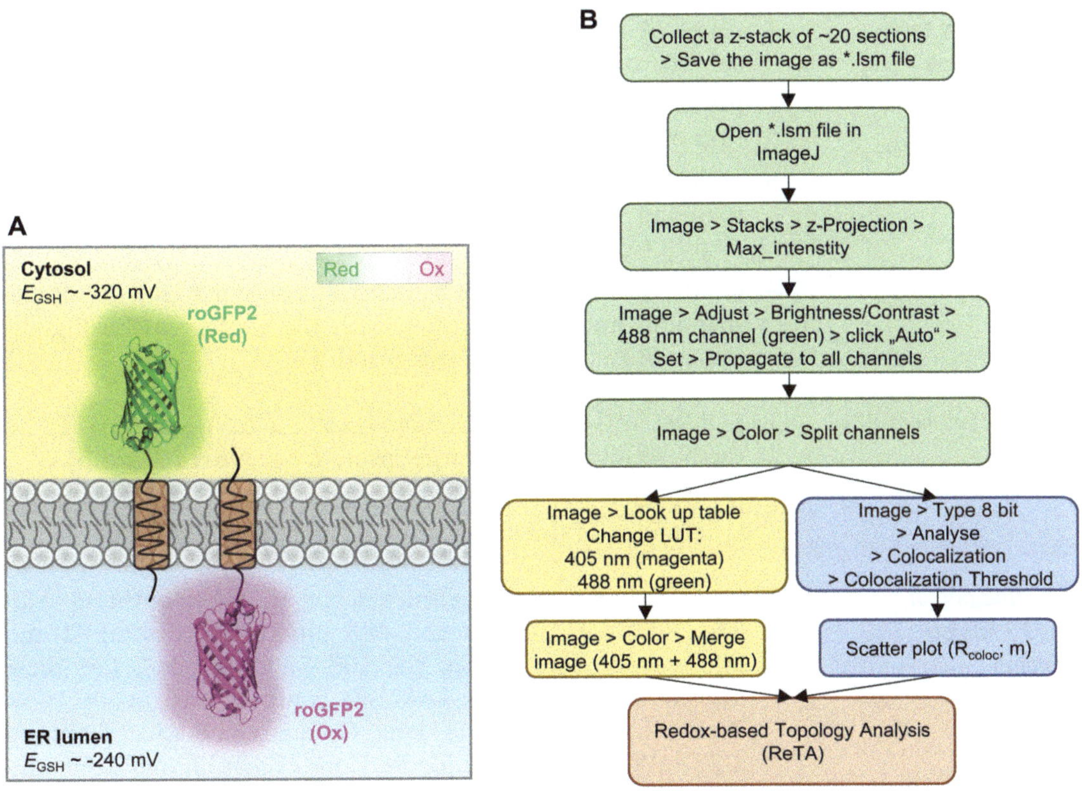


**Fig. 1** Schematic outline of procedures for redox-based topology analysis (ReTA) for ER membrane proteins. (**a**) Illustration of the ReTA assay for a single-spanning ER membrane protein or only the N-terminal transmembrane domain for type II proteins with roGFP2 fused to either the N- or the C-terminus. (**b**) Flowchart of the different steps for ReTA. The analysis concludes in either a merge of the two fluorescence channels of roGFP2 or a colocalization analysis between them

color or a correlative assessment of the two intensity values of pixels in a dual-channel image by applying tools for colocalization analysis [14].

## 2 Materials

### 2.1 Plant Material and Bacteria

1. 2- to 4-week-old tobacco (*Nicotiana tabacum* L.) plants (*see* **Note 1**).
2. *Escherichia coli* DH5α and Top10.
3. *Agrobacterium tumefaciens* AGL1 (*see* **Note 2**).

### 2.2 Culture Media Components

1. Luria–Bertani (LB) medium.
2. Antibiotics for plasmid selection: kanamycin, gentamycin, rifampicin, carbomycin and spectinomycin.
3. Spectrophotometer (BioPhotometer plus, Eppendorf SE, Hamburg, Germany) to measure the $OD_{600}$.

### 2.3 Vectors for roGFP2 Fusion Proteins

#### 2.3.1 Control Vectors

1. pBinCM-roGFP2 (cytosol): control construct for expression of soluble roGFP2 in the cytosol (*see* **Note 3**).
2. pWEN22-CHI-roGFP2-HDEL: control construct for expression of soluble roGFP2 in the lumen of the ER (*see* **Note 3**).
3. pCM01-roGFP2N-SEC22: control construct for a single-spanning membrane protein with the N-terminus facing the cytosol (*see* **Note 3**).
4. pSS01-SEC22-roGFP2C: control construct for a single-spanning membrane protein with the C-terminus facing the lumen of the ER (*see* **Notes 3** and **4**).

#### 2.3.2 Vectors for roFGP2 Fusions

1. pCM01-roGFP2N: Gateway vector for fusion of roGFP2 to the N-terminus of the target protein (*see* **Notes 3** and **5**).
2. pSS01-roGFP2C: Gateway vector for fusion of roGFP2 to the C-terminus of the target protein (*see* **Notes 3** and **6**).

### 2.4 Imaging of roGFP2 Fusion Proteins

1. Confocal laser scanning microscope (CLSM) outfitted with laser lines for 405 nm and 488 nm to excite roGFP2 and fluorescence detection in the 505–530 nm band. For these measurements, a Zeiss LSM780 confocal microscope (Carl Zeiss Microscopy GmbH, Jena, Germany) was used, featuring a 25 mW Ar/ML laser for 488 nm excitation and a 30 mW 405 nm diode laser. Alternative microscopes equipped with these laser lines can also be used, as long as they support sequential excitation in line-switching mode.
2. Glass slides and coverslips (typically #1.5; thickness 0.17 mm).
3. Cork borer (∅ = 6 mm).

4. Featherweight forceps.
5. Adhesive rubber tape.
6. Imaging buffer: 10 mM MES, 10 mM $MgCl_2$, 10 mM $CaCl_2$, 5 mM KCl, pH = 5.8 (KOH) supplemented with the different compounds for specific experiments.
7. 40× lens (C-Apochromat 40×/1.2 W Korr).

### 2.5 Image Analysis Software

1. ZEN (black edition) (Zeiss).
2. ImageJ (https://imagej.nih.gov/ij/download.html) with the Colocalization Threshold plugin Coloc 2 (https://github.com/fiji/Colocalisation_Analysis/releases) installed (*see* **Note 7**).

## 3 Methods

### 3.1 Cloning of roGFP2 Fusion Proteins

1. Design specific primers for PCR amplification of the coding sequence of the protein of interest that include the attB1 and attB2 Gateway® sites (according to the Gateway® manual, Invitrogen, Waltham, MA). For N-terminal roGFP2 fusions, exclude the start codon, while for C-terminal fusions, exclude the stop codon of the gene of interest.
2. Do the BP recombination reaction to clone the CDS of the protein of interest into an entry vector containing the attP1 and attP2 sites, such as pDONR201 or pDONR207. Use 25 fmol of the PCR product (generated in **step 1**) and 75 ng of the entry vector and mix with 1 μL of BP II clonase to a final volume of 4 μL and incubate for 1 h at room temperature.
3. Add 0.5 μL proteinase K and incubate for 10 min at 37 °C.
4. Use the whole volume for transformation into competent *Escherichia coli* cells (DH5α or Top10) and incubate for 20 min on ice (*see* **Note 8**).
5. Heat shock cells for 1 min at 42 °C, and place them back on ice for 2 min. Subsequently, add 600 μL LB medium and let the cells recover for at least 45 min at 37 °C with constant agitation at 120 rpm.
6. After recovering, pellet the cells by centrifugation at 6000 g for 2 min.
7. Discard 450 μL of the supernatant and resuspend the pellet with the residual supernatant.
8. Transfer 200 μL of the cell suspension to a plate with LB medium and agar supplemented with the respective antibiotic and grow colonies overnight at 37 °C (50 μg $mL^{-1}$ kanamycin for pDONR201 or 50 μg $mL^{-1}$ gentamycin for pDONR207).

9. Confirm successful transformation of bacteria via colony PCR and gel electrophoresis to verify the presence of the recombined constructs (*see* **Note 9**).
10. Incubate positive colonies overnight at 37 °C in LB medium containing the respective antibiotic.
11. Purify plasmids and confirm the correct insertion by sequencing.
12. Do the LR recombination reaction for subcloning the constructs in the respective destination vectors pSS01-roGFP2C for C-terminal roGFP2 fusions or pCM01-roGFP2N for N-terminal fusions. Use 25 ng of the entry clone and 50 ng of the destination vector and mix with 1 μL of LR II clonase to a final volume of 4 μL and incubate for 1 h at room temperature.
13. Repeat workflow **steps 3–11** with the appropriate antibiotic selection (50 μg $mL^{-1}$ kanamycin for pSS01-based vectors or 25 μg $mL^{-1}$ spectinomycin for pCM01-based vectors).

### 3.2 Transformation of A. tumefaciens Cells

1. Add 10 ng of the confirmed destination vector to a 100 μL aliquot of frozen electrocompetent *Agrobacterium* cells and incubate them for 45 min on ice.
2. Transfer the *Agrobacterium* cells to a pre-chilled electroporation cuvette and electroporate them according to the manufacturer's recommendations. Here, a MicroPulser Electroporation system (Bio-Rad, Hercules, CA) was used.
3. Directly add 600 μL of prewarmed (28 °C) LB medium into the cuvette, resuspend the cells, and put them back into a 1.5 mL Eppendorf tube. Allow the cells to recover for 1.5 h at 28 °C with constant agitation at 120 rpm.
4. After recovering, pellet the cells by centrifugation at 6000 g for 2 min.
5. Remove 450 μL of the supernatant and resuspend the pellet with the remaining medium.
6. Plate 100 μL of the resuspended cells on a plate containing LB medium with agar supplemented with the respective antibiotics for the *Agrobacterium* cells (50 μg $mL^{-1}$ rifampicin and 50 μg $mL^{-1}$ carbomycin) plus the antibiotics required for selection of the destination plasmid (50 μg $mL^{-1}$ kanamycin for pSS01-based vectors or 25 μg $mL^{-1}$ spectinomycin for pCM01-based vectors).
7. Incubate the plate for 48 h at 28 °C.

### 3.3 Transient Expression of ReTA Constructs in Tobacco Leaves

1. Take a colony of *Agrobacterium* cells from a plate and inoculate 5 mL LB medium supplemented with the respective antibiotics for selection. Grow the culture overnight at 28 °C with continuous agitation at 120 rpm.
2. Measure the $OD_{600}$ of the saturated culture with the spectrophotometer.
3. Dilute in fresh LB medium supplemented with the respective antibiotic to an $OD_{600}$ of 0.3 and let the cells grow until $OD_{600} = 0.8$ is reached.
4. Pellet 4 mL of the culture by centrifugation using 2 mL tubes at 6000 g for 5 min.
5. Wash the cells by resuspending them in 2 mL sterile $dH_2O$ and centrifuge again for 5 min at 6000 g.
6. Discard the supernatant and resuspend the pellet with 2 mL sterile $dH_2O$.
7. Measure the $OD_{600}$ using $dH_2O$ as a blank.
8. If new constructs are to be tested, dilute the cells to $OD_{600}$ of 0.2, 0.5, and 1, to evaluate the transfection efficiency.
9. If additional vectors are required (e.g., fluorescent markers for subcellular compartments), mix the different cells carrying the respective constructs in a 1:1 ratio.
10. Infiltrate 2- to 4-week-old tobacco leaves with a 1 mL syringe from the abaxial side. Carefully wipe excess infiltration medium from the leaf surface with tissue paper before placing them back in the growing chamber.
11. 48 h after the infiltration, image the transiently transformed tobacco using a CLSM.

### 3.4 Image Acquisition for ReTA

1. Set up the confocal microscope and adjust laser intensities such that the maximum dynamic range of roGFP2 can be used. Use appropriate control constructs either with the previously mentioned soluble roGFP2 expressed in either the cytosol or the ER lumen or N- and C-terminal fusions of an established single-spanning membrane protein, e.g., SEC22 [12, 14] (*see* Subheading 2.3 and **Note 4**).
2. Cut leaf disks from a previously infiltrated tobacco leaf expressing the roGFP2 fusion proteins with a cork borer and vacuum infiltrate the disk with imaging buffer inside a 10 mL syringe.
3. Place the leaf disk with the abaxial side up in a drop of imaging buffer on a slide and place a coverslip on top. Use adhesive rubber tape as a spacer between the slide and coverslip to avoid squashing the leaf disk.

4. Set two consecutive tracks for collecting roGFP2 fluorescence at 505–530 nm while exciting at either 405 nm or 488 nm in line-switching mode.
5. Set the power of the 405 nm laser line relative to the power of the 488 nm line such that the maximum dynamic range of roGFP2 can be used (*see* **Note 10**).
6. Take images of epidermal cells expressing roGFP2 fusion proteins with a 40× lens and a 3x zoom factor. Collect images in 12-bit depth with a pixel size of 0.14 $\mu m^2$.
7. Collect z-stacks of about 20 sections with a step width of 0.415 μm. Each z-stack spans from the nucleus to either the upper or bottom cell wall, which is defined in the imaging software by the "first" and "last" image to collect, respectively.
8. Save the images as *.lsm files.

### *3.5 Image Processing in ImageJ*

1. Open the *.lsm file in ImageJ.
2. For the 3-D image (z-stack), go to Image > Stacks > Z-projection > Max_intensity.
3. Adjust the brightness of the image: Image > Adjust > Brightness/Contrast > 488 nm Channel > click "Auto" > Set > click Propagate to the other channels of this image (*see* **Note 11**).
4. Split the channels: Image > Color > Split Channels.
5. Change the color of the 405 nm channel to magenta and the 488 nm channel to green: Image > Look up table.
6. Merge the two images: Image > Color > Merge Channels (keep source images) > add 405 nm and 488 nm Channels. Save 405 nm, 488 nm and the 405 nm + 488 nm color merge image as JPEGs (*see* Fig. 2).
7. Select the split channel images generated in **step 5** and change their type to 8 bits: Image > Type 8 bit.
8. For regression analysis in the dual-channel image: Analyze > Colocalization > Colocalization Threshold. Select the 405 nm image as Channel 1 and the 488 nm image as Channel 2, which represent the X and Y axes of the scatterplot, respectively. Click on 'Show Scatter plot'.
9. Save the scatterplot as a JPEG image and the values for Pearson's correlation coefficient above the threshold ($R_{coloc}$) and the gradient (m) of the linear regression fit, shown in the results window (*see* Fig. 2 and **Note 12**).

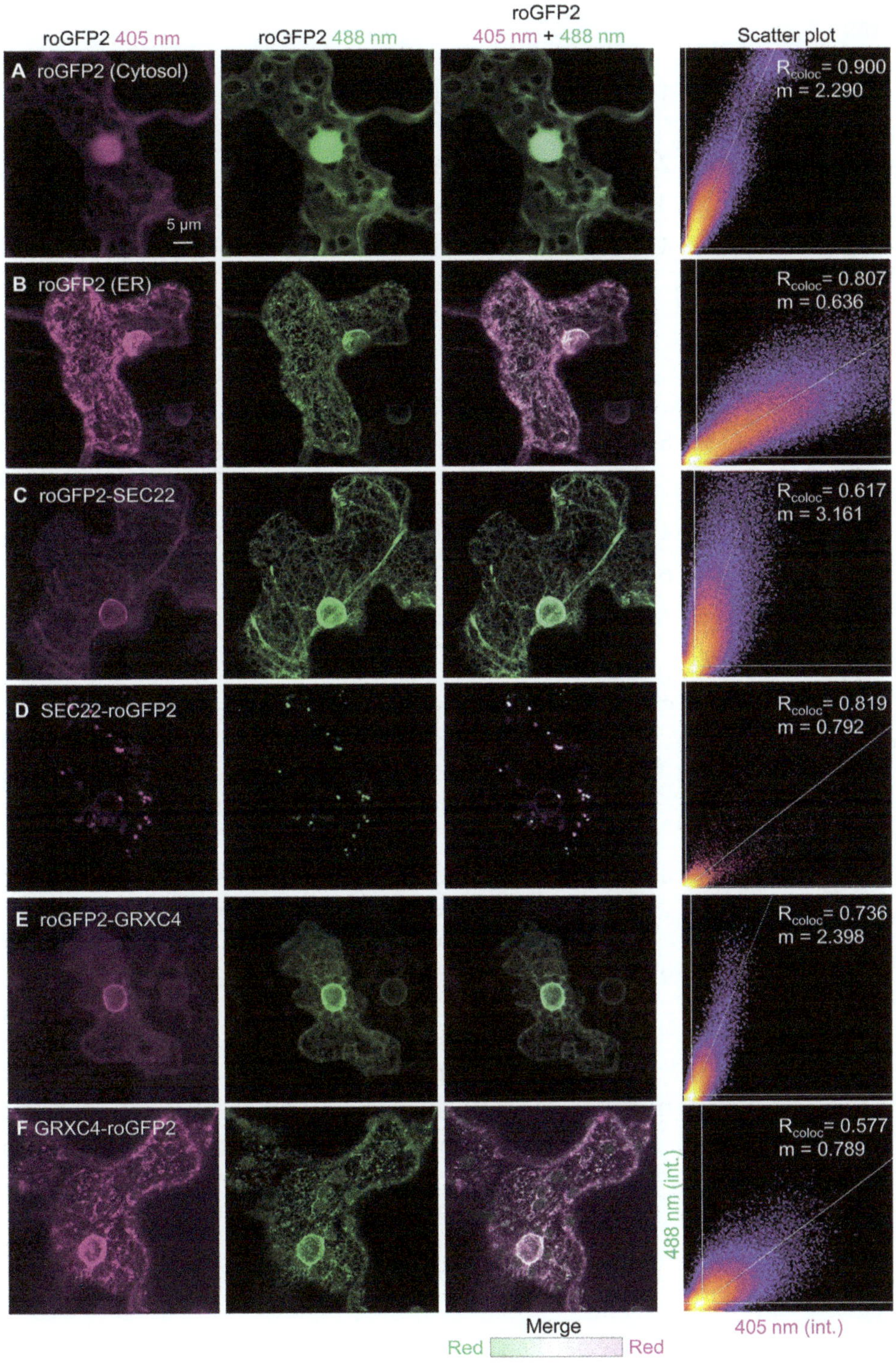


**Fig. 2** Redox-based topology analysis (ReTA) of type II ER membrane proteins in epidermal cells of tobacco leaves. (**a, b**) Soluble roGFP2 in either the cytosol (**a**) or the ER lumen (**b**). (**c, d**) Control constructs with N- (**c**) and C-terminal (**d**) fusion of roGFP2 to SEC22. (**e, f**) Fusion of roGFP2 to the N- (**e**) and the C-terminus (**f**) of glutaredoxin C4 (GRXC4). In all cases, the fluorescence signal was collected at 505–530 nm after excitation at either 405 nm (magenta) or 488 nm (green). To display the ER network, images are shown

## 4 Notes

1. Transient transformation can be also done with *N. benthamiana*.
2. Other *Agrobacterium* strains work as well for transformation.
3. All plasmids are available upon request from the authors' lab.
4. C-terminally tagged type II proteins frequently accumulate in the Golgi (for examples *see* Fig. 2; [14, 15]). Given that the orientation of such proteins is not altered during transfer from the ER to the Golgi and because the lumen of Golgi stacks is also oxidizing, Golgi localization does not impede the binary readout for the protein orientation within the membrane.
5. Fusion of roGFP2 to the N-terminus of a putative ER membrane protein may potentially interfere with correct interpretation of an N-terminal signal peptide. In exceptional cases, the fusion protein may then localize in the cytosol and the nucleus.
6. roGFP2 can also be fused to the C-termini of truncated proteins. For multispanning, roGFP2 may swap sides relative to the membrane if transmembrane domains are deleted stepwise [12].
7. ImageJ is sufficient to open and process Zeiss files (*.lsm). But it is advised to install a plugin such as Bio-formats (http://www.openmicroscopy.org/bio-formats/) or use Fiji to read file formats from other imaging platforms, such as Nikon (*.nd2) or Olympus (*.oib).
8. For high-efficiency cloning and plasmid propagation of the destination clones, used for LR reactions, transform Top10 cells, when the transformation with DH5α is not efficient.
9. For confirmation of the transformation of the pDONR201/207 entry vectors via colony PCR, use a forward primer that is specific for the pDONR201/207 T-DNA and as reverse primer that binds to the gene of interest.
10. For appropriate adjustments of laser power, display the histograms showing the fluorescence intensity distribution for fully reduced and fully oxidized roGFP2 and ensure that the main peaks in the histogram are not clipped [16].

**Fig. 2** (continued) as maximum intensity projections of z-stacks. The 405 nm + 488 nm merge image provides a binary readout for the orientation of roGFP2 relative to the ER membrane (green, cytosolic; magenta/pink, ER lumen). Formal colocalization analysis of these images for the 405 nm and 488 nm intensity values of all pixels in the image results in scatterplots with gradients $m > 1$ (reduced) of $m < 1$ (oxidized)

11. To avoid saturation, perform the brightness and contrast adjustments to the channel with the highest signal first. Then propagate this display range to the other channels of the image.
12. Plotting the intensity values of each pixel for the two channels against each other results in a scatterplot with a cloud of points color-coded depending on their frequency within the respective intensity values, from white/yellow indicating high pixel frequency to purple/blue for intensity values with lower frequencies. Since a dual-channel image is analyzed for fluorescence intensity values from the same pixel, the intercept of the linear regression fit should be close to zero. Provided the full dynamic range of roGFP2 is used with appropriate instrument settings, the gradient (m) will be significantly below 1 for oxidized roGFP2 and significantly above 1 for reduced roGFP2. Because by definition both fluorescence signals should originate from the same pixel or voxel, respectively, Pearson's correlation coefficient nominally should be close to 1. Autofluorescence excited with the respective laser lines might generate a significant noise level that ultimately would lower the Pearson coefficient.

## Acknowledgments

Funding by the DFG and the University of Bonn is gratefully acknowledged.

### References

1. Krogh A, Larsson B, Von Heijne G, Sonnhammer ELL (2001) Predicting transmembrane protein topology with a hidden Markov model: application to complete genomes. J Mol Biol 305:567–580. https://doi.org/10.1006/jmbi.2000.4315
2. Von Heijne G (2006) Membrane-protein topology. Nat Rev Mol Cell Biol 7:909–918. https://doi.org/10.1038/nrm2063
3. Aviram N, Schuldiner M (2017) Targeting and translocation of proteins to the endoplasmic reticulum at a glance. J Cell Sci 130:4079–4085. https://doi.org/10.1242/jcs.204396
4. Mehlhorn DG, Asseck LY, Grefen C (2021) Looking for a safe haven: tail-anchored proteins and their membrane insertion pathways. Plant Physiol 187:1916–1928. https://doi.org/10.1093/plphys/kiab298
5. Von Heijne G (1986) The distribution of positively charged residues in bacterial inner membrane proteins correlates with the transmembrane topology. EMBO J 5:3021–3027. https://doi.org/10.1002/j.1460-2075.1986.tb04601.x
6. Lorenz H, Hailey DW, Wunder C, Lippincott-Schwartz J (2006) The fluorescence protease protection (FPP) assay to determine protein localization and membrane topology. Nat Protoc 1:276–279. https://doi.org/10.1038/nprot.2006.42
7. Zamyatnin AA, Solovyev AG, Bozhkov PV et al (2006) Assessment of the integral membrane protein topology in living cells. Plant J 46:145–154. https://doi.org/10.1111/j.1365-313X.2006.02674.x
8. Swarup R, Kargul J, Marchant A et al (2004) Structure-function analysis of the presumptive Arabidopsis auxin permease AUX1. Plant Cell 16:3069–3083. https://doi.org/10.1105/tpc.104.024737
9. Hanson GT, Aggeler R, Oglesbee D et al (2004) Investigating mitochondrial redox potential with redox-sensitive green fluorescent protein indicators. J Biol Chem 279:

13044–13053. https://doi.org/10.1074/jbc.M312846200

10. Meyer AJ, Dick TP (2010) Fluorescent protein-based redox probes. Antioxid Redox Signal 13:621–650. https://doi.org/10.1089/ars.2009.2948
11. Meyer AJ, Brach T, Marty L et al (2007) Redox-sensitive GFP in *Arabidopsis thaliana* is a quantitative biosensor for the redox potential of the cellular glutathione redox buffer. Plant J 52:973–986. https://doi.org/10.1111/j.1365-313X.2007.03280.x
12. Brach T, Soyk S, Müller C et al (2009) Non-invasive topology analysis of membrane proteins in the secretory pathway. Plant J 57:534–541. https://doi.org/10.1111/j.1365-313X.2008.03704.x
13. Ugalde JM, Aller I, Kudrjasova L et al (2022) Endoplasmic reticulum oxidoreductin provides resilience against reductive stress and hypoxic conditions by mediating luminal redox dynamics. Plant Cell 34:4007–4027. https://doi.org/10.1093/plcell/koac202
14. Schlösser M, Moseler A, Bodnar Y, et al (2023) Localization of four class I glutaredoxins in the cytosol and the secretory pathway and characterization of their biochemical diversification. bioRxiv preprint. https://doi.org/10.1101/2023.09.01.555924
15. Attacha S, Solbach D, Bela K et al (2017) Glutathione peroxidase-like enzymes cover five distinct cell compartments and membrane surfaces in *Arabidopsis thaliana*. Plant Cell Environ 40:1281–1295. https://doi.org/10.1111/pce.12919
16. Meyer AJ, Brach T (2009) Dynamic redox measurements with redox-sensitive GFP in plants by confocal laser scanning microscopy. In: Pfannschmidt T (ed) Plant signal transduction. Humana Press, Totowa, pp 93–107

# Chapter 29

# Organ-Specific Microsomes from Dark-Grown Hypocotyls of *Arabidopsis thaliana*

**Seinab Noura, Jürgen Kleine-Vehn, and Sascha Waidmann**


## Abstract

In this book chapter, we present a method for microsome isolation from the hypocotyl tissue of dark-grown *Arabidopsis thaliana*. Microsomes are heterogeneous, vesicle-like membranes, which are, not exclusively, derived but enriched with membranes of the endoplasmic reticulum (ER). Here, we describe the experimental setup, including sample preparation, homogenization, differential centrifugation steps, and quality control measures after microsome isolation.


**Key words** Microsomes, *Arabidopsis thaliana*, Dark-grown hypocotyl, Endoplasmic reticulum (ER), Membrane-bound protein, Ultracentrifugation

## 1 Introduction

Microsomes are membrane fragments that are formed during tissue homogenization. Microsomes are composed mainly of endoplasmic reticulum (ER) membranes but also display membranes of other organelles, such as the Golgi apparatus, tonoplast, and plasma membranes (PMs). As these endomembranes break into smaller fragments, they have the ability to reseal, creating enclosed vesicle-like structures [1]. Differential centrifugation enables selective isolation and enrichment of specific cellular compartments. In the process of ER-enriched microsome isolation, two centrifugation steps are performed. In the first, slow, centrifugation step, the nuclei, chloroplasts, and mitochondria are removed. The remaining membranes are further separated through ultracentrifugation. Among others, microsomal protein fractions can be used to investigate the function of membrane-bound proteins, protein–protein interactions, and membrane transport processes [2].

In this book chapter, we present a method for isolating organ-specific microsomes from dark-grown hypocotyls of *Arabidopsis thaliana*. We exemplify the usage of this method to enrich

Verena Kriechbaumer (ed.), *The Plant Endoplasmic Reticulum: Methods and Protocols*, Methods in Molecular Biology, vol. 2772, https://doi.org/10.1007/978-1-0716-3710-4_29,

membranes of ER-localized, auxin transport facilitator PILS3 [3, 4]. These microsomes could be subsequently used for microsomal auxin transport assays and/or other procedures. This chapter outlines the various steps of microsome isolation, starting from sample preparation to Western blot analysis.

## 2 Material

### 2.1 Plant Growth Medium, Seedling Growth, and Plant Hypocotyl Harvesting

1. *Arabidopsis thaliana* Col-0 (WT), 35S::GFP-PILS3 [3, 4].
2. Plant growth medium: 2.3 g/L Murashige and Skoog (M0221.0050, Duchefa Biochemie) salt, 0.5 g/L MES (4256.3, Roth), 10 g/L sucrose (S0809.5000, Duchefa Biochemie), 8 g/L plant agar (P1001.0500, Duchefa Biochemie); adjust pH to 5.9, autoclave, and store at RT.
3. Sterile square petri dishes (120 × 120 × 17 mm).
4. Breathable paper tape.
5. 70% and 100% ethanol.
6. 0.1% (w/v) agarose in dd$H_2O$: autoclave and store at RT.
7. 200 μL pipette tips with a wide bore tip.
8. Mortar and pestle.
9. Razor blade or scalpel.
10. Ice box.
11. Ice.

### 2.2 Microsome Isolation of Dark-Grown Hypocotyl from Arabidopsis thaliana

1. Homogenization buffer (20 mM HEPES pH 7.5, 200 mM sorbitol, 1 mM DTT and protease inhibitor mix (11873580001, Roche)); store on ice.
2. Miracloth (475855-1, Calbiochem).
3. Small funnel.
4. 15 mL tubes.
5. 50 mL polypropylene copolymer centrifuge tubes.
6. Ultracentrifuge.
7. Ultracentrifuge tubes.

### 2.3 Immunoblot

1. Specific antibodies, such as α-GFP antibody (anti-GFP antibody, 11814600014, Roche), α-V-ATPase antibody (epsilon subunit of tonoplast $H^+$-ATPase, AS07 213, Agrisera, Sweden), α-$H^+$-ATPase antibody (plasma membrane $H^+$-ATPase , AS07 260, Agrisera, Sweden) and α-RbcL antibody (rubisco large subunit, form I, AS03 037, Agrisera, Sweden)), were used for the detection of transgenic GFP-PILS3 reporter proteins, vacuolar as well as plasma membranes, and rubisco.

2. 10× TBS: 1.5 M NaCl, 100 mM Tris–HCl, pH 8.0. Autoclave and store at RT.
3. TBS-T: 1× TBS + 0.05% Tween 20.
4. Rotator.

## 3 Methods

### 3.1 Plant Material, Growth, and Harvesting

1. Pour 50 mL of media into sterile square petri dishes.
2. Let the plates cool at room temperature for about an hour or until the agar solidifies.
3. Sterilize the seeds in a 1.5 mL Eppendorf tube by adding 700 μL of 70% ethanol. Remove the ethanol and add 700 μL of 100% ethanol.
4. Remove 100% ethanol completely and allow the seeds to dry in a laminar flow hood.
5. Mix the seeds with sterile 0.1% agarose. The volume of agarose should be adjusted to the amount of seeds (roughly four times more).
6. Sow the seeds by pipetting them with a wide bore tip in a horizontal line on the plate. Two separated lines of seeds per plate were sown, with an "agar-seed" volume of approximately 200 μL.
7. Ensure that the seeds containing agar is fully dried before placing the lid onto the plates.
8. Seal the plates with air-permeable breathable paper tape that permits slight aeration while preventing desiccation.
9. Stratify the seeds for 2 days by keeping the plates in darkness at 4 °C.
10. Place the plates in a plant growth chamber, exposing them to light for a duration of around 6 h. Subsequently, wrap the plates with aluminum foil and transfer them to a black box located inside a plant growth chamber.
11. Seedlings are grown in darkness in a plant growth chamber for 3 days.
12. To harvest plant material, carefully cut at the hypocotyl to root junction (refer to Fig. 1). Use a razor blade or scalpel to excise the hypocotyl from a petri dish and collect the tissue using flat tweezers.

### 3.2 Isolation of Total Protein

1. Place the harvested plant material (around 3 g) immediately into the chilled mortar containing 3 mL of cold homogenization buffer (*see* **Notes 1** and **2**).

**Fig. 1** Example of hypocotyl harvesting at 3 days after germination

**Fig. 2** Homogenate after grinding of the hypocotyls using mortar and pestle on ice

2. Grind the plant material thoroughly for at least 3 min (refer to Fig. 2).
3. Transfer homogenate (T) into a pre-cooled 15 mL tube using a funnel lined with two layers of Miracloth.
4. Rinse mortar and pestle with an additional 1.5 mL of homogenization buffer.
5. Squeeze homogenate through the layers.

### *3.3 Isolation of Microsomal Protein*

1. Place the filtered homogenate (T) into a pre-cooled 50 mL polypropylene copolymer centrifuge tube.

**Fig. 3** (**a**) Ultracentrifuge tube containing homogenate separated into supernatant ($S_{38}$) and the pellet ($P_{38}$) after 38,000 rpm spin. (**b**) Resuspended pellet ($P_{38}$)

2. The homogenate (T) is centrifuged for 10 min at 8000× *g*, 4 °C. This will result in the separation of a supernatant fraction ($S_8$) and a pellet fraction ($P_8$) (*see* **Note 3**).
3. In the meantime, pre-cool the ultracentrifuge to 4 °C.
4. Transfer the supernatant $S_8$ into pre-cooled ultracentrifuge tubes.
5. The supernatant fraction ($S_8$) is subsequently centrifuged at 38,000 rpm using a 60 Ti rotor (145,134× *g*) for a duration of 30 min at 4 °C (refer to Fig. 3a).
6. Discard the supernatant (*see* **Note 3**) and proceed to resuspend the pellet in 200 μL of homogenization buffer (*see* **Note 4**). It is recommended to begin with a smaller volume and gradually increase it until reaching the final desired volume (refer to Fig. 4).
7. Keep the samples on ice thereafter.

### *3.4 Western Blot*

1. To assess the success of the isolation of the microsomal protein from soluble/cytosolic proteins (refer to Fig. 4), the protein samples T, $S_8$, $S_{38}$, and M were subjected to an SDS–PAGE [5] using a 10% gel, followed by immunoblot analysis with the antibodies listed in Subheading 2.3.
2. To ensure the effectiveness of your microsomal isolation, it is important to include subcellular compartment marker proteins along with the protein of interest. In this protocol, antibodies against various microsomal marker protein were used such as epsilon subunit of tonoplast $H^+$-ATPase, plasma membrane $H^+$-ATPase, and rubisco large subunit, form I.

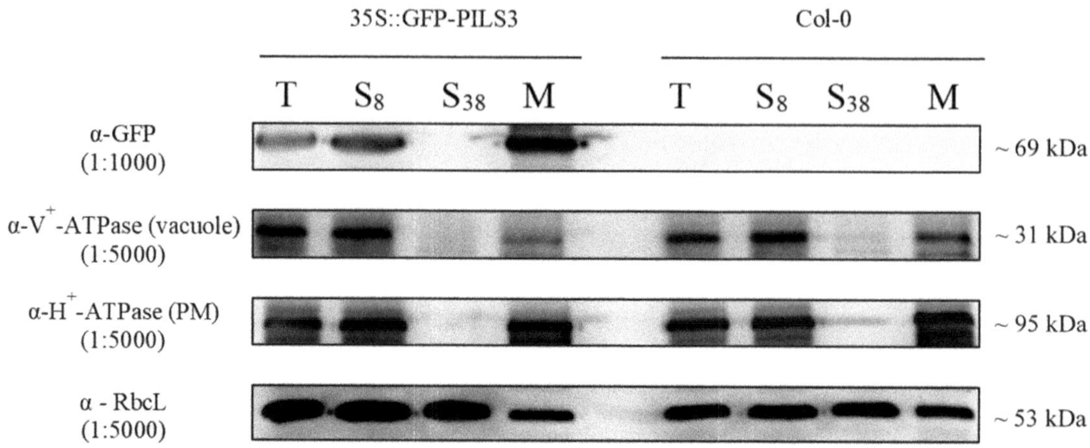


**Fig. 4** Western blot analysis after microsomal isolation by differential centrifugation from dark-grown hypocotyls of *Arabidopsis thaliana* lines 35S::GFP-PILS3 and Col-0. Antibodies used were against GFP and specific subcellular marker proteins. As expected, there is a particular enrichment of ER resident proteins, such as PILS3-GFP, in the microsomal protein (M) fraction when compared to $S_8$, $S_{38}$, and the total protein (T). Please note that the volume of the microsomal protein (M) is tenfold less than $S_8$, $S_{38}$, and T. In this case, 30 μL of the collected $S_8$, $S_{38}$, and T was loaded for this Western blot analysis, whereas only 3 μL of the microsomal protein (M) was used

## 4 Notes

1. The homogenization buffer volume can be adjusted proportionally to the plant material mass.
2. DTT and protease inhibitor mix should be always freshly added.
3. Preserve approximately 50 μL of your sample for the subsequent utilization in a Western blot analysis.
4. Prior to resuspending the pellet, it is possible to observe a white residue along the inner wall of the tube and the surface of the supernatant. It is advisable to remove this residue by gently adding 1 ml of homogenization buffer along the tube wall and subsequently discarding it. Exercise caution to avoid contact with the pellet during this process.

## Acknowledgments

We are grateful to our team members for their helpful discussions. This work was supported by the German Research Foundation (DFG; 470007283 and CIBSS – EXC-2189 – Project ID 390939984 to J.K.-V.) and the Austrian Science Fund (FWF; P29754 to J.K-V., P33497 to S.W.).

## References

1. LaMontagne ED, Collins CA, Peck SC, Heese A (2016) Isolation of microsomal membrane proteins from Arabidopsis thaliana. Curr Protoc Plant Biol 1(1):217–234
2. Peng F, Zhan X, Li MY, Fang F, Li G, Li C, Zhang PF, Chen Z (2012) Proteomic and bioinformatics analyses of mouse liver microsomes. Int J Proteom 2012:832569
3. Barbez E, Kubes M, Rolcik J, Beziat C, Pencik A, Wang B, Rosquete MR, Zhu J, Dobrev PI, Lee Y, Zazimalova E, Petrasek J, Geisler M, Friml J, Kleine-Vehn J (2012) A novel putative auxin carrier family regulates intracellular auxin homeostasis in plants. Nature 485:119–122
4. Béziat C, Barbez E, Feraru MI, Lucyshyn D, Kleine-Vehn J (2017) Light triggers PILS-dependent reduction in nuclear auxin signalling for growth transition. Nature Plants 17(8):1–9
5. Gallagher SR (2012) One-dimensional SDS gel electrophoresis of proteins. Curr Protoc Mol Biol 97:10.2A.1–10.2A.44

# Chapter 30

# Studying Secretory Protein Synthesis in *Nicotiana benthamiana* Protoplasts

**Jing An and Jurgen Denecke**

## Abstract

Transient gene expression in plant protoplasts facilitates the analysis of hybrid genes in a fast and reproducible manner. The technique is particularly powerful when studying basic conserved biochemical processes including de novo protein synthesis, modification, assembly, transport, and turnover. Unlike individual plants, protoplast suspensions can be divided into almost identical aliquots, allowing the analysis of independent variables with uncertainties restricted to minor pipetting errors/variations. Using the examples of protein secretion and ER retention, we describe the most advanced working practice of routinely preparing, electroporating, and analyzing *Nicotiana benthamiana* protoplasts. A single batch of electroporation-competent protoplasts permits up to 30 individual transfections. This is ideal to assess the influence of independent variables, such as point mutations, deletions or fusions, or the influence of a co-expressed effector gene in dose–response studies.

**Key words** Protoplast, Electroporation, Transfection, Dose–response, Expression, Secretion, Protein targeting, Protein turnover

## 1 Introduction

Plant protoplasts have been instrumental in the study of protein secretion in plants [1] as they are non-adherent, form homogeneous cell suspensions, and readily release secreted proteins to the culture medium from which they can be separated routinely. However, their research potential is much broader, as long as the processes under investigation are not cell-type dependent or require intact plant organs and/or highly specific growth conditions. Tobacco protoplasts have been used for the detailed analysis of protein sorting signals mediating ER retention [2, 3] and vacuolar sorting [4, 5], but also for the study of protein assembly and protein–protein interactions [6] and analysis of co-expressed effectors [3, 7–9].

Verena Kriechbaumer (ed.), *The Plant Endoplasmic Reticulum: Methods and Protocols*, Methods in Molecular Biology, vol. 2772,
https://doi.org/10.1007/978-1-0716-3710-4_30,

The use of dual and triple expression plasmids also allows the comparison of a modified test gene against a reference gene and to carry out dose–response assays [10–12]. These advances pave the way for robust experimental strategies to study more enigmatic steps in gene expression, such as the influence of the 5′untranslated end on mRNA translation efficiency. Also the technique can be ideal to investigate the potential role of translational pausing on the efficiency of protein folding as well as the full range of post-translational events contributing to overall gene expression such as modification, assembly, transport, and turnover. Regretfully, the advantages of plant protoplasts are not widely recognized, which is in part due to lack of technical knowledge in how to isolate large quantities of protoplasts routinely and how to carry out reproducible transfections, culturing and harvesting cells, and medium for subsequent analysis.

ER retention of soluble proteins is an excellent example of a process that can be easily studied in protoplasts by comparing secreted with cellular proteins. Since the ER export step is the rate-limiting factor in constitutive secretion [13], secreted proteins in transit to the plasma membrane and beyond are mainly present in the ER. A strong shift in the steady-state distribution can be observed when a secreted protein is supplemented with an ER retention signal [3] and even further when the K/HDEL receptor is co-expressed [11, 12], allowing functional analysis of receptor mutants. This chapter aims at providing a comprehensive guide to the latest transient expression protocol for *Nicotiana benthamiana* protoplasts to allow the wider research community to take advantage of this system.

## 2 Materials

All solutions are prepared with double deionized water (15 MΩ-cm at 25 °C), and either autoclaved or filter-sterilized at room temperature, as specified. Long-term storage is at 4 °C unless indicated otherwise.

### *2.1 Plant Growth Stock Solutions*

1. Macroelements (10×): Add 4000 mL of water to a graded 5 L beaker placed on a magnet stirrer. Weigh out the indicated amounts of the following macroelements and add individually $NH_4NO_3$ (66 g), $KNO_3$ (76 g), $CaCl_2 \cdot 2H_2O$ (17.6 g), $MgSO_4 \cdot 7H_2O$ (14.8 g), and $KH_2PO_4$ (6.8 g). When fully dissolved, bring up to 5 liters and continue to stir for another 5 min. Autoclave in aliquots for 20 min at 120°C and store at 4°C.
2. Fe EDTA (100×): Add $FeSO_4 \cdot 7H_2O$ (2.785 g) and $Na_2$ EDTA (3.36 g) to 950 mL of hot water (>50 °C), dissolve fully, and bring up to 1 L. Autoclave in aliquots for 20 min at 120 °C and store at 4 °C.

3. Microelements (1000×): Add $ZnSO_4 \cdot 4H_2O$ (8.6 g), $H_3BO_3$ (6.2 g), $MnSO_4 \cdot 4H_2O$ (22.3 g), $CuSO_4 \cdot 5H_2O$ (0.025 g), KI (0.83 g), $CoCl_2 \cdot 6H_2O$ (0.025 g), and $Na_2MoO_4 \cdot 2H_2O$ (0.25 g) to 950 mL water, dissolve fully, and bring up to 1 L. Autoclave in aliquots for 20 min at 120 °C and store at 4 °C.
4. MS2 medium for plant growth: Combine 4000 mL of water with 500 mL macroelements (10×), 50 mL Fe EDTA (100×), and 5 mL microelements (1000×), dissolve 100 g sucrose and 2.5 g 2-(N-morpholino)ethanesulfonic acid (MES), add 1.25 mL KOH (5 M), and bring up to 5 L when all solids are dissolved. Check pH to be 5.7. Then divide equally over six 1 L Duran flasks (approximately 833 mL in each), add 3.6 g Duchefa plant agar, and autoclave for 20 min at 120 °C. Allow to cool to 45–50 °C before pouring into glass jars (*see* **Note 1**).

### 2.2 Protoplast Stock Solutions

1. TEX buffer: 10× macroelement 100 mL/L, 100× Fe EDTA 10 mL/L, 1000× micro 1 mL/L, 500 mg/L MES, 750 mg/L $CaCl_2$ ($2H_2O$), 0.4 M sucrose (13.7%, 137 g/L), and 250 μL 5 M KOH per liter to adjust to pH 5.7. When fully dissolved, bring up to 1 L using a volumetric flask. Filter-sterilize through 0.2 μm filter and store in 200 mL aliquots at 4 °C.
2. Leaf digestion mix concentrate (40×): Add 2 g macerozyme R10 and 4 g cellulase R10 to 100 mL of TEX buffer, and allow to resuspend under constant stirring using a magnet stirrer for 2 h. Divide into two equal portions of 50 mL using a 50 mL Falcon tube, and spin for 20 min at 3000 g to remove insoluble fractions. Filter-sterilize the cleared supernatant through 0.2 μm filter and freeze 2.5 mL aliquots in 50 mL Falcon tubes at −80 °C for long-term storage.
3. Electroporation buffer: 0.4 M sucrose (13.7%, 137 g/L), 2.4 g/L HEPES, 6 g/L KCl, 600 mg/L $CaCl_2$ ($2H_2O$). Adjust to pH 7.2 with 5 M KOH. Filter-sterilize through 0.2 μm filter and store in 200 mL aliquots at 4 °C.

### 2.3 Buffers for Protein Extraction and Enzymatic Assays

1. GUS extraction buffer: 50 mM sodium phosphate buffer (pH 7.0), 0.10 mM $Na_2EDTA$, 0.1% sodium lauryl sarcosine, 0.1% Triton X-100, and 10 mM β-mercaptoethanol.
2. GUS dilution buffer: 50 mM sodium phosphate buffer (pH 7.0) and 0.1% Triton X-100.
3. GUS reaction buffer: 50 mM sodium phosphate buffer (pH 7.0), 0.1% Triton X-100, 2 mM 4-nitrophenyl- β-D-glucopyranosiduronic acid, and 10 mM β-mercaptoethanol.
4. GUS stop buffer: 2.5 M 2-amino-2-methyl propanediol.
5. Amy extraction buffer: 50 mM sodium malate, 50 mM sodium chloride, 2 μM calcium chloride, and 0.02% sodium azide.

6. Amy assay reagent: blocked P-nitrophenyl maltoheptaoside (BPNPG7, 54.5 mg), glucoamylase (100 U at pH 5.2), α-glucosidase (100 U at pH 5.2), dissolved in 10 mL distilled sterile water, aliquoted (0.5 mL), and stored at −80 °C
7. Amy stop buffer: 1% Trizma base

### 2.4 Equipment, Materials, and Consumables

Petri dishes, 9 cm diameter.

Petri dishes, 5 cm diameter, extra high.

Forceps and scalpels.

100 micron mesh filter.

1.5 mL Eppendorf tubes.

Conical tubes, 10 mL.

96-well microtiter plates.

Cuvettes.

Alcohol burner.

Weck jars.

Electroporation machine.

Manually controlled peristaltic pump.

## 3 Methods

All sterile work with plants and plant protoplasts is carried out in a laminar flow hood, while protein extraction and enzymatic assays can be carried out at the bench.

### 3.1 Maintaining an In Vitro Plant Forest

1. Surface-sterilize *N. benthamiana* seeds by dispensing 20 microliters of seeds (100–200 seeds) in an Eppendorf tube, adding 1 mL of 20% bleach (sodium hypochloride) and supplement with a single drop (20 microliters) of Tween 20. Invert the tube several times to fully dissolve the viscous Tween 20 and then place on a slow spinning rotary wheel for up to (but not exceeding) 30 min. Wash the seeds 4× by removing all the liquid from the seeds and replacing with 1 mL of sterile distilled water each time, followed by inverting the tubes and letting the seeds settle to the bottom (*see* **Note 2**). At the last wash, the water seed mixture is decanted into a sterile 9 cm Petri dish and further distilled sterile water is added. Seeds are individually sucked up with a 1 mL tip using a P1000 micropipette and expelled onto the MS2 medium surface in a Weck jar, approximately 50 seeds per jar spaced at least 0.5 cm apart (*see* **Note 3**)
2. 2–3-week-old seedlings are transferred to grow individually in the center of a Weck jar. Lift up each seedling by placing forceps underneath the two cotyledons without touching/squeezing

the hypocotyl, and gently pushing upward against the cotyledons. In a single motion, place the recovered seedling into the center of a fresh Weck jar and pull the roots and part of the hypocotyl underneath the agar. 4–6 weeks later, the plants are big enough for leaf harvesting.

### 3.2 Preparation of Protoplasts

1. A 2.5 mL aliquot of leaf digestion mix concentrate (40×) is dispersed in 100 mL TEX buffer and 16 individual plates containing approximately 6 mL of 1× digestion mix. Gently agitate the plates so that the liquid covers the entire surface of the plate.
2. Leaves are removed individually from the Weck jars and cut gently on the lower leaf surface every 1–2 mm (without cutting through the whole surface). This is done by balancing the scalpel so that a fraction of its own weight exerts the pressure on the leaf surface while gliding along, effectively scratching the surface. The leaf is held in place with a pair of forceps at the midnerve to minimize the damage at the leaf surface. An alternative easier method is to punch holes with a small needle bed (*see* **Note 4**, Fig. 1a, b).

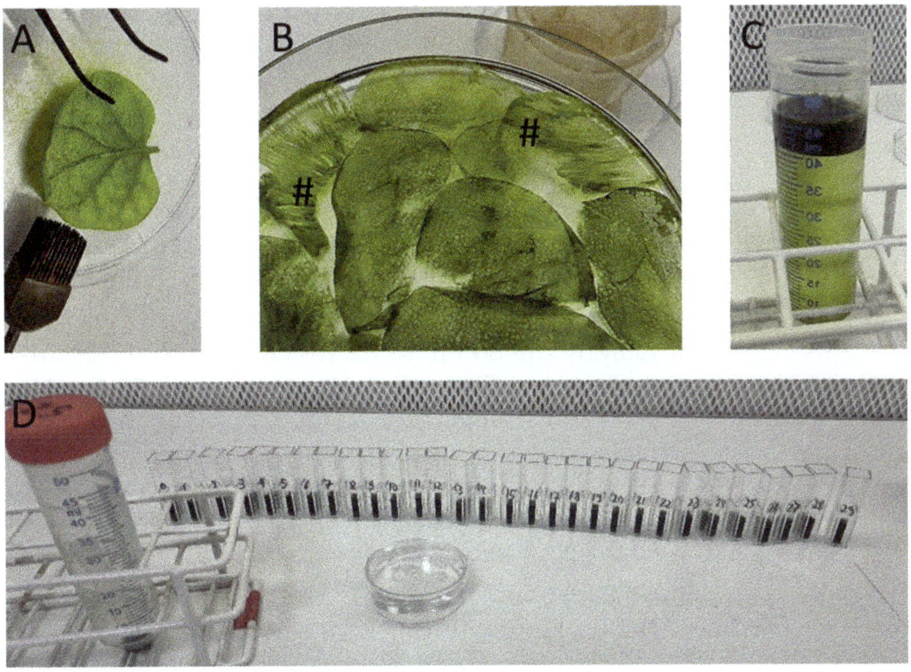


**Fig. 1** (**a**) Needle bed puncher with perforated leaf and forceps resting on a 9 cm Petri dish. (**b**) Overnight digested leaves prior to agitation, prepared the night before with either scalpel or forceps (#) or the needle bed puncher (all other leaves). (**c**) Floating protoplast band after filtration and spinning. 5 mL of floating cells corresponds to 70 million protoplasts. (**d**) Concentrated protoplasts aliquoted (500 μL) into cuvettes for electroporation, ready to receive plasmid DNA mixes in 100 μL of electroporation buffer

3. The midnerve is removed and the two halves of the leaf are transferred to a Petri dish containing digestion mix (1×), with the cut side facing the liquid surface. It is possible to cut leaves and cover the surface of the liquid in approximately 5 min (*see* **Note 5**).
4. The plates are incubated in the dark overnight (16 h) at room temperature, and 30 min before use, it is recommended to gently agitate the plates to release protoplasts from the cuticula. Each plate should provide $5 \times 10^6$ protoplasts after overnight digestion.
5. The digestion mix is filtered through a 100 micron nylon filter and the filter can be washed with electroporation buffer (EB). This will release further protoplasts from the tissue remnants and provide a first step in the adaptation to the new buffer (*see* **Note 6**).
6. The filtered protoplast suspension is then centrifuged in Falcon tubes (50 mL) for 15 min at 100 g at room temperature in a swing-out rotor and stopped without brake. Live intact protoplasts will form a layer on the surface of the solution (Fig. 1c), whereas cell debris will form a pellet or stay in solution.
7. A long-sterilized Pasteur pipette is connected to a peristaltic pump which can pump up to 1 L per minute, and the Pasteur pipette is inserted through the floating cell layer. To avoid that many protoplasts adhere to the pipette and move down with it, a "window" is first created by pushing the cells from the center outward in all directions. As soon as the pipette is pushed through this window, start the peristaltic pump and the window will open further. The pellet and the underlying medium are removed until the band of living protoplasts reaches the bottom. It is important to slow down the pump rate in advance (*see* **Note** 7).
8. Quickly add 30 mL of EB to wash down protoplasts adhering to the sides of the tubes and to resuspend the floating band of protoplasts. Spin again at 100 g for 10 min. Remove the underlying solution as described under **step** 7 above and repeat this procedure twice. The solution below the protoplasts should become clear and there should be hardly any pellet visible after three washes (*see* **Note 8**).
9. After the final wash, protoplasts are resuspended typically by adding 12–13 mL EB, yielding slightly more than 15 mL concentrated protoplast suspension, allowing 30 individual electroporations (*see* **Note 9**). This suspension can be stored for up to 1 h, but the sooner the cells are transfected, the better the final result.

### 3.3 Electroporation

1. Cut the tip of a 1 mL pipette tip for a P1000 micropipette with a sterile scalpel to remove approximately 5 mm, to achieve a wider opening (*see* **Note 10**). Aspirate 500 microliters of homogeneous protoplast suspension and transfer to a 1 mL cuvette. After pipetting five aliquots, gently stir the remaining protoplast suspension to avoid inhomogeneity from floating protoplasts, and continue with five more aliquots. Proceed until you have 30 identical aliquots (Fig. 1d).
2. Dilute plasmid DNA in a final volume of 100 microliters using EB and transfer to the appropriate cuvette. Do not exceed 25 microliters of plasmid volume as this will significantly dilute the osmolarity of the final EB–protoplast mix. Shake the cuvette to mix diluted plasmid DNA with the protoplast suspension. Mock transfection is done by pipetting 100 microliters of EB alone. When mixing a cargo encoding plasmid with an effector plasmid, use a constant amount of cargo plasmid in each transfection. The control transfection is with cargo plasmid alone. The sum of cargo and effector plasmids should not exceed 25 microliters (*see* **Note 11**).
3. Plasmid DNA should be allowed to adhere to the protoplasts for a minimum of 5 min. When protoplasts are floating up, resuspend again by gentle shaking the cuvette. During this time, prepare a shallow jar with 100 mL distilled water, a second jar with ethanol (99.8%), and a Petri dish with EB, and light a small waking flame (an alcohol candle suffices).
4. To electroporate, we use a complete discharge of a 910 microFarad capacitor charged with 160 volts (*see* **Note 12**). Immediately before electroporation, shake the cuvette briefly to achieve a homogeneous protoplast suspension, insert the electrode into the cuvette, press the switch for 3 s to fully discharge the capacitor, gently remove the electrode, and keep the cuvette standing still for at least 15 min, and up to 1 h (*see* **Note 13**).
5. Wash the electrode first by dipping in distilled water to remove any EB buffer with plasmid from the previous electroporation, then dehydrate by dipping in ethanol, shake off excess ethanol and then ignite the alcohol on the electrode with the waking flame, allow 15 s of laminar airflow to ensure all ethanol has burned/evaporated, and then dip in EB buffer to cool down the electrode-plates so they are ready for the next electroporation. This cleaning cycle takes just under 1 min.
6. Repeat **steps 4** and **5** until all transfections are complete. When observing the individual cuvettes, the first electroporated should show a floating band of protoplasts on the liquid surface.

**Fig. 2** After mixing, place the dish on the lid to increase the liquid depth on one side for smooth aspiration of the 500 μL, prior to transfer of the remaining 2 mL to the 10 mL conical tube

7. In the same order in which they have been electroporated, rinse each cuvette with 1 mL of TEX buffer, and pour the suspension into a 5 cm Petri dish. Wash again with 1 mL of TEX buffer, and pour into the same Petri dish, this time making sure most of the liquid is removed from the cuvette (*see* **Note 14**). The resulting protoplast suspension has a reproducible volume of 2.5 mL.
8. Incubate the closed Petri dishes in the dark at room temperature for 20–24 h and do not move the dishes during this incubation period (*see* **Note 15**).

### 3.4 Measurement of the Transfection Efficiency

1. When using dual or triple expression plasmids harboring the transfection marker GUS [10], Petri dishes containing the transfected protoplast suspensions are gently agitated to ensure a homogeneous cell suspension, and exactly 500 microliters is sampled (20% of the transfected cell suspension) and mixed with 500 microliters of GUS extraction buffer and placed on ice (*see* **Note 16**, Fig. 2). Ideally, one researcher carries on with the further steps below, while a second researcher proceeds to harvesting cells and medium from the remaining 2 mL of cell suspension (*see* Subheading 3.5).
2. Sonicate the 1 milliliter samples for 5 s at 60% amplitude, making sure that the probe is not touching the walls of the tube and is inserted deep enough into the liquid to avoid surface foam formation. Place back on ice for further steps.
3. Sonicated samples are vortexed for 5 s and centrifuged in a refrigerated microfuge (4 °C) at 16,000 g for 5 min.
4. Dilute 50 microliters of supernatant with 450 microliters of GUS dilution buffer and place on ice.

5. 100 microliters of the diluted sample is mixed with 100 microliters of fresh GUS reaction buffer in a 96-well microtiter plate, cover the surface to avoid evaporation, and incubate at 37 °C for 20 h.
6. Stop the reaction by mixing 80 microliters of GUS stop buffer. Stop reactions in the same order as they were started to ensure incubation times are identical (*see* **Note 17**).
7. Measure the OD at 405 nm using a plate reader. All transfected protoplast samples without GUS plasmids can serve as blanc (use the average to subtract from all raw ODs).
8. Activity in arbitrary units is calculated as ΔOD per hour and per mL original suspension (*see* **Note 18**).

### 3.5 Harvesting Cells and Medium

1. Immediately after sampling the 500 microliters suspension for GUS analysis (*see* Subheading 3.4, **step 1**), the remaining cell suspension (2 mL) is quantitatively transferred to a 10 mL conical tube (*see* **Note 19**). We recommend that this step is done in conjunction with the GUS sampling by two investigators working side-by-side.
2. When all suspensions have been transferred to individual conical tubes, the tubes are briefly shaken to avoid protoplasts adhering to the sides of the tubes, transferred to a swing-out centrifuge, and centrifuged for 5 min at 150 g (zero break).
3. Refine long Pasteur pipettes to achieve 1 cm ends with an outer diameter of 0.3–0.5 mm (*see* **Note 20**, Fig. 3a). This step is crucial because it permits penetration of the floating protoplast band without taking any protoplasts down with it.

**Fig. 3** (**a**) Refined end of the glass pipette. (**b**) Conical tube secured on a firm surface while removing the lid. (**c**) After aspirating 1 mL of clear medium from underneath the floating protoplast band. (**d**) Clear medium, recovered to be dispensed into an Eppendorf tube

4. Prepare a set of numbered Eppendorf tubes in a rack to recover clear culture medium. Insert the first refined pipette into the rubber sucker and leave it next to the rack.
5. While standing next to the centrifuge, carefully lift the first conical tube out of the rotor with both hands, one to hold the tube lid and the other to secure the tube at the level of the floating protoplast band to avoid any agitation. Secure the bottom of the tube on a firm surface such as the edge of the centrifuge, and then carefully open the lid (Fig. 3b).
6. Using your free hand, pick up the rubber bulb with the refined pipette, press out most of the air and keep the finger pressure fully applied, then slide the glass pipette down toward the protoplast surface across the edge of the tube entrance, penetrate the protoplast band in a single motion until the pipette entrance is at least 1 cm below the protoplast band, and then carefully aspirate approximately 1 mL of clean culture medium (*see* **Note 21**). Watch closely as the floating band comes nearer and stop aspirating before protoplasts come too near to the pipette opening (Fig. 3c).
7. Remove the pipette and while holding it, immediately shake the tube gently with your other hand to resuspend the protoplasts. Place the tube in a rack, and do not yet close the lid.
8. Now pipette the clear supernatant (Fig. 3d) in the appropriate tube, discard the pipette, remove the rubber bulb, and insert a fresh pipette so you are ready to start with the next sample at **step 5**.
9. Repeat **steps 5–8** until all medium samples are recovered in the Eppendorf tubes (continue with **step 13** after completing **steps 10–12** first).
10. Add 9 mL of wash buffer (250 mM NaCl) to the tubes with the cell suspensions and then close all lids. Invert the tubes and gently flicker the conical part to remove all liquid from the narrow end and allow efficient mixing with the wash buffer. Repeat this step and insert into the swing-out rotor. Centrifuge for 5 min at 150 g (no break) to reveal a compact pellet and crystal clear supernatant above.
11. Using a normal Pasteur pipette connected to the peristaltic pump set to maximum aspiration speed, remove the supernatant by holding the pipette close to the liquid surface and moving down with it until you reach the pellet (*see* **Note 22**).
12. Close the lids on the tubes with the cell pellets and freeze at –80 °C for subsequent analysis.
13. Centrifuge the Eppendorf tubes with the culture medium in a refrigerated microfuge (4 °C) at 16,000 g for 15 min. The resulting supernatant has the same concentration (1×) as the

original cell suspension. Remove exactly 500 microliters and transfer to new tubes which can also be frozen at −80 °C for subsequent analysis (*see* **Note 23**).

### 3.6 Measurement of Secreted and Cell-Retained α-Amylase

1. Medium samples are thawed and 500 microliters of Amy extraction buffer is added. Vortex thoroughly to ensure proper mixing. The resulting sample is 2× diluted compared to the original cell suspension.
2. Frozen cell pellets are thawed and resuspended with 950 microliters of Amy extraction buffer and transferred to numbered Eppendorf tubes, and kept on ice.
3. Sonicate the resulting 1 milliliter samples for 5 s at 60% amplitude, making sure that the probe is not touching the walls of the tube and is inserted deep enough into the liquid to avoid surface foam formation. Place back on ice for further steps.
4. Sonicated samples are vortexed for 5 s and centrifuged in a refrigerated microfuge (4 °C) at 16,000 g for 15 min. The resulting supernatant is 2× concentrated compared to the original suspension.
5. Perform a pilot assay to determine the ideal conditions for the enzyme assay. If known, select the samples expected to exhibit the highest and lowest enzyme activities together with a blanc (*see* **Note 24**). Pipette 30 microliters of sample into an Eppendorf tube and complete the pilot series in a rack next to the water bath (45 °C). In 15-s intervals, pipette 30 microliters of substrate into the tube with the samples, mix by aspirating up and down three times, and place in the water bath with the lid closed. When the first sample has incubated 5 min, stop the reaction by mixing 150 microliters of stop buffer. Transfer 200 microliters to a 96-well microtiter plate well, and measure the OD at 405 nm using a plate reader. Subtract the appropriate blanc from raw ODs to determine the ΔOD (*see* **Note 25**).
6. Based on the results, carry out the assay on all samples using the most appropriate conditions. When dose–response assays cause major shifts in activities, it may be necessary to apply different conditions to avoid substrate limitation or extremely low ΔOD values. For instance, co-transfecting a constant amount of cargo plasmid encoding Amy-HDEL with increasing amounts of effector plasmid encoding ERD2 will lead to a strong increase of the Amy-HDEL activity in the cells at the expense of the activity in the medium. This means that some medium samples may have to be assayed with longer enzyme incubation times, and some cell extracts may have to be diluted further. Apply differences in dilutions and incubation times appropriately when calculating the activities as ΔOD per minute and per mL original suspension (*see* **Note 26**).

7. The secretion index (SI) is calculated by simply dividing the activity in the medium by the activity in the cells, to achieve a ratio (without units). A ratio of 1 means that 50% is secreted, and 50% is retained in the cells (the vast majority of which resides in the ER).

## 4 Notes

1. With each bottle, you can pour 12–13 Weck jars (https://www.shop-weck.de/. Conical "Sturz"-form 580 mL Artikel-Nr.: 1116), approximately 65–70 mL medium per jar. Leaving the lid half open to allow steam to dissipate and avoid condense inside the glass jars. Once the agar has set, the lids can be closed, secured with two metal clips, and the jars can be stored in a clean cupboard for up to 6 weeks.
2. The first wash can be facilitated by a very short spin at 1000 rpm but not all seeds will sediment. Remove the liquid and avoid sucking up seeds (touching the bottom of the tube with the tip to limit seed entry into the tip). During the second wash, all seeds effectively pellet down even without spinning.
3. Repeat this step every 2 weeks to maintain a constant level of plants.
4. We use a needle bed with 11 × 11 needles spaced 1 mm of each other and covering 1 square centimeter. The small holes generated allow the digestion mix to enter the spongy parenchyma effectively.
5. Leaves can be cut into smaller tringles to fill spaces between larger leaves.
6. Wash the filters immediately after use with hot water to remove sugars and lipids from the remaining cuticle.
7. A programmable digital pump is not suitable and an analog, manually controlled pump is much preferred.
8. Further washes are not recommended as each time protoplasts are resuspended, a small proportion will rupture, reducing the yield.
9. Approximately two million protoplasts per electroporation.
10. This avoids damaging protoplasts by shearing.
11. It is possible to co-transfect more than two plasmids but then ideally only one should be the independent variable and the others should be constant.
12. We use a homemade electroporation device with a pair of stainless steel plate electrodes embedded in a Teflon insulator, which is directly inserted into the cuvette, masking the two electrodes inside the cuvette. No metal parts are exposed to

fingers or hand when the electroporation takes places. It is not commercially available because our setup does not prevent you from inserting the electrode into your mouth and pulling the trigger. Commercial machines use a more complex setup to make this step fool-proof, but the principle remains the same.

13. Do not immediately resuspend protoplasts after electroporation; the protoplasts need time to repair small holes in the plasma membrane. Usually, a single electroporation cycle takes 1 min, so a series of 30 takes enough time for cell repair after electroporation.
14. Use a whip motion with your wrist but carefully aim at the Petri dish. If done well, only 100 microliters of liquid remains behind in the cuvette. With experience, each transfected protoplast suspension is 2.5 mL with less than 3% variation. Having this reproducible volume is an important asset of this protocol.
15. The incubation time is critical when doing secretion assays and should be kept constant when comparing secretion rates. The longer the incubation time, the higher the amount in the medium relative to the amount in the cells. 20–24 h is the most popular experimental range, although time-courses with sampling as early as 2 h after electroporation can give rise to detecting recombinant proteins encoded by the transfected plasmids.
16. Protoplasts are fragile and unmix by floating to the liquid surface. It is therefore important to aspirate the 500 microliters immediately after mixing the protoplast suspension. It also has to be performed in one single step. If accidentally some air is aspirated, one should not pipette the incomplete sample back to the Petri dish and try again as some of the protoplasts will be damaged. The only option is to transfer the incomplete sample to a separate tube, measure exactly how much is missing to reach 500 microliters, and then go back to the suspension and aspirate the missing amount.
17. Starting and stopping the reactions takes approximately 15 s each. If carried out in the same order, the incubation times will be identical.
18. Each assay contains the equivalent of 10 microliters of original suspension. This means that values of ΔOD per hour must be multiplied by 100.
19. The side of the bottom half of the 5 cm Petri dish hosting the cell suspension is held against the side at the opening of the conical tube at an angle of 90°, and the plate is tilted so that the liquid runs down the inside of the tube in a single motion. It is important to have a continuous stream of liquid on the inner

edge of the tube rather than liquid dripping into the center because the surface tension will allow almost 100% of the suspension to run into the tube.

20. Holding the thin end of the pipette with one hand, and the thick end with the other, place the thin portion (as close as possible to the end where you are holding it) into the edge of the Bunsen burner flame until the flame turns yellow. In a parallel motion, take the heated pipette out of the flame and then pull gently outward to extend by 1–2 cm. Do not pull while the glass is still in the flame. Break off the end to reveal a refined section of approximately 1 cm. It is important to prepare all refined Pasteur pipettes before starting **step 4** of harvesting medium and cells.
21. All work is done at eye level. The protoplast band is not evenly thick on the surface and contains a low-density area; this is where you should penetrate with the pipette. It is essential that once penetrated deep enough, you do not move the pipette at all while aspirating. The exact amount of medium is not critical, as long as it is around 1 mL.
22. Staying close to the surface creates a "slurping" noise and agitates the liquid surface, avoiding the formation of large droplets on the side of the tube. It is important to slow down the descent when you are close to the pellet, allowing you to remove clear liquid up to 1 mm above the pellet without losing any cells. When done correctly, the combined volume of the cell pellet and any remaining supernatant is 50 microliters.
23. It is possible to continue analyzing cells and medium immediately, but in practice, it is better to focus on harvesting cells and medium and to start the GUS assay (*see* Subheading 3.4) on the day of harvesting. The GUS assay is the easiest of the assays, and if any unexpected result suggests that the transfections went wrong, it is possible to abort the experiment (or parts of it) and avoid wasting time.
24. Cell extracts have a higher intrinsic absorbance compared to medium samples; therefore, it is important to use separate blancs to control for auto-absorbance in cell extracts and medium extracts.
25. Ideally, ΔOD values should be between 0.1 and 1.5. Above this range, the reaction will suffer from substrate limitation, and below the range, the reproducibility deteriorates rapidly. The exact assay can now be fine-tuned by changing the dilution of the sample and/or the incubation time.
26. 15–30 min is the ideal incubation time for this assay. If the ΔOD is expected to be much higher than 1.5 for some of the samples, the next approach is to dilute those samples more. In

that case, a portion of the blanc sample (mock transfection) must also be diluted in the same way. If samples are measured for different incubation times, one has to include blancs with the respective incubation times too.

## Acknowledgments

The work was supported by the Biotechnology and Biological Sciences Research Council (BBSRC) project nr. BB/D016223/1 and The Leverhulme Trust (F/10 105/E).

### References

1. Denecke J, Botterman J, Deblaere R (1990) Protein secretion in plant cells can occur via a default pathway. Plant Cell 2:51–59
2. Denecke J, De Rycke R, Botterman J (1992) Plant and mammalian sorting signals for protein retention in the endoplasmic reticulum contain a conserved epitope. EMBO J 11: 2345–2355
3. Phillipson BA, Pimpl P, Pinto L, daSilva P, Crofts AJ, Taylor JP, Movafeghi A, Robinson DG, Denecke J (2001) Secretory bulk flow of soluble proteins is efficient and COPII dependent. Plant Cell 13:2005–2020
4. Dombrowski JE, Schroeder MR, Bednarek SY, Raikhel NV (1993) Determination of the functional elements within the vacuolar targeting signal of barley lectin. Plant Cell 5:587–596
5. Pimpl P, Hanton SL, Taylor JP, Pinto-daSilva LL, Denecke J (2003) The GTPase ARF1p controls the sequence-specific vacuolar sorting route to the lytic vacuole. Plant Cell 15:1242–1256
6. Pedrazzini E, Giovinazzo G, Bollini A, Ceriotti A, Vitale A (1994) Binding of BiP to an assembly-defective protein in plant cells. Plant J 5:103–110
7. DaSilva LL, Taylor JP, Hadlington JL, Hanton SL, Snowden CJ, Fox SJ, Foresti O, Brandizzi F, Denecke J (2005) Receptor salvage from the prevacuolar compartment is essential for efficient vacuolar protein targeting. Plant Cell 17:132–148
8. DaSilva LL, Foresti O, Denecke J (2006) Targeting of the plant vacuolar sorting receptor BP80 is dependent on multiple sorting signals in the cytosolic tail. Plant Cell 18:1477–1497
9. Foresti O, Gershlick DC, Bottanelli F, Hummel E, Hawes C, Denecke J (2010) A recycling-defective vacuolar sorting receptor reveals an intermediate compartment situated between prevacuoles and vacuoles in tobacco. Plant Cell 22:3992–4008
10. Gershlick DC, Lousa Cde M, Foresti O, Lee AJ, Pereira EA, Dasilva LL, Bottanelli F, Denecke J (2014) Golgi-dependent transport of vacuolar sorting receptors is regulated by COPII, AP1, and AP4 protein complexes in tobacco. Plant Cell 26:1308–1329
11. Silva-Alvim FAL, An J, Alvim JC, Foresti O, Grippa A, Pelgrom AJE, Adams TL, Hawes C, Denecke J (2018) Predominant Golgi-residency of the plant K/HDEL receptor is essential for its function in mediating ER retention. Plant Cell 30:2174–2196
12. Alvim JC, Bolt RM, An J, Kamisugi Y, Cuming A, Silva Alvim FAL, Concha JO, daSilva LLP, Hu M, Hirsz D, Denecke J (2023) The K/HDEL receptor does not recycle but instead acts as a Golgi-gatekeeper. Nat Commun 14:1612
13. Crofts AJ, Leborgne-Castel N, Hillmer S, Robinson DG, Phillipson B, Carlsson LE, Ashford DA, Denecke J (1999) Saturation of the endoplasmic reticulum retention machinery reveals anterograde bulk flow. Plant Cell 11 (11):2233–2247

# Chapter 31

# Analyzing Photoactivation with Diffusion Models to Study Transport in the Endoplasmic Reticulum Network


## Matteo Dora, Frédéric Paquin-Lefebvre, and David Holcman



## Abstract

Photoactivation is a paradigm consisting in local molecular fluorescent activation by laser illumination in a chosen region (source) while measuring the concentration at a target region. Data-driven modeling is concerned with the following questions: how from the measurement in these two regions is it possible to infer the properties of molecular propagation? How is it possible to use such responses to infer motions occurring in networks such as the endoplasmic reticulum? In this book chapter, we shall review the data-driven analysis based on diffusion-transport models and numerical simulations to interpret the photoactivation dynamics and extract biophysical parameters. We will discuss modeling approaches to reconstruct local network properties from photoactivation transients.



**Key words** Photoactivation, Endoplasmic reticulum, Modeling, Diffusion, Network


## 1 Introduction

How molecules, ions, or proteins spread across cell compartments remains an active field of cell biology. Molecules can move passively as Brownian motion but can also be transported actively by molecular motors from one region to another. Yet exploring the exact propagation mode is not straightforward as it requires combining molecular labeling with refined live super-resolution microscopy.

More than 20 years ago was developed a key approach called fluorescence recovery after photobleaching (FRAP), which consists in photobleaching a region labeled with a fluorescent marker and observing how this region gets re-populated [1]. This approach in combination with modeling propagation as diffusion provided a first tool to fit experimental recovery curves by a diffusion model, allowing estimating an effective diffusion coefficient and the fraction of immobile molecules. Later on, an opposite paradigm

Matteo Dora and Frédéric Paquin-Lefebvre equally contributed


Verena Kriechbaumer (ed.), *The Plant Endoplasmic Reticulum: Methods and Protocols*, Methods in Molecular Biology, vol. 2772, https://doi.org/10.1007/978-1-0716-3710-4_31,

emerged [2, 3] because of a new photoactivable (PA) molecule PA-GFP (Green Fluorescent Protein), where a population was photoactivated. For that case experimental data are obtained by counting the arrival of activated molecules in a given region. Photoactivation consists in locally activating fluorescent proteins by laser illumination in a chosen region (source) and measuring the propagation at a target region located at a distance apart from the source [4], thus providing information about first arrival times and also biophysical parameters such as diffusion coefficients.

While these experiments are not too difficult to perform in the cytosol of a cell, they remain quite tedious in organelles such as the endoplasmic reticulum which consists of a thin network being constantly reshaped, with tubule size lying under the diffraction limit. Analyzing fluorescence data collected from population photoactivation in the endoplasmic reticulum (ER) remains challenging due to several difficulties associated with noisy signals, segmentation of the ER tubules, and the lack of a defined biophysical model to describe lumen and membrane motion. Various imaging approaches have been developed to denoise these images: the first is the variance stabilizing transformation followed by additive white noise removal [5], algorithms specifically designed for Poisson–Gaussian mixtures [6], and deep learning models based on convolution neural networks [7, 8]. These approaches are not the subject of the current chapter and will be reviewed somewhere else.

However, the experimental approaches mentioned above are in contrast with single-particle trajectories (SPTs) where each molecule is activated at a random time and at a random position, leading to a random sample of trajectories. Interestingly, previous work with SPTs reveal active motion in the ER tubules, while passive diffusion in nodes. But today some discrepancies persist in the description of motion between the population and the individual trajectory approaches [9, 10].

In this book chapter, we shall review modeling approaches used to estimate and to interpret transient rates at which molecules disappear from one region to appear in another. We shall further discuss how propagation in the endoplasmic reticulum can be studied by solving simple planar diffusion models. We will also show how network simulations can be employed to explore the consequence of an active flow between ER nodes. Lastly we will discuss simulations to extract arrival times to ER nodes.

## 2 Principles of Photoactivation and Source–Target Interaction

Photoactivation consists in locally activating fluorescent proteins by laser illumination in a chosen region (source) (Fig. 1a) and measuring the propagation at a target region located at a distance $d$ apart

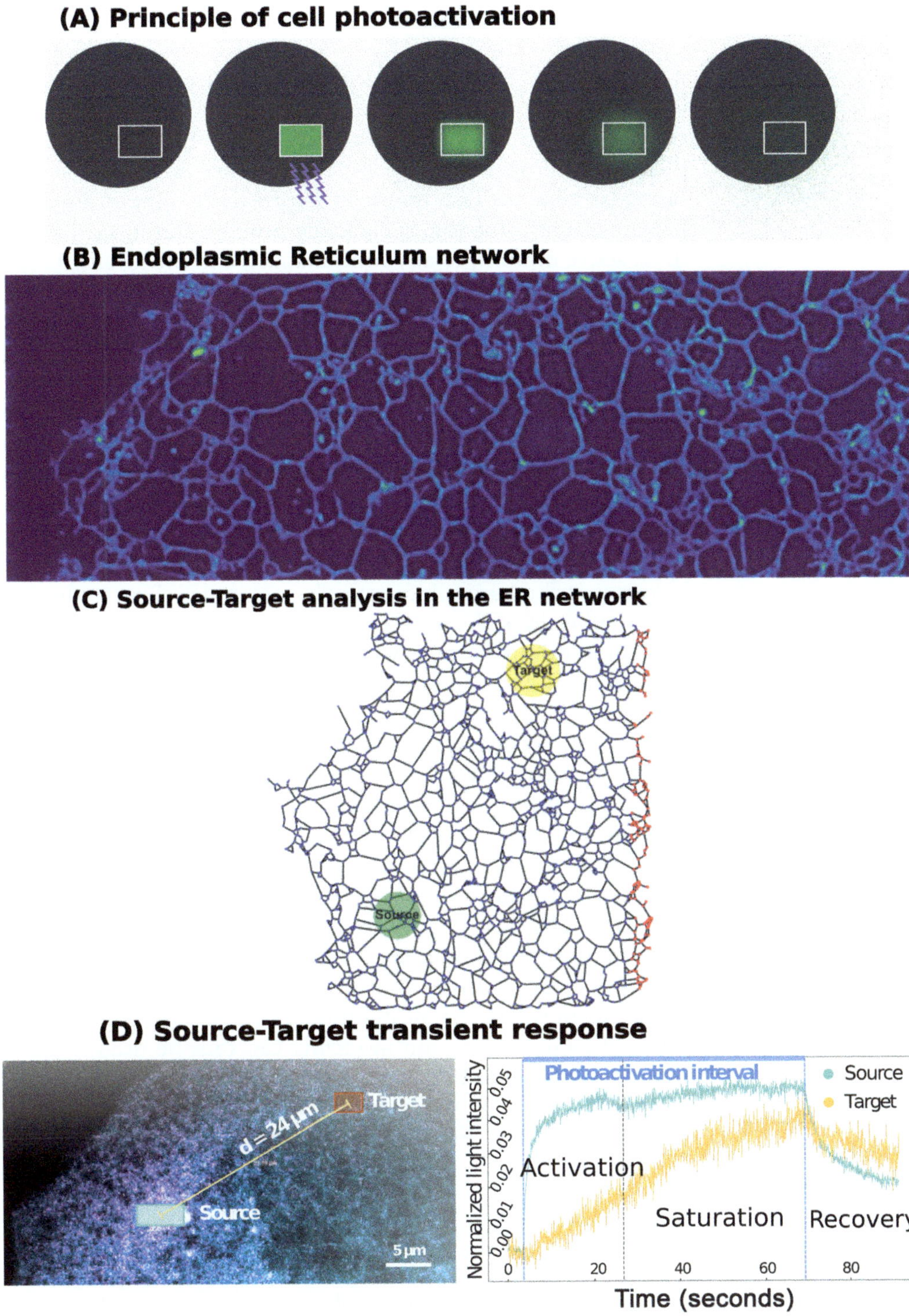


**Fig. 1 Principle of photoactivation in the ER.** (**a**) A subregion of a cell (round scheme) is activated by a laser, followed by the spread of activated particles in the cell and their disappearance from the source region. (**b**) Example of an Endoplasmic Reticulum network. (**c**) Principle of source–target photoactivation analysis: A source region is photoactivated while the fluorescent signal is analyzed at the target region. (**d**) Transient responses of photoactivation at the source (blue) and target (yellow). The dynamics is decomposed into three phases: (1) Transient activation starts when fluorescent molecules are activated, until (2) a transient steady state is achieved (saturation), and (3) the last phase starts when the laser stops, leading to a different decay in both regions

from the source. The source and the target are disks of small radius (few microns), which can be adjusted from the microscope aperture even if located on a network such as the ER (Fig. 1b,c). This procedure allows recording the fluorescence signal $f(d,t)$ on the target at time $t$. The transient fluorescence time course carries biophysical information, such as the propagation mode (diffusion, drift, active motion, or any combination) and the local structure of the underlying network with possible inhomogeneous node-tubule distribution, which should be extracted from the function $f(d,t)$. However, this procedure requires a model for $f$, in addition to a statistical method to extract the relevant parameters.

Photoactivation can be decomposed into three phases. There is first a fast rising phase where dye molecules are photoactivated (Fig. 1d), during which some molecules can be exchanged with the neighboring environment and which lasts until a saturation is reached (plateau in Fig. 1d). In the second phase, the signal saturates until the laser is stopped. During these two phases, the signal increases at the target, interestingly sometimes close to the saturation amplitude of the source.

The third phase starts when the laser excitation stops, and it is characterized by a relaxation of the activated molecule toward the rest of the ER, leading to a decay of the concentration as the activated molecules spread inside the network. The three phases can be recorded at any distance from the source (from few to tens of microns or more). The goal of a modeling approach is to extract from the ensemble of response the type of motion occurring in the ER. The area of the source and target can be chosen containing one or multiple ER nodes. For a size of several microns, the molecule can be locally mixed, and it would then be possible to use a continuous approximation of the ER geometry such as the two-dimensional plane. Another possibility is to simulate the ER as a network where the mean number of nodes (degree) is equal to three [9]. Photoactivation modeling can thus be used to quantify transport within the ER. In brief, we propose to study the local (few microns) and far away transport (tens of microns or more). We summarize below the possible local transport hypotheses inside the ER.

### 2.1 Molecular Transport Hypotheses in the ER

We summarize here various modes of transport that could occur in the ER, which can be explored through modeling and simulations.

1. **Large-scale diffusion.** Molecular transport could be dominated by diffusion at both small and large scales. In that case, computing a single diffusion coefficient is sufficient to characterize the entire dynamics. This hypothesis was initially tested using FRAP and lead to a diffusion coefficient of $D = 1.65\ \mu m^2/s$ [11].

2. **Space-dependent diffusion.** Another possibility is a dynamics characterized by a spatially-dependent diffusion process, with the diffusion coefficient $D(\boldsymbol{x})$ depending on the position $\boldsymbol{x}$ (which could represent the center of a disk of radius $a$, where $a$ characterizes the resolution). The output of this analysis is a diffusion map [12], and in that case the diffusion coefficient $D(\boldsymbol{x})$ characterizes the local transport. Spatial variation of the diffusivity should be explained: nodes could be of different sizes or contain fewer passages, or there could be changes in the density of nodes.
3. **Active drift and diffusion.** An active drift potentially depending on ATP could also be possible. This was observed in tubules using SPTs [9] but was not found at higher spatial scales.
4. **Alternating active drift in tubules.** If tubules can let molecules move in only one direction at a time by an as-yet-unknown mechanism, then it is possible to anticipate a motion where molecules move alternatively and this would lead to molecular packets.
5. **ER membrane transport.** Transport on ER membrane could be different from the lumen motion [9].
6. **Asymmetric transport.** Some asymmetry could exist between diffusion and active transport at regions near the nucleus vs the cell periphery.

We will show below how PA can serve to test the different transport hypotheses.

### 2.2 Fluorescent Mean Square Displacement Analysis

In this section, we recall diffusion models that are used to characterize the type of molecular motion in the ER lumen. We further mention how parameters should be computed to avoid certain pitfalls.

To quantify the nature of the motion from the PA data, one method consists in fitting for each cell the effective Mean Square Displacement (MSD) curve describing the time evolution of the average distance traveled by a protein released from the PA source (Fig. 2a–d). Is it worth stating the assumption about local isotropy that was tested: for collecting data, we use PA responses in $p$ different small regions located at a distance $r$ from the source, and the average fluorescence $r^2 u_i(r,t)$ was recorded for $i = 1, \ldots, p$. After varying the distance $r$, a collection of points is obtained (Fig. 2d).

To evaluate possible deviations from classical diffusion, a fit of thee MSD point distribution curves is used with the function $ct^{\alpha}$ (Fig. 2d), where $\alpha$ is called the anomalous exponent. The exponent $\alpha = 1$ is associated with classical diffusion, while it represents subdiffusion for $\alpha < 1$ and superdiffusion for $\alpha > 1$. This procedure can be repeated for each cell to determine the MSD exponent $\alpha$.

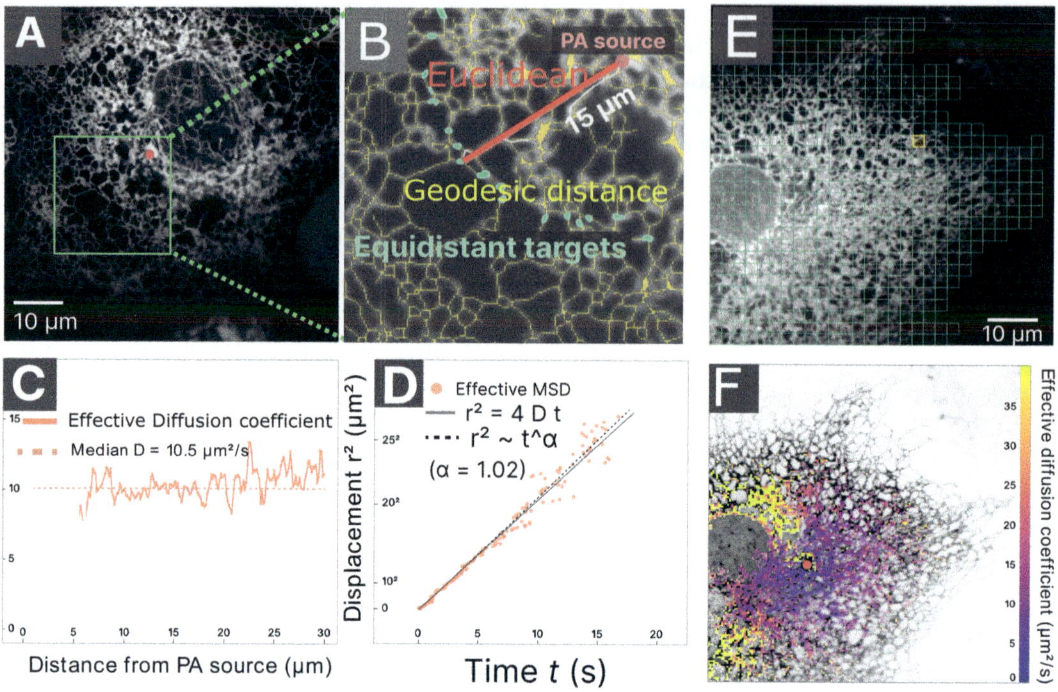


**Fig. 2 Diffusion analysis of large-scale ER luminal dynamics.** (**a**)–(**b**) ER structure of a COS-7 cell showing the location of photoactivation (red dot). ER subdomains (green regions) all equidistant from the photoactivation region (the distance calculated on the underlying ER morphology is $d = 15$ μm). Here the Euclidean distance from the photoactivation region is 15 μm (red line). (**c**) Effective diffusion coefficient vs the distance $d$ from the photoactivation region, obtained from separate fits at each distance, for the model accounting for the ER morphology (orange) and neglecting it (light blue). The error bars represent the approximated confidence interval estimated by observed Fisher information ($\pm 5$ standard deviations). (**d**) Effective MSD obtained by ensemble estimate. (**e**) Subdivision of the ER with a grid of square bins. (**f**) Spatial map describing the effective diffusion coefficient computed from the mean arrival time of HaloTag-KDEL. The red dot indicates the PA source. The grid bin size is 400 nm (3 × 3 camera pixels), adapted from [10]

It is interesting to note that how distances are measured does influence fitting procedure results. The first choice is the Euclidean distance, given by the shortest segment from the source to the target (the red line in Fig. 2b). Another possibility is to instead consider the geodesic distance, which corresponds to the length of the minimal path within the ER network (thus accounting for its morphology) connecting the source to the target (the white path in Fig. 2b). The geodesic distance is longer than the Euclidean distance.

For instance, by calculating the MSD curve with the geodesic distance, population statistics of cells expressing HaloTag-KDEL had an average MSD exponent of $\alpha \approx 1$, thus suggesting a description of luminal dynamics by classical diffusion over this spatiotemporal scale (Fig. 2e, green). However, neglecting the ER morphology and using the Euclidean distance resulted in an apparent super-diffusive behavior with $\alpha \approx 1.2$ [10]. This discrepancy could create a confusion, arguing for a possible drift toward the peripheral ER [13–16].

## 3 Predicting the Rising Phase of the Target from Source Fitting Within the Diffusion Regime

We present in this section various models that can be used to extract parameters from the early phases of photoactivation between two regions called the source (laser target) and a target region (a region located at a certain distance away).

In a first approximation, diffusion models have been used to extract the dynamics occurring in the ER lumen or membrane. The first consists of replacing the complex ER network geometry by a two-dimensional planar geometry, thereby neglecting the local topology (three-way junctions) dynamics of tubules and the possible heterogeneity of ER distribution [10] or the anomalous restricted motion of junctions [17]. For example, the ER structure in COS-7 cells is sufficiently flat to be considered as planar [18].

However two-dimensional geometries are insufficient for three-dimensional ER networks, for which more appropriate approximations could be obtained by considering a fractal description with an average dimension $\alpha = 1.6$ [17]. This representation accounts for the number $n(r)$ of ER networks located in a ball of radius r, changing $n(r) \approx r^{\alpha}$ as $r$ increases [17]. This representation results in an overall anomalous diffusion [19].

### 3.1 Diffusion Model of Photoactivation at the Source

We first describe in this subsection a diffusion method to extract the effective diffusion coefficient using an approximation of the ER network as a two-dimensional plane. In that case the structure of each node/sheet/tubule is not accounted for and simply replaced by a continuum of distances. Also, the distance between the source and the target is computed using the Euclidean norm, an over simplification as ER lengths could be longer and should thus be computed over network paths.

A diffusion model allows to predict the concentration changes at the target site (Fig. 2b) from the dynamics of the source. The fluorescence curve $\phi(t)$ at the source $S$ of the photoactivation region of interest (ROI) is approximated as a sum of exponentially saturating functions

$$\phi(t) = \sum_n c_n \left(1 - e^{-\lambda_n t}\right), \tag{1}$$

where the parameters ($c_n$ and $\lambda_n$) are fitted to the empirical fluorescence (Fig. 2c). In practice a sum of three exponentials is sufficient. The next step is to use the diffusion equation in the plane to compute the concentration at the target (Fig. 2b). An optimal fit of the data allows us to estimate the diffusion coefficient $D$. This approach considers that the recorded luminosity is proportional to the density of photoactivated particles.

### 3.2 Diffusion Model of Photoactivation at the Target

The density of photoactivated particles at any point in space is obtained by solving the diffusion equation

$$\begin{cases} \partial_t u(\mathbf{x},t) - D\nabla_{\mathbf{x}}^2 u(\mathbf{x},t) = 0, & \mathbf{x} \notin S \\ u(\mathbf{x},t) = \phi(t), & \mathbf{x} \in \partial S \\ u(\mathbf{x},t=0) = 0, & \end{cases} \tag{2}$$

where the source effect $\phi(t)$ is imposed at the boundary of the source region $S$, with a zero initial condition. The solution can be expressed using a simpler auxiliary problem with a time-independent boundary condition (called Duhamel's principle [20])

$$\begin{cases} \partial_t v(\mathbf{x},t) - D\nabla_{\mathbf{x}}^2 v(\mathbf{x},t) = 0, & \mathbf{x} \notin S \\ v(\mathbf{x},t) = 1, & \mathbf{x} \in \partial S \\ v(\mathbf{x},t=0) = 0. & \end{cases} \tag{3}$$

The solution of Eq. (2) is the convolution between the auxiliary solution $v(r,t)$ and the time-varying boundary term:

$$u(\mathbf{x},t) = \int_0^t \phi(t-\tau)\ \partial_\tau v(\mathbf{x},\tau) \mathrm{d}\tau. \tag{4}$$

We now solve problem (3) in 2D with radial symmetry, where the source $S$ consists of a disk of radius $a$ (thus the domain is $r > a$) using Laplace's transform $L\{\cdot\}$. Taking then the Laplace transform [21] on both sides of (3) yields the modified Bessel equation of order 0,

$$z^2 \partial_z^2 \hat{v}(r,s) + z\partial_z \hat{v}(r,s) - z^2 \hat{v}(r,s) = 0, \qquad \text{with } z = \sqrt{s/D}\ r\,, \tag{5}$$

where $\hat{v}(r,s) = L\{v(r,t)\}$. The solution is

$$\hat{v}(r,s) = A(s)\ K_0(\sqrt{s/D} r), \tag{6}$$

where $K_0(z)$ is the modified Bessel function of the second kind and order 0, while $A(s)$ is a coefficient determined by imposing the boundary condition on $r = a$, which yields

$$\hat{v}(r,s) = \frac{K_0(\sqrt{s/D}\ r)}{s\ K_0(\sqrt{s/D}\ a)}. \tag{7}$$

Finally the solution of the time-dependent problem is

$$u(r,t) = \int_0^t \phi(t-\tau)\ \left(L^{-1}\{s\ \hat{v}(r,s)\}|_\tau\right) d\tau. \tag{8}$$

Equation (8) is evaluated numerically, thus allowing to fit the photoactivation ROI at the target and recover an approximated diffusion coefficient $D$, which can be computed over the entire ER, leading to a diffusion map representation (Fig. 2e, f).

### 3.3 Numerical Simulations of Photoactivation with Diffusion: Exploring Various Target Local Geometries

To better characterize the photoactivation protocol, various scenarios, each associated with a different target geometry, can be explored: (1) an open local domain, (2) an absorbing domain, and (3) a partially absorbing domain (Fig. 3). Molecules are moving by diffusion inside a domain consisting of a large disk $B(O, R)$ of radius $R$ centered at the origin $O$, in which the source photoactivation region corresponds to the smaller disk $B(O, a)$ of radius $a \ll R$ (Fig. 4). Inside the source, molecules are activated by a laser during the time interval $[0, t_s]$, which can be modeled by a step function. The concentration of activated molecules is then measured over another disk $B(P, a)$ located at a distance $L = |OP|$ away from the source.

#### 3.3.1 Photoactivation Diffusion Model

We now present the diffusion equations that serve as a model of photoactivation. The concentration $u(x, t)$ is modeled by the diffusion equation [22]

$$\frac{\partial u}{\partial t} = D\Delta u\ , \quad x \in B(O, R)\ , \quad t > 0\ , \tag{9}$$

with reflective boundary conditions (no material escapes the domain)

$$\frac{\partial u(x,t)}{\partial n} = 0, \quad x \in \partial B(O, R)\ , \quad t > 0\ . \tag{10}$$

The laser source is activated during the time interval $[0, t_s]$, modeled by

$$u(x,t) \quad = \Phi(x)(H(t) - H(t - t_s))\ , \quad x \in B(O, a), \tag{11}$$

where the profile $\Phi(x)$ can be uniform or Gaussian. Here $H(t)$ is the Heaviside function ($H(t) = 0$ for $t < 0$ and $H(t) = 1$ for $t > 0$). A modified diffusion equation has to be solved after a time $t > t_s$, with $u(x, t_s)$ as an initial condition for again Eq. (9). However this may not be straightforward, and one alternative is to replace equations (9)–(11) by a single equation with an appropriate spatio-temporal source term:

$$\frac{\partial u(x,t)}{\partial t} = D\Delta u(x,t) + \frac{c_0}{4\pi a^2 t_s}\left(1 + \tanh\left(\frac{t_s - t}{\gamma_s}\right)\right)\left(1 + \tanh\left(\frac{a - x}{\gamma}\right)\right),\ x \in B(0, R)\ ,\ t > 0\ , \tag{12}$$

with zero initial condition $u(x, t) = 0$ and reflective boundary condition. The source term acts as a step function from time

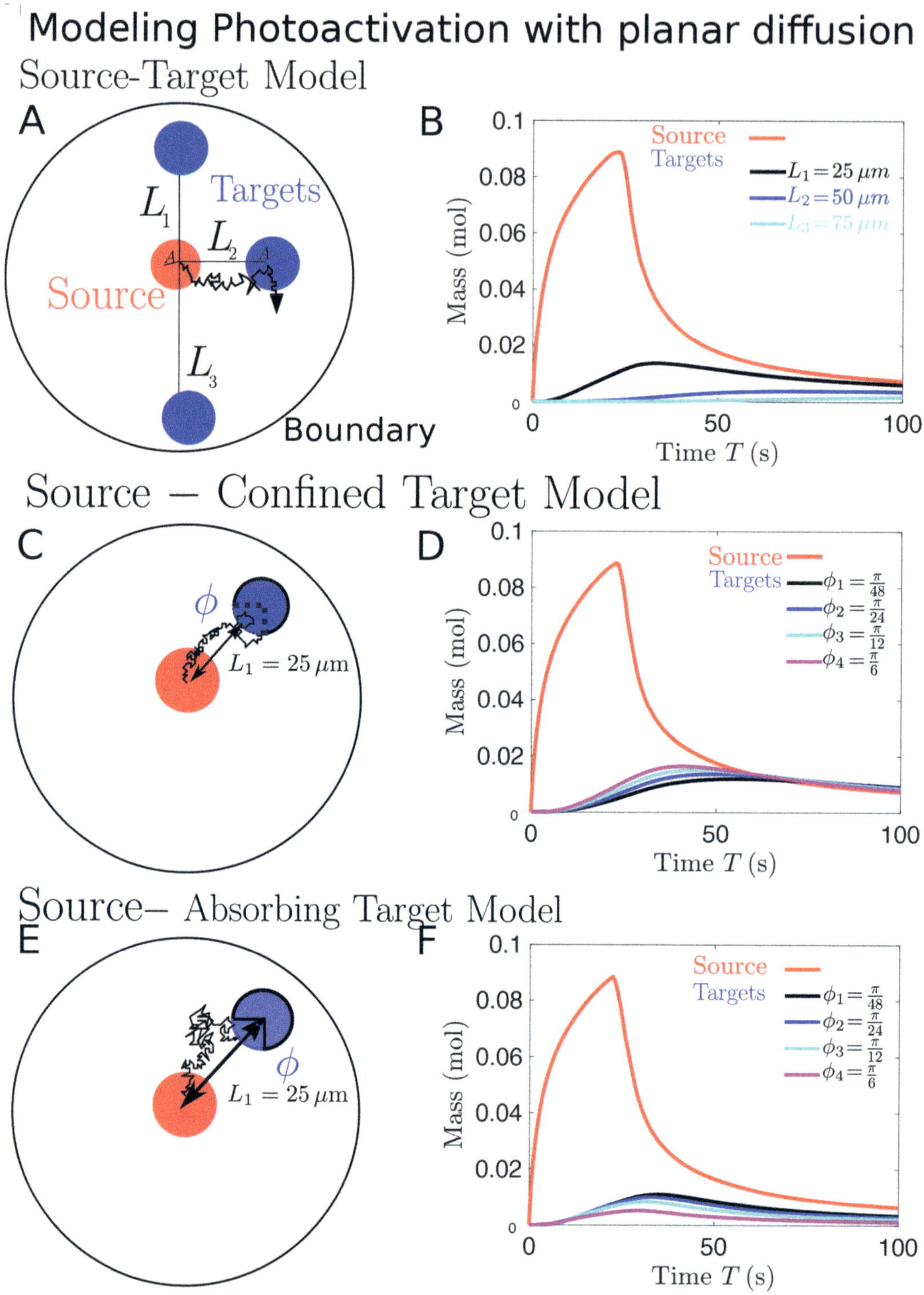


**Fig. 3 Effect of target organization and permeability on source–target responses.** (**a**) No further conditions are imposed on the target boundary $\partial B(0, a)$. The source is simulated (red curve) and compared to the target responses (black, dark, and light blue curves) for three distances $L = 25$, 50, and 75 μm. (**b**) Particles diffuse freely in and out of the small boundary subsection $\partial\Omega_{t_a}$ of size $|\partial\Omega_{t_a}| = \phi a$, where $\phi$ is the free boundary angle, while the large remaining boundary portion $\partial\Omega_{t_r}$ is reflecting. (**c**) Particles hitting the boundary subsection $\partial\Omega_{t_a}$ are absorbed and reinjected in the bulk at a rate $1/\overline{\tau}$, with $\overline{\tau}$ given by Eq. (15). Parameter values are $D = 10\ \mu m^2/s$, $R = 100\ \mu m$, $a = 5\ \mu m$, $c_0 = 1$ mol, $t_s = 25$ s, $\gamma_s = 1$ s, and $\gamma = 1\ \mu m$

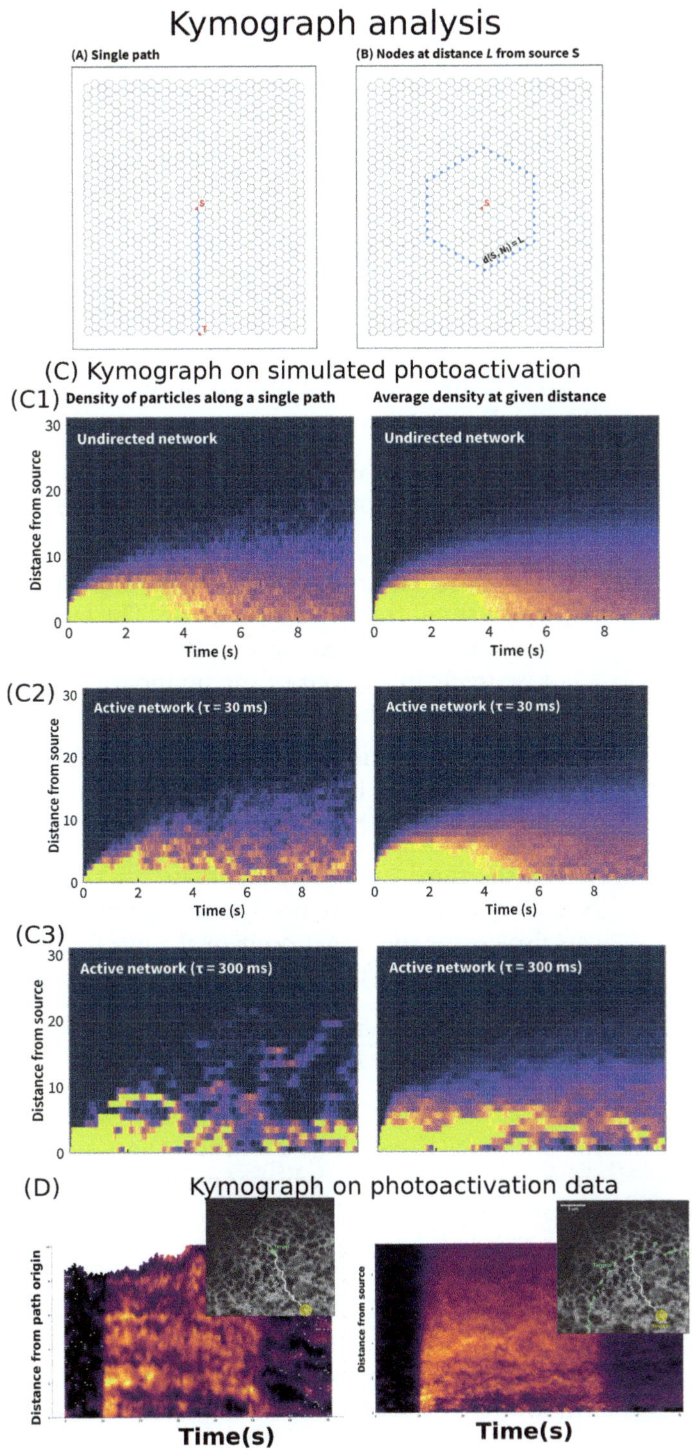


**Fig. 4 Kymograph Analysis of Photoactivation.** (**a**)–(**b**) Representation of the distance computed along a path or by averaging over an equidistance from the center in a tridiagonal network. (**c**) Time–distance kymograph representation of the photoactivation occurring at position $X = (0, 0)$ for three different propagation modes: undirected (**c1**) or modulated by the switching time of the unidirectional flow with $\tau = 30$ ms (**c2**) and $\tau = 300$ ms (**c3**). (**d**) Kymograph of photoactivation data [10]

$0 \leq t \leq t_s$ on the narrow disk $\Omega_s = \{x \mid 0 \leq x \leq a\}$. The concentration at the target is computed by solving Eq. (12) and by computing the local concentration at the target $B(P, a)$:

$$S_{\text{target}}(t) = \int_{B(P,a)} u(x, t) dx \,. \tag{13}$$

Similarly, at the source the total fluorescence is

$$S_{\text{source}}(t) = \int_{B(O,a)} u(x, t) dx. \tag{14}$$

The source–target photoactivation responses are evaluated for three different cases:

1. **No further boundary conditions are imposed at the target site** $B(P, a)$. Molecules can freely diffuse in and out of the target region.
2. **The target boundary consists of a dominant reflective part and a small freely diffusing one.** The target boundary decomposes into two pieces as $\partial B(P, a) = \partial\Omega_{t_r} \cap \partial\Omega_{t_a}$. Particles diffuse freely in and out of the boundary subsection $\partial\Omega_{t_a}$, of size $|\partial\Omega_{t_a}| = \phi a$, where $\phi$ is the angle of the free boundary.
3. **Reflecting boundary condition on** $\partial\Omega_{t_r}$ **and absorbing on** $\partial\Omega_{t_a}$. Particles hitting the boundary subsection $\partial\Omega_{t_a}$ are permanently absorbed.

In the third case, to compensate for the unrealistic absorbing boundary condition, we need to account for absorbed particles that can also escape the target domain at a rate $1/\overline{\tau}$, given by the narrow escape formula [23] on the disk

$$\overline{\tau} = \frac{a^2}{D} \log\left(\frac{1}{\phi}\right) + C + o(1) \,, \tag{15}$$

where $C$ is a known constant [23]. Thus we can derive an equation for the total mass $S_{\text{target}}(t)$ inside the target region $B(P, a)$ by subtracting a loss term, which is proportional to the overall number of particles, from the influx due to the absorbing boundary condition. This balance equation is given by

$$\frac{dS_{\text{target}}(t)}{dt} = J(t) - \frac{S_{\text{target}}(t)}{\overline{\tau}} \,, \tag{16}$$

where $J(t)$ is the time-dependent flux $J(t)$ through the narrow window

$$J(t) = \int_{\partial\Omega_{t_a}} D \frac{\partial u(x, t)}{\partial n} dx \,. \tag{17}$$

The solution for the total mass (case 3) within the target is given by

$$S_{\text{target}}(t) = \int_0^t e^{-(t-s)/\bar{\tau}} J(s)ds \, . \tag{18}$$

The COMSOL software allows exploring how the source–target located at a distance $L$, as well as the target permeability, controlled by the angle $\phi$ of free portion of boundary, affect the responses for the three different cases (Fig. 3). Lower responses are recorded as the source–target distance increases (case 1, Fig. 3a). Then for the second case, smaller angles $\phi$ lead to smoother target responses: it is harder for particles to penetrate the target region, but once inside they stay longer on average (Fig. 3b). Steeper response curves are also found for small angles $\phi$ when the target opening is fully absorbing, with the exit modeled using the classical narrow escape theory (case 3, Fig. 3c).

### 3.4 Modeling Photoactivation Protocol by a Reaction–Diffusion Process at the Source

We now discuss briefly an alternative model of photoactivation that accounts for the decay of activated molecules inside the source. Going back to the diffusion Eq. (2), a source term can be added restricted to the disk of radius $R$:

$$f(r,t) = \begin{cases} \kappa(1 - u(r,t)) & r < R \\ 0 & r \geq R, \end{cases} \tag{19}$$

describing the activation of inactive particles by the laser beam at a rate $\kappa$. The dynamics of active molecules occurring in a two-dimensional plane is thus modeled by the reaction–diffusion equation:

$$\begin{cases} \partial_t u(r,t) - D\left(\partial_r^2 u(r,t) + \frac{1}{r}\partial_r u(r,t)\right) = f(r,t), \quad t > 0 \, , \\ u(r, t = 0) = 0 \, . \end{cases} \tag{20}$$

The equation is solved in two distinct regions, for $r < R$ and $r > R$ with solutions $u_{r<R}(R,t)$ and $u_{r>R}(R,t)$, respectively (*see* Appendix 7.1.1). For short-time ($\frac{R-r}{\sqrt{t}} \gg 1$), we find (*see* also Appendix 7.1)

$$u_{r<R}(R,t) \approx 1 - \exp(-\kappa t) - \kappa t^{3/2}\sqrt{D}\frac{2R}{\sqrt{\pi}r(R-r)}\exp\left(-\frac{(R-r)^2}{4Dt}\right). \tag{21}$$

Thus to leading order, the fluorescence decay at the source can be used to estimate the rate $\kappa$. Outside the source, the fluorescence decay is characterized by

$$u_{r>R}(R,t) \approx \kappa\sqrt{\frac{R}{r}}\left(\frac{2t\sqrt{\pi Dt}}{r-R}\exp\left(-\frac{(r-R)^2}{4Dt}\right)\right.$$
$$\left.+\frac{r-R}{\sqrt{\pi D}}\sqrt{t}\exp\left(-\frac{(r-R)^2}{4Dt}\right)\right). \tag{22}$$

This expression shows that the decay away from the source could be used to extract the diffusion coefficient once the rate $\kappa$ has been determined (*see* also Appendix 7.2).

## 4 Kymograph Construction of Photoactivation

We describe in this section the kymograph approach which is a time–position representation of the fluorescence. A kymograph was originally a device that draws the spatial position of a tracking object over time in which a spatial axis represents time. This time–space representation is used to study the velocity over time (continuous and jump moments) of a tracked moving object, such as cells, organelles, molecules, or molecular motors [24].

A kymograph approach can be adapted to follow the distribution of fluorescent molecules over time and space, using a time–space plot of the fluorescence (color coded by intensity), measured along one dimension [25–27]. The kymograph representation can be used to follow the spread of photoactivation. However due to the network structure, the analysis has to account for the dynamics of tubular paths through the structure and the low copy number of fluorescent molecules. The distance between the PA source and a point $P$ is computed from the segmented ER structure. We can also average the fluorescence extracted at subdomains located at a given distance. This average kymograph summarizes the time evolution of fluorescence versus the distance from the PA source (Fig. 4a).

This representation can be tested on simulated data [28], revealing the differences between classical diffusion and active transport (Fig. 4a–d). In particular packet transport is shown by the heterogeneous concentration distribution. In photoactivation data, the kymograph shows a concentrated signal close to the PA source and a gradual spreading through the ER, moving toward distant regions, as photoactivation increases. The time evolution of the fluorescence can be fitted by a diffusion model, $\langle r^2 u(r,t)\rangle_r \sim 4Dt$ [10]. To conclude, the kymograph allows to quantify the spread of fluorescence across a tubular path, thus allowing to detect possible deviations from diffusion.

## 5 Modeling Active Versus Passive Transport on Graph to Reconstruct the Photoactivation Response

We discuss in this section how network modeling can be used to simulate photoactivation and trafficking inside an ER. Indeed, another possibility to model photoactivation is to use a network (*see* also Appendix 7.3) of nodes connected by edges representing tubules (Fig. 5) [29, 30]. The degree represents the number of connections per node of the graph and thus accounts for its topology. Reconstructing a network from structure illumination microscopy images leads to a homogeneous number of connections of degree $d=3$. It is also possible to use a homogeneous hexagonal lattice (Fig. 5), with various conditions at the periphery nodes: molecules can be reflected or absorbed, or periodic conditions can be imposed.

A network is described by a graph $G=(V,E)$ of vertices and edges. The transition probability between nodes is described by the matrix $M=[m_{ij}]_{i,j\in V}$

$$m_{ij}=\begin{cases} 1/d_i, & (i,j)\in E \\ 0, & \text{otherwise.} \end{cases} \tag{23}$$

The time evolution of a transition process describes the number of molecules occupying a node at time $t$. In a discrete time representation, we have

$$\boldsymbol{X}_{t+1}=M\ \boldsymbol{X}_t, \qquad t=1,2,\ \dots\ , \tag{24}$$

and consequently the time evolution depends on the initial distribution

$$\boldsymbol{X}_t=M^t\ \boldsymbol{X}_0, \tag{25}$$

where $\boldsymbol{X}_0$ is the initial state at time 0.

### 5.1 Diffusion-Network Models

Diffusion in a network consists in splitting at each node the number of particles between the number of connected nodes at either random time or synchronously in case of a simulation with a clock time $\Delta t$. The time in each tubule is zero (infinite velocity). For a rate $\lambda$ and a size $l$, the effective diffusion coefficient is $D=\lambda l^2$. Diffusion is characterized by a uniform spread of molecules until steady state where the final density is uniform $p=\frac{N}{n_{\text{nodes}}}$, where $N$ is the initial number of particles and $n_{\text{nodes}}$ the number of nodes [28].

### 5.2 Active-Network Model

In an active-network model, the direction of the flow in each tubule alternates: edges have a directionality that changes in time according to a Poissonian rate. In an active-network model, a particle moves from one node to another by selecting a random incident edge from those which are outward directed and jumps through it. When all edges are inward directed (capture state), particles at the nodes wait a random time and then retry to escape until one of the edges becomes outward (Fig. 5). Thus the stochastic motion in

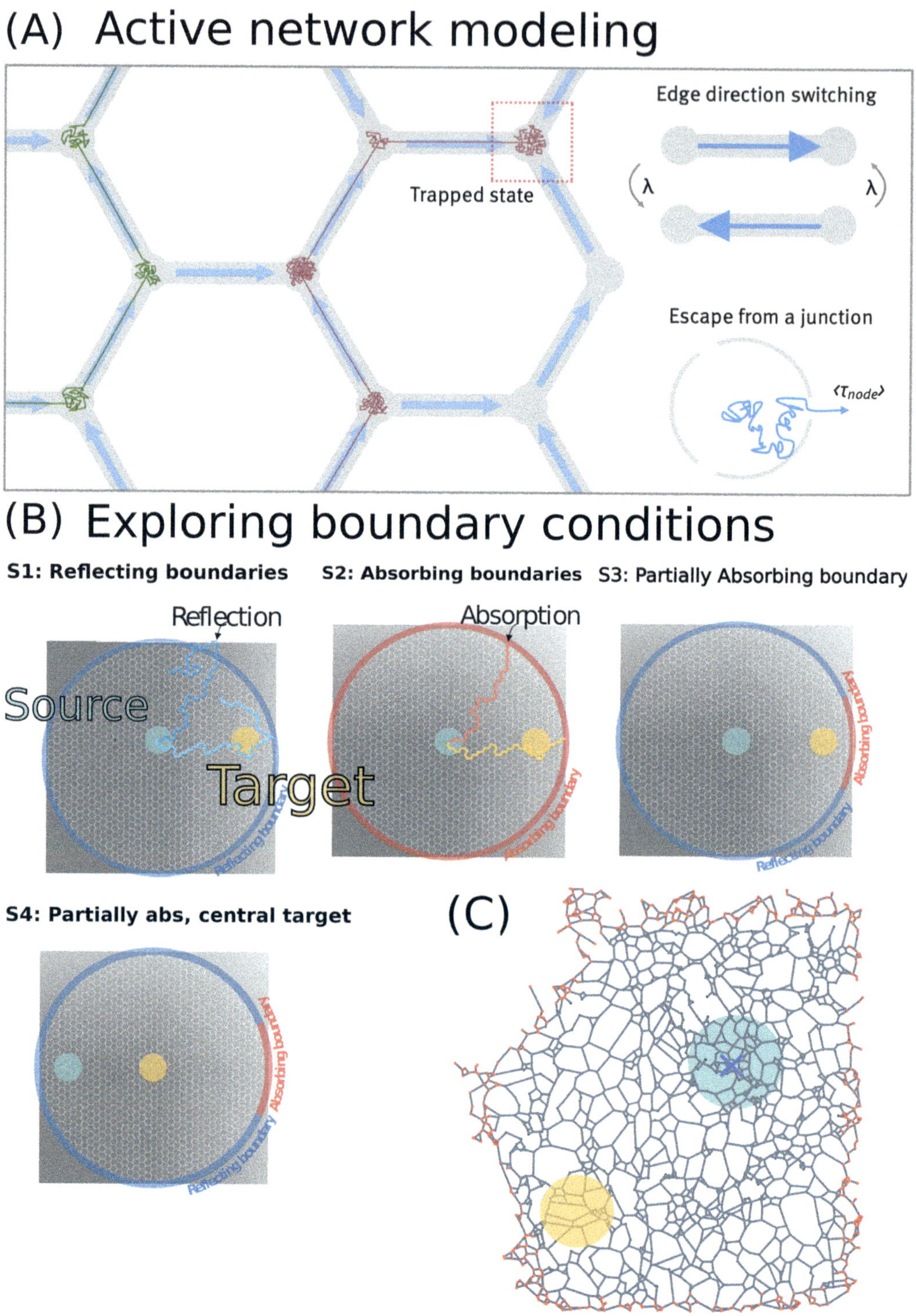


**Fig. 5 ER network modeling and conditions at boundary nodes.** (**a**) Active-network modeling consisting of alternating flows in tubules occurring at a rate $\lambda$. Escape from a junction occurs with a probability which is the reciprocal of the number of connected nodes. (**b**) Examples of source–target representation in a hexagonal lattice network with various boundary conditions imposed at a disk of radius R. Different boundary conditions modulate the time course of the simulated PA, allowing to explore the optimal configuration for a given PA dynamics. (**c**) Reconstruction of a partial ER network from SIM (structure illumination microscopy), where any boundary conditions can be given at external nodes (red), adapted from [28]

an active network is characterized by two parameters: a tubule switching rate $\lambda$ for tubular flow to change directions and a waiting rate $\lambda$ for particles to move to the chosen neighboring node.

### 5.3 Deconvolution of the Injection Rate

The first step for simulating PA responses is to reproduce the fluorescence dynamics or the number of molecules at the source. To simulate the PA process, we need to convert the measured photoconverted molecular dynamics into an injection rate. Concentration is not an injection rate due to molecules entering and exiting the source region. The injection rate $r(t)$ should be found from the measured concentration c(t). Thus this step relies on finding the injection rate or the number of photoactivated particles. When particles are released gradually over time at a single source node $s$, then we need to replace the initial vector $\boldsymbol{X}_0$ such that $x_{0,i=s}=1$ and $x_{0,i\neq s}=0$ by a vector accounting for the injection rate constants. For an injection rate $\mu_t$, we obtain a classical convolution for the distribution of particles at time $t$ as

$$\boldsymbol{\rho}_t=\sum_{k=0}^{t}\mu_k\left(M^{t-k}\,\boldsymbol{X}_0\right). \tag{26}$$

When the density $\phi_t$ of particles is known in an ROI and the injection node, the rate $\mu_t$ at which particles were injected can be deconvolved from Eq. (26) using the vector $\boldsymbol{\pi}$ for obtaining the total density in the ROI.

$$\pi_i=\begin{cases}1, & i\in\text{ROI}\\ 0, & \text{otherwise.}\end{cases} \tag{27}$$

Thus,

$$\phi_t=\boldsymbol{\rho}_t\cdot\boldsymbol{\pi}=\sum_{k=0}^{t}\mu_k\left(M^{t-k}\,\boldsymbol{x}_0\right)\boldsymbol{\pi}=\sum_{k=0}^{t}\mu_k\left(\boldsymbol{x}_{t-k}\,\boldsymbol{\pi}\right). \tag{28}$$

Using the first time $T$ for a particle to reach the ROI from the source node, we have $\boldsymbol{x}_t\,\boldsymbol{\pi}=0$ for any $t<T$, leading to

$$\phi_t=\sum_{k=0}^{t-T}\mu_k\left(\boldsymbol{X}_{t-k}\,\boldsymbol{\pi}\right), \tag{29}$$

and we can recover the coefficients $\mu_t$ recursively, starting from $\mu_0=\frac{\phi_T}{x_T\,\boldsymbol{\pi}}$ using

$$\mu_t=\frac{1}{\boldsymbol{x}_T\,\boldsymbol{\pi}}\left(\phi_t-\sum_{k=0}^{t-T}\mu_k\left(\boldsymbol{x}_{t-k}\,\boldsymbol{\pi}\right)\right). \tag{30}$$

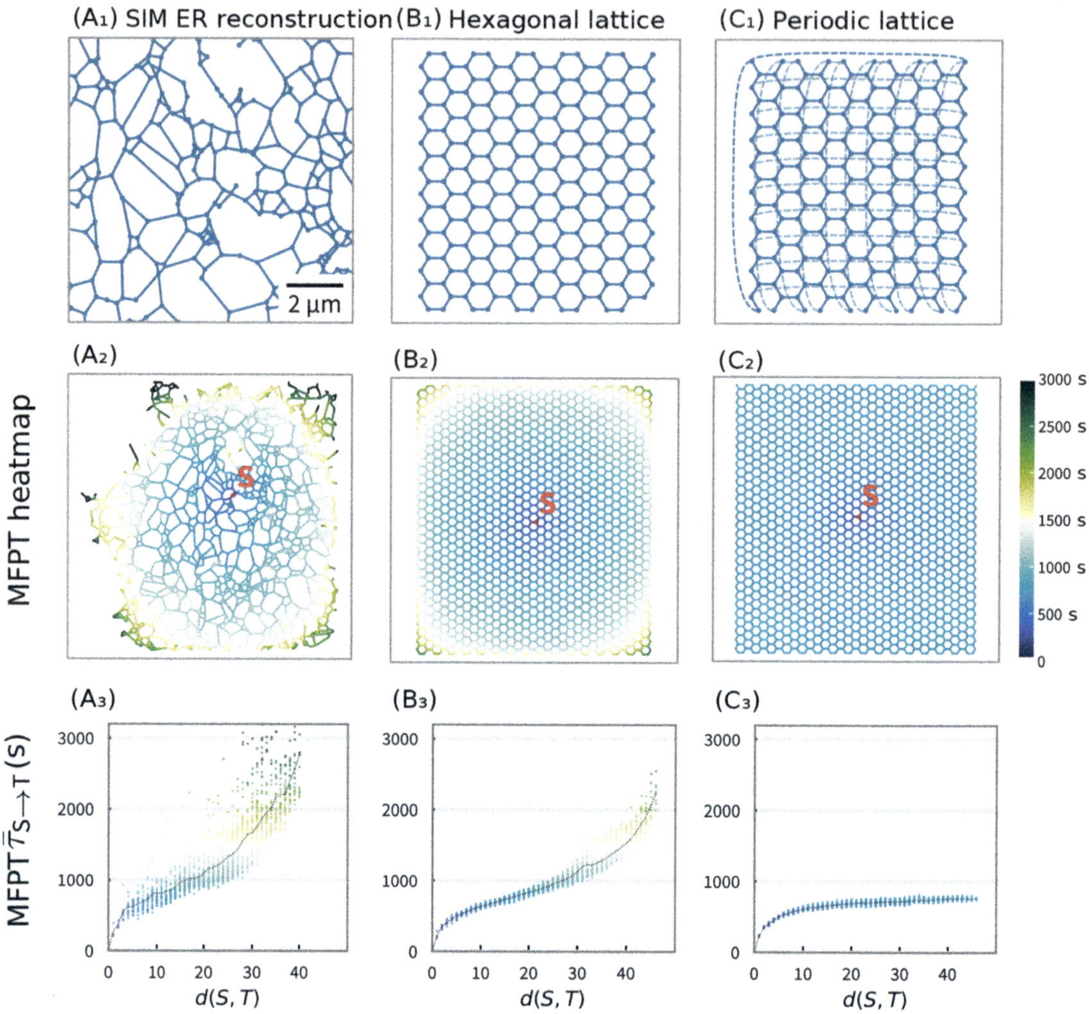


**Fig. 6 Stochastic simulations of arrival times in three graphs.** (**a**) Three examples of network reconstruction: $a_1$ from structure illumination microscopy data, $b_1$ hexagonal lattice with reflecting boundary on the edge nodes, and $c_1$ hexagonal lattice with periodic boundary. (**b**) Probability maps of the first arrival time depending on the distance from the source $S$ for the three reconstructed graphs. (**c**) Distribution of the Mean First-Passage Times (MFPTs) vs distances. Parameters are summarized in Tables 1 and 2

### 5.4 Simulating PA in Hexagonal or Reconstructed Network

The flow of material is often only available in a small portion of a graph. To model photoactivation, the first step is to choose a network (hexagonal or empirical, Fig. 6a1-b1-c1). The first-passage time analysis reveals how geometry can influence trafficking as shown in Fig. 6a2/a3-b2/b3-c2/c3, where simulation parameters are summarized in Tables 1 and 2.

Once a model is selected, which can either be diffusion-based or with an active flow controlling directionality [28], the PA in the source region can be used to deconvolve the injection rate and thus to obtain a first photoactivation approximation. Once the source is well calibrated, the model parameters should be adjusted such that

**Table 1**
**Network structure characteristics**

| Structure | Nodes | Edges | Radius | Degree (avg) |
|---|---|---|---|---|
| Hexagonal lattice | 2046 | 3006 | 47 | 2.94 |
| Periodic hex. lattice | 1984 | 2976 | 47 | 3 |
| Reconstructed ER | 1920 | 2787 | 41 | 2.90 |

**Table 2**
**Simulation parameters**

| Param | Value | Distribution |
|---|---|---|
| $\tau_{edge}$ | 0 | Constant |
| $\tau_{node}$ | 100 ms | Exponential |
| $\tau_{switch}$ | 30 ms | Exponential |

the simulations and the PA responses at the source match. Diffusion on a graph is simulated as a jump process between nodes: with a mean distance between neighboring nodes of around $L_{ER} = 1$ μm [9, 18] with an exponential waiting time with a mean $\tau = \frac{L_{ER}^2}{D_e}$.

Two examples are shown in Fig. 7, with such an analysis allowing to recover the diffusion coefficient. Alternatively for an active flow model, the two parameters to adjust are the $\mu$ and $\lambda$ switching rates. Another free parameter of the simulation is the flux condition at nodes located on the boundary of the simulation region. These boundary conditions can influence both the rising and the decay phases. Several scenarios are possible: the flux cannot (resp., can) return to the domain when reaching a boundary point, described as purely absorbing (resp., reflecting) points. In an intermediate case, a fraction $P$ of the flow is reflected and the rest $1 - P$ is absorbed.

Transients during the initial Activation-Saturation phases are not sensitive to the boundary condition imposed on the partially simulated network, because it depends mostly on the local properties of the network and laser conversion rate. Interestingly, the boundary condition influences drastically the recovery phase and the dynamics at the target (Fig. 7). It remains difficult to recapitulate the PA dynamics with the present diffusion network simulations. Another example concerns an active network with a switching time of 0.01 s and waiting time of 0.01 s, when the distance between the source and the target is $d = 23$ μm, with a target radius of 2.2 μm. Results are shown in Fig. 7.

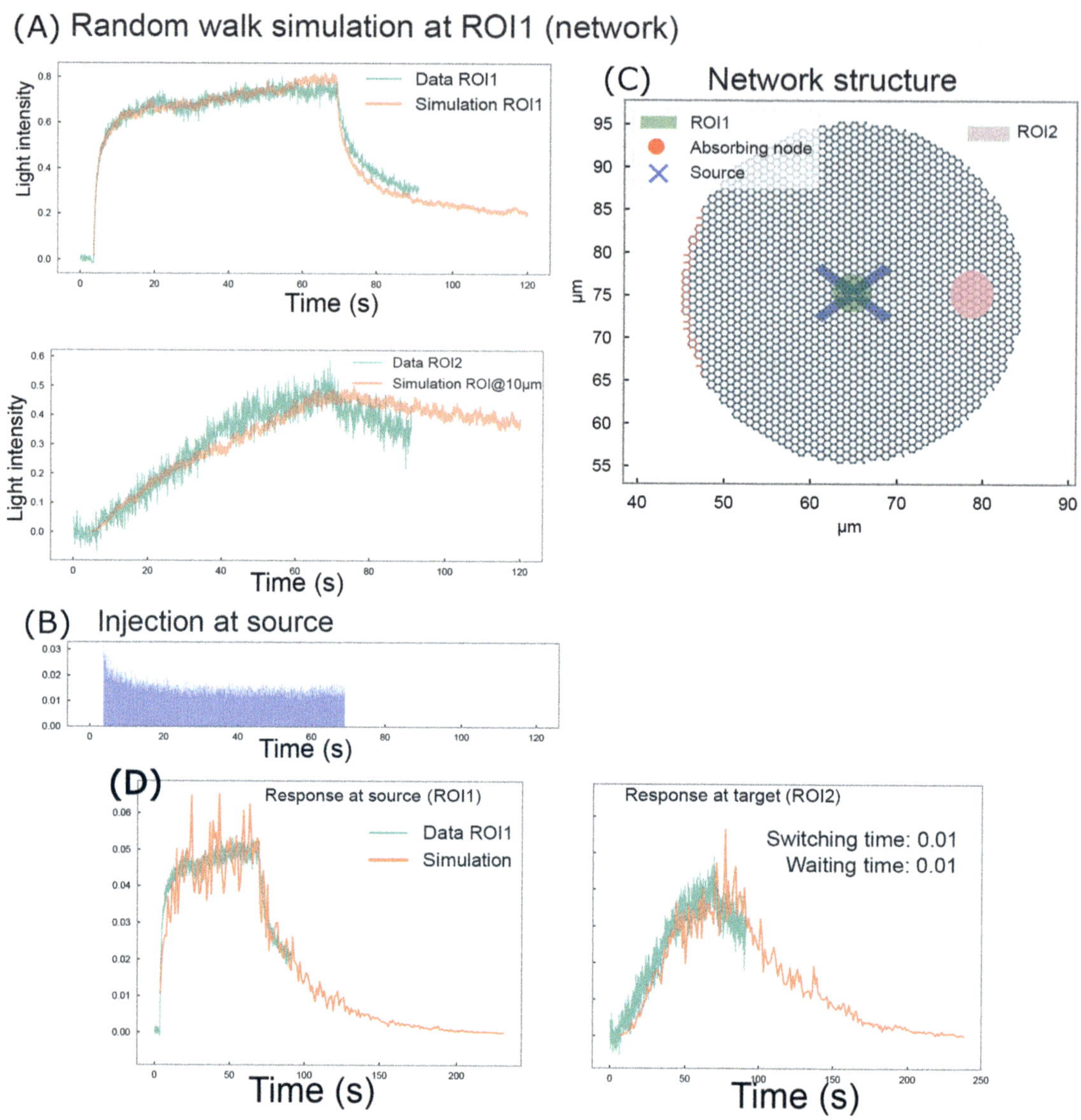


**Fig. 7 Stochastic simulations of photoactivation responses.** (**a**)–(**b**) Photoactivation at the source (ROI1) and the target (R02) in green versus random walk simulations for an injection rate described by Eq. (30). (**c**) Brownian simulations in a hexagonal network, where the source (green) is at the center and the target is ROI2 (purple). The boundary conditions on the nodes are either reflecting (black) or absorbing (red). (**d**) Source–target responses within an active flow network with similar switching and waiting times $\frac{1}{\mu} = \frac{1}{\lambda} = 0.01s$

## 6 Discussion

We presented in this book chapter computational approaches based on solving diffusion equations, computing Mean Square Displacement (MSD) or Kymograph analyses [10, 28], in order to analyze the first phases of photoactivation. These approaches are used to characterize molecular transport in the endoplasmic reticulum.

They account for fluorescence data at a source and a target site. Biophysical-based models rely on classical and anomalous diffusion. We reviewed here how these methods are used to extract various parameters, such as the anomalous exponent, the diffusion coefficient, and the diffusion map.

It is possible to adapt these methods to study the decay phase of the PA, of which it would be interesting to extract the associated biophysical parameters. In particular using stochastic simulations in a network allows exploring different boundary conditions that should be imposed at each peripheral node to recover the PA decay: this could be a mix of absorbing or partially absorbing conditions with reflection. These conditions affect the decay rate at short time and thus could reveal hidden control mechanisms of local transport.

Previous stochastic modeling [28] explored the possibility that molecules in the ER lumen could traffic together in packets. This transport is in contrast with classical diffusion laws, characterized by molecular dispersion. So far, local transiently persistent fluorescence was replicated by solving the diffusion equation in a heterogeneous network, accounting for local differences in molecular accumulation in nodes [10]. This approach can rely on an image processing procedure to remove noise and to account for the continuously changing ER morphology during lumen trafficking, an approach that will be reviewed somewhere else. Once key parameters are extracted, computations and numerical simulations can be used to estimate arrival rates to any regions of interest and predict the accumulation of molecules at exit sites. Another application of the computational methods presented here would be to develop simulations of misfolded protein trafficking. A possible model consists in diffusion aggregation, where the diffusion coefficient could depend on the size of aggregates [12]. It would also be possible to apply the present method to quantify the effect of ATP depletion on PA to explore the role of possible active components.

## 7 Appendix

### 7.1 Short-Time Behavior of Reaction–Diffusion Approximation to Photoactivation

Equation (20) is solved by separately considering the two cases $r < R$ and $r \geq R$. For $r \geq R$, the homogeneous partial differential equation is

$$\partial_t u(r,t) - \alpha\left(\partial_r^2 u(r,t) + \frac{1}{r}\partial_r u(r,t)\right) = 0, \tag{31}$$

where the diffusion coefficient is now denoted by $\alpha$. By then taking the Laplace transform, we obtain the modified Bessel equation

$$z^2 \partial_z^2 \hat{u} + z \partial_z \hat{u} - z^2 \hat{u} = 0, \quad \text{with } z = \sqrt{s/\alpha}\, r, \tag{32}$$

which has a general solution

$$\hat{u}_{r \geq R}(r, s) = c_1(s)\, I_0(\sqrt{s/\alpha}\, r) + c_2(s)\, K_0(\sqrt{s/\alpha}\, r). \tag{33}$$

In the case $r < R$, we need to solve the nonhomogeneous equation:

$$\partial_t u(r,t) - \alpha\left(\partial_r^2 u(r,t) + \frac{1}{r}\partial_r u(r,t)\right) = \kappa(1 - u(r,t)). \tag{34}$$

By taking the Laplace transform, we obtain

$$z^2 \partial_z^2 \hat{u} + z \partial_z \hat{u} - z^2 \hat{u} = -\frac{\kappa}{s(s+\kappa)} z^2, \quad \text{with } z = \sqrt{\frac{s+\kappa}{\alpha}}\, r. \tag{35}$$

The general solution of Eq. (35) is the sum of the solutions of the corresponding homogeneous equation and a particular solution $\psi_{\text{part}}$. The corresponding homogeneous equation is a modified Bessel equation with solutions $I_0(z)$ and $K_0(z)$. The particular solution is constant in $r$ and $\psi_{\text{part}} = \frac{\kappa}{s(s+\kappa)}$. The solution for the case $r < R$ is then

$$\hat{u}_{r<R}(r,s) = c_3(s)\, I_0\left(\sqrt{\frac{s+\kappa}{\alpha}}\, r\right) + c_4(s)\, K_0\left(\sqrt{\frac{s+\kappa}{\alpha}}\, r\right) + \frac{\kappa}{s(s+\kappa)}. \tag{36}$$

Putting the two solutions together, the solution of Eq. (20) in the Laplace domain is

$$\hat{u}(r,s) = \begin{cases} \hat{u}_{r<R}(r,s), & 0 \leq r < R \\ \hat{u}_{r \geq R}(r,s), & r \geq R. \end{cases} \tag{37}$$

The coefficients $c_1, c_2, c_3$, and $c_4$ are determined by imposing constraints on the solution: it has to be bounded in $r$, leading to $c_4 = 0$ to prevent a diverging behavior at $r = 0$. Similarly, we have $c_1 = 0$ to prevent the solution to diverge for $r \to \infty$. The remaining coefficients are found by using the continuity of the solution and its first derivative at $r = R$:

$$\hat{u}_{r<R}(R,s) = \hat{u}_{r \geq R}(R,s), \tag{38}$$

$$\partial_r \hat{u}_{r<R}(r,s)|_{r=R} = \partial_r \hat{u}_{r \geq R}(r,s)|_{r=R}. \tag{39}$$

Finally, we get

$$c_2 = \frac{\kappa\, I_1\left(\sqrt{\frac{s+\kappa}{\alpha}}R\right)}{s\,\sqrt{s+\kappa}\left(\sqrt{s}\,K_1\left(\sqrt{\frac{s}{\alpha}}R\right)\,I_0\left(\sqrt{\frac{s+\kappa}{\alpha}}R\right)+\sqrt{s+\kappa}\,K_0\left(\sqrt{\frac{s}{\alpha}}R\right)\,I_1\left(\sqrt{\frac{s+\kappa}{\alpha}}R\right)\right)}, \tag{40}$$

$$c_3 = -\frac{\kappa\, K_1\left(\sqrt{\frac{s}{\alpha}}R\right)}{\sqrt{s}(s+\kappa)\left(\sqrt{s}\,K_1\left(\sqrt{\frac{s}{\alpha}}R\right)\,I_0\left(\sqrt{\frac{s+\kappa}{\alpha}}R\right)+\sqrt{s+\kappa}\,K_0\left(\sqrt{\frac{s}{\alpha}}R\right)\,I_1\left(\sqrt{\frac{s+\kappa}{\alpha}}R\right)\right)}. \tag{41}$$

#### 7.1.1 Short-Time Asymptotics of Reaction–Diffusion Equation

$$\begin{aligned} u_{r<R}(R,t) = {} & \frac{\kappa}{s(s+\kappa)} \\ & - \frac{\kappa}{\sqrt{s(s+\kappa)}(s+\kappa)}\frac{R}{r}\exp\left(-\sqrt{\frac{s+\kappa}{D}}(R-r)\right) \end{aligned} \tag{42}$$

and

$$\begin{aligned} u_{r>R}(R,t) = {} & \kappa\sqrt{\frac{R}{r}}\left(\left(t/2+\frac{(r-R)^2}{4D}\right)\operatorname{erfc}\left(\frac{r-R}{2\sqrt{Dt}}\right)\right. \\ & \left. -\frac{r-R}{2\sqrt{\pi D}}\sqrt{t}\exp\left(-\frac{(r-R)^2}{4Dt}\right)\right). \end{aligned} \tag{43}$$

### 7.2 Short-time Behavior of Diffusion Approximation to Photoactivation

When photoactivation stops at the source, we can use as an approximation the solution of a diffusion equation where a concentration $C_1$ is maintained constant on the disk of radius $r = a$. The radial solution is well known [20] (p. 87) to be

$$\frac{C(r,t)}{C_1} = 1 + \frac{2}{\pi}\int_0^{\infty}\exp\left(-Du^2t\right)\frac{J_0(ur)\Upsilon_0(ua)-J_0(ua)\Upsilon_0(ur)}{J_0^2(ua)+\Upsilon_0^2(ua)}\frac{du}{u}, \tag{44}$$

where $J_0$ and $\Upsilon_0$ are the Bessel functions. For short time, the decay at the source is approximated by

$$\begin{aligned} \frac{C(r,t)}{C_1} & = \left(\frac{a}{r}\right)^{1/2}\operatorname{erfc}\left(\frac{r-a}{2\sqrt{Dt}}\right) + o((r-a)t^{1/2}) \\ & = \left(\frac{a}{r}\right)^{1/2}\frac{2\sqrt{Dt}}{\sqrt{\pi}(r-a)}\exp\left(-(r-a)^2/(4Dt)\right). \end{aligned} \tag{45}$$

This expression could be used to fit the diffusion coefficient for a dense ER network, so that the possible ER does not impact the two-dimensional planar approximation.

### 7.3 Modeling Diffusion and Active Transport in a Graph

We recall here the general theory about transport in a network. The network is represented by the matrix $A_{pq}$ accounting for the topology (specific nodes to be connected), where $A_{pq}=1$ if the nodes $p$ and $q$ are connected and $A_{pq}=0$ otherwise. The transition probability density function $P_p(t+\Delta t)$ for a particle ending at time $t+\Delta t$ at node $p$ is given by the sum of the transition rates $\lambda_{p\to q}(t)$ from $p$ to $q$ at time $t$ multiplied by the probability that a particle is at node $q$ at time $t$:

$$\begin{aligned} P_p(t+\Delta t) &= \sum_{\{q\neq p, A_{pq}=1\}} \lambda_{q\to p}(t)\Delta t P_q(t) \\ &+ \left(1 - \sum_{\{q\neq p,\ A_{pq}=1\}} \lambda_{p\to q}(t)\Delta t\right) P_p(t). \end{aligned} \tag{46}$$

The transition rates are

$$\lambda_{q\to p}(t) = \begin{cases} \lambda U_{p\to q}(t), & A_{pq}=1 \text{ and connected;} \\ 0, & \text{otherwise,} \end{cases}$$

and the random Poissonian variable satisfies

$$U_{p\to q}(t) = \begin{cases} 1, & \text{if the flow is going from p to q, when } A_{pq}=1; \\ 0, & \text{otherwise.} \end{cases}$$

The pdf of $U_{p\to q}(t)$ reflects the switching flow between two connected nodes. It is Poissonian with rate $\lambda_{sw}$, so that the Markov chain for the pdf $P_{p\to q}$ is independent of the node and tubule and is given by $P_{p\to q}+P_{q\to p}=1$ and

$$\frac{dP_{p\to q}}{dt} = \lambda_{sw}P_{q\to p} - \lambda_{sw}P_{p\to q}. \tag{47}$$

If a particle is ready to escape from a node, but all tubules are characterized by converging flow direction, the particle will have to wait that the flow in one of the tubules changes direction in order to be expelled from the node to the next one. Thus the escape rate $\lambda$ is such that

$$\frac{1}{\lambda} = \langle\tau_{node}\rangle + \frac{1}{\lambda_{sw} n_{tub-node}}, \tag{48}$$

where $\langle\tau_{node}\rangle$ is the mean residence time in a node, and $n_{tub-node}$ is the number of tubules per node (equal to 3 here). When the switching time is much larger than $\langle\tau_{node}\rangle$, this condition models extra-long confinement. In the opposite regime, where the

$\langle \tau_{node} \rangle \gg \frac{1}{\lambda_{sw} n_{tub-node}}$, the process is equivalent to a random walk on a graph. In that case, the mean first-passage time from a node to a given one can be computed asymptotically for a hexagonal lattice [28].

## Acknowledgements

We thank Chris Obara for critical feedback on this manuscript. This project has received funding from the European Research Council (ERC) to D.H. under the European Union's Horizon 2020 research and innovation program (grant agreement No. 882673). D.H. also gratefully acknowledges the support from the Agence Nationale de la Recherche via the grants ANR NEUC 00001 and ANR AstroXcite. F.P.-L. received funding from the Fondation ARC (grant No. ARCPDF12020020001505).

### References

1. Lippincott-Schwartz J, Snapp EL, Phair RD (2018) The development and enhancement of frap as a key tool for investigating protein dynamics. Biophys J 115(7):1146–1155
2. Patterson GH, Lippincott-Schwartz J (2002) A photoactivatable GFP for selective photolabeling of proteins and cells. Science 297(5588): 1873–1877
3. Patterson GH, Lippincott-Schwartz J (2004) Selective photolabeling of proteins using photoactivatable GFP. Methods 32(4): 445–450
4. Lippincott-Schwartz J, Snapp E, Kenworthy A (2001) Studying protein dynamics in living cells. Nat Rev Mol Cell Biol 2(6):444–456
5. Foi A, Trimeche M, Katkovnik V, Egiazarian K (2008) Practical Poissonian-Gaussian noise modeling and fitting for single-image raw-data. IEEE Trans Image Process 17(10):1737–1754
6. Luisier F, Blu T, Unser M (2011) Image denoising in mixed poisson-Gaussian Noise. IEEE Trans Image Process 20(3):696–708
7. Zhang K, Zuo W, Chen Y, Meng D, Zhang L (2017) Beyond a gaussian denoiser: residual learning of deep CNN for image denoising. IEEE Trans Image Process 26(7):3142–3155
8. Lehtinen J, Munkberg J, Hasselgren J, Laine S, Karras T, Aittala M, Aila T (2018) Noise2-Noise: learning image restoration without clean data. In: Proceedings of machine learning research, vol 80. PMLR, New York, pp 2965–2974
9. Holcman D, Parutto P, Chambers JE, Fantham M, Young LJ, Marciniak SJ, Kaminski CF, Ron D, Avezov E (2018) Single particle trajectories reveal active endoplasmic reticulum luminal flow. Nat Cell Biol 20(10):1118–1125
10. Dora M, Obara CJ, Abel T, Lippincott-Schwartz J, Holcman D (2023) Simultaneous photoactivation and high-speed structural tracking reveal diffusion-dominated motion in the endoplasmic reticulum. bioRxiv, pp 2023–04
11. Nehls S, Snapp EL, Cole NB, Zaal KJ, Kenworthy AK, Roberts TH, Ellenberg J, Presley JF, Siggia E, Lippincott-Schwartz J (2000) Dynamics and retention of misfolded proteins in native er membranes. Nature Cell Biol 2(5): 288–295
12. Hozé N, Holcman D (2017) Statistical methods for large ensembles of super-resolution stochastic single particle trajectories in cell biology. Annu Rev Stat Appl 4:189–223
13. Ölveczky BP, Verkman A (1998) Monte Carlo analysis of obstructed diffusion in three dimensions: application to molecular diffusion in organelles. Biophys J 74(5):2722–2730
14. Sbalzarini IF, Mezzacasa A, Helenius A, Koumoutsakos P (2005) Effects of organelle shape on fluorescence recovery after photobleaching. Biophys J 89(3):1482–1492
15. Sbalzarini IF, Hayer A, Helenius A, Koumoutsakos P (2006) Simulations of (an) isotropic diffusion on curved biological surfaces. Biophys J 90(3):878–885
16. Sun Y, Yu Z, Obara CJ, Mittal K, Lippincott-Schwarz J, Koslover EF (2022) Unraveling Single-Particle Trajectories Confined in Tubular Networks. arXiv preprint arXiv:2112.05884

17. Speckner K, Stadler L, Weiss M (2018) Anomalous dynamics of the endoplasmic reticulum network. Phys Rev E 98(1):012406
18. Nixon-Abell J, Obara CJ, Weigel AV, Li D, Legant WR, Xu CS, Pasolli HA, Harvey K, Hess HF, Betzig E, et al (2016) Increased spatiotemporal resolution reveals highly dynamic dense tubular matrices in the peripheral ER. Science 354(6311):aaf3928
19. Metzler R, Redner S, Oshanin G (2014) First-passage phenomena and their applications, vol. 35. World Scientific, Singapore
20. Crank J (1979) The mathematics of diffusion. Oxford University Press, Oxford
21. Abramowitz M, Stegun IA, Romer RH (1988) Handbook of mathematical functions with formulas, graphs, and mathematical tables
22. Carslaw HS, Jaeger JC (1959) Conduction of heat in solids, 2 edn. Oxford University Press, Oxford
23. Holcman D, Schuss Z (2015) Stochastic narrow escape in molecular and cellular biology. Anal Appl Springer New York 48:108–112
24. Nitzsche B, Bormuth V, Bräuer C, Howard J, Ionov L, Kerssemakers J, Korten T, Leduc C, Ruhnow F, Diez S (2010) Studying kinesin motors by optical 3d-nanometry in gliding motility assays. Methods Cell Biol 95:247–271
25. Zhou HM, Brust-Mascher I, Scholey JM (2001) Direct visualization of the movement of the monomeric axonal transport motor unc-104 along neuronal processes in living caenorhabditis elegans. J Neurosci 21(11):3749–3755
26. Mangeol P, Prevo B, Peterman EJ (2016) Kymographclear and kymographdirect: two tools for the automated quantitative analysis of molecular and cellular dynamics using kymographs. Mol Biol Cell 27(12):1948–1957
27. Jakobs MA, Dimitracopoulos A, Franze K (2019) Kymobutler, a deep learning software for automated kymograph analysis. Elife 8:e42288
28. Dora M, Holcman D (2020) Active flow network generates molecular transport by packets: case of the endoplasmic reticulum. Proc R Soc B 287(1930):20200493
29. Lovász L, et al (1993) Random walks on graphs: A survey. Combinatorics, Paul Erdős is Eighty 2(1):1–46
30. Masuda N, Porter MA, Lambiotte R (2017) Random walks and diffusion on networks. Phys Rep 716:1–58

# INDEX



Verena Kriechbaumer (ed.), *The Plant Endoplasmic Reticulum: Methods and Protocols*, Methods in Molecular Biology, vol. 2772, https://doi.org/10.1007/978-1-0716-3710-4,

GPSR Compliance
The European Union's (EU) General Product Safety Regulation (GPSR) is a set of rules that requires consumer products to be safe and our obligations to ensure this.

If you have any concerns about our products, you can contact us on

ProductSafety@springernature.com

In case Publisher is established outside the EU, the EU authorized representative is:

Springer Nature Customer Service Center GmbH
Europaplatz 3
69115 Heidelberg, Germany

www.ingramcontent.com/pod-product-compliance
Ingram Content Group UK Ltd.
Pitfield, Milton Keynes, MK11 3LW, UK
UKHW052000051026
15624UKWH00003B/490

* 9 7 8 1 0 7 1 6 3 7 0 9 8 *